MIND-ALTERING AND POISONOUS
PLANTS OF THE WORLD

Papavera

Papaver somniferum L.

MIND-ALTERING AND POISONOUS
PLANTS OF THE WORLD

Michael Wink
Ben-Erik van Wyk

TIMBER PRESS
Portland • London

Published in 2008 by
Timber Press, Inc.

The Haseltine Building   2 The Quadrant
133 S.W. Second Avenue, Suite 450   135 Salusbury Road
Portland, Oregon 97204-3527   London NW6 6RJ
U.S.A.   United Kingdom
www.timberpress.com   www.timberpress.co.uk

First edition, first impression, 2008

WARNING & DISCLAIMER

This book contains general information about poisonous and psychoactive plants and fungi. It is intended as a scientific overview and does not encourage the use of mind-altering material. Many of the plants and fungi described in this book are highly toxic and may cause severe allergic reactions, serious injury or even death. Although care has been taken to be as accurate as possible, neither the authors nor the publisher make any expressed or implied representation as to the accuracy of the information contained in this book and cannot be held legally responsible or accept liability for any errors or omissions. Neither the authors nor the publisher can be held responsible for claims arising from the mistaken identity of plants or their inappropriate use. This book provides information of first aid treatment for intoxifications. Many of the therapies can only be administered by doctors. In case of poisoning always consult a medical professional or qualified practitioner.

Note that some of the terms used in this book may refer to registered trade names even if they are not indicated as such.

ISBN-13: 978-0-88192-952-2

Proof reading: David Pearson
Typesetting: Jacqueline Huisman, Lizmarnet Drukkershuis
Printed and bound by Tien Wah Press (Pte.) Ltd, Singapore

Cover photographs: Front, main image; *Papaver somniferum* by
Milos Luzanin, courtesy of iStock international, Inc.
Front, inset from left to right; *Abrus precatorius, Strophanthus
speciosus, Amanita muscaria*
Back, from left to right; *Atropa belladonna, Ricinus communis,
Brugmansia candida*

Contents

handwritten: 1,69 in

Preface

The aim of this book is to give the reader a bird's eye view of the better known poisonous and mind-altering plants of the world. The book is presented as a compact, colourful and scientifically accurate reference text. It covers more than 1 200 of the most important plants from Europe and North America and some of the African, Asian and Australian plants, if they are widely known. There are probably thousands of toxic plants in rain forests and savannas which are not even known to be toxic. It has been suggested that only 30 % of all plants have been studied for their chemical composition. This book should be useful to gardeners, pharmacists, doctors or members of Poison Centres to identify and learn about poisonous and mind-altering plants and to be able to help in case of poisoning and intoxication.

The biological activity of many poisonous and mind-altering plants has become known through scientific research and any literature search (via the internet, for example) will reveal that numerous new publications are added to the scientific literature every day. One of the reasons for the extensive scientific research is the fact that many of the poisonous and mind-altering plants are also used as medicinal plants in many countries of the world. There is simply no space to allow for a comprehensive literature list for the more than 1 200 plants and their relatives treated and illustrated in this book. The reader can consult the list of key references (p. 433) for further information. Our texts are based on published information and research carried out in our laboratories.

The book gives a relevant background on how hazardous and mind-altering plants work (i.e., their modes of action). An overview of the various active ingredients (secondary metabolites) is provided in an attempt to clarify the complexity of metabolic and physiological effects caused by toxic plants. Plants often contain a mixture of substances that have additive or even synergistic effects, so that the toxicological and pharmacological properties are difficult to test and verify. This should be kept in mind when we deal with the toxicology of isolated chemicals; the effects of the complex mixture (e.g. the whole plant or extract) can differ. The book provides information on toxicity, symptoms of poisoning and first aid treatment. The scope of the book does not allow full coverage of these topics and readers should consult textbooks of toxicology for a more detailed account of the pharmacological and medical background.

This guide contains more than 200 illustrated monographs of important mind-altering and poisonous plants. For further information, a short checklist of over 1 200 relevant plant species is provided, giving the correct scientific name, common name(s), family, origin, main compounds and main actions. Although not plants, some poisonous and mind-altering mushrooms have been included, since they are usually discussed in such a context.

We could not avoid using medicinal and pharmacological terminology, although we are aware that it is not used in everyday language and readers might not be familiar with it. Otherwise the texts would have become much too long or if shorter or simpler, without useful information. We refer the reader to a glossary of pharmacological and toxicological terms in the back of the book, that should help to clarify the technical language.

Abbreviations

AAPCC	Association of the American Poison Control Centres	i.m.	intramuscular
		i.v.	intravenous
BW	bodyweight	i.p.	intraperitoneal
CA	chemical abstracts	mAChR	muscarinic acetylcholine receptor
CNS	central nervous system	MW	molecular weight
Da	Dalton	nAChR	nicotinic acetylcholine receptor
DW	dry weight	NPAA	non-protein amino acid
FW	fresh weight	PA	pyrrolizidine alkaloid
GABA	gamma aminobutyric acid	p.o.	per os (oral)
GI tract	gastro-intestinal tract	s.c.	subcutaneous
GLC	gas-liquid chromatography	SM	secondary metabolite
HPLC	high performance liquid chromatography	TLC	thin-layer chromatography
		VOD	veno-occlusive disease

Introduction

What are toxins and toxic plants?

A toxin is a substance that exerts negative effects on an organism and its metabolism. The scientific discipline that deals with the study of toxins, their modes of action and the treatment of intoxications is toxicology, a subdiscipline of pharmacology. A toxin can cause transient disturbances, long-term effects or even death. Depending on the severity of the effects we distinguish between **poisons** that can kill in minute amounts, **toxins** that are less toxic than poisons and **toxicants** that are toxic in high concentrations only.

Viruses and bacteria are infectious agents and are not considered as toxins, although some of them produce toxic substances. Substances or materials that injure an organism mechanically or by irradiation are not toxins and are not considered in this book.

It is important to note that most substances, both natural and synthetic, can harm our body if they are given at an elevated dose. This is even true for salt, vitamins, nutrients and water. Already in 1537 Paracelsus (1493–1541) observed that *"sola dosis facet venenum"* ("It is the dose that makes a poison" or "All things are poison and nothing is without poison, only the dose permits something not to be poisonous".) The toxic dose depends, however, on many variables, such as:
- route of administration (the intravenous or intraperitoneal application is usually more effective than the oral route)
- solubility of poisons and toxins in body fluids
- frequency of intoxication (acute, subacute and chronic)
- health and age of a person (sick persons are more susceptible than healthy ones, babies and children more than adults; old more than younger people; women more than men)

In order to compare the toxicity of different poisons and toxins, acute toxicity data have been determined in laboratory animals (mostly rats, mice, rabbits and guinea pigs). Although humans are often more susceptible than rodents, the experimental data provide a good indication of what to expect in humans. The value often used is the LD_{50} (lethal dose) value; it indicates that dose which kills 50 % of test animals. This value is easier to assess than the LD_{100}, which is that dose that kills all animals in an experiment.

The World Health Organisation (WHO) recognises **four toxicity classes**, which are used in this book, as follows:

Class Ia: extremely hazardous
Class Ib: highly hazardous
Class II: moderately hazardous
Class III: slightly hazardous

The system is based on LD_{50} determination in rats, thus an oral solid agent with an LD_{50}
- at 5 mg or less/kg bodyweight is Class Ia
- at 5–50 mg/kg bodyweight is Class Ib
- at 50–500 mg/kg bodyweight is Class II
- and at more than 500 mg/kg bodyweight is Class III

Compounds or plants that fall into classes Ia and Ib have traditionally been termed "highly poisonous", those of class II (LD_{50} between 50 and 200 mg/kg bodyweight) as "poisonous" and the others as "less poisonous".

Toxins interfere with central functions of an organism. In animals and humans, the most poisonous substances are neurotoxins that affect the brain and the nervous system, followed by cytotoxins and metabolic poisons that disturb the liver, kidneys, heart or respiration (see p. 240).

All plants produce secondary metabolites as defence compounds against herbivores, micro-organisms and viruses (see p. 12). Among them are several groups of compounds which are poisonous or deadly poisonous for humans and animals when ingested even in small amounts. As a rough estimate, there are about 750 very poisonous substances occurring in more than 1 000 species.

Plants with alkaloids, cardiac glycosides, phorbol esters, lectins and cyanogenic glycosides are often classified as extremely toxic and hazardous plants (see Toxins). Some examples of extremely hazardous plants include species of the genera *Abrus, Acokanthera, Aconitum, Arum, Atropa, Brugmansia, Cicuta, Colchicum, Conium, Daphne, Datura, Hyoscyamus, Laburnum, Nerium, Ricinus, Strophanthus, Strychnos, Taxus, Thuja* and *Veratrum*. We have also discussed plants that are considered moderately hazardous and slightly hazardous if they are frequently the subject of consultations in Poison Centres. Also treated are a number of potentially harmful exotic pot plants, which have been introduced in gardens or sold for decorative use in the home, because most people are unaware of their properties.

Paracelsus – His real name was Theophrastus Bombastus von Hohenheim (born 1493 near Einsiedeln, Switzerland; died 24 September 1541 in Salzburg). Painting by Jan van Scorel.

Aconitum napellus

Acokanthera oblongifolia

Daphne mezereum

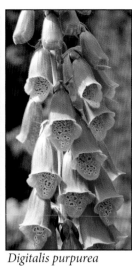
Digitalis purpurea

Nerium oleander

Conium maculatum

Strophanthus gratus

Taxus baccata

Examples of extremely hazardous plants

9

What are mind-altering substances and psychoactive plants?

A subclass of toxins affect the performance of the brain and the central and peripheral nervous system. At low concentrations these substances are often mind-altering or stimulating, whereas they can be deadly poisons at higher doses. For example, atropine (an alkaloid from *Atropa belladonna* and several other plants of the family Solanaceae) inhibits the muscarinic acetylcholine receptor in the brain. This induces hallucinations and deep sleep with vivid, often erotic dreams. This action must have been experienced by humans since antiquity. An early account can be found in the Odyssey: When Ulysses visited Circe, he and his comrades were offered wine adulterated with an extract of plants with atropine or scopolamine. The comrades turned into pigs whereas Ulysses remained conscious because he had taken the herb Moly as an antidote before. In medieval times, witches used ointments with tropane alkaloids, which were administered onto the skin. The alkaloids were absorbed and caused a deep sleep with the sensation of being able to fly. This is why witches are often depicted as flying on broomsticks. High doses of atropine, however, lead to cardiac and respiratory arrest, since acetylcholine regulates the activity of cardiac and smooth muscles.

Depending on the effects, mind-altering or psychoactive substances or plants are sometimes classified as:
- **Stimulants:** Plants or substances which alert and stimulate the mind or activity without changing the perception. This group includes plants such as coffee (*Coffea arabica*), tea (*Camellia sinensis*), cacao (*Theobroma cacao*), guarana (*Paullinia cupana*), mate (*Ilex paraguariensis*), Ephedra (*Ephedra* species), khat (*Catha edulis*) or coca (*Erythroxylum coca*).
- **Sedatives, hypnotics or narcotics:** Plants or substances with sedating and sleep-inducing or anxiolytic or narcotic properties which can change perception and cause vivid dreams and euphoria. Important in this group are opium poppy (*Papaver somniferum*), valerian (*Valeriana officinalis*), hop (*Humulus lupulus*) and kava kava (*Piper methysticum*).
- **Hallucinogens:** Plants or substances with substantial influence on perception of space and time and on emotional feelings. These substances have also been termed psychomimetics, psychotics, psychedelics, empathogens, entheogens and others. Hallucinogens cause illusions that appear real to the person who is under the influence of these drugs. Hallucinogens have been used for centuries during human history (see p. 19) by shamans, sorcerers or witches as magic plants, gifts of the gods or holy drugs in their spiritual and healing rituals. Many of these drugs are used today as recreational drugs to elicit the daily "high" and are of great concern for most societies. Important plants include yopo (*Anadenanthera peregrina*), wormwood (*Artemisia absinthium*), deadly nightshade (*Atropa belladonna*), ayahuasca (*Banisteriopsis caapi*), tree datura (*Brugmansia* species), marijuana (*Cannabis sativa*), thorn-apple (*Datura stramonium*), henbane (*Hyoscyamus niger*), morning glory (*Ipomoea tricolor*), peyote (*Lophophora williamsii*), mandrake (*Mandragora officinarum*), wild tobacco (*Nicotiana rustica*), African rue (*Peganum harmala*), iboga (*Tabernanthe iboga*), ololiuqui (*Turbina corymbosa* and some other Convolvulaceae), and epena (*Virola* species). Also included are a few hallucinogenic mushrooms, which are treated in this book (although they are not plants but a kingdom of their own). Important fungi include fly agaric (*Amanita muscaria*), ergot (*Claviceps purpurea*) and magic mushroom (*Psilocybe mexicana, P. cubensis*).

The number of mind-altering molecules in nature appears to be limited and their modes of action have been elucidated in many instances already. These substances may have similar structures as endogenous neurotransmitters that transfer an electric signal (action potential) from one nerve cell to another or to the neuromuscular plate. If a synapse becomes activated by an action potential, neurovesicles that are filled with neurotransmitters fuse with the cell membrane of the presynapse and release the neurotransmitters into the synaptic cleft. The neurotransmitter binds to a neuroreceptor at the postsynaptic target cell, which in turn activates a downstream signal transduction pathway. Thus neurotransmitters convert an electric signal into a biochemical signal. Mind-altering compounds often enhance or inhibit the activity of the neurotransmitters acetylcholine, serotonin, dopamine, noradrenaline, glutamate or gamma aminobutyric acid (GABA). Some psychoactive compounds affect ion channels. A more detailed explanation is given on pages 242 to 251.

Amanita muscaria

Brugmansia suaveolens

Hyoscyamus aureus

Mandragora officinarum

Papaver somniferum

Datura stramonium

Claviceps purpurea

Atropa belladonna

Examples of some mind-altering plants and fungi

Why do poisons and mind-altering substances exist in nature?

Plants, fungi and non-mobile invertebrates cannot run away when attacked by enemies such as herbivores or predators, nor do they have an immune system against attacking bacteria, fungi or viruses (which humans have as a defence system). Plants and other sessile organisms (such as marine animals) have developed various biologically active chemicals during evolution that help them to defend themselves against predators (insects, molluscs, vertebrates), microbes, viruses and competing other plants. Some animals, such as spiders or snakes, use poisons for both defence and prey acquisition; these poisons often consist of toxic polypeptides (enzymes, ion channel modulators) and neurotransmitters. Mind-altering compounds appear to be an exception at first glance, since they make the consumer dependent and thus they would feed even more on a given plant. But mind-altering secondary metabolites (SMs) also function as defence compounds: If a herbivore becomes intoxicated and "high" it is no longer alert and becomes unable to watch its surroundings. The likelihood is high that it will fall out of trees or from rocks or be killed by predators. Furthermore, most psychoactive compounds are deadly poisons at higher doses. Since animals cannot determine the dose, the chance of a lethal poisoning is quite high. Therefore, the production of attractive mind-altering substances is another trick of plants to get rid of herbivores.

Whereas primary metabolites are present in all species and are essential for life, secondary metabolites (SMs) occur in varying mixtures that differ between species and systematic units. Because some SMs are typical for specific plant groups, scientists have used them as taxonomic markers ("chemotaxonomy"). SMs are not essential for primary or energy metabolism but are important for the ecological fitness and survival of the organisms producing them. SMs are adaptive traits and their occurrence in plants reflects common descent as well as environmental selection.

In order to be effective, SMs need to be present at the right site, time and concentration in the plant producing them. This means that biosynthesis, transport, storage and even turnover must be highly regulated and coordinated processes. The biosynthesis of several SMs is usually constitutive whereas in many plants the synthesis can be induced and enhanced by biological stress conditions, such as wounding or infection. This activation can be biochemical, e.g. through hydrolysis of glycosides that are stored as "prodrugs" (Table p. 14) or via the activation of genes responsible for synthesis, transport or storage of the SM. Signal transduction pathways that cause gene activation in plants include the pathway leading to jasmonic acid or salicylic acid that have been found to trigger defence reactions in plants. Defence chemicals are usually stored in tissues that are important for survival and reproduction (such as fruits and seeds); very often they are located in strategically advantageous sites, such as epidermal or bark tissues. For these reasons, seeds often represent the most toxic part of a plant. Whereas hydrophilic compounds are preferably stored in vacuoles, lipophilic SMs are stored in resin ducts, lacticifers, oil cells, trichomes or in the cuticle. The site of storage need not be identical to the site of biosynthesis. Therefore, long-distance transport via xylem or phloem is a regular theme in secondary metabolism.

Plants use SMs (such as volatile essential oils and coloured flavonoids or tetraterpenes) not only for defence but also to attract insects for pollination or other animals for seed dispersal. In the case of flowers, pollinators are attracted by colour or scent and are then rewarded by nectar, which usually contains sugar, but sometimes amino acids and SMs. In the case of fruits, mature fruits attract dispersing animals by offering sweet fruit pulp, whereas unripe fruits are loaded with repellent SMs. Whereas the ripe pulp is edible, the seeds contain toxic SMs. The strategy of plants is obvious: An animal should eat the fruit pulp but not the seeds, which pass the GI tract intact and are dispersed with the faeces. In the case of pollinators and seed dispersers, SMs serve as **signal and defence compounds**. Animals that store toxic SMs for defence or prey acquisition often advertise this property by red and yellow warning colours (the pigments are also SMs), a phenomenon called **aposematism**.

The table on page 13 gives an overview of known classes and numbers of SM structures and their function as poisons or mind-altering substances. A second table on page 15 lists the occurrence of poisons and mind-altering substances in the main groups of producers. Whereas small molecules dominate in plants, bacteria, fungi, algae and sponges, toxic polypeptides are common in animals. Some groups of SMs are often defined and named after their bioactivities, such as antibiotics or mycotoxins, although they can also be grouped according to their structural types.

Structural types of plant secondary metabolites and abundance of poisons and psychoactive substances

Class	Number of structures	Poisons	Psychoactive substances
With nitrogen			
Alkaloids	20000	most	many
Non-protein amino acids (NPAAs)	700	some	few
Amines	100	some	some
Cyanogenic glucosides	60	most	few
Glucosinolates	100	some	none
Alkamides	150	most	none
Lectins, peptides, polypeptides	2000	most	none
Without nitrogen			
Monoterpenes (including iridoids)	2500	some	some
Sesquiterpenes	5000	many	few
Diterpenes	2500	many	few
Triterpenes, steroids, saponins	5000	most	few
Tetraterpenes	500	none	none
Phenylpropanoids, coumarins, lignans	2000	some	few
Flavonoids, tannins	4000	some	none
Polyacetylenes, fatty acids, waxes	1500	some	few
Polyketides (anthraquinones)	750	most	none
Carbohydrates	200	few	none

Functions of secondary metabolites in plants

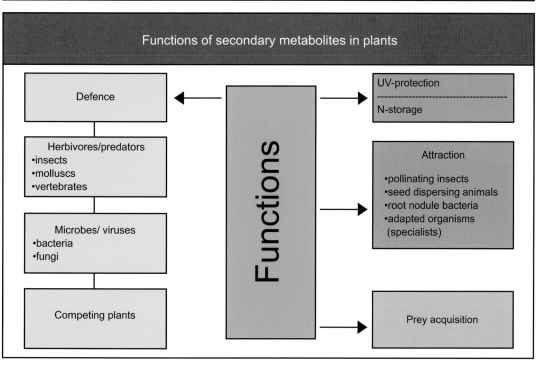

13

Typical "prodrugs" present in plants that are activated by wounding, infection or in the human body	
Secondary metabolites of undamaged tissue	**Active metabolite**
Cyanogenic glucoside	hydrocyanic acid (HCN)
Glucosinolate	isothiocyanate
Alliin	allicin
Coumaroylglucoside	coumarin
Arbutin	benzoquinone
Salicin, methylsalicylate	saligenin, salicylic acid
Gein	eugenol
Bi-desmosidic saponins	mono-desmosidic saponins
Cardiac glycosides with terminal glucose residues	cardiac glycosides without terminal glucose residues
Cycasin	methylazoxymethanol (MAM)
Ranunculin	protoanemonin
Tuliposide	tulipalin
Crocetin	safranal
Glycosylated cucurbitacins	free cucurbitacins

Occurrence of toxins and mind-altering substances in nature

A number of common poisons are man-made or of inorganic origin, such as ammonia, arsenic, KCN, DDT, E 605, CO, H_2S, heavy metals, phenols or phosphins. Unknown to most modern people, nature also provides a rich variety of noxious substances; indeed the most toxic of compounds, the Botulinus toxin, is a natural product. Although this book mainly deals with plants and a few fungi, this chapter gives an overview of the occurrence of toxic and mind-altering substances throughout nature, i.e. including microorganisms and animals.

Especially rich in hazardous compounds are plants, fungi, mushrooms, certain algae and microorganisms. Marine animals, especially sessile organisms, are also a rich source of poisons. Spiders, scorpions, insects and snakes employ toxins for both defence and hunting.

Many toxins are small molecules that are called "**secondary metabolites**" (unlike primary metabolites, they are not important for metabolism or energy) or **natural products**. Secondary metabolites (SMs) are often molecules with a molecular weight below 1 000 Dalton. We can distinguish between N-containing and N-free secondary metabolites (N = nitrogen). More than 120 000 SM structures have been determined (Table p. 13) by NMR, MS and X-ray analysis although only 20–30 % of relevant organisms have been studied in some depth so far.

Toxins are not always small molecules; some plants produce toxic macromolecules (peptides and proteins), such as lectins (ricin from *Ricinus communis* is one of the most potent toxins found in plants). Especially rich in toxic proteins are bacteria and animals.

It is likely that far more than 150 000 secondary metabolites exist in nature and most of them exert negative effects on other organisms. Several are highly poisonous or mind-altering and will be discussed in this book in more detail.

Occurrence of poisons, toxins and mind-altering SMs in nature

Producer	Type of secondary metabolite or toxin
Bacteria	
Bluegreen algae (Cyanobacteria)	polyketides (aplysiatoxin, brevetoxin B, dinophysistoxin-1), alkaloids (lyngbyatoxin, anatoxin-a)
Bacteria	antibiotics, polyketides, alkaloids, terpenoids, phenylpropanoids, peptides
Algae, plants and other groups	
Algae	polyphenols, terpenoids, polysaccharides
Dinoflagellates (Dinophyceae)	paralytic shellfish poisoning, alkaloids (saxitoxin, gonyautoxin, ciguatoxin)
Lichens	anthraquinones, polyphenols, phenylpropanoids
Higher plants (ferns, angiosperms, gymnosperms)	alkaloids, amines, glucosinolates, cyanogenic glucosides, cyanolipids, glucosinolates, non-protein amino acids, terpenoids, saponins, phenylpropanoids, tannins, lignans, anthraquinones, fatty acids, polyenes, phloroglucinols, alkylphenols, lectins
Fungi	
Fungi (Basidiomycetes)	organic acids, alkaloids and peptides (hydrazine derivatives, amatoxins, bufotonin, ibotenic acid, muscarine, phallotoxins, psilocin, psilocybin, virotoxin), nonprotein amino acids, cyanogenic glucosides, phenolics, sesquiterpenes, triterpenes
Moulds (Deuteromycetes)	mycotoxins (aflatoxins, phenolics, furanocoumarins, citrinin, citreoviridin, cytochalasin, penicillium toxins, trichothecens, anthraquinones, ergot alkaloids, fusarium toxins, aspergillus toxins, ochratoxins, patulin, penitrem A, rugolosin, rubratoxins, zearalenon)
Animals	
Sponges (Porifera)	terpenoids, sesquiterpenes, diterpenes, steroids, alkaloids, halogenated SMs, histamine derivatives
Corals, jelly fishes (Cnidaria)	toxic polypeptides in nematocysts (cytolysins; haemolysins; neurotoxins; physaliatoxin, esterase, hyaluronidase, proteases)
Sea anemones (Anthozoa)	toxic polypeptides in nematocysts (neurotoxins)
Worms (Nemertini)	alkaloids (anabaseine; nereistoxin)
Scorpions (Scorpiones)	toxic polypeptides (neurotoxins: α-, β-, γ-toxin; phospholipase A, hyaluronidase)
Spiders (Araneae)	polyamines (argiotoxin, argiopin, NSTX, ISTX), polypeptides (robustoxin, latrotoxin, sphingomyelinase D)
Scolopender (Chilopoda)	toxic peptides, neurotransmitters (histamine, serotonin)
Insects (Insecta)	SMs acquired from host plants (alkaloids: PAs, quinolizidines, aconitine; cardiac glycosides, cyanogenic glucosides, phorbol esters); quinones, terpenoids; toxic polypeptides (phospholipase A2, hyaluronidase, haemolysins, melittin, apamin, wasp kinins, neurotransmitters (histamine), pyridine and piperidine alkaloids (solenopsine), iridoids (dolichodial, iridodial), sesquiterpenes, diterpenes, cantharidin
Cone shells (Mollusca)	toxic polypeptides, conotoxins; saxitoxin, gonyautoxin in *Turbo* and *Tectus* species
Octopuses (Cephalopoda)	toxic polypeptides (hyaluronidase, cephalotoxin, eledoisin); tetrodotoxin, neurotransmitters (serotonin, tyramine, octopamine, noradrenaline)
Starfish (Asteroidea)	toxic polypeptides (phospholipase A2), steroid saponins
Sea urchins (Echinoidea)	toxic polypeptides, amines
Sea cucumbers (Holothuroidea)	steroidal saponins (holothurin A, B)
Stingrays (Chondrichthyes)	toxic polypeptides (phosphodiesterase)
Bone fish (Osteichthyes) (weever, scorpion, fire fish)	toxic polypeptides, neurotransmitters (serotonin, acetylcholine), alkaloids (pahutoxin)
Puffer fish, molluscs, amphibia	marine bacteria produce toxins that accumulate in the food chain: tetrodotoxin, palytoxin
Toads, salamanders, frogs (Amphibia)	alkaloids (bufotonin, samandarine, tetrodotoxin, batrachotoxin, pumiliotoxin, histrionicotoxin), bufadienolides (bufotoxin)
Gila monster (Reptilia)	toxic polypeptides (hyaluronidase, kallikrein, gilatoxin)
Snakes (Reptilia)	toxic polypeptides (neurotoxins: neurotoxic phospholipase A2, α-neurotoxins, choline esterase, inhibiting proteins; enzymes: hyaluronidase, phosphatases, phospholipase A2, proteases, oxidases)

Foaming grasshopper

Monarch caterpillar

Poison frog

Scorpion

Firefish

Scolopender

Cobra

Examples of a few poisonous animals

Poisonous and mind-altering plants in human history – murder, magic and medicine

Because humans are at least partly herbivores, they had to cope with the toxic properties of plants from early times. Due to their intelligence, humans learned to avoid or to use them as poisons (for murder, euthanasia, executions, suicide, abortion and hunting) or for medicine. Medicine is only the other side of the coin: A natural product which functions as a protective poison for the plant at a high concentration, often exhibits medicinally useful pharmacological properties at lower doses.

In developing civilisations, knowledge about the medicinal and toxic properties of plants was closely interwoven with mythology. The god of medicine and therapy was Asclepias (with a snake as insignia, which is still in use today). Knowledge of the intrinsic activities of poisonous and mind-altering plants was often restricted to a small elite: priests, medicine men, shamans, magicians or witches. Since overdosing could be lethal for a patient or the shaman himself, the use of plant extracts demanded expertise in dose effects. At least some of the substances and plants that were used as poisons or medicines in antiquity have been successfully converted into clinically acceptable drugs today.

Conium maculatum

Digitalis purpurea

Colchicum autumnale

Examples of famous poisonous plants

Murder and magic

Poisonous and psychoactive plants were used by hunters, priests, witches and magicians. Some secondary metabolites, which block the neuronal signal transduction to the muscles and cause immediate paralysis when injected, played a significant role as **arrow poisons** for hunting and warfare. Examples are aconitine, *Taxus* alkaloids, atropine, toxiferine, tubocurarine and most importantly various cardiac glycosides. The ancient Greek word "toxicon" originally meant "poison for arrows". In Central and South America, frogs of the genera *Phyllobates* and *Dendrobates* secrete toxic alkaloids, such as batrachotoxin or pumiliotoxins. Batrachotoxin is five times more potent than tetrodotoxin and achieves its toxicity by activating Na^+-channels, thus disturbing neurotransmission. Natives used, and still use, these toxins as potent arrow poisons – the secretion of one frog provides poison for 50 arrows. In the Kalahari desert, San hunters use the larvae of the poison beetle (*Diamphidia* species) to prepare very potent arrow poison. However, the main arrow poisons in Africa are fast-acting cardiac glycosides from genera such as *Acokanthera, Boophone, Strophanthus* and *Adenium*. Crushed plant materials are also widely used as fish poisons to stun fish.

Before and during the Middle Ages, several other alkaloids and toxins were used for **suicide, murder and executions**; infamous are coniine, aconitine, atropine, strychnine, colchicine, cardiac glycosides, or cyanogenic glycosides. Extracts of *Conium maculatum* were frequently used for murder, suicide (which was not stipulated in antiquity) and execution (according to Dioscorides, Juvenal, Libanius, Theophrastos and Plinius). The most famous victim was Socrates, who was forced to drink hemlock juice in 399 BC, because – in the eyes of the people of Athens – he had taught his students about the gods in a contemptuous way and had spoiled the youth. According to Tacitus, *Conium* extracts were used by Nero to murder his stepbrother Britannicus, after the effectiveness of the poison had been previously tested on a billy-goat. In ancient Greece and Rome, hemlock was usually mixed with aconite or oleander for a quicker execution. A famous example of this type of execution in Africa is the trial by ordeal concept, where a poisonous decoction is administered to someone accused of a serious offence. The person is deemed innocent if he or she vomits and thus survives the poisonous potion; if not, then the verdict is guilty and death is the automatic punishment (see notes under the ordeal tree *Erythrophleum suaveolens* and *Physostigma venenosum*).

Death of Socrates (J. L. David, 1787)

The yew tree (*Taxus baccata*) occurs in Eurasia (other *Taxus* species occur in America) and provides an interesting example of a highly poisonous plant known in antiquity. Nikander (200 BC) described its use as a poison which renders a death very painful because the throat becomes contracted. According to Dioscorides, chickens die by suffocation and humans by diarrhoea. Yew toxins were used for suicide; Caesar describes in "de bello gallico" the Celtic chieftain Catuvolcus (53 BC) who preferred to ingest yew rather than become a Roman slave. The ancient Celts employed extracts from bark and needles to prepare potent arrow and spear poisons. Also the goddess Artemis used arrows poisoned with yew extracts for hunting and murder. According to Plinius, the extracts were first called "taxica" which later developed into "toxica" (from which modern "toxins" thus obtained their name). The yew was dedicated to the furies, the Erinyes (the goddesses of vengeance), who pursued evil-doers with their anger and punished them with the toxic yew. Theophrastos stated that animals die when ingesting needles, but that the fruits are not toxic. This is true; the red and fleshy aril is indeed without toxins, but the seed (which is eliminated intact via the faeces) is (see *Taxus baccata*). Another use of extracts made from needles was for abortion, which often had a fatal outcome for both mother and embryo, since 50 to 100 needles might be lethal. The wood of the yew tree, which can become up to 2 000 years old, has been esteemed for woodwork since ancient times, and is a traditional material for the manufacture of bows.

In addition to utilisation for suicide, murder, executions and hallucinogens, a number of plants were also used as **abortifacients** in antiquity and medieval ages. These were mostly plants rich in cytotoxic alkaloids, essential oils and sesqui- and triterpenes (such as cucurbitacins). Examples are *Bryonia dioica, Claviceps purpurea, Helleborus viridis, Juniperus sabina, Petroselinum crispum* and *Taxus baccata*. Many women have died because they used the wrong plant or inadequate doses. The development of birth control with the Pill or condoms has made abortifacient plants obsolete and our awareness and knowledge about these properties have gone. Who thinks about abortion when eating parsley at a buffet?

Many alkaloids and alkaloid-producing plants have been discovered by mankind as possessing **stimulating, euphoric and hallucinogenic properties**. Each continent (especially South America, Europe and Africa) has its own set of psychoactive plants, but common to them all is that the alkaloids or terpenoids involved usually affect signal transduction in neurons of the brain: either by modulating the receptors of neurotransmitters or hormones (e.g. hyoscyamine, scopolamine, nicotine, muscarine, morphine, psilocin, psilocybin, mescaline, N,N-dimethyltryptamine, harmaline, ergot alkaloids, etc.), the degradation of neurotransmitters (MAO, cholinesterase: e.g. β-carboline alkaloids, physostigmine, galanthamine) or their re-uptake into the presynaptic neuron and/or secretory vesicles (e.g. cocaine, ephedrine, reserpine). Some psychoactive agents affect ion channels.

Mind-altering plants of the Old World

Unfortunately, very few of the ancient customs in Africa (the cradle of mankind) have been documented, so that we have to look at current practices to get a glimpse of the past. Mind-altering plants range from mild stimulants such as coffee (*Coffea arabica*), khat (*Catha edulis*) and kanna or sceletium (*Mesembryanthemum tortuosum*) to aphrodisiacs and euphorants (e.g. blue water lily, *Nymphaea nouchali*), powerful sedatives (e.g. wild yam, *Dioscorea dregeana*) and even hallucinogens, such as poison bulb (*Boophone disticha*) or *Tabernanthe iboga*. In southern Africa (the home of the oldest of human cultures, dating back to 70 000 years or more), *Boophone* bulb scales were traditionally used to embalm Khoisan bodies and are also strongly associated with the traditional "trance dance" (an ancient form of healing and divination).

The mind-altering properties of plants must have puzzled early man and it is plausible that corresponding descriptions found their way into mythology. Female gods have been described very early in this context. One of them is Kybele, the goddess of animals, mountains and medicine. It is likely but not documented that the Kybele cult employed hallucinogenic alkaloid drugs. Near the Black Sea, Hecate was worshipped as the goddess of witchcraft. Helpers of Hecate were the Pharmakides ("pharmacists"), witches experienced in medicinal and toxic plants. It has been said that Hecate was the first to use *Aconitum* as a deadly poison and tested it when having guests for a party. Hecate had a "botanical garden" on the island of Colchis where the following alkaloidal plants were kept: Akoniton (*Aconitum napellus*), Diktamnon (*Dictamnus albus*), Mandragores

(*Mandragora officinarum, M. autumnalis*), Mekon (*Papaver somniferum*), Melaina (*Claviceps purpurea*), Thryon (*Atropa belladonna*) and *Colchicum* species.

Our knowledge about drug use in antiquity is also derived from the Odyssey by Homer. He might be the author of the epic poems the Iliad and the Odyssey. However, Homer is apparently a legendary figure rather than a historical person. The Iliad and the Odyssey are probably the products of a long tradition of orally composed poems. The poems can be the product of a single person or they rather slowly evolved towards their final form over a period of centuries; in this view, they are the collective work of generations of poets. In the Odyssey we can discover a number of episodes which describe the utilisation of psychoactive plants.

The most important hallucinogens of antiquity came from plants with tropane alkaloids (*Atropa, Mandragora, Scopolia, Hyoscyamus*), ergot alkaloids (*Claviceps*) and opium (*Papaver somniferum*). A famous description of tropane alkaloid intoxication can be found in the story of Circe and Ulysses (see p. 10). Ulysses had apparently used an antidote in order to remain sober. We can speculate that Ulysses had ingested bulbs of some lilies (such as *Narcissus, Galanthus* and *Leucojum*), which contain the alkaloid galanthamine, before he met Circe. This alkaloid is an active inhibitor of cholinesterase and would counteract the activity of tropane alkaloids (see hyoscyamine). It should be noted that witches' potions also contained extracts from several plants containing tropane alkaloids, such as *Datura, Hyoscyamus, Scopolia* and *Mandragora* and additional compounds with hallucinogenic properties, such as skins from toads (*Bufo bufo*) with *N*-methyltryptamine and bufotenin (besides bufotoxine, an esterified bufadienolide). Three *Hyoscyamus* species were well known to ancient Greeks and Romans. According to Plinius, henbane was known in Greece as "Herba Appolinaris" and taken by the priestesses of Appollon for preparing their oracles. The priestesses of the Delphic oracle were said to have inhaled smoke from smouldering henbane. In this context it is remarkable that scopolamine (which is abundant in henbane), was used in modern times for "brain-washing". Because of the higher scopolamine content in henbane, the hallucinogenic properties are more pronounced than in *Atropa*. During medieval times, tropane alkaloids (especially scopolamine) were applied as ointments and resulted in the sensation of flying. Because of massive abuse of henbane and other drugs, many people became conspicuous and were either burnt as witches or directly executed. It has been reported that the condemned were given a drink with henbane to sedate and tranquillise them, to make death less agonising.

Aphrodisiacs were also famous and sought after. Most of them contain extracts of *Mandragora* and of other plants with tropane alkaloids, such as *Hyoscyamus, Atropa* and *Datura*. The goddess of love, Aphrodite was therefore called "mandragoritis". Another historic remark: extracts of *Atropa* or *Hyoscyamus* were used by ancient doctors to anaesthetise their patients during surgery (sometimes combined with alkaloid extracts of *Papaver somniferum*). This application came to a stop during the Middle Ages when witch-hunting started. Since patients would talk about the erotic dreams (induced by the alkaloids) they had after surgery, the risk of being accused of being a sorcerer became too great for surgeons. For many years thereafter, no anaesthetic was available until the advent of modern compounds such as ether (1846), chloroform and other products.

Poppy is a Eurasian plant that has played a special role in the history of humans. Its latex is rich in benzylisoquinoline and morphinane alkaloids, such as morphine, codeine, thebaine, noscapine and papaverine (see *Papaver somniferum*). The dried latex, prepared from unripe fruits, is better known as opium (the alcoholic solution of which is the famous "laudanum"). The poppy was known in antiquity and the Sumerians used it as early as 4000 BC. It was also employed in ancient Egypt, Persia, Arabia and Greece. In Greece, poppy was dedicated to the gods of death "Thanatos", of sleep "Hypnos" and of dreams "Morpheus". Dioscorides described the harvest of opium by cutting the unripe poppy fruits. He gave an excellent description of the medicinal properties of opium: a pea-sized piece of opium kills pain effectively, induces sleep, activates digestion, silences cough and stomach troubles. Applied on the head (together with rose oil) opium helps against headache, and mixed with almond oil, saffron and myrrh helps against earache when brought into the auditory channel. But an overdose of opium will produce extreme lethargy and death. Dioscorides also described the use of poppy seeds (which are rich in oil but almost devoid of alkaloids) for baking bread (as is done today) and the adulteration of opium with latex of *Papaver rhoeas, Glaucium flavum* (these plants were mentioned by Theophrastos as a purgative) or other lactiferous plants. More colourful descriptions can be found in the Odyssey, where opium ("nepenthes") taken with wine made people peaceful and let them forget all sorts of sorrows and pain. When Persephone was kidnapped by Hades to the "underworld", her sorrows were alleviated when Demeter provided her with opium. Vergilius reports in the "Aeneas" that "nepenthes" would even calm the extremely dangerous three-headed dog Cerberus.

The Romans used opium medicinally and for murder in the same way as the Greeks. According to Plinius (50 BC) the imperator Nero was an ardent user of toxins and probably murdered his step-brother Britannicus with a mixture of hemlock and poppy. Although Roman doctors were officially not allowed to dispense opium for people wishing to commit suicide, there are several descriptions from Plautus and other writers suggesting that they did. The Romans were fond of "catapotiae", which had strong effects as a painkiller or sleeping potion since they contained extracts from *Papaver somniferum*, *Conium maculatum*, *Mandragora officinarum* or *Hyoscyamus niger*. According to our present knowledge, the use of these plants was based on pharmacologically active alkaloids, such as morphine, coniine, hyoscyamine and scopolamine. Hannibal's caring relatives, according to Silius in "de bello punico", regularly helped him to sleep with opium, because he suffered from insomnia. Opium in combination with tropane alkaloids was the main narcotic during the Middle Ages, but was more or less abandoned during the times of witch-hunting. It should be mentioned that the use of morphine underwent a renaissance in the 20th century. Because of its exceptional pain-killing properties, it is increasingly used to treat cancer patients. Opium smoking was introduced to China in the early 17th century, where it became a major problem. Wars were fought to maintain a free trade in opium. At the beginning of the 20th century, many countries agreed to limit opium production, but only with limited success. The misuse of opium or morphine is still a problem today, less in the form of "morphinism" (a problem formerly common with doctors) but more with heroinism: Since heroin (a diacetyl derivative of morphine, which was first marketed by Bayer Co. in 1898) is more lipophilic, it can pass the blood-brain barrier of the brain more easily, and produces stronger sensations and addiction. Because the market value of opium is still high, poppy is illicitly cultivated in many parts of the world. Opium is processed and distributed through several "dark channels" across the world, from Afghanistan, Pakistan and other countries in SE Asia to Europe, Asia or America.

Boophone disticha

Nymphaea caerulea

Catha edulis

Cannabis sativa

Examples of mind-altering plants of the Old World

Mind-altering plants of the New World

In Central and South America, the use of hallucinogenic plants followed a different (although in essence similar) route: sources indicate that medicinal and psychotropic plants as well as mushrooms have been used by Indians for more than 4 000 years. Written evidence originates from the sixteenth century, for example from Hernandez ("Rerum Medicarum Novae Hispanieae Thesaurus") and Bernardino de Sahagun ("Historia general de las cosas de la Nueva Espana").

A rich knowledge had been gathered by the Aztecs, who used toaloatzin (*Datura*), ololiuqui (*Turbina corymbosa*), teonanacatl (*Psilocybe mexicana*), peyotl (*Lophophora williamsii*) and several other drugs for divination and magic-religious purposes. Further famous mind-altering plants of the New World include *Banisteriopsis caapi, Anadenanthera peregrina, Brunsfelsia, Brugmansia* and *Datura, Calea ternifolia, Ipomoea, Nicotiana, Theobroma cacao,* and *Turnera diffusa*. Besides shamans, brujos (witch-doctors), yerberos (herbalists) and curanderos (healers) still exist today. Whereas the former two are more concerned with magic and divination, the herbalists and healers, who are lower in social status, collect plants for medicinal applications. Although Central and South America were Christianised four centuries ago, some of the magic-religious rites of the Aztecs still exist today (albeit in Christian "guise", i.e. connected with the Virgin Mary, or the Saints Peter and Paul).

An interesting story concerns the narcotic alkaloid cocaine. The stimulant properties of coca leaves must have been known for a long time; the mummy of a Peruvian king in the Nazca area, dated AD 500, was accompanied by several bags of coca leaves. During the Inca civilisation (ca. 10th century), the utilisation of coca leaves was well established but restricted to a few and used to silence hunger, to provide vigour and to help to forget the miseries of life. Coca was an integral part of religious ceremonies and initiation rites and the shamans used it to induce trance-like states to communicate with the gods. After the Incas were conquered by Pizarro in 1533, coca use was no longer restricted to the shamans and Indians started coca-chewing everywhere to enjoy its stimulating effects. The Conquistadores brought coca back to Europe and it was even celebrated as an "elixir of life". In the 1860s Angelo Mariani successfully invented the "Vin Mariani", which contained coca extracts and aphrodisiac damiana extracts (*Turnera diffusa*), which was said to have analgesic, anaesthetic and carminative properties. The wine was officially approved by the Vatican under Pope Leo XIII. In the USA, a "Peruvian wine of Coca" was sold from Sears, Roebuck and Co. with similar indications. Even more famous was Coca-Cola, which originally contained true extracts of *Erythroxylum coca* together with caffeine-containing *Cola nitida* and wine and was the

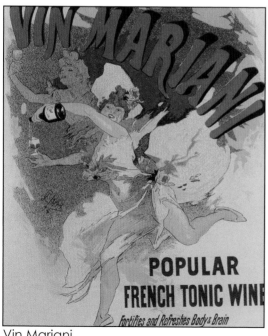

Vin Mariani

Absinthe was a famous drink in Europe

22

Argemone mexicana   *Psilocybe mexicana*   *Datura innoxia*

Erythroxylum novogranatense Lobelia tupa   *Lophophora williamsii*

Examples of mind-altering plants of the New World

equivalent to the Vin Mariani. In 1886, the wine was replaced by sugar syrup and the beverage was called Coca-Cola. In 1904, coca extracts had to be eliminated from Coca-Cola, but the trade name has been maintained until today. In Cola Lite of present time, even the sugar has been removed.

Cocaine was isolated in the 1860s and became available in pure form. In 1884, Sigmund Freud experienced the stimulating and euphoric properties of cocaine and in the same year his assistant C. Köller demonstrated its anaesthetic properties and introduced cocaine as a local anaesthetic for eye surgery. In 1915, cocaine was purified and used in large quantities for medicinal purposes by the Merck company in Darmstadt, for example. It was in the 20th century that cocaine became an infamous drug which was first experienced by the "intelligentia", such as Aldous Huxley. At present, cocaine is illegally produced in South America but exported for illicit purposes to North America and Europe. Cocaine is applied by snorting (via the mucous membranes of the nose – facilitating an immediate passage to the brain), by smoking or intravenously. While cocaine provides a powerful stimulating effect on the pleasure centres of the brain, it also leads to addiction and the long-term consequences of drug abuse. It has been estimated that more than 30 million Americans have taken cocaine at least once and cocaine abuse is a severe problem in many countries of the world at present. More than 15 million Indians, the so-called "coquereos", regularly chew coca leaves (ca. 50 g fresh leaves per day, equivalent to 0.4–2 g cocaine) together with a little lime, which converts cocaine into its free base for better absorption. An increased endurance, a suppression of hunger and an easing of life's difficulties are experienced. Cocaine does not primarily affect the dopamine receptor but interferes with the re-uptake of noradrenaline and dopamine.

Early medicine

The earliest evidence that humans used medicinal, poisonous and mind-altering plants is derived from Assyrian clay tablets, written in cuneiform characters. On these 6 000 year old plates, about 250 different plants are mentioned which include a number of poisonous plants, e.g., *Papaver somniferum, Atropa belladonna,* and *Mandragora.* As in Mesopotamia, empirical knowledge of medicinal plants developed in China; the Emperor Shen Nung described 365 drugs in about 3000 BC. In India, the traditional medicine was documented in the Ayurveda about 900 BC. Its main remedy was "soma", which had to be collected under moonlight and had to be extracted while praying. The effect of soma included slight intoxication and a feeling of strength, courage and sexuality. The identity of the source plant has been disputed: its main components could have been the hallucinogenic plants *Cannabis sativa, Sarcostemma acidum, Sarcostemma brevistigma* or more likely the mushroom *Amanita muscaria.* From these older cultures, some of the knowledge eventually reached the Mediterranean countries through traders and migrations. About 1550 BC medicinal plants were described in the Ebers Papyrus (the German Egyptologist, Professor Ebers, had obtained the papyri in 1872 in Luxor and had recognised its content and value). In addition to recipes for purgatives (including castor oil, rhubarb, aloe and senna), some toxic plants were mentioned: henbane (*Hyoscyamus*), pomegranate (*Punica granatum*) and poppy (*Papaver somniferum*).

The ancient Greeks could rely on Sumerian, Egyptian and Indian traditions. Early "pharmacognosy" reached a summit with Hippocrates (460–377 BC), who critically reviewed more than 200 medicinal plants (overall, more than 1 000 plant species were known to ancient Greeks and Romans). Hippocrates based his medicine on empirical data and freed medicine from the mythical past. He influenced Western medicine substantially and the "oath of Hippocrates" is still relevant today. Besides Aristoteles (384–322 BC), it was Theophrastos of Eresos (371–287 BC), who produced a first "Historia Plantarum". In this major text, the available knowledge on more than 450 plant species was systematically summarised. During the next centuries, Greek science concentrated on astronomy and mathematics and not on botany. But the Greek knowledge was passed to the Romans Polybos and Plinius (AD 23–79). Pedanios Dioscorides (ca. AD 40–90) produced the famous "De Materia Medica" in AD 78 which described more than 500 medicinal plants (roughly 8 % of the plants that occur in the eastern Mediterranean) and their uses in detail, even mentioning the various synonyms. Dioscorides recognised plants for over 50 medical indications. This *Materia Medica* represents a rather modern approach, since the importance of efficacy and doses was already recognised. The work was used as the standard textbook until the Middle Ages and served as the basis for various herbals. The Greek Claudius Galen (AD 131–199) combined the wisdom of Aristoteles, Theophrastos and Hippocrates with the anatomical and physiological knowledge developed in Perganum, Smyrna, Corinth and Alexandria. Besides 304 medicinal plants, Galen

Dioscorides and his Materia Medica

employed complex mixtures, which were specially devised for each therapy; he thus became the founder of "galenics".

In Arabia, Avicenna (AD 930–1036) based his medicine on Hippocrates, Dioscorides and Galen. He was followed by Ibn al-Baitar (1197–1248) who recognised 1 400 drugs and medicinal plants, combining the work of Dioscorides and that from the Far East. In Europe, several herbals appeared from the Middle Ages onwards, which eventually led to modern medicine and pharmacy, which often forget their roots.

In modern medicine, several isolated natural products from poisonous and mind-altering plants are still in use to treat various diseases and disorders including ajmaline, anthraquinones, atropine, berberine, caffeine, camptothecin, cardiac glycosides, cinchonine, codeine, colchicine, ergobasine, ergotamine, ephedrine, lobeline, morphine, noscapine, papaverine, quinine, sparteine, thebaine, yohimbine, reserpine, scopolamine, serpentine, sanguinarine, strychnine, taxol, tubocurarine, vinblastine, vincristine, etc. A more detailed description can be found in "Medicinal Plants of the World" another book in this BRIZA series.

Importance of poisonous plants in modern life

This book describes many highly poisonous or mind-altering plants. Poisoning with plants or mushrooms represent usually less than 5 % of all poisoning incidents that are recorded in Western countries. However, the number of consultations has been increasing over the years. Thus, reliable information about toxins and poisonous plants becomes increasingly more important. If the plants in question can be identified in time, an adequate and necessary therapy can be endorsed to avoid serious poisoning or even death.

Ingestion of hazardous material

Babies and young children explore their environment and do not hesitate to lick, chew or eat plants or mushrooms. Coloured plant parts such as flowers, seeds and fruits appear to be especially attractive. A bad or bitter taste is often not as repellent to infants as it is to adolescents and adults. Many young children eat plant leaves or brightly coloured berries collected from plants in gardens, near playing grounds, in parks or from ornamental pot plants. Therefore, most consultations of poison control centres concern children or pet animals. In Germany, for example, about 90 % of all cases were children and adolescents. Consultations are more numerous than actual poisoning. Serious cases involving plants are rare; only 9 % of all cases needed clinical therapy. More substantial and abundant are intoxication through chemicals, cosmetics, medicines and cigarettes. During summer and autumn, when many plants produce fruits in the northern hemisphere, the number of ingestions especially by children increases; at Christmas time, problems can occur when babies or small children chew on Christmas decorations, which sometimes are made of toxic plant materials. Necklaces may be made of attractive plant seeds, such as seeds from *Abrus precatorius* and *Ricinus communis*. Since the seeds contain highly toxic lectins, poisoning can occur in the people who produce the necklaces and in children or teenagers who start to chew on the beads.

The chances that **adults** will suffer from accidental poisoning are small, except when collecting and eating wild mushrooms. It has become a fashion in Europe to collect wild herbs and to eat them as a salad. If collectors are not able to distinguish wild garlic (*Allium ursinum*) from *Convallaria majalis* or *Colchicum autumnale* (see pictures on p. 29), which grow at similar places and have similar leaves at the same time of the year, serious poisoning is likely. Such cases, and even fatalities, are more common in Africa, where wild herbs are still important food items.

Poisonous and mind-altering plants are used in phytomedicine in most countries of the world, so that poisoning can also occur if isolated compounds or plant extracts are overdosed during medical treatment. For example, cardiac glycosides are widely used in cardiac medicine to treat heart insufficiency. Since these compounds have a small therapeutic window, accidental poisoning by overdosing is not a rare event. The extensive drinking of teas with coumarins (*Galium odoratum, Melilotus officinalis, Dipteryx odorata*) can lead to liver failure and blood clotting problems. This also applies to several other recipes from traditional medicine, including TCM (Traditional Chinese Medicine) and African Traditional Medicine.

25

Poisoning in adults is often the result of a deliberate action; the plants or substances are either ingested with suicidal intentions or as hallucinogens. The last decades have seen a number of serious intoxications in teenagers experimenting with hallucinogenic plants, such as those with tropane alkaloids (*Atropa, Datura, Hyoscyamus, Brugmansia*). In addition, poisonous plants or plant extracts have been and are still employed for murder or to induce abortions (today less often than in historic times). In general, intoxication in adults is more serious and about 70 % of all poisoning instances in Germany needed medical therapy.

Symptoms of human poisoning may vary and depend on the type of plant/toxin and quantity ingested. The following symptoms may occur:
- Nausea and vomiting
- Stomach pains or cramps; diarrhoea
- Local irritation, with burning, itching or local pain (mouth, skin or eyes)
- Difficult breathing
- Disturbed vision; delirium or hallucinations
- Convulsions; loss of consciousness
- Palpitations or quick and irregular pulse/heartbeat

Plants that affect the skin

Some plants have compounds that are aggressive and **irritating** to skin and mucosal tissue, such as those with oxalate crystals, capsaicin in *Capsicum*, phorbol esters in Euphorbiaceae and Thymelaeaceae, protoanemonine in Ranunculaceae, tulipalin in tulips, alkylphenols in Anacardiaceae (e.g. poison ivy), polyacetylenes in Asteraceae and Araliaceae, isothiocyanates in Brassicales, podophyllotoxin in Berberidaceae and several monoterpenes in Lamiaceae, Cupressaceae and Myrtaceae. Especially painful is contact of these substances with the eye.

Several secondary metabolites are **phototoxic**, such as furanocoumarins from various Apiaceae and Rutaceae or naphthodianthrones in *Hypericum* and *Fagopyrum*. If skin that came into contact with these compounds is exposed to UV-rich sunlight, the compounds become activated and interact with proteins and DNA (see chapter on Toxins). Affected cells often die from apoptosis and slow healing burns can be a result (e.g. after contact with *Heracleum mantegazzianum*).

A number of compounds can cause allergic reactions; infamous are contact dermatitis caused by alkylphenols of Anacardiaceae, Hydrophyllaceae or Proteaceae, tulipalins in tulips or sesquiterpene lactones in Asteraceae and Lauraceae. Since the number of allergenic plant species is very large, only a few of the more common and potent species are covered in this book. A comprehensive handbook has been published by Hausen et al. (1997).

Some plants can cause wounds mechanically by **thorns and spikes** or stinging hairs that are filled with noxious chemicals (sometimes neurotransmitters and histamine). Such a contact can lead to local inflammation. This group of hazardous plants is also generally not covered in our book, but see *Urtica dioica*.

Statistics of plant poisoning

The Poison Centres in Europe and North America have long lists indicating which plants have been involved in poisoning cases. Some examples of these statistics are given here. A cursory look at the list indicates that the plants which turn up in consultations are not necessarily the most poisonous plants known to mankind. Instead, they include many plants that produce attractive fruits and that are found in gardens and parks as ornamental plants. As can be seen in the table, there are many regional differences but also examples of cosmopolitan plants. Some of the "poisonous" plants are harmless, such as *Mahonia aquifolium*, but are nevertheless the subject of enquiries. Species considered as less hazardous are **printed in bold** and do not require a thorough detoxification. Some of the pot plants and potted flowers that are commonly sold in nurseries and florist shops are poisonous. Customers are usually not aware of a potential danger. The monographs in this handbook will describe and characterise most of these potentially harmful pot plants.

Statistics of plant poisonings				
Scientific name	Number of consultations			Poison class
	Southern Germany	Switzerland	USA	
Arum maculatum	212	646		Ia
Atropa belladonna		474		Ia
Berberis species	145	221		III
Brugmansia species	635			Ib
Capsicum annuum			4095	II
***Chrysanthemum* species**			998	II
Clivia miniata	1074			Ia
Colchicum autumnale		118		Ia
Convallaria majalis	455	904		Ib
***Cotoneaster* species**	645	1190		III
***Crassula* species**	459		1293	II
Daphne mezereum	120	401		Ia
Dieffenbachia species	545	777	1435	Ia
Digitalis species	129			Ib
Epipremnum species	206		1087	Ib-II
***Eucalyptus* species**			820	II
Euonymus europaeus	127	216		Ib
***Euphorbia pulcherrima* and other Euphorbiaceae**	499	1069	3073	II
***Ficus* species**	2519		1697	III
Hedera helix	516	245	936	II
Ilex species	474	454	3091	Ib
Laburnum anagyroides	467	364		Ia
Ligustrum vulgare	830	354		II
***Lonicera* species**	792	775		II
Mahonia aquifolium	940	713		III
Narcissus species		289		II
Nerium oleander	238		894	Ib
Phaseolus vulgaris	257			Ib (raw)
***Philodendron* species**	222	318	3280	II-III
Physalis alkekengi		352		II-III
Phytolacca americana			2502	II
Prunus laurocerasus	1074	1428		II
Pyracantha coccineus	364	694	701	II-III
Rhododendron species			939	II
Sambucus nigra	373	690		III-II
Schlumbergera species	125		816	II
Solanum species		640		II
Sorbus aucuparia	1150	617		III
Spathiphyllum species			3557	Ib
Symphoricarpos albus	344	184		II
Taxus baccata	2015	1022		Ia
Thuja species	221	141		Ib
Tulipa species	130	226		II
Viscum album	297	276		II

Although many plants are somewhat hazardous, only a small number are actually deadly. Usually a large quantity of berries, pods, leaves, flowers or seeds must be eaten from most plants before serious symptoms occur. Since children and pets are usually mostly affected, the following list provides suggestions of how to avoid poisoning or to help the poison centre in case of intoxication.

- If you have small children or curious pets, keep all plants (including a bouquet of flowers in a vase) out of their reach.
- Know the names (including scientific name) of all your plants both inside and outside the house. Make sure that all adult members of the household have this information. Show grandparents and babysitters where to find the plant names. Poison Centre staff cannot identify a plant simply from a description given over the phone.
- Since mushrooms and berries are very attractive to children, teach your child not to put any part of a plant (leaves, berries, stems, bark, seeds, flowers, nuts, pods and bulbs) into their mouth. Teach children that sucking the nectar from flowers can be dangerous.
- Never let children chew on necklaces made of decorative seeds or beans.
- When you are camping, hiking, picnicking or travelling with young children, take syrup of ipecac or medicinal charcoal with you. If an ingestion occurs, you will be prepared and can induce vomiting or detoxification.
- If you store bulbs and seeds over the winter, label them and place them safely out of the reach of children and pets. Do not confuse bulbs with onions or shallots.
- Wear gloves and protective clothing when handling plants that can be irritating to the skin, e.g. *Heracleum mantegazzianum*. Wash hands and clothes well afterwards.
- Smoke from burning poisonous plants (especially poison oak, *Rhus toxicodendron*) can irritate the eyes, nose, throat and lungs.

Euphorbia pulcherrima (poinsettia)

Mahonia aquifolium (Oregon grape)

Physalis alkekengi (winter cherry)   *Sorbus aucuparia* (quickbeam)   *Symphoricarpos albus* (snowberry)

Non-toxic plants, berries and fruits

Animal poisoning

As a general rule, plants that are listed as toxic to humans should also be considered toxic to animals. There are a few cases of plants considered non-toxic or mildly toxic to humans causing problems to animals. If you keep cats or dogs indoors, make sure that they cannot feed on your toxic pot or garden plants. The ingestion of a few leaves from oleander (*Nerium oleander*) can easily kill a cat or rabbit.

Farm animals and game often depend entirely on wild plants for their daily food intake. Both grazers and browsers may nibble on poisonous plants but few species are sufficiently toxic to cause harm in small mouthfuls. Under normal circumstances there are rarely any problems, because the bulk of the daily intake of plant material is non-toxic. Furthermore, small amounts of toxins can be broken down or rendered harmless by microorganisms in the rumen. For this reason, monogastric animals (pigs, ostriches, chickens etc.) are often more sensitive than ruminants (cattle, sheep, goats etc.). However, when drought and other factors (such as fires) cause temporary or permanent shortages of suitable grazing, hungry animals may be forced to eat poisonous plants that they would normally avoid. Poisoning may also occur when animals are moved from one area to another (or even from one camp to another) so that they come into contact with toxic plants that they have not yet learnt to avoid.

Farmers should take care that livestock do not feed on garden clippings as they may contain hazardous plants. There are several reports of horses having died after feeding on clippings from yew trees. Although most farm animals avoid eating toxic plants, they may do so if other food becomes rare, for example in times of drought. It is therefore advisable to get rid of highly poisonous plants in places where cattle, horses, sheep or goats feed. This book also deals with livestock poisoning in cases where it is relevant.

Allium ursinum   *Convallaria majalis*   *Colchicum autumnale*

Whereas *Allium* can be used as a vegetable, leaves of *Convallaria* and *Colchicum* are deadly poisonous.

Seeds of *Ricinus communis*   Seeds of *Abrus precatorius*

Toxic seeds which are used as beads in necklaces

First aid treatment

Emergency treatment of poisoning should be handled with great care and caution by unskilled persons. Supportive therapy (see below), as in other medical emergencies, is the most important aspect of handling a case of plant poisoning. The saying "treat the patient, not the poison" remains the most basic rule. It is critically important to get the poisoned person to the nearest hospital with emergency facilities as soon as possible. Stay calm and think logically and systematically.

The purpose of induction of vomiting is to prevent further absorption of the poison from the stomach. However, although many textbooks on poisoning or first aid state that vomiting should be induced, it must be remembered that in the case of plant poisoning, vomiting should not be tried as a first aid measure, because of the danger of inhaling (aspiration) plant material into the lungs during the process. Also, vomiting frequently occurs spontaneously in any case, due to the irritation caused by ingestion of toxic plants.

Fortunately, because of the unpleasant, often bitter taste of poisonous plants and the accompanying pronounced irritation of the inside of the mouth and throat, most people will not eat enough plant material to develop serious signs of poisoning.

Diagnosis

Before any treatment can start, one should try to identify the poisonous plant, and find out which part of it has been eaten. Keep any small pieces of the plant (leaves, seeds, berries) that you have found to show to the doctor or pharmacists and send these with the casualty to the hospital or poison centre. If a person has vomited, information can be gained from plant residues or substances found in the vomit. Morphological characters as described in this handbook can help to identify toxic plants; consult other botany texts for plant identification guides if you do not find the plant in this book or if you are not certain about the correct species (see Further reading). It is important to identify the species; a wrong identification can either lead to unnecessary treatment (if a harmless plant is concerned) such as gastric lavage or to inadequate treatment when a more comprehensive treatment was actually required, as in the case of a highly hazardous species.

To extract toxic substances, various solvents are typically used. Shredded or powdered plant material (or urine, blood) is shaken up in the solvent and left for several hours (with or without heat), so that the soluble plant compounds can dissolve in the solvent. Solvent extraction can also be done in a special glass perculator known as a soxhlet apparatus. Non-polar compounds are usually extracted with methanol or dichloromethane, while more polar compounds such as sugars, glycosides, proteins, amino acids and lectins are extracted with water. The unique property of alkaloids to form water-soluble salts under acidic conditions, allow their easy separation from other compounds. A weak acid is used, and the alkaloids are extracted as salts (other, non-polar compounds will not dissolve). After careful filtration of the liquid, a base (such as ammonia) is added so that the solution becomes alkaline (pH of 9 or more). Under alkaline conditions, the alkaloidal salt becomes a free base once again and can now be extracted from the water phase with a solvent such as chloroform or dichloromethane – all non-alkaloidal compounds remain water-soluble. After extraction, the solvents are removed by evaporation to leave behind a solid residue (the crude extract) that will contain the required toxic compounds.

Chemical methods of identification often require that the compound to be identified must be pure – it must therefore be separated from all other compounds in the crude extract. A process called column chromatography is used. A glass tube fitted with a valve at the base is filled with a solid stationary material. The mixture or crude extract is applied on top of the column and a specially prepared solvent mixture, the so-called mobile phase, is slowly and carefully washed through the column. Different compounds will move at different rates down the column, depending on their solubility in the mobile phase. Silica gel is the most commonly used stationary phase solid for small molecules. The silica gel is polar and the more polar a compound is, the stronger it will be adsorbed onto the silica. The less polar constituents will elute much faster from the column than more polar ones. Fractions are collected from the valve at the bottom of the column. The first fractions will therefore contain the least polar and most soluble compound or compounds, while the later fractions will contain more polar constituents. For larger molecules such as proteins and ectins, the column is filled with a special stationary phase that forms countless minute pores of a specific (required) size. Small molecules are trapped inside the pores, while the larger ones move faster down the column, a process known as size-exclusion chromatography. Fractions containing the single pure compound can then be dried by evaporating the mobile phase, often leaving the compound as small crystals or as an oily residue in the bottom of the test tube. The chemical structure of the toxin can now be determined using mass spectrometry (MS) and nuclear magnetic resonance spectroscopy (NMR). Biological activity tests and toxicity tests can also be done.

Chromatographic methods include thin layer chromatography (TLC), high performance liquid chromatography (HPLC) and high-resolution capillary gas-liquid chromatography (GLC). Whereas TLC is a quick and inexpensive method, it might be difficult to identify less common or less abundant toxins. HPLC and GLC are more sensitive and offer a better resolution of individual compounds. Both have the advantage that they can be coupled with mass spectrometers (HPLC-MS, GLC-MS). Results are recorded in the form of chromatograms; the compounds are shown as a series of peaks, each with a particular retention time (i.e. the time it took for the molecules to move through the HPLC or GLC column). The larger the peak, the higher the concentration of the compound. By comparing the chromatographic behaviour of the unknown compound with that of an authentic reference sample of a known compound, the unknown can be identified. Using mass spectrometry, one can easily identify known compounds on account of their known mass fragmentation patterns. If isolated substances are available, NMR (both ^{1}H-NMR, ^{13}C-NMR and various coupling methods) is the method of choice to elucidate the structure of known and unknown compounds. For a number of mind-altering substances, specific and quick assays are available including ELISA tests and colour reactions.

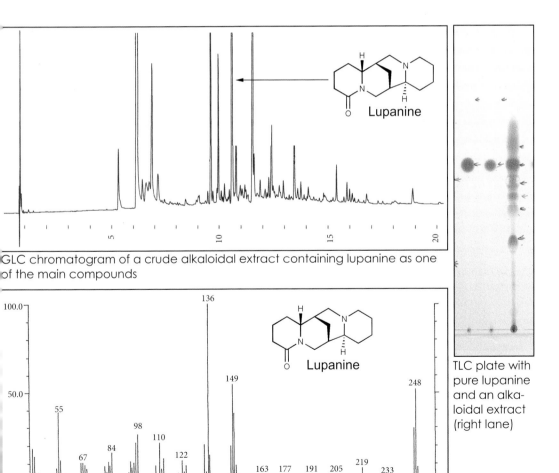

GLC chromatogram of a crude alkaloidal extract containing lupanine as one of the main compounds

Mass spectrum of lupanine

TLC plate with pure lupanine and an alkaloidal extract (right lane)

31

Therapy

A comprehensive therapeutic treatment would be completely unnecessary if a non-hazardous plant has been ingested. On the other hand, one should not wait for the symptoms of poisoning before starting treatment, as this may result in severe illness or even death in case of highly hazardous plants or substances. Therefore, as mentioned above, identification of the ingested material is of prime importance.

In most instances, only small amounts of plant material are eaten or the plant was not really hazardous. If symptoms occur, they usually include gastrointestinal disturbances such as nausea, vomiting, diarrhoea and abdominal pains. In mild cases, patients should be given ample amounts of liquids to drink (tea, juices, water) and possibly powdered medicinal charcoal (children 0.5–1.0 g/kg bodyweight; adults 50–100 g) to absorb the toxins. Activated charcoal is a deep black very fine, odourless and flavourless powder, capable of adsorbing (binding) various substances including toxins, poisons and medicines, because of its large surface area and porosity. The dangerous substances are thus inactivated. Activated charcoal is available from most pharmacies and administering it to a person who has ingested poisonous plant material could be an important first aid action. If successful intake and retention of this mixture has taken place, the chances are good that most of the toxic or poisonous ingredients from the plant will be bound and inactivated in the gut and excreted harmlessly from the body via the faeces (stool).

A laxative, such as Glauber's salt (sodium sulphate), could be helpful for a rapid elimination of toxins and medicinal charcoal. Charcoal tablets are less effective, since their adsorption capacity is lower than that of charcoal powder. Drinking milk is no longer recommended, because milk can enhance the uptake of lipophilic toxins.

In case only small amounts of a known hazardous plant have been ingested (e.g. if a child has only sucked on a leaf or has eaten fewer than three berries), the patient should be observed but not treated. If large amounts of a slightly hazardous plant have been consumed (resulting from confusion with a food plant), or if significant amounts of known hazardous plant material has been ingested, a full therapy, as outlined below, might be necessary in order to avoid severe poisoning and even death. **Remember: As mentioned in the introduction – the dose makes the poison.**

First aid – removal of toxins

Skin-irritating substances
If the skin has been exposed to irritating or noxious plant juices or extracts, it should be washed with warm water and soap as soon as possible. Lipophilic (fat-soluble) substances, such as essential oils or milk juice from Euphorbiaceae can be removed with a polyethylene glycol 400 solution (e.g. Roticlean). Inflammation from chemical burns should be treated several times a day with a cortisone ointment.

Ingested plant parts or toxins
A general tool to prevent toxins from being absorbed in the gastrointestinal tract into the blood is pulverised medicinal charcoal, which can bind most substances with a high capacity. Usually 30–100 g **medicinal charcoal**, suspended in water, is given to the patient to drink. Children get 0.5–1 g/kg body weight. Medicinal charcoal adsorbs most lipophilic drugs efficiently. If the patient vomits, apply medicinal charcoal again. Provide at least 10 times more medicinal charcoal as the ingested toxin.

If medicinal charcoal is not available, a primary toxin elimination can be achieved by inducing vomiting. In a first step, the patient should drink large amounts of any liquid (except alcohol or milk). Then the throat must be tickled so that a vomiting reflex is initiated. The head should be lower than the body during this procedure. The process should be repeated until plant residues can no longer be detected in the vomit. Do not use hypertonic sodium chloride (NaCl) to induce vomiting, since this can lead to NaCl poisoning. In children, the application of "**syrup of ipecac**" induces spontaneous vomiting (see antidotes). Medicinal charcoal will inactivate emetine and thus the emetic activity. Since ipecac contains the toxic alkaloid emetine, this treatment should be carried out with care. It should not be applied on a regular basis as this can lead to severe poisoning and even death. Medicinal charcoal should be given as soon as possible after vomiting in order to

32

bind any toxin that has already been transported down the intestines. As a prophylactic measure against shock symptoms, patients should receive warm tea and sweet juice (e.g. raspberry).

Vomiting should not be provoked when patients are unconscious or dizzy, or in case of spasms or shock. This treatment is also contraindicated when patients have ingested strong irritants, organic solvents or detergents. If the patient is already showing symptoms of severe poisoning (dizziness, sleepiness or excitation) or is no longer conscious, gastric lavage and further clinical treatment should be carried out in a clinic.

Clinical therapy

Elementary help from trained personnel is necessary:
- To support and maintain respiration (clear respiratory tract from vomit; supply oxygen and artificial respiration; control lung oedema)
- To maintain circulation (treatment or avoidance of shock; reduction of arrhythmia and cardiac arrest; stabilisation of cardiac activity)
- To stop spasms, colic and convulsions

Given that the poison is known, a causal therapy should address the organ-specific disturbances.
Gastric lavage: This is a safe and efficient way to eliminate ingested poisons and should be carried out as soon as possible. Gastric lavage may be indicated in certain cases, but this depends on a medical doctor's discretion. The procedure must only be performed in a hospital or emergency room by trained medical staff. Patients are firstly treated with atropine and are then positioned on a bed lying on their abdomen. Lavage entails "washing out" of the stomach contents, by means of a thick tube (18 mm diameter) that is placed into the patient's stomach and through which a certain volume of water is administered and then sucked out immediately afterwards. The process is repeated several times until only clear liquid is sucked out, meaning that all harmful material in the stomach has been removed. In case of alkaloids, potassium permanganate is added (colour resembling burgundy red wine) to inactivate the toxins. It is usually of no use to perform gastric lavage if more than four hours have elapsed since intake of the poisonous plant, because by then most of the ingested material will have passed into the intestines.

After lavage, 30–100 g (adults) or 1 g/kg (children) of medicinal charcoal and sodium sulphate (2 spoonfuls, 15–20 g) are instilled. In case of lipophilic poisons, 150 ml polyethylene glycol 400 can be applied.

Further clinical treatments
- **Acidosis treatment:** In order to stimulate kidney function and to avoid shock, an infusion of 1 M sodium bicarbonate solution is often adequate to make the blood more alkaline.
- **Electrolyte substitution:** After heavy vomiting and diarrhoea, the loss of electrolytes should be overcome by either drinking mineral-rich juice or by infusions with K^+, Na^+, Cl^--ions.
- **Provision of plasma expander:** Infusion of plasma expander (Haes 3 %) or of plasma can counteract shock due to poisoning.
- **Colic:** Intestinal spasms or colic can be treated intravenously with atropine or metamizol.
- **Spasms:** In case of epileptic spasms inject diazepam (Valium) or a barbiturate (Luminal) intravenously. Monitor the occurrence of cerebral bleeding when spasms occur several times.
- **Fever:** Fever can be treated physically by cooling with icy water or pharmacologically with metamizol (Novalgin) or acetylsalicylic acid (Aspirin).
- **Electrocardiogram:** In case of poisons that cause arrhythmia, ECG should be recorded as soon as possible to detect bradycardia and tachycardia. Bradycardia can be treated with atropine or Alupent; tachycardia with Xylocain or quinidine. If cardiac arrest occurs, the patient needs immediate cardiac massage and artificial respiration.
- **Artificial respiration:** In case of shallow breathing, blue lips or loss of consciousness, immediately provide artificial respiration, including intubation and mechanical respiration.
- **Reanimation:** Initial cardiac arrest can be overcome by cardiac massage, artificial respiration, injection of adrenaline, acidosis treatment and infusion of plasma expander.
- **Blood dialysis:** In a few cases of severe poisoning, blood dialysis, peritoneal dialysis, blood transfusion or forced diuresis may be indicated.

Relevant antidotes and solutions

- **Atropine:** Inject i.v. or i.m. 1 mg (adults), 0.2 mg (babies), 0.4–0.6 mg (children) atropine in case of colics and bradycardia. Repeat every 2 hours if necessary.
- **Chibro-Kerakain:** Apply 1–2 drops into the eye in case of painful chemical burns before washing the eye with a washing solution.
- **Cortisol ointment:** In case of chemical burns apply several times daily to the affected parts of the skin.
- **Dexametasone spray:** In case of burns in mouth and throat or inhalation or irritants, apply a dexametasone-21-isonicotinate spray frequently.
- **Diazepam (Valium):** Inject 2–4 mg (children) or 10–20 mg (adults) in case of spasms.
- **4-Dimethylaminophenol (4-DMAP):** Inject 3 mg/kg i.v. in case of cyanide poisoning, followed by sodium thiosulphate (50–100 mg/l 10 % i.v.)
- **Medicinal charcoal:** Suspend 30–100 g (adults) or 0.5–1 g/kg (children) powdered medicinal charcoal (carbo medicinalis) in water.
- **Naloxon:** Inject 0.4–0.8 mg in case of adults and 0.01 mg/kg for children (i.v., i.m., s.c.) if respiratory arrest is caused by opiates.
- **Physostigmine:** Inject (i.m., i.v.) 0.03 mg/kg bodyweight in case of poisoning with anticholinergic substances (e.g. atropine).
- **Polyethylene glycol 400 (Roticlean, lutrol):** Use this to clean the skin from lipophilic toxins or for gastric lavage (1.5 ml/kg).
- **Sodium sulphate (Glauber's salt):** For induction of diarrhoea after the application of medicinal charcoal, instil 2 teaspoons (adults) or 1 teaspoon (children).
- **Potassium permanganate:** Dissolve a few crystals to make a 0.05 to 0.1 % solution (colour resembling red wine). Use for gastric lavage in case of alkaloid, cyanide and glycoside ingestion.
- **Syrup of ipecac:** A recipe to induce spontaneous vomiting in children is given here. Handle with care, because ipecac is hazardous on its own (see *Psychotria ipecacuanha*). The syrup can be stored in a fridge for up to 3 months. Dose: Children under 1.5 years: 2 teaspoons (10 ml); 1.5–2 years: 1 tablespoon (15 ml); 2–3 years: 1.5 tablespoons; above 3 years: 2 tablespoons (20 ml). Provide plenty of water to drink after vomiting.

Recipe

Dry extract of ipecac (*Psychotria ipecacuanha*)	1.37 g
Glycerol	10.0 g
p-Hydroxybenzoic acid propylester	0.025 g
p-Hydroxybenzoic acid methylester	0.075 g
Sugar syrup	84.4 g

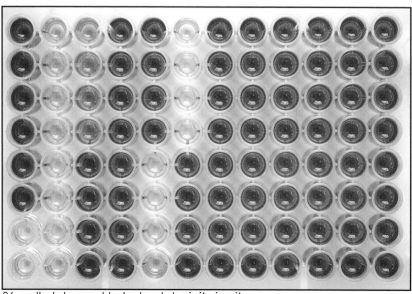

96-well plate used to test cytotoxicity in vitro

Methods of testing for toxicity in humans

There are no foolproof ways of testing the toxicity of plant extracts or pure compounds for humans. There are two main approaches: *in vivo* tests (in live animals) and *in vitro* tests (using laboratory models). In animal experiments, several different species are used (often rodent and nonrodent models). The LD_{50} toxicity test is done to determine that dose which kills 50 % of the test animals. It is important to note that the route of administration (oral, dermal or inhalation) may lead to very different LD_{50} values. Some Menispermaceae alkaloids, for example, are very poisonous when injected but harmless when ingested (so that the meat of animals killed with poisoned arrows or blow darts can safely be consumed). Some substances are considered non-toxic except when inhaled and absorbed through the lungs. Animal experimentation has become controversial but is still considered valuable to predict possible toxic effects that a new chemical compound may have in humans. The modes of action of toxic substances, as well as their absorption, retention and excretion can be studied by quantifying the levels of toxins in blood and urine. Special forms of toxicity, such as carcinogenicity, mutagenicity and teratogenicity, are also tested in carefully designed tests that may last for extended periods. The same is true for substances intended for use on human skin and mucosa (e.g. the throat or eye). Prolonged experiments using animal organs may be necessary. All new or unknown substances intended for use in humans should be carefully tested for safety and toxicity as a matter of principle. Ethical committees at research institutes and universities are responsible for ensuring that animals are not subjected to unnecessary suffering.

For routine testing, laboratory models using isolated human cells (*in vitro* studies) are often used to test for cytotoxicity. The method involves suspended, live human cells that are transferred to a test plate with numerous small wells or hollows of equal size. Different concentrations of a plant extract or pure compound are added to the wells (in order of increasing concentration). After a fixed period of time, a colour reagent (indicator) is added to all the wells. Those cells that have been killed by the plant substances remain uncoloured, while live cells become coloured (often bright pink or purple). Large numbers of compounds (and concentrations) can be tested in a single experiment on human cells without the need to sacrifice animals. However, individual cells may not react in the same way as cells that are part of an organ system. *In vivo* tests therefore only give an indication of toxicity and may lead to unrealistic conclusions if not properly interpreted.

In laboratories all over the world, there are ongoing efforts to gain a deeper understanding of the adverse and curative effects of plant compounds. Valuable plant-based drugs are explored for their toxicological and pharmacological properties, and in the process, animal and laboratory studies play an important role.

Information

This book provides basic information about plant characters, toxic plant substances, symptoms, modes of actions and possible therapeutic treatment. Other reference books are listed on pages 433–435.

In case of emergency, consult your MD or local pharmacist or directly phone the nearest poison centre. Since this book is distributed in many countries, we cannot give detailed addresses. If you have small children, obtain this sort of information from your doctor or the internet and store it in a place that can be easily reached in case of emergency.

TOXLINE is produced and supplied by the National Library of Medicine (USA). It is an extensive collection of online bibliographic information covering the biochemical, pharmacological, physiological and toxicological effects of, amongst others, poisonous plants. It is available, free of charge, on the Specialized Information Services (SIS) (http://toxnet.nlm.nih.gov).

MEDLINE is also produced and supplied by the National Library of Medicine (USA). The service offers a broad coverage of medicine and life sciences. It covers, to some extent, medicinal and poisonous plants. It can be accessed, free of charge, from Pubmed (http://www.ncbi.nih.gov/entrez).

Colchicum autumnale L.

Plants in alphabetical order

Structure of monographs

About 200 of the most relevant poisonous and mind-altering plants are characterised by concise monographs. Essential facts about another 1 000 species are provided in the "Quick guide to poisonous and mind-altering plants" at the end of the book.

Each monograph is organised in the following 12 sections:

NAME: Monographs are listed alphabetically according to the scientific plant name; commonly used English names are given in the next line.

PLANTS WITH SIMILAR PROPERTIES: The monograph usually focuses on one or a few species within a genus as a representative. In cases where several species occur in a genus, we can very often assume that the other species will have similar ingredients and thus similar toxicological properties. Therefore, we mention how many similar species are in a given genus. Sometimes we refer to other genera with similar properties, especially when there is no space for a separate monograph.

PLANT CHARACTERS: Short description of the growth form and morphological characters of the plant; one or two photos illustrate the general appearance of the species (and sometimes close relatives).

OCCURRENCE: Information on the origin and present distribution of the species.

CLASSIFICATION: Short classification of the type of biological activity that is exhibited by the plant, such as "neurotoxin"; classification of a poison class, according to the WHO recommendations (see Introduction):
- Class Ia: extremely hazardous
- Class Ib: highly hazardous
- Class II: moderately hazardous
- Class III: slightly hazardous

ACTIVE INGREDIENTS: Information on the main and toxicologically relevant secondary metabolites (SMs) found in the species.

UTILISATION: Information on the historic and present uses of the plant or substances isolated from it.

TOXICITY: Information on toxicological importance. If available, relevant LD_{50} and LD_{100} values are presented for plant material or isolated compounds. More information is provided in the Toxin monographs.

SYMPTOMS: Description of reported symptoms of poisoning or psychic effects in humans, or sometimes in animals. More information can be found in the Toxin monographs.

PHARMACOLOGICAL EFFECTS: Information on the possible mode of action of SMs present in the plant to understand the symptoms. More information is provided in the Toxin monographs, which cover the topic in more detail.

FIRST AID: Brief information for immediate first aid and, if necessary, the next steps of clinical therapy. More information is given in the Toxin monographs.

SYSTEMATICS: Scientific name, abbreviated authority, major synonyms (if any); and plant family, followed by French, German, Italian and Spanish names, if available.

Left, from Köhler's Medizinalpflanzen (1887)

Abrus precatorius

crab's eye vine • coral pea

Abrus precatorius leaves

Abrus precatorius seeds and pods

PLANTS WITH SIMILAR PROPERTIES *Abrus* has 17 pantropical species (see also *Ricinus*).

PLANT CHARACTERS A woody climber with pinnate leaves and clusters of pale purple, inconspicuous flowers. Each pod has 4–5 brightly coloured, red and black seeds (diameter: 5–9 mm) with a hard, water-impermeable seed coat.

OCCURRENCE *A. precatorius* has a wide distribution in the African and Asian tropics.

CLASSIFICATION Cell toxin; extremely hazardous, Ia.

ACTIVE INGREDIENTS Seeds contain the lectin abrin which is a mixture of 4 lectins, abrin a–d with a MW of 63–67 kDa. Abrin can be inactivated by high temperatures, i.e. cooking in water at more than 65 °C. Roots and leaves contain sweet-tasting triterpene saponins (abrusosides).

UTILISATION *Abrus* seeds are widely used as beads in necklaces, toys, musical instruments, masks and decorations. Necklaces are known to cause dermatitis. Despite their toxicity, the seeds are traditionally (e.g. in India) used as aphrodisiacs, oral contraceptives, abortifacient, or emetics. Powdered seeds are employed to treat eye infections, snakebite, ulcers and intestinal worms.

TOXICITY Abrin is one of the most deadly plant toxins known. Fatal poisoning in both humans and domestic animals has been reported. Seeds become highly poisonous when the tough seed coat is damaged. When chewed well, even a single seed is apparently sufficient to cause death (especially in children). The brightly coloured seeds are attractive to children and they may be poisoned if damaged or punctured seeds are swallowed. The oral LD_{50} of ground seeds in mice is only 2 mg/kg body-weight. About 0.5 g of seeds are fatal for humans (adults). Abrin: LD_{50} in mice: i.p. 10–31 μg/kg; i.v. 0.7 μg/kg.

SYMPTOMS Symptoms appear only after a latent period of several hours. Typical are a loss of appetite, bloody diarrhoea and vomiting, followed by intestinal inflammation (acute gastroenteritis), haemorrhage, diabetes due to pancreas damage, delirium, chills, convulsions and coma. Death is caused by heart failure after 3–4 days. Serious toxic effects occur when abrin comes into contact with wounds or eyes.

PHARMACOLOGICAL EFFECTS Abrin has haemagglutinating properties and inhibits ribosomal protein synthesis (see ricin). It consists of two polypeptide chains connected by disulphide bonds. One of the peptides (B-chain) binds to the cell surface (haptomer), the other (A-chain) inactivates ribosomal protein biosynthesis (effectomer).

FIRST AID Treatment is indicated if damaged or chewed seeds, e.g. from necklaces have been ingested. Treatment involves the induction of vomiting and detoxification as outlined under ricin.

Abrus precatorius L.

family: Fabaceae

pois rouge (French); *Paternostererbse* (German)

Acokanthera oppositifolia

bushman's poison bush

Acokanthera oblongifolia flowers

Acokanthera oppositifolia fruits

PLANTS WITH SIMILAR PROPERTIES A genus of 5 species in Arabia and tropical East and South Africa, including *A. oblongifolia* and *A. schimperi*.

PLANT CHARACTERS *Acokanthera* species are evergreen shrubs or small trees of up to 5 m in height. The elliptical leaves have a thick texture and are glossy green. Clusters of small, sweetly scented, white or pale pink flowers with 5 petals are followed by red to purple, oblong, plum-like fruits of up to 30 mm long.

OCCURRENCE *A. oppositifolia* is widely distributed in C, S and E Africa. *A. oblongifolia* and *A. schimperi* occur mainly in southern Africa.

CLASSIFICATION Heart poison; extremely hazardous, Ia

ACTIVE INGREDIENTS Several cardenolides such as acovenoside A, B, C, acolongoflorodis A, glucoavenocoside B.

UTILISATION *Acokanthera* species have been used throughout Africa as a source of very effective arrow poisons. Ouabain is known to be the active principle of E African arrow poisons. The poison is a gum-like substance, obtained by boiling the leaves, roots or wood in water for a long period. This extract is mixed with *Euphorbia* latex and *Acacia* gum and then applied to arrows. Humans are known to have died within 15 or 20 minutes after being wounded with a poison arrow.

Murderers coat the very prickly fruits of *Tribulus terrestris* with the arrow poison and place them in the path of barefoot victims. *A. oppositifolia* (wood with 1.1 % cardenolides) has also been used as an ordeal poison. Traditional medicinal uses include the treatment of abdominal pain, headache and snakebite. Ouabain is used medicinally as a pure compound to treat acute heart failure.

TOXICITY Poisoning of animals is surprisingly rare but cattle are sometimes at risk during droughts. Cases of murder and suicide have been recorded. Lethal dose for ouabain in cats (i.v.): 0.1 mg/kg; for acovenoside A: 0.2 mg/kg.

SYMPTOMS Accidental or deliberate poisoning with strong decoctions cause heart failure within a few minutes. Symptoms that have been recorded are salivation, nausea and retching, gastrointestinal disturbance, purging and exhaustion. As is typical for heart glycosides, respiratory and cardiac symptoms (arrhythmia, hypertension, coma, cardiac arrest) may occur.

PHARMACOLOGICAL EFFECTS Inhibition of the Na^+, K^+-ATPase (see cardiac glycosides).

FIRST AID First aid treatment is indicated if plant material from *Acokanthera* or isolated cardiac glycosides have been ingested or if persons are injured by poisoned arrows. Treatment involves the induction of vomiting and detoxification as outlined under cardiac glycosides.

Acokanthera oppositifolia (Lam.) Codd (= *A. ouabaio*, *A. venenata*)   family: Apocynaceae

acokanthéra (French); *Wachsbaum* (German)

Aconitum napellus

aconite • monkshood • wolfsbane

Aconitum napellus   *Aconitum napellus* flowers   *Aconitum lycoctonum* flowers

PLANTS WITH SIMILAR PROPERTIES *Aconitum* comprises about 300 species with similar properties such as *Aconitum chinense*, *A. reclinatum*, *A. ferox*, *A. lycoctonum*, *A. carmichaelii*, *A. kuznezoffii*, *A. uncinatum* and *A. columbianum*.

PLANT CHARACTERS An erect perennial herb (up to 1.5 m) with a tuberous rootstock. It has deeply dissected leaves and erect stems with clusters of hood-shaped, purple to blue flowers.

OCCURRENCE *Aconitum* species are widely distributed in the northern hemisphere and are commonly used as ornamentals and cut flowers.

CLASSIFICATION Neurotoxin, mind-altering; extremely hazardous, Ia

ACTIVE INGREDIENTS Mixtures of diterpene alkaloids such as aconitine, mesaconitine; lycoctonine and other terpene alkaloids; aerial parts 0.2–2 %, tubers up to 3 %, and seeds up to 2 %.

UTILISATION Extracts of several species have been used by early humans in Europe, Alaska and Asia as an arrow poison to kill vermin and enemies. Aconite is infamous for murder, suicide and as a deadly poison to remove criminals and unwanted persons. It was also used, with raw meat as a bait, to kill wolves, foxes (reflected in *A. "vulparia"* and "wolfsbane"); even as insecticide. Roots have been smoked together with cannabis in India. During medieval times, aconite extracts were part of witch ointments, inducing the feeling of having wings and being able to fly. *Aconitum* is used as a psychoactive drug in modern days and medicinally as a painkiller for neuralgic pains.

TOXICITY Accidental poisoning of humans is rare but fairly common in livestock and pet animals. 5 g of dried tubers are lethal for dogs; 10–15 g of dried tubers for humans. Aconitine: 3–6 mg are lethal for humans. LD_{50}, mouse, i.v. 0.16 mg/kg, i.p. 0.32 mg/kg, p.o. 1.8 mg/kg; rat, i.v. 0.08–0.14 mg/kg; cat, i.v. 0.07–0.13 mg/kg.

SYMPTOMS Aconite toxication causes burning and paraesthesia of the mouth and throat, followed by insensitivity of fingers and toes, accompanied by cold sweat, general coldness, nausea, severe diarrhoea, bradycardia, arrhythmia, serious pain, and spasms. Death is caused after 30 min to 3 h by respiratory or cardiac arrest (see aconitine). Aconite is strongly psychedelic when smoked or absorbed through the skin.

PHARMACOLOGICAL EFFECTS Aconitine is a Na^+-channel agonist that blocks neuronal and neuromuscular signal transduction. This causes rapid paralysis and anaesthesia (see aconitine).

FIRST AID Indicated if any parts of *Aconitum* have been ingested. It involves the induction of vomiting; the application of sodium sulphate or potassium permanganate and medicinal charcoal; for **clinical therapy**, which includes gastric lavage and artificial respiration, see aconitine.

Aconitum napellus L.   family: Ranunculaceae

aconit napel (French); *Blauer Eisenhut* (German); *aconito* (Italian); *acónito* (Spanish)

Adenia digitata

wild granadilla

Adenia digitata leaves and flowers

Adenia digitata fruits

PLANTS WITH SIMILAR PROPERTIES A genus of 94 species in tropical Africa and Asia.

PLANT CHARACTERS A perennial climber with soft stems growing from an underground tuber. The presence of climbing tendrils are characteristic, as are the deeply lobed to digitate leaves and tubular, almost bell-shaped flowers. The fruits are yellow to bright red, fleshy capsules. Other species known to be poisonous are *A. glauca* (which lacks tendrils and has a distinctive bright green, bottle-shaped stem) and *A. gummifera* (a climber with thick, woody, green stems).

OCCURRENCE *A. digitata* and *A. glauca* occur in subtropical regions of southern Africa, while *A. gummifera* has a wide distribution in southern and eastern Africa. *A. digitata* and *A. glauca* are cultivated by succulent plant enthusiasts.

CLASSIFICATION Cell toxin; extremely hazardous, Ia

ACTIVE INGREDIENTS *Adenia* species contain two toxins – a very poisonous lectin called modeccin and cyanogenic glycosides (the latter are typical of the family). In *A. digitata* and *A. glauca*, tetraphyllin B has been identified as the main glycoside. Modeccin is similar to other lectins in structure and activity (see abrin).

UTILISATION *A. gummifera* is used in traditional medicine as an emetic and for treating leprosy and malaria. It is said to be an effective disinfect-

ant for use around the house. *Adenia* species have also been used as sources of fish poison.

TOXICITY *Adenia* is involved in stock losses, homicide and suicide. *A. digitata* is known to be extremely poisonous and death may occur within a few hours after plant material has been consumed. Accidental and fatal poisoning may occur if the tuberous base of the plant is confused with edible tubers. Adults may apparently be killed by as little as 30 g of fresh root. Children often fall victim to the brightly coloured fruits (several fatalities have been recorded). Modeccin is extremely poisonous – it is claimed that a mere 0.01 mg/kg is lethal to a rabbit (when injected).

SYMPTOMS Typical of *A. glauca* poisoning are nausea, vomiting, diarrhoea and fits. Damage to the kidneys and liver is typical.

PHARMACOLOGICAL EFFECTS Tetraphyllin B releases HCN after cleavage by β-glucosidase; HCN is rapidly inactivated in the human body at low concentrations. Higher doses stop cellular respiration; the resulting ATP shortage leads to general paralysis and causes respiratory arrest (see amygdalin). Modeccin appears to work in a similar way to ricin and abrin, which are very potent inhibitors of cellular protein biosynthesis (see abrin).

FIRST AID Treatment involves the induction of vomiting and detoxification as outlined under abrin and amygdalin.

Adenia digitata (Harv.) Engl.

family: Passifloraceae

Adenium multiflorum

impala lily

Adenium multiflorum flowers

Adenium boehmianum flowers

PLANTS WITH SIMILAR PROPERTIES A genus traditionally accepted to include 6 species from tropical and subtropical Africa and Arabia: *A. obesum* (desert rose), *A. boehmianum*, *A. oleifolium*, *A. swazicum*, *A. socotranum* and *A. somalense*. In this system, *A. multiflorum* is considered to be merely a variety of *A. obesum* and the two are indeed very similar. The alternative system recognises only one species (*A. obesum*), with 6 subspecies: subsp. *obesum*, subsp. *boehmianum*, subsp. *oleifolium*, subsp. *socotranum*, subsp. *somalense* and subsp. *swazicum*.

PLANT CHARACTERS *A. multiflorum* is a thick-stemmed xerophyte up to 3 m in height. When broken, the stems exude a watery, milky sap. The leaves are glossy green and borne at the branch ends. Spectacular clusters of white and pink or white and red flowers are borne in spring. The oblong capsules contain numerous hairy seeds that are wind-dispersed.

OCCURRENCE *A. multiflorum* is widely distributed in the tropical parts of Africa. *A. obesum* occurs from E Africa to S Arabia. *A. boehmianum* is known from N Namibia and Angola. Plants are also cultivated as garden ornamentals, especially some spiny-stemmed species from Madagascar.

CLASSIFICATION Heart poison; extremely hazardous, Ia

ACTIVE INGREDIENTS More than 30 different cardiac glycosides have been isolated from the stems, including cardenolides and pregnane glycosides. The major cardiac glycoside in *A. multiflorum* is obebioside B, but 16-deacetylanhydrohongheloside A (0.6 %) and 16-anhydrostropeside (0.08 %) are also major compounds.

UTILISATION The impala lily and desert rose are known in Africa as sources of fish, arrow and ordeal poisons, prepared from the bark and soft trunk. It is used in combination with other poisons. *A. boehmianum* is used as a source of a highly toxic arrow poison.

TOXICITY Concentrated decoctions or substances from poisoned arrows may cause death within a few minutes due to heart failure.

SYMPTOMS Nausea, salivation, retching, gastrointestinal disturbance, purging and exhaustion, with the usual respiratory and cardiac symptoms (arrhythmia, hypertension, coma, cardiac arrest) associated with heart glycoside poisoning.

PHARMACOLOGICAL EFFECTS Cardenolides inhibit Na^+, K^+-ATPase and are therefore strong general poisons (see cardiac glycosides).

FIRST AID First aid treatment is indicated if plant material or isolated cardiac glycosides have been ingested or if persons are injured by poisoned arrows. Treatment involves the induction of vomiting and detoxification as outlined under cardiac glycosides.

Adenium multiflorum Klotzsch (= *A. obesum* var. *multiflorum*) family: Apocynaceae

lis des impalas, rose du désert (French); *Impalalilie* (German)

Adonis vernalis

yellow pheasant's eye • spring adonis

Adonis vernalis

Adonis vernalis flower

PLANTS WITH SIMILAR PROPERTIES The 26 species of the genus *Adonis* occur in temperate Eurasia and includes *A. annua* (pheasant's eye), *A. aestivalis* and *A. microcarpa*.

PLANT CHARACTERS *Adonis vernalis* is a perennial herb of up to 0.4 m in height with compound leaves that have a feathery appearance. Also typical are the large, bright yellow flowers that are 40–80 mm diameter (in early spring).

OCCURRENCE The species occurs naturally in Europe (excluding the British Isles), West and East Siberia. *A. vernalis* and *A. aestivalis* are indigenous to large parts of the continent, particularly the central and northern regions. *A. annua* occurs in S Europe and SW Asia. *Adonis* species, including *A. aestivalis*, *A. annua* and *A. vernalis* are cultivated as ornamental plants in cold temperate regions of the world.

CLASSIFICATION Heart poison; highly hazardous, Ib

ACTIVE INGREDIENTS Leaves and roots of *A. vernalis* and the other *Adonis* species contain several cardenolides in concentrations of 0.2–1 %. Cymarin is the main active compound (aglycone k-strophantidin), with smaller amounts of adonitoxin (aglycone adonitoxigenin) and 26 further cardenolides.

UTILISATION *Adonis vernalis* is sometimes used as a heart stimulant, in the same way as more familiar sources of heart glycosides such as digitalis. It is considered particularly useful when cardiac conditions are accompanied by nervous symptoms. Several traditional uses are known, including the treatment of bladder and kidney stones.

TOXICITY Concentrated decoctions or i.v. injections may cause rapid death due to heart failure (see cardiac glycosides). LD_{50} for adonitoxin in cats: 0.191 mg/kg, i.v.; cymarin: 0.11–0.13 mg/kg, i.v. Poisoning of horses by *A. aestivalis* or *A. microcarpa* has been reported.

SYMPTOMS Symptoms include nausea, salivation, colic, gastrointestinal disturbance, purging and exhaustion, with the usual respiratory and cardiac symptoms (arrhythmia, hypertension, coma, cardiac arrest) associated with heart glycoside poisoning (see cardiac glycosides).

PHARMACOLOGICAL EFFECTS In general, cardiac glycosides inhibit the Na^+, K^+-ATPase, which is one of the most critical targets in animal cells (see cardiac glycosides).

FIRST AID First aid treatment is indicated if large amounts of plant material from *Adonis* or isolated cardiac glycosides have been ingested. Treatment involves the induction of vomiting, administration of medicinal charcoal, sodium sulphate and detoxification, including gastric lavage as outlined under cardiac glycosides.

Adonis vernalis L.

family: Ranunculaceae

adonis du printemps (French); *Frühlings-Adonisröschen* (German); *adonide gialla* (Italian)

Aesculus hippocastanum

horse chestnut

Aesculus hippocastanum

Aesculus hippocastanum seeds and capsules

PLANTS WITH SIMILAR PROPERTIES A genus of about 13 species in Europe, Asia and N America.

PLANT CHARACTERS A tall, deciduous tree (up to 30 m) bearing large, palmately compound leaves. The flowers are usually white with pink spots and are borne in attractive, oblong clusters. They are followed by spiny capsules containing the large, shiny brown seeds.

OCCURRENCE Horse chestnut has a wide natural distribution from E Europe (the Balkans) to C Asia (the Himalayas). It is a popular garden and street tree and several cultivars and hybrids have been developed.

CLASSIFICATION Cell poison; slightly hazardous, III

ACTIVE INGREDIENTS The main ingredient of seeds (up to 10 % of dry weight) is known as aescin (which is actually a mixture of saponins). The main saponins are glycosides of monodesmosidic protoaescigenin and barringtogenol C. Procyanidin B_2 is the main proanthocyanidin found in the seed coat (oligomers of catechol and epicatechol occur in horse chestnut).

UTILISATION In contrast to edible chestnut (*Castanea sativa*), the seeds of horse chestnut are poisonous. Extracts from the ripe seeds are used in traditional and modern medicine to treat symptoms of venous and lymphatic vessel insufficiency (including varicose veins, and ulcerations of the legs and also piles). Seed, leaf or bark tincture are either taken orally or gargled to treat mouth ulcers. They are also used as an ingredient of skin creams for their venotonic activity.

TOXICITY Human poisoning with deadly outcome has been described. Overdosing of extracts during medicinal treatment can cause anaphylactic shock.

SYMPTOMS Severe gastrointestinal disturbance which is caused by the saponins, can be seen after ingestion of only one seed. The symptoms include flushing (reddened skin), thirst, mydriasis, oedema, vomiting, diarrhoea, hypotension, unconsciousness and collapse.

PHARMACOLOGICAL EFFECTS Saponins strongly interfere with cellular membranes and can cause cytotoxic effects. On skin and mucous membranes, irritation may occur at low doses (see saponins). Since some of the saponins are only weakly haemolytic, the toxic effects of horse chestnut and its extracts are moderate but care should nevertheless be taken to avoid overdosing.

FIRST AID Provide medicinal charcoal and plenty of fluids to drink. In severe cases of allergic reactions, the application of antihistaminics and corticoids can help to overcome serious symptoms (see saponins).

Aesculus hippocastanum L.

family: Sapindaceae or Hippocastanaceae

châtaigner de cavalle, (French); *Rosskastanie* (German); *ippocastano* (Italian); *castaña de Indias* (Spanish)

Agrostemma githago

corn cockle

Agrostemma githago

Agrostemma githago flower

PLANTS WITH SIMILAR PROPERTIES A genus of 2 European and Mediterranean species, including *A. githago* and *A. brachylobum*.

PLANT CHARACTERS A tall hairy annual with narrowly lanceolate leaves and pale reddish-purple flowers of 30–50 mm in diameter. The calyx tube is very hairy and ribbed. Sepals are hairy, lanceolate and longer than the petals. Many small black seeds are borne in capsules.

OCCURRENCE. *A. githago* and *A. brachylobum* originated from the Mediterranean but were introduced as weeds in C and NW Europe. Corn cockle is grown as an ornamental plant.

CLASSIFICATION Cell toxin; highly hazardous, Ib

ACTIVE INGREDIENTS All parts, but especially the seeds, contain triterpene saponins such as githagin (aglycone githagenin = gypsogenin), agrostemmic acid. Furthermore, a non-protein amino acid, orcylalanin (0.4 %) and a toxic lectin "agrostin" = RIP1) are present in seeds.

UTILISATION As a weed of corn fields, the seeds of corn cockle can contaminate grain. It was formerly a widespread weed in cereal fields but has become rare as a result of extensive herbicide spraying. The plant is also a traditional medicine for treating skin problems, gastritis and cough.

TOXICITY In former times, consumption of grain contaminated with *Agrostemma* seeds resulted in poisoning of humans and livestock. Horses, pigs and calves are very sensitive, while chickens are less susceptible. LD_{50} of triterpene saponins in rats: 2.3 mg/kg i.v.; 50 mg/kg p.o. Lethal dose in pigs: 2–5 g seeds/kg body weight. Doses higher than 5 g seeds are lethal for humans.

SYMPTOMS Irritation of mucosal tissues in the mouth and throat is followed by tear formation, nausea, vertigo, headache, weak but enhanced pulse, spasms, abdominal pain, shock and respiratory arrest, coma and death.

PHARMACOLOGICAL EFFECTS The triterpene saponins enhance the uptake of the lectin by cells. Both of them are toxic; the saponins affect the permeability of the cell membrane whereas the lectin inhibits protein biosynthesis.

FIRST AID Give medicinal charcoal and paraffin oil. Contact with eyes or skin: wash with polyethylene glycol or water; in case of skin blisters, cover with sterile cotton and locally with a corticoid ointment. **Clinical therapy:** After ingestion, gastric lavage (possibly with potassium permanganate), instillation of medicinal charcoal and sodium sulphate, as well as electrolyte substitution is required. Control acidosis (with sodium bicarbonate, urine pH 7.5) and kidney function; in case of spasms give diazepam; against colic atropine; in severe cases supply intubation and oxygen.

Agrostemma githago L.

family: Caryophyllaceae

nielle des champs (French); *Kornrade* (German); *gittaione* (Italian)

Aleurites fordii

tung oil tree

Aleurites fordii flowers

Aleurites fordii capsules

PLANTS WITH SIMILAR PROPERTIES A genus of 5 species in the Indomalayan and W Pacific region, including *A. montana* in C Asia and Burma and *A. moluccana* in SE Asia. Less well known are *A. cordata* and *A. trisperma*. All of them are poisonous.

PLANT CHARACTERS A large tree, 6–15 m in height, with thick branches and large, heart-shaped leaves borne on long petioles. The flowers are white with dark pink veins. The seeds or nuts are borne in rounded, pendulous capsules that split open at maturity.

OCCURRENCE The tree occurs naturally in C Asia (China); it is widely cultivated.

CLASSIFICATION Cell toxin, co-carcinogen, skin irritant; moderately hazardous, II

ACTIVE INGREDIENTS Several phorbol esters of the tigliane type, such as 12-decanoyl-13-O-acetyl-16-hydroxyphorbol (0.01 %) occur in various parts of the plant, as well as toxic saponins in the fruits. In some parts of the tree, Welensali factor F1 has been found. Also present are polyphenols, such as ellagitannins (geraniin).

UTILISATION The tree is used as an ornamental garden tree to some extent but is grown commercially as a source of so-called tung oil (obtained from ripe seeds). Tung oil has a characteristic, but unpleasant scent and is used as ingredient in paints and quickly-drying varnishes. *A. moluccana* has been cultivated for at least 3 000 years as a source of China wood oil, candle-nut oil or lumbang oil which is included in curries and shampoos. Roots have been used for tattooing on Tonga and on Hawaii. The oil is used medicinally to treat ulcers, bruises and burns.

TOXICITY Poisoning may occur if the fruits, nuts or oil are ingested.

SYMPTOMS If tung oil or tung meal is ingested, it may cause vomiting, diarrhoea, abdominal pain and colic, and irritation of the skin and internal organs. The oil is a strong purgative.

PHARMACOLOGICAL EFFECTS. Phorbol esters activate protein kinase C, which is a key enzyme in several signalling pathways. They are co-carcinogens (see phorbol esters). Welensali factor F1 can activate certain tumour viruses, such as Epstein-Barr virus, which is associated with nasopharynx carcinoma in China and Burkit lymphoma in Africa.

FIRST AID Give medicinal charcoal and sodium sulphate. Provide plenty of tea. Contact with eyes or skin: wash with polyethylene glycol or water in case of skin blisters, cover with sterile cotton and locally apply corticoid medication. **Clinical therapy:** Gastric lavage (possibly with potassium permanganate), instillation of 10 g charcoal and sodium sulphate; for further treatment see phorbol esters.

Aleurites fordii Hemsl. (= *Vernicia fordii* (Hemsl.) Airy Shaw)　　　　　　　　　　　　　　family: Euphorbiaceae

alévrite, bois de Chine (French); *Tungölbaum* (German)

Amanita muscaria

fly agaric

Amanita muscaria

Amanita pantherina

FUNGI WITH SIMILAR PROPERTIES *A. pantherina* (panther cap), *A. gemmata* (crenuated amanita), *A. regalis*, *Tricholoma muscarium*.

FUNGAL CHARACTERS Spectacular and unmistakable; the large red head (50–200 mm in diameter) has several white dots. The stem is 80–200 mm long. Panther cap is similar but has a brown head.

OCCURRENCE The fly agaric and panther cap are common in Europe, Asia and N America.

CLASSIFICATION Neurotoxin; hallucinogen; moderately hazardous, Ib–II

ACTIVE INGREDIENTS Both mushrooms accumulate active nitrogen-containing compounds, such as ibotenic acid (0.2–1 %), muscimol, and traces of muscarine. Ibotenic acid is metabolised in the body to the more active muscimol and excreted in the urine. The red colour is due to betalaines (muscaflavine, muscapurpurine).

UTILISATION Fly agaric is a natural insecticide. It might be the oldest of all hallucinogens, ingested by nomadic people in Europe, Asia (Siberia) and N American Indians (perhaps a relict of their Siberian origin, some 10 000 years ago?). Shamans in Siberia are said to use it dried or suspended in milk. Since urine becomes hallucinogenic, poor people recycled it up to 5 times. The fierceness of the Vikings and the soma of ancient India are ascribed to fly agaric. *A. pantherina* is also used as a psychoactive drug.

TOXICITY 100 g of fresh mushroom are lethal for humans. About 10 mg muscimol induce dizziness and psychic excitation in humans; 15 mg cause mental disorientation, hallucination and tantrums. Higher doses can be lethal. LD_{50} of muscimol in rats 4.5 mg/kg i.v.; 45 mg/kg p.o.; in mice 2.5 mg/kg i.p., 3.8 mg/kg s.c.; ibotenic acid: LD_{50} in mice 15 mg/kg i.v., 38 mg/kg p.o.

SYMPTOMS Similar to ethanol intoxication, setting in after 30–90 min: unusual visions, euphoria, disorientation and dizziness. Hallucinations are both pleasant and horrible. Some people dance, jump in the air or have a serious tantrum. Also nausea, vomiting, diarrhoea, ataxia, paralysis, epileptic seizures, tachycardia, arrhythmia, mydriasis and dry mouth, followed after 12 h by a deep sleep. Higher doses cause coma and death.

PHARMACOLOGICAL EFFECTS Whereas muscarine affects the "muscarinic" AChR as an agonist, muscimol activates the $GABA_A$ receptor leading to inhibition of motoric functions. Ibotenic acid is an agonist at glutaminergic receptors, especially of NMDA receptor. Muscimol and ibotenic acid are easily absorbed and can pass the blood brain barrier.

FIRST AID Induce vomiting, give medicinal charcoal and sodium sulphate. In serious cases, provide gastric lavage, charcoal, plasma expander, sedation with neuroleptics; artificial respiration.

Amanita muscaria (L. ex Fr.) Secr.

family: Amanitaceae

agaric mouchet (French); *Roter Fliegenpilz* (German); *moscario, tignosa dorata* (Italian)

Amanita phalloides

deadly agaric

Amanita phalloides

Amanita phalloides

FUNGI WITH SIMILAR PROPERTIES *A. virosa, A. verna, A. bisporigera, A. ocreata, A. tenuifulia, Galerina marginata* and *Lepiota brunneo-incarnata* also produce amanitin. *A. phalloides* and *A. virosa* accumulate phalloidin.

FUNGAL CHARACTERS Head 40–120 mm in diameter, greenish white. Stems white, 80–120 mm long, with a bulbous base surrounded by a whitish sheath. Spores are white.

OCCURRENCE Europe and N America, in forests and parks under trees.

CLASSIFICATION Cell toxin; extremely hazardous, Ia

ACTIVE INGREDIENTS Main toxic groups include cyclic peptides (amatoxins, phallotoxins), and other peptides (antamanide, cycloamanide). Amatoxins are bicyclic octapeptides, among them α- and β-amanitin (up to 0.5 %). Phallotoxins (up to 0.5 %) are bicyclic hexa- and heptapeptides such as the neutral phalloidin, phalloin, phallisin, prophalloin and the acidic phallotoxins phallacin, phallacidin and phallisacin.

UTILISATION Deadly agaric looks superficially similar to edible mushrooms; poisoning occurs regularly in areas where wild mushrooms are collected. About 90 % of fatal mushroom poisonings in Europe are caused by *A. phalloides*.

TOXICITY α-Amanitin: LD_{50} (mouse, i.p.) 0.1 mg/kg; LD_{50} (rat, i.p.) 2 mg/kg; LD_{50} (guinea pig, i.p.) 0.05 mg/kg. Phallotoxins: LD_{50} (mouse i.p., i.m.) 1.8–3.3 mg/kg. Lethal dose for humans amanitin: 0.1 mg/kg; phallotoxins 1–2 mg/kg Half a mushroom (50 g) is a lethal human dose 18–22 % of *A. phalloides* poisonings are fatal.

SYMPTOMS After 6–24 h: bloody diarrhoea vomiting, abdominal pain and colic. In sever cases, serious liver and kidney damage set in after 2–5 days; also bleeding of the GI tract, icterus liver swelling and kidney failure. The heart, CNS and other organs are affected. Death after 7 days is due to liver failure and circulatory arrest.

PHARMACOLOGICAL EFFECTS Amanitin are strong inhibitors of protein biosynthesis, especially in the liver and kidneys. They specifically inhibit RNA polymerase II and thus the formation of mRNA, causing cell death. Amanitins are recycled via the enterohepatic cycle and therefore remain toxic for a prolonged period. Phalloidin binds to F-actin and stabilises microfilaments.

FIRST AID Intensive medicinal care is immediately required: lavage, charcoal treatment electrolyte substitution, acidosis control, haemoperfusion, peritoneal dialysis, haemodialysis and forced diuresis. Control circulation and support heart (e.g. by cardiac glycosides). Useful are liver protecting drugs, such as silibinin (from *Silybum marianum*), penicillin, and vitamins. A strict diet is required after poisoning.

Amanita phalloides (Vaill.) Secr.

family: Amanitaceae

amanite phalloide (French); *Grüner Knollenblätterpilz* (German)

Amaryllis belladonna

march lily • belladonna lily

Amaryllis belladonna flowers

PLANTS WITH SIMILAR PROPERTIES A monotypic genus. The name *Amaryllis* is also used for *Hippeastrum* hybrids, e.g. *H.* x *hortorum* (see p. 276).

PLANT CHARACTERS The March lily is a bulbous plant bearing long, strap-shaped leaves that die back completely in summer. The leafless bulb produces a single, thick, bright red flowering stalk in late summer and autumn. Very large, pale to dark pink, trumpet-shaped flowers are borne in a many-flowered umbel.

OCCURRENCE *A. belladonna* occurs naturally only in South Africa, while *Hippeastrum* (see p. 276) originates from tropical America. Both are popular as garden ornamentals and house plants.

CLASSIFICATION Cell poison; highly hazardous, Ib

ACTIVE INGREDIENTS All parts of *A. belladonna* contain toxic Amaryllidaceae alkaloids, the bulbs being most toxic. Ambelline is the major alkaloid, with lycorine, caranine, acetylcaranine and undulatine. Reports claiming that lycorine is the major alkaloid can perhaps be explained by geographical or seasonal variation.

UTILISATION Extracts have been employed as arrow poisons. The bulb is used in horticulture.

TOXICITY *Amaryllis* and *Hippeastrum* bulbs and seeds are very poisonous. School children are known to have died from eating bulbs of Amaryl-lidaceae but there are no clearly recorded cases of fatal *Amaryllis* poisoning in humans. Care should be taken to protect infants and pet animals from bulbs growing in the garden. When injected into mice, the LD_{50} of ambelline was found to be 5 mg/kg body-weight. Only 200 g of fresh bulb is a fatal dose for sheep, while 2–3 g are lethal to humans. Lycorine is highly toxic – the LD_{50} in dogs is only 41 mg/kg.

SYMPTOMS Symptoms of human poisoning include nausea, vomiting, dizziness, strong perspiration, diarrhoea and kidney trouble. In severe cases, respiratory arrest may occur.

PHARMACOLOGICAL EFFECTS Ambelline has pain-killing properties not unlike that of morphine but is unfortunately too toxic to be used as analgesic in modern medicine. Acetylcaranine shows activity against some forms of leukaemia and is also a uterine stimulant. Caranine is deadly to animals and causes respiratory paralysis. Lycorine is cytotoxic, emetic and diuretic.

FIRST AID Give medicinal charcoal and sodium sulphate. Provide plenty of tea to drink and shock prophylaxis. **Clinical therapy:** After ingestion, gastric lavage (possibly with potassium permanganate), electrolyte substitution, control acidosis with sodium bicarbonate (urine pH 7.5) and kidney function; in case of spasms give diazepam; in severe cases supply oxygen and intubation.

Amaryllis belladonna L.

family: Amaryllidaceae

amaryllis belladonna (French); *Belladonnalilie* (German)

Anadenanthera peregrina

cohoba • yopo • niopo

Anadenanthera peregrina

Anadenanthera peregrina pods

Anadenanthera peregrina seeds

PLANTS WITH SIMILAR PROPERTIES
A. colubrina (cebil, villca) is the only other species. Psychoactive *N,N*-DMT is also found in several other legumes (*Acacia, Mimosa, Mucuna*).

PLANT CHARACTERS A shrub or tree up to 20 m high with black bark and pinnate leaves of 0.3 m long. Rounded terminal racemes of 35–50 small white flowers are followed by pods of 50–350 mm long, each containing 3–15 thin, black, shiny seeds (10–20 mm in diameter).

OCCURRENCE S America (Brazil, Andes). *A. colubrina* mainly in Argentina and the S Andes.

CLASSIFICATION Hallucinogen; mind-altering; moderately hazardous, II

ACTIVE INGREDIENTS Seeds and bark are especially rich in tryptamine derivatives such as *N,N*-dimethyltryptamine (*N,N*-DMT) and *N,N*-dimethyl-5-methoxytryptamine (5-MeO-DMT), as well as other serotonin mimics such as bufotenin. Also present are traces of β-carboline alkaloids (such as 2-methyltetrahydro-β-carboline). Total alkaloids: bark: 0.4 %; seeds: 0.2 %.

UTILISATION Indians of the Orinoco basin of Colombia and Venezuela, in the northern part of the Brazilian Amazon and in northern parts of Argentina have used yopo as psychotropic drug. Seed powders act as strong hallucinogens. Free bases, produced by mixing seed powder with alkaline ash or chalk from snails, are applied nasally as a snuff (called yopo, niopo) or rectally as enema (both more effective than the oral route). The free bases are readily absorbed and pass the blood-brain barrier. Sometimes the drug is smoked together with tobacco or ingested with honey.

TOXICITY Applied intravenously, about 0.5–1.0 mg *N,N*-DMT are enough to induce a state of high intoxication. Higher doses cause toxic effects. Extracts of *Virola* (Myristicaceae) are rich in tryptamines and are used by Indians as arrow poisons because they are extremely strong neurotoxins after injection (see bufotenin).

SYMPTOMS Yopo intoxication sets in after 10–15 min; symptoms include a complete change of consciousness, multidimensional visions and strong and diverse psychedelic hallucinations (conversion into animals, erotic ecstasy, and sensation of flying). Intoxication starts with headache, salivation, vomiting and continues with trance-like conditions including dancing, singing and shouting (see bufotenin).

PHARMACOLOGICAL EFFECTS Bufotenin, *N,N*-DMT and 5-MeO-DMT activate serotonin receptors. β-carboline alkaloids are potent inhibitors of MAO and also serotonin receptor agonists. These activities explain the observed hallucinogenic effects (see bufotenin).

FIRST AID Instillation of sodium sulphate and medicinal charcoal (see bufotenin).

Anadenanthera peregrina (L.) Spegazzini (= *Piptadenia peregrina*)　　　　　　　　　　　　　　　　　family: Fabaceae
yopo (French); *Yopo* (German)

Anchusa officinalis

bugloss

Anchusa officinalis

Anchusa arvensis

PLANTS WITH SIMILAR PROPERTIES A genus with 40 species in Europe, N & S Africa and W Asia, including *A. arvensis* and *A. capensis*. A related genus is *Alkanna*, with up to 30 species in S. Europa, the Mediterranean and W Asia. The latter includes *A. tinctoria* which produces a red dye in the roots, that has been used as a natural dye.

PLANT CHARACTERS Biennial or perennial plant of up to 1.2 m high with rough hairs on all aerial parts. The leaves are lanceolate and the flowers are funnel-shaped and starlike, with 5 deep blue petals. Fruits are light brown in colour, with four unsymmetrical nutlets.

OCCURRENCE *A. officinalis* is common in SE and C Europe, where it occurs in sandy and rural places. A few species are grown in gardens, such as *A. azurea*.

CLASSIFICATION Liver poison, neurotoxin, mutagen; slightly hazardous, II-III

ACTIVE INGREDIENTS All parts of these plants contain pyrrolizidine alkaloids (PAs); especially rich in PAs are the roots and flowers. *Anchusa* accumulates PA monoesters such as lycopsamine, acetyllycopsamine, cynoglossine and consolidine; also triterpene saponins and allantoin. *Alkanna* has several PAs in roots, such as triangularine and 7-angeloylretronecine as well as coloured naphthoquinones (alkannin). In most plants PAs accumulate mainly as PA-N oxides,

which are water soluble.

UTILISATION *A. officinalis* has been used in traditional phytomedicine for various indications and as a vegetable. *Alkanna* roots have been used as colourants of food or drinks.

SYMPTOMS AND TOXICITY Herbs and extracts from members of the Boraginaceae should not be used internally if PAs are present (see senecionine).

PHARMACOLOGICAL EFFECTS When PA plants are ingested, the PA-N oxides are reduced to free PAs in the intestine by the gut microorganisms. The free PAs can then diffuse across the gut epithelia and are transported to the liver. In the liver, PAs are oxided by the enzymes of drug metabolism; reactive pyrroles are generated, which can alkylate DNA bases and proteins (see senecionine). As a result, they cause liver damage, and after prolonged intake are mutagenic, teratogenic and carcinogenic. Consolidine inhibits cholinergic neuronal activity, similar to curare. The saponins have cytotoxic and haemolytic properties.

FIRST AID Give medicinal charcoal to absorb the alkaloids; administer plenty of warm black tea, and sodium sulphate. Provide shock prophylaxis (keep patient quiet and warm). **Clinical therapy:** In case of ingestion of large doses: gastric lavage and further treatments described under senecionine.

Anchusa officinalis L. (= *A. arvalis*)   family: Boraginaceae

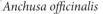

buglosse (French); *Ochsenzunge* (German); *buglossa* (Italian)

Apocynum cannabinum

Indian hemp

Apocynum cannabinum

PLANTS WITH SIMILAR PROPERTIES The 12 species that have been described in the genus *Apocynum* occur from Russia to China and also temperate America. They are similar to species of *Asclepias* and *Gomphocarpus*. Many members of the families Apocynaceae and Asclepiadaceae contain milky latex and heart glycosides. The two families are very closely related. In modern classification systems, the Asclepiadaceae is usually included in the Apocynaceae (as the subfamily Asclepiadoideae).

PLANT CHARACTERS Indian hemp is an erect, multi-stemmed shrub of up to 1 m in height. The long, thin stems grow from a woody underground rhizome (which has yellowish peeling bark and many latex canals) and bear opposite, narrow, bright green leaves of up to 100 mm long. All parts of the plant produce white latex when damaged. White flowers are borne in clusters, followed by typical *Asclepias* fruits (pairs of pointed capsules). The capsules burst open when ripe and the seeds are dispersed by means of silky seed hairs.

OCCURRENCE Indigenous to N America; plants are widely cultivated in gardens.

CLASSIFICATION Heart poison; highly hazardous, Ib

ACTIVE INGREDIENTS *Apocynum cannabinum* is known to contain cymarin, k-strophan-

toside; apocannoside, cynocannoside and several other cardenolides.

UTILISATION Indian hemp has been used in traditional medicine. Roots have the pharmaceutical name of *Apocyni cannabini radix* and have been employed as emetic and cardiac stimulant. The bark is a traditional source of fibre for making ropes, nets and sails (hence the common name Indian hemp). The silky hairs on the seeds are suitable for use as tinder.

TOXICITY All parts of the plant, but especially the roots, are known to be toxic.

SYMPTOMS Concentrated decoctions may cause rapid death due to heart failure. Symptoms include nausea, salivation, retching, gastrointestinal disturbance, diuresis, purging and exhaustion. Also typical are the usual respiratory and cardiac symptoms (arrhythmia, hypertension, coma, cardiac arrest) associated with heart glycoside poisoning.

PHARMACOLOGICAL EFFECTS In general, cardiac glycosides inhibit the Na^+, K^+-ATPase (see cardiac glycosides).

FIRST AID First aid treatment is indicated if plant material from *Apocynum* or isolated cardiac glycosides have been ingested. Treatment involves the induction of vomiting, supply of medicinal charcoal, and detoxification as outlined under cardiac glycosides.

Apocynum cannabinum L.   family: Apocynaceae

chanvre du Canada (French); *Hundswürger* (German); *canapa acquatica* (Italian)

Areca catechu

areca nut

Areca catechu palm

Areca catechu fruits

Areca catechu seeds

PLANTS WITH SIMILAR PROPERTIES A genus of 60 spp., e.g. *A. concinna* and *A. vestiaria*.

PLANT CHARACTERS A slender palm tree (up to 25 m) with 2 m long leaves. Male and female flowers occur in cylindrical inflorescences, forming 150–200 fruits. The fruits are oval and 70 mm long, each containing a brown seed of 3–10 g.

OCCURRENCE *Areca catechu* is a cultigen and is grown throughout tropical SE Asia, in E Africa, and on Madagascar. It was probably domesticated by the Hoabhinians around 8 000–3 000 BP.

CLASSIFICATION Neurotoxin; stimulant; moderately hazardous, II

ACTIVE INGREDIENTS Mainly alkaloids (0.3–0.6 %) with arecoline as main alkaloid, furthermore arecaidine, arecolidine, guvacoline and guvacine. When betel is being chewed some arecoline is hydrolysed to arecaidine.

UTILISATION *Areca* seeds are ingested as "betel", a combination of leaves of the vine *Piper betle*, slices of the seeds and lime (to convert the alkaloids to their free base, which can easily be taken up). For several hundred years, betel chewing has been a common habit of more than 450 million people in Asia and East Africa. This causes saliva to turn red because of the phlobatannins formed after alkaline treatment. The alkaloid free bases are directly absorbed and quickly pass the blood-brain barrier. Arecoline is used in veterinary medicine against intestinal worms of cattle and dogs (formerly also in European tooth powders).

TOXICITY The alkaloids and possibly eugenol from betel are mutagenic and can cause oral cancer. 8–10 g of seeds can be fatal. Arecaidine: LD_{50} mouse: 850 mg/kg p.o.; 520 mg/kg i.v.; arecoline: LD_{50} mouse: 100 mg/kg s.c., 34 mg/kg i.v.

SYMPTOMS Betel increases salivation and dulls the appetite but as stimulant gives a relaxed feeling (similar to alcohol). Side effects are profound sweating, burning in mouth and throat, and nausea. High doses cause bradycardia, tremors, vomiting, central disorganisation, spasms, dilated pupils, diarrhoea, respiratory or cardiac arrest.

PHARMACOLOGICAL EFFECTS Arecoline acts as a parasympathomimetic; it activates mAChR such as pilocarpine. The enhanced secretory activities of glands (salivation) are a consequence. Arecaidine is a mild narcotic; it inhibits the uptake of GABA into inhibitory neurons. It has stimulating and sedating properties.

FIRST AID Give medicinal charcoal. **Clinical therapy:** (large doses) Gastric lavage, instillation of medicinal charcoal and sodium sulphate; 1–2 mg atropine i.v. as antidote. Control of acidosis with sodium bicarbonate. In severe cases provide intubation, artificial respiration, cardiac massage and shock treatment. Check liver and kidney function.

Areca catechu L.   family: Arecaceae

aréquier (French); *Betelnusspalme* (German); *avellana d'India* (Italian)

Argemone mexicana

prickly poppy • Mexican poppy

Argemone mexicana

Argemone ochroleuca

PLANTS WITH SIMILAR PROPERTIES 23 species occur in N and S America and Hawaii.

PLANT CHARACTERS A prickly annual of up to 0.9 m high, with spiny stems and leaves that exude a bright yellow juice when damaged. The flowers are yellow and the spiny fruit capsules contain many black seeds of 2.5 mm in diameter.

OCCURRENCE Originally C & S America; *A. mexicana* and *A. ochroleuca* spread to Africa, Europe and Asia as weeds and ornamental plants.

CLASSIFICATION Neuro- and cell toxin; mind-altering; moderately hazardous, II

ACTIVE INGREDIENTS The isoquinoline alkaloids berberine and protopine (whole plants, up to 0.13 % DW); sanguinarine and derivatives are concentrated in the seeds.

UTILISATION The Aztecs regarded this poppy as food for the dead. The bright yellow sap has medicinal uses (insomnia, eye and kidney complaints). The seeds are narcotic, emetic, sedative, and purgative. A condition known as epidemic dropsy occurs in India when they are used to adulterate mustard seed and mustard seed oil. In Mexico, Chinese immigrants have used the plant as "chicalote opium" as an addictive stimulant. This poppy is smoked as a marijuana substitute and aphrodisiac.

TOXICITY Wheat that was contaminated with *Argemone* seeds has caused lethal poisoning in S Africa. In Nepal and India, edible oil has been adulterated with *Argemone* seed oil, causing epidemic illness (dropsy) with partly deadly outcome. Stock animals avoid the prickly and unpalatable plant but are sometimes poisoned by contaminated grain. About 8.8 ml/kg seed oil is required to produce toxic effects in humans. The alkaloid sanguinarine has an LD_{50} of about 18 mg/kg in mice (i.p.).

SYMPTOMS Seeds or seed oil may cause vomiting, diarrhoea and blurred vision. Swollen legs and other serious symptoms indicate chronic poisoning. Berberine is moderately toxic but has pronounced effects on the respiratory system. Protopine is a powerful stimulant of the heart and uterus and has been linked to glaucoma in India in people who consumed adulterated mustard oil (see chelidonine).

PHARMACOLOGICAL EFFECTS Sanguinarine and berberine are strong DNA intercalating substances. These interactions can lead to frameshift mutations and very likely to malformations and cancer if the substances are ingested over prolonged periods (see chelidonine).

FIRST AID Give medicinal charcoal. **Clinical therapy:** (only after ingestion of large doses) Gastric lavage, instillation of medicinal charcoal and sodium sulphate; in severe cases provide intubation and oxygen.

Argemone mexicana L.

family: Papaveraceae

pavot épineux (French); *Stachelmohn* (German); *pavero messicano* (Italian)

Aristolochia clematitis

birthwort

Aristolochia clematitis   Aristolochia macrophylla flower

PLANTS WITH SIMILAR PROPERTIES An Old World genus of 120 species in the tropics and warm climates of Europe and Asia. Many of the species are perennial creepers (vines) with attractive flowers. All of the species produce toxic aristolochic acid. *Asarum*, with 70 species in N temperate regions, is a related genus which accumulate phenylpropanoids such as α-asarone and also aristolochic acid. The best-known species is *Asarum europaeum* (s. p. 305).

PLANT CHARACTERS *A. clematitis* is an erect perennial herb with stems branching at ground level and large heart-shaped leaves. Tubular yellow flowers, that resemble a human embryo, are borne in the leaf axils.

OCCURRENCE The plant occurs in S Europe and has been naturalised in C and W Europe.

CLASSIFICATION Cell toxin, mutagen, moderately hazardous, II

ACTIVE INGREDIENTS Aristolochic acids and magnoflorine, an alkaloid derived from tyrosine.

UTILISATION *Aristolochia* was known and used in ancient times. Dioscorides, who recognised several *Aristolochia* species, described the following applications: extracts of *Aristolochia* were used by women in childbed, against snake bites and intoxication. In addition, it was said to help against internal inflammations and in wound-healing. *Asarum* and *Aristolochia* were used as abortifaciens. Because of the carcinogenic properties of aristolochic acid, extracts of *Aristolochia* are no longer allowed in pharmaceutical preparations.

TOXICITY In the last decade, several instances of serious *Aristolochia* poisoning (including kidney tumours) have been reported when TCM drugs were confused with *Aristolochia*. At least 70 cases of fibrous interstitial nephritis were recorded in Belgium. The likely cause was substitution of *Stephania tetrandra* (*hanfangji*) with *Aristolochia fangchi* (*guangfangchi*) (both official in China).

SYMPTOMS High doses lead to vomiting, spasms, tachycardia, serious kidney damage, hypotension and convulsions; death is caused by respiratory failure.

PHARMACOLOGICAL EFFECTS Aristolochic acids can be activated metabolically in the liver to reactive intermediates that can alkylate DNA bases. Mutagenic and carcinogenic effects are a consequence.

FIRST AID Give medicinal charcoal. **Clinical therapy:** (only after ingestion of large doses) Gastric lavage, instillation of 0.1 % potassium permanganate with medicinal charcoal and sodium sulphate; electrolyte substitution; control acidosis with sodium bicarbonate (urine pH 7.5); check kidney function; in severe cases provide intubation and oxygen respiration.

Aristolochia clematitis L.   family: Aristolochiaceae

arrasine (French); *Osterluzei* (German); *aristolochia clematite* (Italian)

Armoracia rusticana

horseradish

Armoracia rusticana leaves

Armoracia rusticana flowers

PLANTS WITH SIMILAR PROPERTIES A genus of 4 species in Europe, Asia and N America; *Capparis spinosa* and other members of the Capparaceae are used in food (capers); they also release mustard oil. *Nasturtium officinale* and *Tropaeolum majus* are other plants with glucosinolates, which could cause problems after overdosing.

PLANT CHARACTERS A stemless perennial herb with large, dark green leaves growing from a thick taproot. The small white flowers have 4 petals and are borne on leafless stalks (up to 1 m).

OCCURRENCE Uncertain, possibly SW and C Asia. Fruits do not form viable seeds and the plant is probably a sterile hybrid. It has been cultivated since ancient times from root cuttings.

CLASSIFICATION Cell toxin, slightly hazardous, III (at higher doses)

ACTIVE INGREDIENTS Roots contain glucosinolates (mustard oil glycosides), of which gluconasturtiin and sinigrin are the main compounds, respectively forming phenylethyl isothiocyanate and allylisothiocyanate after hydrolysis.

UTILISATION Horseradish is traditionally used to treat bronchial conditions and urinary infections, and externally for relief of rheumatism and inflammation. It is a popular food flavourant and commercial source of the enzyme peroxidase.

TOXICITY Allylisothiocyanate is a toxic substance that may cause allergic reactions and irritation to mucous membranes. Even when used as a food flavouring, horseradish should be taken sparingly. The pure essential oil is considered a hazardous substance. Poisoning of livestock (especially horses and ruminants) can occur if they are fed on rape cake or other food items with mustard oil. Ingestion of fresh plants leads to gastroenteritis, inflammation of kidneys, colic and diarrhoea. 3 g allylisothiocyanate is lethal for cattle.

SYMPTOMS The oil causes rapid reddening of the skin and sharp pain lasting up to 48 hours. High oral doses of mustard oil cause gastrointestinal disorders, such as stomach pain, nausea, vomiting and diarrhoea. Severe intoxication result in paralysis of central nervous system, low heart and respiratory activity, even coma and death.

PHARMACOLOGICAL EFFECTS The iso thiocyanates can form covalent bonds with proteins and thus alter their properties (see mustard oils). The compounds affect TRP channels, similar to capsaicine. They have proven antibacterial effects against both gram-positive and gram-negative bacteria. The oil acts as skin irritant.

FIRST AID Medicinal charcoal; possibly artificial respiration, shock treatment; externally: wash skin, mucosa or eye thoroughly with water or polyethylene glycol. **Clinical therapy:** Only in severe case, gastric lavage and other treatment (see mustard oils).

Armoracia rusticana P. Gaertn., Mey. & Scherb. (= *Cochlearia armoracia* L.)   family: Brassicaceae

grand raifort (French); *Meerrettich* (German); *cren* (Italian); *rábano picante* (Spanish)

Arnica montana

arnica

Arnica montana

Arnica montana flower heads

PLANTS WITH SIMILAR PROPERTIES A genus with 32 species in the N hemisphere. Examples of other commercially relevant species are *A. chamisssonis* and *A. fulgens*.

PLANT CHARACTERS *Arnica montana* is a small perennial herb with broad, hairy leaves and large, deep yellow flower heads borne on slender talks. *A. chamissonis* has a more erect habit and smaller flower heads.

OCCURRENCE The well-known *A. montana* occurs naturally in mountainous regions of C and N Europe; *A. chamissonis* and *A. fulgens* in N America. The N American species are nowadays commercial sources.

CLASSIFICATION Cell toxin; moderately hazardous, II

ACTIVE INGREDIENTS The main active principles, in concentrations of 0.2–0.5 %, are helenalin and related sesquiterpene lactones. Flavone glycosides may play a synergistic role. Also present are a volatile oil with thymol, thymolmethylether and azulene as main ingredients, several triterpenoids, phytosterols, phenolic acids, coumarins and polysaccharides.

UTILISATION Arnica is well known as a traditional topical treatment for bruises and sprains, as well as burns, sunburn and diaper rashes. It should not be taken internally but can be used as a mouthwash to treat inflammation of the mucous membranes of the mouth or throat. Arnica has been used as an abortifacient and is considered toxic when ingested.

TOXICITY Poisoning has been reported when *Arnica* plant parts or extracts were taken orally. Several instances of allergic reactions have been observed when products with arnica, such as cosmetics, hair or bathing lotions, were applied.

SYMPTOMS Poisoning with arnica results in irritation of mucosa, disturbance of the gastrointestinal tract, diarrhoea, bleeding, headache, vertigo, anxiety, enhanced pulse, palpitations, dyspnoea, and even death.

PHARMACOLOGICAL EFFECTS Helenalin and other sesquiterpenes are reactive molecules that can form covalent bonds with proteins containing a free SH-group. Since many such interactions are possible in the body, several properties and adverse (e.g. cytotoxic) and beneficial effects have been described (see helenalin). When sesquiterpene lactones couple to proteins, they can become antigens and induce the selection of antibodies. These antibodies can cause allergic reaction upon further contact with sesquiterpenes.

FIRST AID Give medicinal charcoal; externally: wash skin, mucosa or eye extensively with water or polyethylene glycol. **Clinical therapy:** Only in severe cases, use gastric lavage and other treatments (see helenalin).

Arnica montana L.

family: Asteraceae

arnica (French); *Arnika* (German); *arnica* (Italian); *arnica* (Spanish)

Artemisia absinthium

common wormwood

Artemisia absinthium

Artemisia absinthium flower heads

PLANTS WITH SIMILAR PROPERTIES A large genus with more than 350 species in Europe, Asia, Africa and America; among them several medicinal and ornamental plants, such as *A. abrotanum, A. annua, A. arborescens, A. afra, A. cina, A. dracunculus, A. glacialis, A. lactiflora, A. tilesii, A. vulgaris* and *A. mexicana*.

PLANT CHARACTERS A robust perennial herb of up to 1 m in height. The silver-coloured leaves are pinnately compound and deeply dissected. Tiny pale yellow florets are borne in small heads along axillary and terminal stalks.

OCCURRENCE Europe, N Africa and W Asia. It is widely cultivated, especially in E Europe.

CLASSIFICATION Neurotoxin; mind-altering; slightly to moderately hazardous, II–III

ACTIVE INGREDIENTS Absinthin and several other sesquiterpene lactones are thought to be responsible for the medicinal value and bitter taste of wormwood. The main compounds in the essential oil include α-thujone, β-thujone and chrysanthenyl acetate.

UTILISATION Wormwood has been used since antiquity for pest control and abortion. Other traditional uses include the stimulation of appetite and the treatment of dyspepsia and biliary dyskinesia. Essential oil has been an ingredient of absinthe, a liqueur favoured by many artists in the 19th century. Absinthe also contained anise,

fennel, other aromatic herbs and 75 % ethanol. It is a sexual and spiritual stimulant but causes unpleasant hallucinations and neurotoxic effects. Chronic consumers were pale and mentally disturbed; several artists died young or committed suicide, such as Vincent van Gogh. Some of his most expressionistic pictures were probably influenced by his absinthe addiction. Wormwood was banned as an ingredient of absinthe early in the 20th century. *A. mexicana* is smoked as a euphoric stimulant.

TOXICITY The monoterpene thujone appears to be responsible for the neurotoxic and hallucinogenic effects. The LD_{50} of essential oil is 0.96 g/kg (p.o.) in rats. LD_{50} α-thujone (mouse): 87.5 mg/kg s.c.; β-thujone (mouse) 442.4 mg/kg s.c.

SYMPTOMS Acute poisoning: vomiting, strong diarrhoea, dizziness, headache and cramps. Thujone enhances the effects of alcohol but is a convulsing and paralysing poison at high doses which can cause coma and death. Chronic poisoning causes hallucinations, delirium and seizures.

PHARMACOLOGICAL EFFECTS Thujone is highly reactive and can alkylate proteins. It stimulates the autonomous nervous system at lower doses whereas high doses lead to spasms and unconsciousness.

FIRST AID Give medicinal charcoal. Severe cases: gastric lavage (see thujone).

Artemisia absinthium L.

family: Asteraceae

grande absinthe, herbe d´absinthe (French); *Wermut* (German); *assenzio* (Italian)

Artemisia cina

wormseed

Artemisia cina seeds

Artemisia cina pills

Artemisia cina

PLANTS WITH SIMILAR PROPERTIES A cosmopolitan genus of more than 350 species with several species of medicinal and ornamental plants (see *A. absinthium*). *A. maritima* also accumulates santonin.

PLANT CHARACTERS Wormseed is a robust perennial shrub of up to 0.6 m in height. The leaves are pinnately compound. Numerous small, brownish-green flower heads are borne along the branch ends.

OCCURRENCE Steppe of Asia; areas such as Iran and Turkestan. The main commercial source is Kazakhstan.

CLASSIFICATION Neurotoxin; mind-altering; highly hazardous, Ib

ACTIVE INGREDIENTS Several sesquiterpene lactones (up to 6.5 % in flowers), of which α- and β-santonin ($C_{15}H_{18}O_3$; MW 246.31) and artemisin are the main compounds.

UTILISATION Wormseed has been employed to kill intestinal worms (santoninum), since santonin stimulates the muscles of *Ascaris* worms. Overdosing frequently caused poisoning.

TOXICITY The sesquiterpene lactone santonin appears to be responsible for the neurotoxic and hallucinogenic effects. About 10 g of dried flower heads are lethal for adults. The lethal dose of santonin for humans (adults) is about 15 mg/kg; 60–300 mg can be lethal for children. LD_{50} (mouse; p.o.) 900 mg/kg; (i. p.) 130 mg/kg; (i.v.) 180 mg/kg.

SYMPTOMS Santonin causes loss of consciousness and epileptic seizures. Symptoms start with distorted vision (patients see violet, then yellow), followed by intense gall formation, icterus, abdominal pain, diarrhoea, nausea, vomiting, cold skin, salivation, tear production, mydriasis, dyspnoea, haematuria through kidney damage, convulsions of facial muscles, paralysis of legs, lowered body temperature, delirium, coma and death from respiratory arrest. In addition, hallucinations, dizziness and vertigo have been recorded.

PHARMACOLOGICAL EFFECTS Santonin affects the brain and spinal cord; it stimulates motoric centres. Santonin is converted to hydroxysantonin in the body, which turns the urine bright yellow.

FIRST AID Give medicinal charcoal and sodium sulphate; allow large quantities of fluids to drink.
Clinical therapy: Gastric lavage (possibly with 1 % potassium permanganate), instillation of medicinal charcoal, sodium sulphate, polyethylene glycol 400, electrolyte substitution, control of acidosis with sodium bicarbonate. Check kidney function. In case of spasms provide diazepam or thiopental, for colic give atropine; in case of severe intoxication, provide intubation and oxygen respiration.

Artemisia cina O.C.Berg & C.F. Schmidt

family: Asteraceae

barbotine (French); *Zitwerbeifuß* (German); *semenzine* (Italian)

59

Arum maculatum

lords and ladies

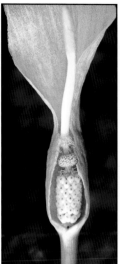

Arum maculatum   *Arum maculatum* flower   *Arum italicum* fruits   *A. maculatum* mature fruits

PLANTS WITH SIMILAR PROPERTIES
A genus of 26 species, including *A. italicum* and
A. rupicola. The related *Arisarum* (3 species) and
Dracunculus (2 species) have similar properties.

PLANT CHARACTERS A perennial herb with
arrow-shaped, often purple-spotted leaves that
emerge from tuberous rhizomes in spring. The
spectacular inflorescence has a greenish-white
spathe and dark brown spadix, which produces
heat when mature and releases amines to attract
pollinating flies. The insects are imprisoned until
the male flowers mature. Fruits are green when
unripe but an attractive red when mature.

OCCURRENCE Forests of C and S Europe;
A. italicum is often planted as an ornamental.

CLASSIFICATION Cell toxin; extremely hazardous, Ia

ACTIVE INGREDIENTS All parts, especially
the red fruits, are poisonous. Toxic principles include oxalate crystals (40–100 μm long needles)
(0.2–0.4 % in fruits); other toxic components are
aroin (0.005 %), saponins, cyanogenic glucosides
(triglochinin; up to 100 ppm in all parts, especially in fruits) and nicotine (traces).

UTILISATION Cooked rhizomes are edible.
They were formerly used to starch linen. Rhizomes have been used in traditional medicine.

TOXICITY The red berries of *Arum* are attractive to children. Fatalities are rare because the intense burning sensation usually limits the quantity ingested. Livestock may eat the aerial parts.

SYMPTOMS The first symptoms are burning o
mouth, throat and oesophagus, followed (withi
1 hour) by a swollen tongue, salivation, stron
convulsions, nausea, bloody vomiting and sever
gastrointestinal disorder. Higher doses lead t
strong clonic spasms, bleeding of gums, stomach
gut and uterus, arrhythmia, intense thirst wit
heavy salivation, and finally paralysis of the CNS
Externally, the sap causes severe dermatitis.

PHARMACOLOGICAL EFFECTS Acidit
and the reduction power of oxalic acid can harn
mucosa of the mouth and throat. The shar
oxalate crystals of Araceae are potent irritant
of skin and mucosal tissues; they penetrate cell
and may mediate the entry of other toxins (in
cluding proteases and lectins). Oxalic acid form
insoluble salts with calcium. If calcium oxalat
is deposited in kidney tubules, kidney tissue be
comes damaged and other consequences follow
(see oxalic acid).

FIRST AID Give medicinal charcoal and sodiun
sulphate. Provide plenty of tea. After contac
with eyes or skin: wash with polyethylene glyco
400 or water; in case of skin blisters, treat locall
with corticoid ointment. **Clinical therapy:** Afte
ingestion, apply gastric lavage and instil medici
nal coal; for further treatments see oxalic acid.

Arum maculatum L.   family: Aracea

gouet maculé, arum tacheté (French); *Aronstab* (German); *aro gigaro* (Italian)

Asclepias tuberosa

butterfly-weed • butterfly milkweed

Asclepias tuberosa

Asclepias curassavica

PLANTS WITH SIMILAR PROPERTIES About 100 members of the genus *Asclepias* are mainly known from N and C America, including *A. syriaca*, *A. curassavica* and *A. eriocarpa*. *A. fruticosa* and *A. physocarpa* are African. The related *Cynanchum* (about 200 species) are vines from warm regions that are rich in cardiac glycosides.

PLANT CHARACTERS A sparsely branched perennial herb (up to 0.6 m) with somewhat fleshy stems and narrowly oblong leaves that exude milky latex when damaged. The bright orange flowers are borne in terminal clusters and are followed by tapering capsules that split open when ripe to release seeds with silky seed hairs.

OCCURRENCE N America; cultivated as ornamental plants. A number of *Asclepias* species have become naturalised in the Old World, such as *A. curassavica*. A few species have an Old World origin, such as *A. fruticosa* and *A. syriaca*. Several species are ornamental garden plants.

CLASSIFICATION Heart poison; moderately hazardous, II

ACTIVE INGREDIENTS *Asclepias* species accumulate cardenolides (especially in the latex), such as afroside, asclepiadin, gomphoside, coroglaucigenin and others. The major compounds of *A. fruticosa* are glucofrugoside, frugoside and oroglaucigenin. The cardenolide fraction had been called "asclepiadin".

UTILISATION Dried roots of *A. tuberosa* are a traditional pleurisy treatment in America but are also used for other respiratory ailments. In Africa, *A. crispa*, *A. fruticosa* and *A. physocarpa* are popular in traditional medicine to treat various ailments, ranging from stomach complaints to tuberculosis and headache. Indians have used fibres for cordage and latex for chewing gum. Latex contains 75 % refinable oil.

TOXICITY *Asclepias* is toxic to livestock but only in large quantities. About 300 g of *A. physocarpa* leaves are said to be a lethal dose for sheep.

SYMPTOMS In humans: nausea, salivation, retching, gastrointestinal disturbance, diuresis, purging and exhaustion, with the usual respiratory and cardiac symptoms (arrhythmia, hypertension, coma, cardiac arrest) associated with heart glycoside poisoning. Animals develop respiratory problems, severe gastroenteritis, fever, paralysis, and a rapid, weak heartbeat.

PHARMACOLOGICAL EFFECTS In general, cardiac glycosides inhibit the Na^+, K^+-ATPase (see cardiac glycosides).

FIRST AID First aid treatment is indicated if plant material of *Asclepias* or isolated cardiac glycosides have been ingested. Treatment involves the induction of vomiting, supply of medicinal charcoal, and detoxification as outlined under cardiac glycosides.

Asclepias tuberosa L.   family: Apocynaceae (Asclepiadoideae)

asclépiade tubéreuse (French); *Knollige Seidenpflanze* (German); *esculapia* (Italian)

Atropa belladonna

deadly nightshade

Atropa belladonna flowers

Atropa belladonna fruits

PLANTS WITH SIMILAR PROPERTIES There are 4 species with similar properties known from the genus *Atropa*.

PLANT CHARACTERS A perennial herb with soft stems and bright green, simple leaves. The tubular flowers are usually yellowish brown, and are followed by attractive, shiny, big black berries, each surrounded by a persistent calyx.

OCCURRENCE Deciduous forests in Europe, Asia and the Mediterranean region of N Africa; plants are widely cultivated.

CLASSIFICATION Neurotoxin, hallucinogen; extremely hazardous, Ia

ACTIVE INGREDIENTS Dry leaves contain 0.3–1.5 % alkaloids; roots up to 2 %. Alkaloids are also present in fruits and seeds. Hyoscyamine is the main and scopolamine a minor component.

UTILISATION Leaves or roots, but especially the isolated alkaloids atropine and scopolamine are medicinally useful as a spasmolytic (see hyoscyamine). The alkaloids cause pupils to dilate. Since wide pupils appeared to make women more attractive (*"belladonna"*), the drug was used as a cosmetic until the Renaissance. Historically important are the hallucinogenic and aphrodisiac properties of *Atropa*. Surprisingly, the feeling of being able to fly is produced when free tropane alkaloids combined with fat and oil were applied on skin (especially of armpits, external genitalia but also into the vagina and rectum). This is reflected in the well-known pictures of witches flying on a broomstick. The dream of being an animal (remember the description in the Odyssey when the comrades of Ulysses were converted into pigs by Circe?) is caused when extracts with tropane alkaloids are taken orally. *Atropa* has been used as arrow poison in the past.

TOXICITY Ingestion of *Atropa* berries is a main cause of poisonings (10–20 berries will kill an adult, 2–5 a child; 0.3 g leaves are toxic). Leaves ingested as salad or as intoxicant are another source of poisoning. LD_{50} values: see hyoscyamine.

SYMPTOMS At low concentrations, the alkaloids have a depressant and sedative effect, but high doses (>3 mg) lead to mostly pleasant hallucinations, euphoria, confusion, insomnia, mydriasis, dry mucosa, red face and tachycardia. Increasing paralysis and damage to the respiratory system can lead to coma and death from central respiratory arrest (see hyoscyamine).

PHARMACOLOGICAL EFFECTS Hyoscyamine is an antagonist at the muscarinergic acetylcholine receptor and therefore works as a parasympatholytic (see hyoscyamine).

FIRST AID Induce vomiting; treat with sodium sulphate and medicinal charcoal. **Clinical therapy:** Provide gastric lavage and physostigmine as an antidote (see hyoscyamine).

Atropa belladonna L.

family: Solanaceae

belladonne (French); *Tollkirsche* (German); *belladonna* (Italian)

Banisteriopsis caapi

ayahuasca

Banisteriopsis caapi

Banisteriopsis caapi flowers

PLANTS WITH SIMILAR PROPERTIES 92 species; *B. inebrians* and *B. rusbyana* also have hallucinogenic properties; S. American *Virola* species (Myristicaceae) are chemically similar.

PLANT CHARACTERS A large climber (vine) with smooth brown bark and opposite, lanceolate leaves (up to 170 x 80 mm). The pink flowers are 14 mm in diameter and are borne in racemes.

OCCURRENCE Plants grow wild (or cultivated from cuttings) in tropical S America (Amazon and Orinoco River, Ecuador, Colombia, Peru).

CLASSIFICATION Hallucinogen; moderately hazardous, II

ACTIVE INGREDIENTS β-Carboline alkaloids, such as harmine (main alkaloid with 2.2 % in the bark), harmaline, tetrahydroharman and several others. Total alkaloids (upper values): stems 0.8 %, twigs 0.4 %, leaves 0.7 %, roots 1.9 %. Furthermore, saponins (0.7 %) and tannins. *B. rusbyana* contains bufotenin and *N,N*-DMT.

UTILISATION Bark infusions, known as *caapi* in Brazil, *ayahuasca* in Peru and *yagé* or *yajé* in Colombia, are of traditional importance as an intoxicant for mental health and religious rites. Indians have called ayahuasca the "light of their souls" and use it as a general remedy against illness. Infusions are often mixed with other plant extracts (*Psychotria viridis* and others). Sometimes bark and leaves are smoked.

TOXICITY Overdosing can induce loss of consciousness and even death. In humans, injection (i.v.) of 100–200 mg harmaline causes nausea, trembling and vomiting, bradycardia and hypotension; about 300–400 mg (p.o.) leads to hallucination. LD_{50} (mouse; i.v.) of harmine 38 mg/kg; LD_{50} of harmaline (rat; s.c.) 120 mg/kg.

SYMPTOMS Starting with dizziness and nausea, mydriasis, salivation, followed by euphoric ecstasy, sometimes with aggressive behaviour. Typical are intense colour visions, dreams, psychedelic (similar to LSD and mescaline) and aphrodisiac effects, but also vomiting and an urge to defecate. A deep sleep usually follows intoxication.

PHARMACOLOGICAL EFFECTS Harmine, harmaline and tetrahydroharmane activate serotonin receptors and inhibit MAO in the brain. This increases the serotonin level and thus further stimulates 5HT receptors. The alkaloids also affect dopamine, GABA, mACh and adrenergic receptors; they inhibit Na^+, K^+-ATPase, intercalate DNA and thus exhibit cytotoxic effects.

FIRST AID Administer medicinal charcoal; try to sedate patient by talking. In case of strong excitation apply doxepin (better not diazepam); keep respiratory tract free. Keep patient warm. **Clinical therapy:** In severe cases, gastric lavage, medicinal charcoal, sodium sulphate, plasma substitution; provide intubation.

Banisteriopsis caapi (Spruce ex Griseb.) Morton

family: Malpighiaceae

ayahuasca (French); *Ayahuasca* (German); *ayahuasca* (Spanish)

Baptisia australis

blue wild indigo • blue false indigo

Baptisia australis

Baptisia tinctoria

Thermopsis caroliniana

PLANTS WITH SIMILAR PROPERTIES A genus with 17 species, including *B. tinctoria* and *B. leucantha*. Plants with similar ingredients are found in the genus *Caulophyllum* (3 species), with *C. thalictroides* in N America (family Berberidaceae). Plants of the legume genus *Thermopsis* (false lupin; 13 species in E Asia, 10 in N America, including *T. lanceolata, T. rhombifolia*) resemble *Baptisia* in habit and chemistry.

PLANT CHARACTERS Perennial plant with an extensive rhizome; stems up to 1 m in height; typical large trifoliolate leaves with leaflets 20 x 10 mm. Blue flowers are borne in a long raceme. The seeds are brown and about 2 mm in diameter.

OCCURRENCE N America; commonly cultivated in gardens as ornamental plant.

CLASSIFICATION Neurotoxin; moderately hazardous, II

ACTIVE INGREDIENTS Cytisine, *N*-methylcytisine, anagyrine, baptifoline and other quinolizidine alkaloids. The total level of alkaloids in aerial parts are below 1 % dry weight.

UTILISATION *Baptisia* species were formerly used as dye plants. *B. tinctoria* is part of immune stimulating herbal prescriptions and has been used in N America to treat infections. It has been widely used in N American traditional medicine for various indications, including inflammation, skin disorders, wounds, septic processes and fever

(diphtheria, influenza, malaria, angina, typhus).

TOXICITY Ingestion of more than 30 g of th root drug or more than 6 seeds or flowers caus toxic effects (see cytisine). Livestock poisonin has also been reported.

SYMPTOMS The first effects of cytisine poisor ing set in after 0.25–1 h and include burning c the mouth and throat, intensive salivation, thirs and nausea followed by long hours of sometime bloody vomiting, cold sweat, mydriasis, vertigc excitation and confusion (with hallucinations, de lirium), anxiety, tachycardia, trembling muscles spasms and collapse. Further symptoms can b liver pains, abdominal pains, constipation an general weakness. Death is caused by respirator or circulatory arrest. Symptoms are weaker i persons who use large amounts of tobacco.

PHARMACOLOGICAL EFFECTS Cytisin and other oxidised quinolizidine alkaloids ar strong agonist at nAChR (similar to nicotine and thus affect the central nervous system, espe cially the *medulla oblongata*, centres of vomiting vasomotor and breathing. Effects are stimulatin, at first but paralysing later on (see cytisine).

FIRST AID Induce vomiting; administration c sodium sulphate, medicinal charcoal, and plent of warm tea or fruit juice to drink. **Clinical ther apy:** After intake of large doses, gastric lavage an further treatments (see cytisine).

Baptisia australis (L.) R.Br. ex Ait.f.

lupin indigo (French); *Blaue Färberhülse* (German)

family: Fabacea

Berberis vulgaris

common barberry

Berberis vulgaris fruits

Mahonia aquifolium

PLANTS WITH SIMILAR PROPERTIES A large genus with more than 500 species; it shows much diversity in Asia and in S and N America. Chinese species such as *B. soulieana* and *B. wilsoniae* are traditional medicines. A related genus with similar secondary metabolites is *Mahonia,* with more than 100 species in the N hemisphere.

PLANT CHARACTERS A spinescent woody shrub (up to 3 m) with small, firm-textured leaves congested on short lateral branches. Small yellow flowers are followed by edible, bright red berries.

OCCURRENCE Barberry is indigenous to Europe and Asia. *Berberis* species have become popular garden shrubs in many parts of the world.

CLASSIFICATION Cell toxin; moderately hazardous, II

ACTIVE INGREDIENTS Isoquinoline and protoberberine alkaloids are present in root and stem bark of *B. vulgaris* and other *Berberis* species in concentrations of up to 13 %, including berberine (the main compound), columbamine, jatrorrhizine and palmatine. Berberine has a yellow colour. The main site of berberine accumulation is the bark of roots and stems. The ripe berries of *B. vulgaris* are practically alkaloid-free. However, the seeds of most other *Berberis* species have alkaloid levels of .5–4.3 %; the fruit pulp is alkaloid-free.

UTILISATION Barberry bark and roots are traditional remedies against disorders of the digestive tract, liver, kidneys and urinary tract, including hepatitis, gall bladder inflammation, jaundice and gallstones. Pure berberine has been used as an ingredient of eye drops to treat conjunctivitis.

TOXICITY Berberine is easily absorbed after oral ingestion. LD_{50} in mice: berberine 0.5 g/kg p.o. Consultations in poison centres often involve ingestion of *Berberis* and *Mahonia* fruits. Most cases are harmless; only 4 % of patients showed symptoms of GI tract disturbance.

SYMPTOMS Berberine increases gut motility and respiration. Other symptoms that have been recorded include dizziness, nosebleed, vomiting, diarrhoea and kidney irritation. High doses cause primary respiratory arrest and haemorrhagic nephritis.

PHARMACOLOGICAL EFFECTS Protoberberine alkaloids intercalate DNA and inhibit various enzymes and neuroreceptors. They show broad antibacterial, antifungal, amebicidal and cytotoxic activities. Berberine activates smooth muscles. The alkaloids exhibit hypotensive and cholekinetic properties.

FIRST AID Provide plenty of warm tea or fruit juice to drink. Administer sodium sulphate and medicinal charcoal. **Clinical therapy:** After intake of large doses, gastric lavage and further treatments are recommended (see chelidonine).

Berberis vulgaris L.   family: Berberidaceae

épine-vinette (French); *Berberitze, Sauerdorn* (German); *crespino* (Italian)

65

Blighia sapida

akee

Blighia sapida

Blighia sapida fruits

Blighia sapida flowers

PLANTS WITH SIMILAR PROPERTIES A genus of 3 species in tropical Africa.

PLANT CHARACTERS Akee is an attractive tree of up to 12 m high, with large compound leaves, each comprising 3 to 5 pairs of oblong leaflets. The small white flowers are borne in sparse clusters. Akee fruits are leathery capsules, oblong in shape, yellow to bright red, with three shiny black seeds surrounded by cream-coloured, fleshy arils.

OCCURRENCE West Africa, especially Cameroon; introduced to the West Indies by slave merchants. Akee is cultivated in Jamaica and other parts of C America.

CLASSIFICATION Cell poison; highly hazardous, Ib

ACTIVE INGREDIENTS The unripe or over-ripe aril, as well as the raphe that connects the aril to the seed, together with the seed itself and the rest of the fruit contain toxic non-protein amino acids (NPAAs) known as hypoglycins, of which hypoglycin A and hypoglycin B are the main compounds. Hypoglycin A is the most toxic compound. It is present in the seed (0.94 %), unripe aril (0.7 %) and fruit wall (0.04 %).

UTILISATION The ripe seed arils are regularly consumed in W Africa and in the West Indies. The fruit has been banned in the USA. The arils are parboiled in salted water and then fried in o or butter. They resemble scrambled eggs and ar commonly known as "vegetable brains". It is use to prepare the famous "akee and saltfish", the na tional dish of Jamaica.

TOXICITY If the arils from unripe or over-rip fruits or the pink base of the aril of ripe fruits (i. the raphe) are eaten, toxic symptoms have bee recorded. The seeds and the rest of the fruit ar also poisonous but are unlikely to be consumed The symptoms are called Jamaican vomitin disease or akee poisoning and regularly caus human deaths. Children are more commonly a fected than adults.

SYMPTOMS Symptoms include uncontrollab vomiting, dizziness, severe hypoglycaemia, con vulsions, coma and even death.

PHARMACOLOGICAL EFFECTS Hypogly cin A is a mimic of glutamic acid. It is metabo lised to methylene cyclopropylacetic acid by de samination, which can bind the coenzyme FAI and butyryl-CoA dehydrogenase. This inhibitio leads to the accumulation of short chain fatty aci which apparently causes the neurotoxic effects.

FIRST AID Administration of sodium sulphat medicinal charcoal, and plenty of warm tea o fruit juice. **Clinical therapy:** After intake of larg doses, gastric lavage and further symptomati treatments will be necessary.

Blighia sapida Koenig

family: Sapindacea

akée d'Afrique, arbre fricassé (French); *Akeepflaume* (German); *arbol de seso, seso vegetal* (Spanish)

Boophone disticha

candelabra flower • bushman poison bulb

Boophone disticha leaves

Boophone disticha flowers

PLANTS WITH SIMILAR PROPERTIES A genus of 2 species in S and E Africa.

PLANT CHARACTERS A perennial with a large bulb (150–220 mm in diameter) covered in papery scales that grows partly above the ground. Two rows of strap-shaped leaves in the form of a fan (up to 0.5 m) emerge after flowering. The pink to reddish flowers are borne in a rounded inflorescence. When the seeds mature, the globular inflorescence breaks away from the stem and is rolled by the wind, thereby dispersing the seeds.

OCCURRENCE Open grassland from S Africa northwards into tropical Africa.

CLASSIFICATION Neurotoxin, hallucinogen; highly hazardous, Ib

ACTIVE INGREDIENTS Buphanidrin is the main toxic compound in the bulb. Also present are several other isoquinoline alkaloids (total alkaloids 0.3 %), including buphanamine, buphanidine, buphanisine, crinine and distichanine.

UTILISATION *B. disticha* bulbs are very well known as a source of arrow poison. The bulbs scales have been used by Khoisan people in southern Africa to mummify bodies. The plant is very poisonous and several cases of murder and suicide have been recorded. Accidental poisoning from overdosing may occur when the bulbs are used in traditional medicine as hallucinogen. Another important traditional use is the effective treat-

ment of painful wounds. The bulb scales are simply applied to the affected area.

TOXICITY Buphanidrine is known as a powerful analgesic, hallucinogen and neurotoxin. Lethal dose in mice: less that 10 mg/kg; in rabbits (s.c.) 15 mg/kg, and in guinea pigs (s.c.) 8 mg/kg.

SYMPTOMS In rabbits, ingestion of bulb material result in restlessness, dyspnoea, dizziness, distorted vision, loss of coordination, dry mouth, enhanced pulse followed by slow pulse, blood and water accumulation in lungs and bleeding of intestinal mucosa. Symptoms of human poisoning are well documented: dizziness, restlessness, impaired vision, unsteady gait and visual hallucinations, leading to coma and death.

PHARMACOLOGICAL EFFECTS Alkaloids apparently interfere with several neuroreceptors and ion channels; they are probably cytotoxic and mind-altering.

FIRST AID Induce vomiting; administration of sodium sulphate and medicinal charcoal; provide plenty of warm tea or fruit juice to drink. **Clinical therapy:** Gastric lavage (possibly with potassium permanganate), instillation of medicinal charcoal and sodium sulphate, electrolyte substitution, control of acidosis with sodium bicarbonate. In case of spasms give diazepam and atropine; in case of severe intoxication, provide intubation and oxygen therapy.

Boophone disticha (L.f.) Herb. (= *Boophane disticha*)   family: Amaryllidaceae or Asparagaceae
Fächerlilie (German)

Bowiea volubilis

climbing potato

Bowiea volubilis flowers

Bowiea volubilis bulbs

PLANTS WITH SIMILAR PROPERTIES The genus is nowadays considered to be monotypic, with a single species divided into two subspecies: the widespread subsp. *volubilis*, and the southern African endemic subsp. *gariepensis*.

PLANT CHARACTERS *B. volubilis* is a very unusual xerophytic plant with greenish-white, fleshy bulbs (150 mm in diameter) borne partly above the ground. The bulb scales are fleshy (even when old), never papery or fibrous. What is unusual is that the main body of the plant is formed by succulent, twining, leafless flowering stems that continue to grow. The small yellowish green flowers are followed by small capsules that split open to release numerous black seeds.

OCCURRENCE *B. volubilis* is widely distributed in S and E Africa. Plants are cultivated in gardens and greenhouses for their curiosity value.

CLASSIFICATION Heart poison; extremely hazardous, Ia

ACTIVE INGREDIENTS Bovoside A is the main cardiac glycoside. It occurs with structurally similar bufadienolides (total content is 0.4 mg/g dry weight). The main aglycones are bovoruboside, bovokryptoside and bovogenine A.

UTILISATION The bulbs are regularly available from traditional medicine markets and are used for a variety of ailments, including headache, oedema, ascites, sterility, infertility and bladder complaints. It is also used as a purgative medicine. As a result, fatal poisoning through accidental overdosing has been reported.

TOXICITY *Bowiea* bulbs and green tissues are extremely toxic (30 x more toxic than *Digitalis* and have been the cause of numerous fatalities in African humans and animals. Death may occur within a few minutes. Cases of livestock poisoning are rare. The main bufadienolides have LD in cats (i.v.) between 0.11 and 0.19 mg/kg. Half an ounce (about 15 g) of fresh bulb can kill a sheep.

SYMPTOMS Concentrated decoctions may cause heart failure and death within a few minutes. Symptoms include nausea, vomiting, salivation, spasms, gastrointestinal disturbance, purging and exhaustion, with the usual respiratory and cardiac symptoms (arrhythmia, hypertension, coma, cardiac arrest) associated with heart glycoside poisoning. Skin irritation has been recorded.

PHARMACOLOGICAL EFFECTS Bufadienolides inhibit the Na^+, K^+-ATPase (see cardiac glycosides) and cause disturbances of atrio-ventricular conduction, including complete AV block.

FIRST AID First aid treatment is indicated if plant material from *Bowiea* or isolated cardiac glycosides have been ingested or if persons are injured by arrow poisons. Treatment involves the induction of vomiting and detoxification as outlined under cardiac glycosides.

Bowiea volubilis Harv. ex Hook. f. (= *B. kilimandscharica*; = *B. gariepensis*)   family: Hyacinthacea

Zulukartoffel (German)

Brassica nigra

mustard • black mustard

Brassica nigra flowers

Brassica nigra fruits

PLANTS WITH SIMILAR PROPERTIES A genus of 35 Eurasian species, many of which are used in human nutrition, such as cabbage (*B. oleracea*; white mustard (*Sinapis alba*) and Indian mustard (*Brassica juncea*).

PLANT CHARACTERS An upright, robust, leafy annual that grows to a height of about 1 m. It has soft, lobed leaves, panicles of small, yellow flowers and smooth (hairless) seed capsules.

OCCURRENCE Europe and Asia (black mustard); Europe (white mustard); S and E Asia (Indian mustard). The plants are all grown as seed crops on all continents.

CLASSIFICATION Cell toxin, irritant; moderately hazardous, II (high doses)

ACTIVE INGREDIENTS Glucosinolates (mustard-oil glycosides) are the main compounds: sinigrin in black mustard; sinalbin in white mustard. When hydrolysed by the enzyme myrosinase, sinigrin and sinalbin yield volatile allylisothiocyanate and non-volatile *p*-hydroxybenzoyl isothiocyanate, respectively.

UTILISATION Seeds have been used in traditional medicine to stimulate circulation and digestion. Mustard oil (in the form of mustard spirits – *Spiritus sinapis* – or mustard plasters) is a traditional topical treatment (counter-irritant) for rheumatic pain. Internal uses are against bronchitis, influenza and urinary tract infection.

TOXICITY Overdosing leads to skin inflammation and skin blisters. LD_{50}: Allylisothiocyanate: Mouse: 108.5 mg/kg p.o., 69.5 mg/kg after inhalation; rats (p.o.) 25 to 200 mg/kg. 3 g of allylisothiocyanate is lethal for cattle (present in 400 g mustard press cake). *B. napus* oil contains erucic acid that is toxic to animals. Modern Canola varieties are low in erucic acid and glucosinolates.

SYMPTOMS High oral doses cause stomach pain, nausea, vomiting and diarrhoea and in severe cases, paralysis of central nervous system, low heart and respiratory activity and even coma and death. On skin, mustard oils cause rapid reddening and sharp pain lasting up to 48 hours. Thiocyanates can disturb thyroid hormones (goitre formation) and nitriles are liver toxins.

PHARMACOLOGICAL EFFECTS The isothiocyanates of mustard oil can form covalent bonds with proteins and thus alter their properties. They activate TRP channels of sensitive nerves and are antibacterial against both gram-positive and gram-negative bacteria. The oil acts as skin irritant; a stimulant to increase peripheral blood flow (see mustard oils).

FIRST AID Administer medicinal charcoal. After intake of large amounts give artificial respiration and shock treatment if necessary; externally: wash skin, mucosa or eyes thoroughly with water or polyethylene glycol (see mustard oils).

Brassica nigra (L.) Koch

family: Brassicaceae

moutarde noire (French); *Schwarzer Senf* (German); *senapa vera* (Italian); *mostaza negra* (Spanish)

Brugmansia suaveolens

angel's trumpet

Brugmansia suaveolens leaves and flowers   Brugmansia x candida flowers

PLANTS WITH SIMILAR PROPERTIES 14 species are recognised, such as *Brugmansia* x *candida* (a hybrid between *B. aurea* and *B. versicolor*; sometimes called *B. arborea*), *B. sanguinea*, *B. versicolor* and *B. aurea*. *Brugmansia* species were formerly included in the genus *Datura*, but they are woody perennials with pendulous flowers, while *Datura* species are annuals with erect flowers.

PLANT CHARACTERS *B. suaveolens* is a shrub or small tree with large velvety leaves and enormous, pendulous, trumpet-shaped flowers (length up to 300 mm) that produce a pleasant smell in the evening to attract pollinators.

CLASSIFICATION Neurotoxin, hallucinogen; extremely hazardous, Ia

OCCURRENCE The natural distribution area of *Brugmansia* species is S and C America but the plants are widely grown as garden ornamentals.

ACTIVE INGREDIENTS *Brugmansia* species accumulate tropane alkaloids such as scopolamine (hyoscine), hyoscyamine, norhyoscine and tigloyl esters of tropine. The alkaloid content of leaves is 0.3–0.6 % dry weight.

UTILISATION *Brugmansia* has been used as medication (as narcotic) and powerful hallucinogen (causing excited visions) by American Indians. The form "Methysticodendron" has been vegetatively propagated by shamans. Similar to the experiences with tropane plants in the Old World (see *Atropa belladonna*), Indians also experienced the feeling of being able to fly after using *Brugmansia* ointments.

TOXICITY Leaves, flowers and especially the seeds are considered to be very poisonous. Cases of severe poisoning are becoming more regular in the Western world as young people increasingly experiment with hallucinogenic plants. In Florida alone, in 1994, 112 teenagers were admitted to hospital suffering the side effects of *Brugmansia* intake. Gardeners may discover the disturbing mydriatic effect of tropane alkaloids when handling *Brugmansia* and touching their eyes later.

SYMPTOMS At low concentrations, the alkaloids have a depressant and sedative effect, but high doses lead to violent intoxications with hallucinations (that may last for several hours), euphoria, confusion, insomnia and even death from respiratory arrest. Mydriatic effects are especially long lasting (up to 6 days).

PHARMACOLOGICAL EFFECTS Tropane alkaloids are antagonists at the muscarinic acetylcholine receptor and therefore work as parasympatholytics. Scopolamine has stronger central effects than hyoscyamine (see hyoscyamine).

FIRST AID Induce vomiting; treat with sodium sulphate and medicinal charcoal. **Clinical therapy:** Physostigmine can be used as an antidote (see hyoscyamine).

Brugmansia suaveolens (Willd.) Bercht. & C.Presl   family: Solanaceae

stramoine odorante, Brugmansia (French); *Engelstrompete* (German)

Brunfelsia pauciflora

yesterday, today and tomorrow

Brunfelsia pauciflora

Brunfelsia uniflora

PLANTS WITH SIMILAR PROPERTIES A genus of about 40 species in tropical America, some of which are grown as ornamental plants. These include *B. americana, B. australis, B. grandiflora, and B. uniflora.*

PLANT CHARACTERS A perennial evergreen shrub of 1.5–2 m in height. The stems bear oval-lanceolate leaves and large discus-shaped flowers with five rounded petals. The flowers characteristically change their colour from dark violet to blue and then to white, so that three different flower colours are present simultaneously (hence the common name). Fruits are dark green to black berries with 20 small seeds.

OCCURRENCE Tropical C and S America (Brazil); it is an ornamental garden plant in the subtropics and a pot plant in Europe.

CLASSIFICATION Neurotoxin, mind-altering; moderately hazardous, II

ACTIVE INGREDIENTS Scopoletin (a coumarin), oleanolic acid (a triterpene), and an amidine, pyrrole-3-carboxamidine (also known as brunfelsamidine), which appears to be the toxic principle. *B. uniflora* and *B. pauciflora* both contain the alkaloids manaceine and mancine.

UTILISATION Extracts from *Brunfelsia uniflora* roots were used by Amazon Indians as arrow poison. *Brunfelsia* plants are favourite garden and pot plants. *B. uniflora* was used against syphilis

and as abortifacient. Several *Brunfelsia* species (including *B. grandiflora, B. chiricaspi, B. uniflora* and *B. maritima*) are known for hallucinogenic effects and were used by S American Indians as intoxicating drugs, sometimes together with ayahuasca (*Banisteriopsis caapi*).

TOXICITY Berries have caused serious poisoning in dogs and 5 berries were a lethal dose. Lethal effects have been seen in animal experiments with mice and rats. Human poisoning has not been recorded, but care should be taken when this plant is cultivated in a household with small children and pets.

SYMPTOMS 15 to 60 minutes after ingestion, symptoms of intoxication become visible, including anxiety, restlessness, tachycardia, enhanced respiration, salivation, urination, vomiting, tremors, convulsions and death from respiratory arrest.

PHARMACOLOGICAL EFFECTS The active compounds and their mode(s) of action are still not clear. Pyrrole-3-carboxamidine (brunfelsamidine) is a known convulsant that also occurs in *Nierembergia hippomanica* (Solanaceae) a plant known for causing stock losses.

FIRST AID After ingestion of berries: Instil medicinal charcoal and sodium sulphate. In serious cases, provide gastric lavage with 0.1 % potassium permanganate. In case of spasms, inject diazepam or pentobarbital.

Brunfelsia pauciflora (Cham. & Schldl.) Benth.   family: Solanaceae

Brunfelsia, plante caméléon (French); *Brunfelsie* (German)

71

Bryonia dioica

red bryony

Bryonia dioica flower

Bryonia dioica fruits

PLANTS WITH SIMILAR PROPERTIES A genus of 12 Eurasian and N African species, including *B. alba* (white bryony).

PLANT CHARACTERS The plants are tuberous perennials with creeping and climbing stems. *B. dioica* is dioecious with red berries, while *B. alba* is monoecious and has black berries. The plants exude acrid milky juice when damaged.

OCCURRENCE C Europe and the Mediterranean (*B. dioica*); eastern Europe (*B. alba*).

CLASSIFICATION Cell toxin; highly hazardous, Ib

ACTIVE INGREDIENTS *Bryonia* species contain mixtures of bitter triterpenoids (0.4 % dry weight) known as cucurbitacins, which occur mainly in glycosidic form (bryonin). Cucurbitacins B, D, E, I, J, K and L, together with di- and tetrahydro-cucurbitacins occur in bryony root. The aglycones most likely represent the active toxic principle. Mature fruits accumulate a toxin protein (brydiofin) with a MW of 66 kDa.

UTILISATION The drug is nowadays only used orally in diluted form (D3 to D6) in homoeopathic medicine. It was formerly employed as drastic purgative, emetic and abortifacient. For topical use it is included in creams and ointments to treat rheumatism and muscle pain.

TOXICITY Bryony is highly poisonous and cucurbitacins are amongst the most bitter and acrid

of all substances known to man. LD_{50} values in mice for cucurbitacin B,D,E & I (i.p.) 1 to 2 mg/kg; cucurbitacin E & I (p.o.) 5 mg/kg. Lethal dose of fruit extract in mice was 0.4 mg/kg i.p. About 6–8 berries induce vomiting [about 40 berries c *B. alba* (adults) or 15 (children)]. Livestock po soning has been recorded.

SYMPTOMS Symptoms after oral intake includ vomiting, gastroenteritis, drastic, even bloody di arrhoea, kidney damage, spasms and death from respiratory arrest; in pregnant women root ex tracts cause abortion. Exposure to skin may caus blister formation and inflammation.

PHARMACOLOGICAL EFFECTS Cucurbita cins are spindle poisons and inhibit mitotic cel division and are therefore cytotoxic (see cucur bitacins); they have antitumour activity but ar too toxic to be used in medicine. Bryonin activat intestinal peristaltic movement and cause drasti diarrhoea with spasms and colic.

FIRST AID Provide plenty of liquid (tea) and me dicinal charcoal. After ingestion of more than fruits: Induce vomiting; administration of sodiun sulphate and medicinal coal. Apply sterile plaster on blisters (skin and mucosa). **Clinical therapy** Gastric lavage (possibly with 0.1 % potassiun permanganate), instillation of medicinal charcoa and sodium sulphate; further detoxification (se cucurbitacins).

Bryonia dioica Jacq.   family: Cucurbitacea

bryone dioïque (French); *Rote Zaunrübe* (German); *brionia, vite bianca* (Italian); *alfesira* (Spanish)

Buxus sempervirens

common box

Buxus sempervirens flowers

Buxus sempervirens fruits

PLANTS WITH SIMILAR PROPERTIES A genus with 50 species in Europe, Asia, West Indies and C America, including *B. macowanii* in S Africa and *B. microphylla* in Japan. Similar toxins occur in *Pachysandra terminalis* and *Sarcococca humilis* and related species (Buxaceae).

PLANT CHARACTERS Evergreen shrub or small tree (up to 5 m) with variable growth forms and small oval leaves. Flower clusters each have a single terminal female flower and several male flowers. The fruit is a capsule.

OCCURRENCE Mediterranean area; today it is distributed in S and W Europe. It is widely cultivated as an ornamental and hedge plant.

CLASSIFICATION Cell toxin; highly hazardous, Ib

ACTIVE INGREDIENTS Box contains a complex mixture of toxic steroidal alkaloids ("buxine"); young leaves up to 2.4 %, older leaves and bark up to 1 % total alkaloids. The alkaloids with 2 exocyclic nitrogen atoms derive from the cholesterol skeleton, and include cyclobuxine ($C_{25}H_{42}N_2O$; MW 386.63), buxanine, buxandrine, cyclobuxoviridine, and many others.

UTILISATION During medieval times, extracts from box were used against gout, rheumatism, skin disorders and malaria. The box-tree was devoted to Kybele and Hades in antiquity; Circe held a branch of *Buxus* in her hands when she welcomed Ulysses. It is a symbol of immortality (planted in cemeteries). Dioscorides mentions the "Lykion" of ancient Greece as treatment for wounds, jaundice and intoxication.

TOXICITY Livestock may be poisoned by twigs used as bedding or hedge clippings. 500 g (pigs) or 750 g (horses) were partly lethal. The lethal dose (p.o.) of the alkaloid mixture "buxine" in dogs is 0.1 g/kg, equivalent to 5–10 g/kg of dry leaves.

SYMPTOMS An initial excitation of the CNS is followed by growing immobilisation and paralysis. Poisoning causes irritation of the GI tract, diarrhoea, vomiting, dizziness, epileptiform seizures, hypotension, bradycardia, and finally death caused by paralysis (respiratory failure). 500 g of leaves fed to pigs caused paralysis, breathing difficulties, colic, spasms and eventually death.

PHARMACOLOGICAL EFFECTS Extracts have cytotoxic, anti-inflammatory and antimicrobial properties.

FIRST AID Administer sodium sulphate and medicinal charcoal. **Clinical therapy:** Gastric lavage (possibly with 0.1 % potassium permanganate); medicinal charcoal; sodium sulphate; electrolyte substitution; glucose infusion. In case of spasms provide intubation and antiepileptics (inject small doses of diazepam or barbital to reduce spasms; no phenothiazines); oxygen respiration if necessary.

Buxus sempervirens L.   family: Buxaceae

buis bénit, buis commun (French); *Buchsbaum* (German); *bossolo* (Italian)

Calea ternifolia

dream herb

Calea ternifolia

Calea ternifolia leaves and dried flower heads

PLANTS WITH SIMILAR PROPERTIES A genus of about 60 species of shrubs restricted to montane regions of tropical and subtropical America. A few species have mind-altering properties, among them *C. ternifolia* (mainly known as *C. zacatechichi)* and *C. urticifolia.*

PLANT CHARACTERS An herbaceous plant of 1.5 m height with small, ovate, alternately arranged leaves. Characteristic are the three main veins. Young leaves are violet underneath. The leaf margins have a few short spiny teeth. Tiny white florets occur in small, few-flowered heads borne along the branch ends.

OCCURRENCE Mainly Mexico, but also C America and Jamaica. The plants tend to become weedy.

CLASSIFICATION Hallucinogen; moderately hazardous, II

ACTIVE INGREDIENTS The plant contains a mixture of bitter tasting sesquiterpene lactones, with zacatechinolide, germacrene and caleocromene. Also present are furanone-type heliangolides such as 8β-angeloyloxy-9α-hydroxy-calyculatolide. The presence of flavones and possibly an alkaloid have been reported.

UTILISATION Indians in Oaxaca call this plant thle-pela-kano ("leaf of god"). The Curanderos produce an extract from 60 g of the herb to induce visions and dreams, and to see the future.

Furthermore, the drug is used in traditional medicine to treat skin disorders, against fever and as laxative. This is one of several plants used by Indian communities in Mexico to obtain divinatory messages from dreams. Under the old name *Calea zacatechichi*, and common names such as zacatechichi, dream herb, leaf of God, bitter leaf, dog's grass or cheech, the plant has become popular as a recreational drug. It is taken as a tea and is smoked. In addition to increasing the number and/or recollection of dreams during sleep, it is said to produce a slight hypnotic effect with feelings of well-being and improved mental clarity.

TOXICITY Not known. No serious side effects have been reported.

SYMPTOMS Dream and sleep inducing symptoms are apparent, also in animal experiments. Large doses administered to cats resulted in salivation, ataxia, retching and sporadic vomiting.

PHARMACOLOGICAL EFFECTS The mode of action of the *Calea* sesquiterpene lactones has not been determined yet. The plant is alleged to have oneirogenic (dream-enhancing) effects. Extracts were found to increase the superficial stages of sleep and the number of spontaneous awakenings.

FIRST AID In case of overdosing, instil medicinal charcoal and provide plenty of tea or fruit juice to drink.

Calea ternifolia Kunth. (= *C. zacatechichi* Schlecht.) . family: Asteraceae
Aztekisches Traumgras (German)

Callilepis laureola

ox-eye daisy

Callilepis laureola

Callilepis laureola flower head

PLANTS WITH SIMILAR PROPERTIES A genus with 5 species in S Africa.

PLANT CHARACTERS The ox-eye daisy is a perennial herb of about half a metre in height. It has a thick, strongly tapering, woody taproot with erect branches bearing elliptic to broadly lanceolate leaves. The large and attractive flower heads are borne at the branch tips. They have pale yellow to white ray florets surrounding the contrasting dark purple, almost black disc florets.

OCCURRENCE *C. laureola* is restricted to the eastern parts of S Africa.

CLASSIFICATION Cell poison; highly hazardous, Ib

ACTIVE INGREDIENTS A kaurene glycoside known as atractyloside is the main toxic compound. It is accompanied in the plant by three other kaurenoid glucosides, all of them derivatives of atractyloside.

UTILISATION The characteristic woody tubers, widely known by the Zulu name *impila* (meaning "health"), are popular items of trade on traditional medicine markets. It is used to treat cough in adults but in a very specific way to ensure safety. Traditionally, impila was never given to children under the age of 10, and is never administered as an enema. It should only be given as a weak decoction that should be completely expelled immediately after being drunk. Unfortunately, it

has become a common cause of human fatalities, because people are not aware of the dangers of the medicine and use it in the wrong way.

TOXICITY In the period 1958 to 1977, a total of 263 deaths were recorded in a single hospital in Durban, S Africa. Patients suffer from liver and kidney damage and usually die within 5 days. Of the recorded deaths, about half were children below the age of 15. The LD_{50} of atractyloside in rats (i.m.) is reported to be 431 mg/kg.

SYMPTOMS The symptoms are severe vomiting, abdominal pain, headache, strong convulsions, and rapid progression into coma and death.

PHARMACOLOGICAL EFFECTS Atractyloside is used experimentally because it is a specific inhibitor of ATP transport at the mitochondrial membrane. It is very toxic and produces strychnine-like symptoms in mammals. Blocking ATP transport dramatically disturbs the supply of energy within cells and leads to immobility and cell death.

FIRST AID Quick action is required. Induce vomiting and instil medicinal charcoal and sodium sulphate. **Clinical therapy:** Perform gastric lavage (possibly with 300 ml of 0.2 % potassium permanganate). Instil medicinal charcoal and sodium sulphate; control of acidosis with sodium bicarbonate and provide intubation and oxygen respiration.

Callilepis laureola DC.   family: Asteraceae

Camptotheca acuminata

camptotheca • cancer tree • happy tree

Camptotheca acuminata leaves

Camptotheca acuminata fruits

PLANTS WITH SIMILAR PROPERTIES A monotypic genus. Camptothecin, an important anticancer substance, was discovered in *C. acuminata*, but is also found in other plants: *Nothapodytes foetida, Pyrenacantha klaineana, Merrilliodendron megacarpum* (Icacinaceae), *Ophiorrhiza pumila, O. mungos* (Rubiaceae), *Tabernaemontana* (= *Ervatamia) heyneana* (Apocynaceae) and *Mostuea brunonis* (Gelsemiaceae).

PLANT CHARACTERS A large tree of up to 25 m high. The glossy green leaves have prominent, parallel lateral veins. Small white flowers are borne in rounded heads. The seeds are narrowly oblong.

OCCURRENCE The tree is restricted to China, where it is known as the "happy tree" (*xi shu*). It is cultivated in India, Japan, Germany and the USA. Cultivars with high levels of camptothecin are being developed.

CLASSIFICATION Cell toxin; highly hazardous, Ib

ACTIVE INGREDIENTS The main toxic compound is camptothecin, a pentacyclic quinoline alkaloid. It occurs in levels of about 0.01 % in stem bark, 0.02 % in root bark and 0.03 % in fruits (DW).

UTILISATION The tree is used in traditional medicine in China to treat various ailments of the gall bladder, liver and spleen, as well as psoriasis and the common cold. *Camptotheca* alkaloids or their derivatives have become important in treating colorectal, ovarian and pancreatic cancer. Because camptothecin is poorly soluble and has serious side effects, various semisynthetic analogues have been developed, including 9-amino-20S-camptothecin, irinotecan (also known as irinotecan hydrochloride trihydrate, CPT-11 or Camptosar®) and topotecan (Hyacamptin®).

TOXICITY LD_{50} values for camptothecin: mouse (p.o.) 50.1 mg/kg; rat: (p.o.) 153 mg/kg.

SYMPTOMS Typical are tiredness, asthenia, neutropenia (reduction of white blood cells), hair loss, stomatitis, nausea, vomiting, diarrhoea and haemorrhagic cystitis.

PHARMACOLOGICAL EFFECTS Camptothecin is a very toxic compound but has proven cytostatic and antitumour activity. The activity against cancer is due to the unique ability of camptothecin and its derivatives to inhibit the nuclear DNA topoisomerase I enzyme (the enzyme involved in the uncoiling of DNA) so that replication and transcription are inhibited. As a result, cells die through apoptosis. This activity not only affects cancer cells but all dividing cells.

FIRST AID Administration of sodium sulphate and of medicinal charcoal is recommended. **Clinical therapy:** Gastric lavage (possibly with potassium permanganate), instillation of medicinal charcoal and sodium sulphate, electrolyte substitution; oxygen therapy if necessary.

Camptotheca acuminata Decne — family: Cornaceae (or Nyssaceae)

camptotheca (French); *Camptotheca* (German); *camptotheca* (Italian)

Cannabis sativa

marijuana • Indian hemp

Cannabis sativa

Cannabis sativa flowers

PLANTS WITH SIMILAR PROPERTIES Only 1 species with 2 subspecies (*sativa* and *indica*).

PLANT CHARACTERS A dioecious annual with an erect main stem of up to 4 m. The characteristic leaves are palmately compound (hand-shaped), with serrated margins. Tiny male and female flowers occur on separate plants.

OCCURRENCE Indigenous to Asia but widely cultivated (often illegally) in temperate regions.

CLASSIFICATION Mind-altering; slightly hazardous, III

ACTIVE INGREDIENTS Phenolic terpenoids (cannabinoids). The psychotropic effect is ascribed only to Δ⁹-tetrahydrocannabinol (THC).

UTILISATION Subsp. *indica* is the source of the intoxicant drug: marijuana is the dried flowering twigs of female plants; hashish (which is the stronger stimulant) is the resin collected from leaves and flowers. Since ancient times, cannabis has been used to treat pain, rheumatism and asthma. Modern uses include the treatment of glaucoma, nausea in chemotherapy and the depression and lack of appetite associated with AIDS. Commercial hemp (subsp. *sativa*) has low levels of THC and is grown for its fibres and seed oil.

TOXICITY THC is easily absorbed when inhaled. LD_{50} values for THC: mouse (i.v.) 43 mg/kg; (i.p.) 455 mg/kg; (p.o.) 482 mg/kg; rat: (i.v.) 28 mg/kg; (i.p.) 373 mg/kg; (p.o.) 600 mg/kg.

SYMPTOMS THC is a strong stimulant of the central nervous system. Symptoms set in after consumption of 0.3–1 g marijuana, equivalent to 4–20 mg THC and include euphoria, relaxation, dilated pupils, enhanced pulse, loss of coordination, slurred speech, pleasant sleepiness, and various degrees of visions and hallucinations. THC has bronchodilatory and hypotensive effects. High doses can cause anxiety, panic fears, tremors, psychotic aggressiveness, nausea, vomiting and tachycardia. Chronic use may result in personality changes, a lack of motivation, concentration and memory, as well as psychosis and schizophrenia. As "gateway drug", cannabis may lead to the use of other, more harmful narcotics.

PHARMACOLOGICAL EFFECTS THC binds to CB1 and CB2-receptors in the brain and other organs, which use anandamide (a derivate of arachidonic acid) as a natural ligand. THC appears to increase the levels of the neurotransmitters noradrenaline, dopamine and serotonin. The receptors regulate motoric coordination, memory, appetite and pain. THC can induce neuronal apoptosis. Marijuana and smoke contain mutagenic, carcinogenic and teratogenic substances.

FIRST AID In case of severe intoxication, give medicinal charcoal and tranquilliser (aponal). **Clinical therapy:** Gastric lavage; intubation and artificial respiration if necessary.

Cannabis sativa L.

family: Cannabaceae

chanvre (French); *Hanf* (German); *canapa indiana* (Italian); *cánamo* (Spanish)

Capsicum annuum

hot pepper

Capsicum annuum

Capsicum frutescens

PLANTS WITH SIMILAR PROPERTIES A genus of 10 species in tropical America.

PLANT CHARACTERS *C. annuum* (paprika, hot pepper, Spanish pepper) is an annual plant of up to 0.5 m in height with dark green, stalked leaves, white flowers and oblong, green or red fruit whereas *C. frutescens* (chilli) is a perennial herb. Within *C. annuum*, 5 groups are distinguished: 1. Cerasiforme group, with small and very pungent fruits; 2. Conoides group, with erect, conical fruits; 3. Fasciculatum group with erect, slender, very pungent fruits; 4. Grossum group with green or yellow bell-shaped and scarcely pungent fruits; 5. Longum group, with drooping, 300 mm long, pungent fruits.

OCCURRENCE Originally S America; widely cultivated as spice.

CLASSIFICATION Cell poison; slightly hazardous, III

ACTIVE INGREDIENTS The pungent compounds in peppers (especially concentrated in the seeds and placentas) are alkaloid-like compounds known as capsaicinoids (0.4–0.9 % in *C. frutescens* and 0.1–0.5 % in *C. annuum*). More than half of the total capsaicinoids are usually represented by capsaicin, the main compound.

UTILISATION *Capsicum* preparations are topically applied for effective pain relief in cases of rheumatism, arthritis, neuralgia, itching, lumbago and spasms of the upper body. It may be taken orally as a traditional treatment for colic, dyspepsia and flatulence. Chilli extracts are also gargled to treat chronic laryngitis.

TOXICITY *Capsicum* is not really toxic and the amounts that are normally used in foods are considered safe. However, high concentrations may cause very painful local reactions and severe irritation of the skin and mucous membranes. Exceptional cases are known where infants have died after being treated with chilli-based solutions.

SYMPTOMS Cayenne and paprika may cause discomfort and pain or even blisters and necrotic ulcers when exposed in high concentrations to skin and muscosa. Typical symptoms are erythema (redness), together with a sensation of pain and heat. Nerve ends may become insensitive for long periods, but they recover, apparently without any lasting damage. Toxic doses cause hyperthermia and symptoms similar to anaphylactic shock. Chronic overdosing leads to appetite loss, chronic gastritis, and liver and kidney damage. Paprika can induce food allergies.

PHARMACOLOGICAL EFFECTS Capsaicin activates TRP channels (transient receptor potential family of ion channels).

FIRST AID In case of serious overdosing: Instil medicinal charcoal and sodium sulphate; wash skin and treat with corticoid ointment.

Capsicum annuum L.

family: Solanaceae

poivron, poivre de Guinée (French); *Paprika* (German); *capsico, peperone* (Italian); *pimiento picante* (Spanish)

Catha edulis

khat

Catha edulis leaves

Catha edulis flowers

PLANTS WITH SIMILAR PROPERTIES A genus of 1 species. *Ephedra* species are chemically and pharmacologically similar.

PLANT CHARACTERS Usually a shrub or small tree but sometimes up to 15 m high. The reddish stems bear glossy green leaves with toothed margins and small white flowers in rounded clusters.

OCCURRENCE The distribution area extends from S and E Africa to Arabia and Afghanistan. Main sources of khat are Ethiopia, Somalia and especially commercial plantations in Yemen.

CLASSIFICATION Mind-altering; slightly to moderately hazardous, II–III

ACTIVE INGREDIENTS The stimulants in khat leaves are natural amphetamines known as phenethylamines or khatamines. Cathinone is the main compound; norpseudoephedrine and pseudoephedrine are also present. Cathinone is unstable, hence the need to use fresh leaves. Furthermore, 15 sesquiterpene polyester alkaloids (catheduline) have been reported.

UTILISATION Leaves are either chewed or taken as hot water infusions. About 200–400 g of leaves are used per day, equivalent to 60–120 mg alkaloids. Khat chewing gives stimulating and euphoric effects and is an ancient, socially accepted tradition in Africa and Arabia. Long-term usage leads to addiction, as do other amphetamines. In Yemen, the high proportion of bachelors may be the consequence of anaphrodisiac effects. Traditional uses include the treatment of malaria, cough and asthma. In Europe it is sold as an appetite suppressant but because of its addictive nature this use should be discouraged. Khat became known as a recreational drug in the USA after American soldiers were exposed to it in Somalia. Many countries have imposed legal restrictions.

TOXICITY In mice, the lethal oral dose (extract) was 2 g/kg; LD_{50} (racemic cathinone) 263 mg/kg i.p.; norpseudoephedrine 275 mg/kg s.c.

SYMPTOMS Khat stimulates the mind and increases mental power and sociability while suppressing feelings of fatigue, hunger and thirst. Khat and amphetamines can induce a psychosis resembling schizophrenia. Users often suffer from constipation, caused by tannins in the drug.

PHARMACOLOGICAL EFFECTS Cathinone stimulates α- and β-adrenergic and dopaminergic receptors by increasing the release of noradrenaline and dopamine from catecholic synapses and inhibiting their re-uptake. Khat shows sympathomimetic properties.

FIRST AID Administer medicinal charcoal. **Clinical therapy:** After overdose: Gastric lavage, instillation of medicinal charcoal and sodium sulphate, electrolyte substitution; artificial respiration and oxygen supply if needed; in case of excitation and spasms inject diazepam.

Catha edulis (Vahl) Endl.

family: Celastraceae

qat (French); *Kathstrauch* (German); *catha* (Italian)

Catharanthus roseus

Madagascar periwinkle

Catharanthus roseus

Catharanthus roseus

PLANTS WITH SIMILAR PROPERTIES A genus with 7 species on Madagascar and 1 species in India and Sri Lanka.

PLANT CHARACTERS A short-lived perennial herb of up to 0.4 m in height, with dark green, shiny leaves and attractive purple, pink or white flowers, followed by oblong seed capsules.

OCCURRENCE The plant originates from Madagascar but has been introduced as an ornamental garden subject into many tropical regions of the world where it has become naturalised.

CLASSIFICATION Cell toxin; highly hazardous, Ib–II

ACTIVE INGREDIENTS More than 95 monoterpene indole alkaloids. Catharanthine and vindolinine are hypoglycaemic, while vincristine and vinblastine (two dimeric indole alkaloids) have anticancer activity. They are present in low concentrations (e.g. less than 3 g vincristine per metric ton of plant material).

UTILISATION Leaves were traditionally used to treat diabetes and rheumatism; roots are still used in the supportive treatment of diabetes. It is a source of vinblastine and vincristine (or semisynthetic derivatives, such as vindesine and vinorelbine) used to treat breast cancer, uterine cancer and Hodgkin's and non-Hodgkin's lymphoma.

TOXICITY All parts of the plant are poisonous but human poisoning is rare. Animals may be at risk. Pure vincristine and vinblastine are both extremely poisonous and may cause gastrointestinal and neurological disorders. LD_{50} in rats (i.v.) 1.3 mg/kg, mice 5.2 mg/kg i.p., 2 mg/kg i.v.

SYMPTOMS Vinca alkaloids are irritants of the skin and respiratory tract and cause damage to the cornea of the eye. The dimeric alkaloids can elicit neurological disturbances (they block microtubule formation in the axon of neurons), nausea and vomiting. Paraesthesia of hands and feet and disturbance of vision and hearing are possible. They can inhibit the formation of white blood cells in bone marrow. Also other fast dividing cells are blocked, e.g., hair root cells; this leads to hair loss. In the digestive tract, constipation and diarrhoea have been observed.

PHARMACOLOGICAL EFFECTS The dimeric alkaloids act as antimitotics (spindle poisons) by blocking cell division – they bind to tubulin, the protein that forms the microtubules (spindle) during metaphase. As a DNA intercalator, vincristine has mutagenic properties.

FIRST AID Administration of sodium sulphate and medicinal charcoal. Keep patient in shock position and warm and provide warm tea or coffee

Clinical therapy: Gastric lavage, control of acidosis, electrolyte substitution; in cases of spasms diazepam or atropine, provide intubation and respiration with oxygen if necessary.

Catharanthus roseus (L.) G. Don (= *Vinca rosea* L.)   family: Apocynaceae

pervenche de Madagascar (French); *Madagaskar-Immergrün* (German); *vinca* (Italian)

Cerbera odollam

suicide tree • odollam tree

Cerbera odollam

Cerbera odollam flowers and fruits

PLANTS WITH SIMILAR PROPERTIES There are about 4 species, including *C. odollam* with a wide distribution in Asia. *C. odollam* is sometimes considered to be the same species as *C. manghas* or yellow-eyed cerbera, which is indigenous to Madagascar.

PLANT CHARACTERS Trees and shrubs with the branches and leaves arranged in whorls. The leaves are leathery, lance-shaped and glossy green. Attractive, fragrant white flowers are followed by red, egg-shaped fruits.

OCCURRENCE India, SE Asia, Australia and some Pacific islands. Plants are sometimes cultivated in gardens and parks.

CLASSIFICATION Heart poison; extremely hazardous, Ia

ACTIVE INGREDIENTS Seeds contain several heart glycosides of the cardenolide type such as neriifolin, cerberin, cerberoside, tanghinin and tanghinoside.

UTILISATION *C. odollam* is used by more people to commit suicide and murder than any other plant (although many plants in this category are plants with cardiac glycosides). More than 500 fatal cases have been recorded in the SW Indian state of Kerala in the 10 year period between 1989 and 1999. One tenth of fatal poisoning (death after 6 h) are attributed to this plant. It is also used for murder but doctors, pathologists and coroners often fail to detect the cause of death. *Cerbera* seeds

are crushed and mixed with spicy or sweet food. About 75 % of fatal poisoning victims are women, especially young wives who do not meet the exacting standards of some Indian families. *Cerbera manghas* (= *C. venenifera* or *C. tanghin*), a related species found in Madagascar, has a long history as an ordeal poison, and was responsible for the death of 3 000 people per year in previous centuries. Wood is used for the colourful painted masks typically found in southern Sri Lanka. Decoctions have been employed as a purgative on the Fiji islands.

TOXICITY Concentrated decoctions may lead to death within a few minutes due to heart failure.

SYMPTOMS Symptoms include nausea, salivation, retching, gastrointestinal disturbance, purging and exhaustion, with the usual respiratory and cardiac symptoms (arrhythmia, hypertension, coma, cardiac arrest) associated with heart glycoside poisoning.

PHARMACOLOGICAL EFFECTS Cardiac glycosides inhibit Na^+, K^+-ATPase, one of the most important ion pumps of the body (see cardiac glycosides). Its inhibition blocks neuronal and neuromuscular activities (see cardiac glycosides).

FIRST AID First aid treatment is indicated if plant material from *Cerbera* or isolated cardiac glycosides have been ingested. Treatment involves the induction of vomiting and detoxification as outlined under cardiac glycosides.

Cerbera odollam Gaertn.

family: Apocynaceae

Zerberusbaum, See-Mango (German)

Cestrum parqui

green cestrum

Cestrum parqui

Cestrum elegans flowers

PLANTS WITH SIMILAR PROPERTIES A genus with more than 175 species in tropical America, including *C. elegans*, *C. aurantiacum* and *C. laevigatum*.

PLANT CHARACTERS A 2–3 m shrub with narrow (25 mm wide), pungent-smelling leaves and small, tubular, yellow flowers aggregated at the branch ends. The fruits are small, oval, 10 mm long berries which turn black when they ripen. *C. laevigatum* is related to *C. parqui* but has broader leaves (50 mm wide). *C. aurantiacum* has orange-yellow flowers and white berries.

OCCURRENCE S America. *Cestrum* species are grown as ornamentals but have become troublesome weeds in many parts of the world.

CLASSIFICATION Cell toxin, mind-altering; highly to moderately hazardous, Ib–II

ACTIVE INGREDIENTS The toxic compounds are extremely toxic terpenoids (kaurene glycosides) of which carboxyparquin and the less toxic parquin are the main compounds. These constituents are structurally similar to atractyloside. Saponins are also present.

UTILISATION *C. parqui* has been used as medicine and smoked as stimulant and intoxicant since pre-Columbian times. Fishermen of S Brazil appear to have smoked the leaves of *C. laevigatum* together with or as a substitute for marijuana.

TOXICITY Poisoning of farm animals by *C. par-*qui, *C. aurantiacum* and *C. laevigatum* are quite common. The terpenoids are stable, so that dried material (e.g. hedge clippings) can cause fatalities. Fruits are 10 times as toxic as leaves – ingestion of green berries has resulted in the death of children, but human poisoning is not common. *C. laevigatum* causes death within a few hours. Cattle are killed by 200 g of dried leaves; 15 g per day are fatal to goats. The LD_{50} of carboxyparquin in mice is 4.3 mg/kg.

SYMPTOMS In cattle, these include salivation, arched back, staggering gait and abdominal pain. Symptoms in humans include severe vomiting, abdominal pain, headache, convulsions, and rapid progression into coma and death.

PHARMACOLOGICAL EFFECTS Atractyloside and carboxyparquin are highly toxic in mammals and produce strychnine-like symptoms. Atractyloside is a specific inhibitor of ATP transport at the mitochondrial membrane. Blocking ATP transport dramatically stops the supply of energy within cells. This leads to cell death and paralysis of CNS and muscles.

FIRST AID Instil medicinal charcoal and sodium sulphate. **Clinical therapy:** After overdosing perform gastric lavage (possibly with 300 ml of 0.2 % potassium permanganate). Instil medicinal charcoal and sodium sulphate; control of acidosis with sodium bicarbonate.

Cestrum parqui L'Hér.

family: Solanaceae

cestrum (French); *Chilenischer Hammerstrauch* (German)

Chelidonium majus

greater celandine

Chelidonium majus

PLANTS WITH SIMILAR PROPERTIES A monotypic genus (1 species with 2 varieties).

PLANT CHARACTERS A perennial herb (up to 0.8 m) with yellow-green, deeply lobed leaves and orange-yellow flowers (20–25 mm in diameter). The mutant *C. majus* var. *laciniatum* arose at Heidelberg around 1590. Typical is an orange latex that exudes from damaged stems and leaves.

OCCURRENCE Ruderal places in Europe, Asia and N Africa (naturalised in N America).

CLASSIFICATION Cell poison; neurotoxin; moderately hazardous, II

ACTIVE INGREDIENTS The plant accumulates numerous protoberberine and benzophenanthridine alkaloids at levels of up to 1.3 % in aerial parts and up to 3 % in roots. The main alkaloids are coptisine and chelidonine; also chelerythrine, berberine and sanguinarine. They are complexed with chelidonic acid, resulting in the distinctive yellow or orange colour of the latex.

UTILISATION Extracts have been used in traditional and modern phytomedicine for a wide range of indications, including liver complaints, to stimulate bile flow in cases of hepatitis, jaundice and gall stones, for eye complaints, tinea (ringworm) and eczema. In addition, the latex was used to treat warts of all sorts and was even used against tumours. Plinius recorded a legend of swallows that successfully treated the blindness of chicks in their nest by applying the juice from a twig of *Chelidonium*. It has been suggested that the legend inspired the Latin name, since "chelidon" was the old name for swallow.

TOXICITY The alkaloids have cytotoxic and neuroreceptor activities, so that high doses have adverse effects and can cause death in rare cases. Chelerythrine: LD_{50} mouse 95 mg/kg s.c.; chelidonine: LD_{50} mouse 34.6 mg/kg i.v.; guinea pig 2 g/kg s.c.

SYMPTOMS Ingestion of large doses leads to a burning sensation in mouth, vomiting, paralysis, an urge to urinate, dizziness, arrhythmia, gastrointestinal disturbance with bloody diarrhoea, slow pulse, hypotension and collapse. Chronic intake causes hepatotoxic effects.

PHARMACOLOGICAL EFFECTS The alkaloids affect neuroreceptors and are antispasmodic, weakly analgesic and sedative. They relax smooth muscles in various organs. Berberine, chelidonine and sanguinarine intercalate DNA and inhibit a number of enzymes, including RNA and DNA polymerases, resulting in cytotoxic, antiviral and antibacterial effects.

FIRST AID If latex comes into contact with eyes, wash with water. Ingestion: instil medicinal charcoal and sodium sulphate. **Clinical therapy:** After overdose: gastric lavage (possibly with potassium permanganate); for further detoxification see isoquinoline alkaloids.

Chelidonium majus L.

family: Papaveraceae

chélidoine, grande-éclaire (French); *Schöllkraut* (German); *chelidonia, erba rondinella* (Italian); *célidonia* (Spanish)

Chenopodium ambrosioides

wormseed goosefoot

Chenopodium ambrosioides

Chenopodium quinoa

PLANTS WITH SIMILAR PROPERTIES A genus of 100 species in the Old and New World.

PLANT CHARACTERS A weedy annual or short-lived perennial (up to 1 m high) with lanceolate, toothed leaves and spikes of tiny flowers. It was previously known as *C. anthelminticum* but is now classified as a variety of *C. ambrosioides*. *C. mucronatum* (known in Europe as stinking goosefoot, *C. vulvaria*); a spreading herb with green or red stems and small, greyish green leaves. Quinoa (*Chenopodium quinoa*) contains bitter triterpene saponins and is used in S America as a food plant.

OCCURRENCE C and S America. It is a cosmopolitan weed but is also cultivated in Europe and elsewhere, mainly for the essential oil.

CLASSIFICATION Cell toxin; (oil) highly hazardous, Ib

ACTIVE INGREDIENTS The activity is due to an essential oil (up to 0.6 %) which contains mainly ascaridol (60–70 %), together with *p*-cymol (20–40 %), α-terpinene, limonene and camphor. Amines, such as trimethylamine, were found in *C. mucronatum*.

UTILISATION Wormseed and its oil are traditionally used to treat intestinal worms, especially maw worm (*Ascaris*) and hookworm. It was used as a tonic and also had other uses, including the treatment of snake bite. In the 18th century, Jesu-its brought the herb from Mexico to Europe. The so-called "Jesuit tea" or *tinctura botryos mexicana* has been used for abortion ever since.

TOXICITY Both the plant and the essential oil have caused human poisoning. 0.6 ml/kg of the oil were lethal for rabbits. *C. mucronatum* appears to be toxic to livestock and has killed sheep.

SYMPTOMS The oil causes disturbance of the brain, resulting in insanity. Symptoms also include ringing in the ears, spasms and convulsions, loss of consciousness, paralysis, hypotension, internal bleeding, gastrointestinal problems, coma and even death by respiratory arrest. Trimethylamine poisoning in farm animals is similar to nitrate poisoning and leads to cramps, convulsions, rapid breathing, high blood pressure, respiratory failure and death.

PHARMACOLOGICAL EFFECTS Ascaridol is an effective anthelmintic. As a reactive peroxide it is likely that it can bind to essential proteins and DNA.

FIRST AID Administration of sodium sulphate and medicinal charcoal. Provide warm tea or coffee to drink. **Clinical therapy:** Gastric lavage, instillation of sodium sulphate and medicinal charcoal; control of acidosis, electrolyte substitution, check kidney function; in cases of spasms treat with diazepam or atropine; provide intubation and respiration with oxygen if necessary.

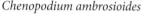

Chenopodium ambrosioides L. var. *anthelminticum* (L.) A. Gray   family: Amaranthaceae or Chenopodiaceae

chénopode anthelmintique (French); *Amerikanisches Wurmkraut, Wurmtreibender Gänsefuß* (German); *chenopodio* (Italian)

Chondrodendron tomentosum

curare vine

Chondrodendron tomentosum

Chondrodendron tomentosum fruits

PLANTS WITH SIMILAR PROPERTIES A genus of 10 tropical species in C & S America, including *C. tomentosum*, *C. platyphyllum* and *C. microphyllum*. The family Menispermaceae is well known for its rich diversity of several types of isoquinoline alkaloids.

PLANT CHARACTERS A climber (vine) with hairy stems and large, leathery leaves (up to 150 mm wide) that are borne on slender petioles of up to 120 mm long. The leaves are characteristically white-hairy on the lower surfaces. Inconspicuous flowers are borne in racemes followed by fleshy black fruits containing halfmoon-shaped seeds.

OCCURRENCE In western parts of Bolivia, Peru, Ecuador, C Colombia and Panama.

CLASSIFICATION Neurotoxin; when applied i.m. or i.v. extremely hazardous, Ia

ACTIVE INGREDIENTS A dimeric isoquinoline alkaloid, D-tubocurarine, is a major ingredient. Furthermore, curine, chondocurarine, chondocurine, cycleanine and norcycleanine have also been found in the plant.

UTILISATION Pure tubocurarine has been used as a muscle-relaxant in surgery. It is a component of curare arrow poison used by S American Indians. The use of a muscle-relaxant in arrow poison (more specifically for blow darts) is a clever way to ensure that the prey animal can easily be retrieved from high forest trees. Since the alkaloid relaxes the muscles, the hunted animal (monkey or bird) does not cling to the branch on which it was perching but falls to the ground. The poison is only toxic when injected, so that the meat of the hunted animal can safely be consumed.

TOXICITY Tubocurarine is only highly toxic when it has been introduced into the bloodstream. Oral and dermal applications are virtually inactive. Acute toxicity: rabbit (s.c.) 3–5 mg are lethal; LD_{50} (i.v.) ranges in different mammals between 0.23 and 0.70 mg/kg.

SYMPTOMS Symptoms, after injection or after being wounded by a poisoned arrow include vertigo, tachycardia, hypotension, bronchospasms, weakness, nausea, pains, fever and thirst, followed by paralysis, coma and death.

PHARMACOLOGICAL EFFECTS D-Tubocurarine is an antagonist at presynaptic nicotinic acetylcholine receptors at neuromuscular plates and is therefore a muscle-relaxant. Furthermore, it stimulates the release of histamine, which causes hypotension, tachycardia, circulatory collapse and bronchospasms.

FIRST AID Overdosing causes breathing difficulties; apply 2–3 mg neostigmine together with 0.5–1 mg atropine in order to antagonise cholinergic effects on circulation, bronchia and glands. In severe cases, provide intubation and artificial respiration.

Chondrodendron tomentosum Ruiz et Pav.

family: Menispermaceae

vigne sauvage, curare (French); *Behaarter Knorpelbaum, Pareira* (German); *pareira brava* (Spanish)

Chrysanthemum vulgare

tansy

Chrysanthemum vulgare

Chrysanthemum vulgare

PLANTS WITH SIMILAR PROPERTIES A group of 150 species (N Hemisphere, Old World), formerly known as *Tanacetum*, but now included in the genus *Chrysanthemum*. Also of interest is the pyrethrum plant, *C. cinerariifolium*. It has dissected, glaucous leaves and large flower heads with yellow disc florets and white ray florets.

PLANT CHARACTERS An aromatic perennial herb (up to 1.5 m), with erect stems and compound, feathery leaves. The small, bright yellow, button-like flower heads are characteristic (they lack ray florets).

OCCURRENCE Common plants of dry and ruderal places in Europe and Asia (*C. vulgare*) or eastern Europe (*C. cinerariifolium*). Both are commercially cultivated and grown in herb gardens.

CLASSIFICATION Cell and neurotoxin, mind-altering; moderately hazardous, II

ACTIVE INGREDIENTS The toxicity is due to the essential oil, which contains β-thujone (up to 95 %), together with smaller quantities of camphor and other monoterpenoids. Tansy also produces sterols (β-sitosterol), terpenoids (α-amyrin, β-amyrin) and sesquiterpenoid lactones (including tanacetin, germacrene D and tanacetol A).

UTILISATION Tansy is traditionally used as anthelmintic to treat tapeworm, roundworm and threadworm; also as carminative, antispasmodic and stimulant to treat migraine, liver problems and loss of appetite. Flower heads of pyrethrum are traditionally used against lice and extracts are now included in commercial insecticides. It contains toxic monoterpenoids (pyrethrins I and II)

TOXICITY Large doses of tansy can be fatal. α-thujone (mouse): 87.5 mg/kg s.c.; β-thujone (mouse) 442.4 mg/kg s.c. The oil from 15–30 flowers can be a lethal dose in humans. Ingestion over a long period may be particularly harmful due to cumulative effects (see thujone).

SYMPTOMS The oil (and thujone) may cause abortion, vomiting, abdominal pain, serious gastroenteritis, rapid breathing, convulsions, mydriasis, rapid heartbeat, arrhythmia, kidney and liver damage, loss of consciousness and death.

PHARMACOLOGICAL EFFECTS The sesquiterpene lactones may cause allergic skin reactions. Thujone is a known neurotoxin.

FIRST AID Induce vomiting; instillation of sodium sulphate and medicinal charcoal. **Clinical therapy:** In case of overdose: provide gastric lavage, instillation of medicinal charcoal, sodium sulphate and polyethylene glycol. In case of excitation inject diazepam. Provide intubation and oxygen respiration in case of respiratory arrest and plasma expander when shock occurs. In case of thujone overdose: treat with atropine against colic and diazepam against spasms; control liver

Chrysanthemum vulgare (L.) Bernh. (= *Tanacetum vulgare* L.)

family: Asteraceae

tanaisie (French); *Rainfarn* (German); *tanaceto* (Italian); *tanaceto, atanasia* (Spanish)

Cicuta virosa

cowbane • water hemlock

icuta virosa   *Aethusa cynapium*

LANTS WITH SIMILAR PROPERTIES *Cita* comprises 8 species in Europe, Asia and America, among them beaver poison (*C. macata* and *C. douglasii*). *Aethusa* is a monotypic geus. Toxic polyenes have also been found in other piaceae; water dropwort (*Oenanthe crocata* and *. aquatica*).

LANT CHARACTERS A perennial herb, 0.5–3 m high, with thick rhizomes and ridged, holw stems bearing bi- or tripinnate leaves that are iangular in outline and 0.3 m long. Umbels are)–130 mm in diameter; fruits rounded, 2 mm ng, ridges blunt. *Aethusa cynapium* is a hairless nnual, 0.5 m high, bearing compound leaves ith oblong, pointed segments. Flowers in bractss umbels; bracteoles are 10 mm long. The fruits e oval and 4 mm long, with broad ridges.

CCURRENCE *C. virosa* occurs in wet places N Europe, N Asia and N America. *Aethusa cypium* occurs in Europe, W Asia and N Africa.

LASSIFICATION Cell toxin; highly hazardous, Ib

CTIVE INGREDIENTS Both species produce ixtures of polyacetylenes: *A. cynapium*: aeusin, aethusanol A & B; leaves contain 0.2 %, ots 1 % polyacetylenes. Some sources mention e alkaloid coniine as a constituent. *C. virosa*: ciItoxin and cicutol; fresh tissue contains 0.2 %, ried rhizomes up to 3.5 % polyacetylenes. Activy remains partly intact in dried plant material.

UTILISATION Both species have been used in traditional medicine. *Aethusa* (false parsley) has sometimes been confused with parsley.

TOXICITY About 15 kg fresh plant material of *A. cynapium* is lethal for cattle. 2–3 pieces of rhizomes and 500 g dried leaves can kill cattle and horses. Toxicity in mice: cicutoxin: LD_{50} 9.2 mg/kg, i.p.; aethusin: LD_{50} 93.3 mg/kg, i.p.; aethusanol A: LD_{50} 100.8 mg/kg, i.p.

SYMPTOMS Ingestion of both plants causes strong epileptic spasms and seizures. Additional symptoms include burning sensation in mouth and throat, nausea and long-lasting vomiting, headache, abdominal pain, dilated pupils, red face, bradycardia, coma, dyspnoea, paralysis starting from legs and arms; death from respiratory arrest (see polyacetylenes).

PHARMACOLOGICAL EFFECTS Polyacetylenes are very reactive compounds that can form covalent bonds with important macromolecules of the cell, such as proteins (including important receptors, ion channels, structural proteins and enzymes). These interactions can cause CNS symptoms and cell death.

FIRST AID Induce vomiting; give medicinal charcoal and sodium sulphate. **Clinical therapy:** Gastric lavage, medicinal coal, in case of spasms diazepam, penthotal, or trapanal, to reduce spasms (see polyacetylenes).

icuta virosa L.   family: Apiaceae

gué aquatique (French); *Giftiger Wasserschierling* (German); *cicuta aquatica* (Italian)

Cinchona pubescens

Peruvian bark tree • red cinchona

Cinchona pubescens

Cinchona pubescens

PLANTS WITH SIMILAR PROPERTIES
A genus of about 40 species distributed from the Andes to Costa Rica. *C. succirubra* is usually considered to be a mere variety or cultivar of *C. pubescens*. The second main source, the well-known quinine tree, are species alternatively known as *C. officinalis* or *C. calisaya* (= *C. ledgeriana*).

PLANT CHARACTERS Forest trees with bright green, simple leaves with prominent white veins.

OCCURRENCE Colombia, W Ecuador and N Peru (*Cinchona pubescens* and *C. officinalis*). Commercial cultivation occurs mainly in India, Indonesia and Africa (Congo).

CLASSIFICATION Cell toxin, neurotoxin; moderately hazardous, II

ACTIVE INGREDIENTS Several very bitter-tasting quinoline alkaloids are present (quinine, quinidine, cinchonine, cinchonidine and others).

UTILISATION The bark contains quinine, an important antimalarial drug that has saved millions of lives. It is a bitter tonic and stimulates appetite (often taken as the famous gin-and-tonic before a meal). Standardised isolated quinine and quinidine are available with antispasmodic, uterotonic and anti-arrhythmic properties. Bark is used as an ingredient of some herbal teas that are taken for flatulence and loss of appetite.

TOXICITY Several cases of human poisoning with quinine and quinidine have been reported (as side effects, overdosing or chronic toxicity).

SYMPTOMS Side effects of medicinal use: extensive skin allergies, fever, shaking chills, icterus, haematuria and accommodation problems (even blindness). In pregnant patients and in those with malaria, quinine can cause severe haemolysis and abortion: symptoms are cold sweat, back pain, vomiting, fever, diarrhoea, haematuria (red urine), dyspnoea, internal bleeding, icterus and death in the second week from respiratory and cardiac arrest. Chronic quinine use (1 g p.o.) can also cause central intoxication with tinnitus, deafness, dizziness, headache and vomiting.

PHARMACOLOGICAL EFFECTS Quinine is a protoplasma poison that selectively kills the asexual stages of *Plasmodium* parasites. Quinidine inhibits sodium channels. Both alkaloids also modulate several neuroreceptors, and are substrates for MDR proteins.

FIRST AID Administration of medicinal charcoal and sodium sulphate. **Clinical therapy:** After overdose: Gastric lavage (possibly with potassium permanganate), instillation of medicinal charcoal and sodium sulphate, electrolyte substitution; check blood coagulation and acidosis. In case of tachycard arrhythmia, give phenytoin or lidocaine; for bradycardia, give atropine or alupent. Intubation and oxygen respiration may be necessary.

Cinchona pubescens Vahl (= *Cinchona succirubra* Pav. ex Klotsch)   family: Rubiaceae

quina, quinquina (French); *Roter Chinarindenbaum* (German); *china rossa* (Italian); *quina* (Spanish)

Citrullus colocynthis

colocynth • bitter apple

Citrullus colocynthis

PLANTS WITH SIMILAR PROPERTIES A genus of 4 species in tropical and southern Africa, including the well-known commercial water melon, *C. lanatus* and its wild relative (*tsama* melon).

PLANT CHARACTERS A perennial creeping climber (vine) with heart-shaped, stalkless leaves and branched climbing tendrils. The yellow flowers are followed by round, apple-like, green fruits (100 mm in diameter) with spongy, white flesh and many seeds.

OCCURRENCE W Africa, naturalised in Arabia, the Mediterranean and in India.

CLASSIFICATION Cell toxin, highly hazardous, Ib

ACTIVE INGREDIENTS Bitter apple contains mixtures of bitter triterpenoids, so-called cucurbitacins (up to 3 %). Cucurbitacins B, E, J and their glycosides are present; fruit tissue: 0.22 %; seeds: 0.18 %, stems: 0.17 %, leaves: 0.15 %.

UTILISATION Colocynth or bitter apple is a traditional purgative medicine of the pumpkin family containing free and glycosylated cucurbitacins, including cucurbitacin E. It has been cultivated since Assyrian times and was also used for rodent control in ancient Rome. The fruit (*Colocynthidis fructus*) is no longer used medicinally because of its toxicity and undesirable side effects. Domesticated water melons are without cucurbitacins and a popular fruit used in many parts of the world.

TOXICITY *C. colocynthis* has been confused with melons and pumpkin or zucchini and thus caused poisoning. Animal poisoning has also been reported. Cucurbitacins are strong cell poisons at higher doses. A dose of 3 g of *Citrullus colocynthis* is lethal.

SYMPTOMS Symptoms after oral intake include vomiting, drastic, even bloody diarrhoea through extensive inflammation of gut tissues, severe kidney damage, and spasms. Death results from respiratory arrest after intake of large doses; in pregnant women, root extracts can cause abortion (see cucurbitacin E). Cucurbitacins are transferred into the milk of lactating women. Exposure to skin may cause inflammation with blister formation.

PHARMACOLOGICAL EFFECTS Cucurbitacins inhibit mitotic cell division and are therefore cytotoxic (see cucurbitacins); they have antitumour activity but are too toxic to be used in medicine. Cucurbitacins activate intestinal peristaltic motility (therefore strong purgatives) and cause spasms with colic.

FIRST AID Induce vomiting; administer sodium sulphate and medicinal charcoal; provide plenty of liquid (tea) to drink. **Clinical therapy:** After overdose: Gastric lavage (possibly with 0.1 % potassium permanganate), instillation of medicinal charcoal and sodium sulphate; for further detoxification see cucurbitacin E.

Citrullus colocynthis (L.) Schrad.   family: Cucurbitaceae

coloquinthe (French); *Koloquinte, Bittermelone* (German); *coloquintide* (Italian)

Claviceps purpurea

ergot

Claviceps purpurea

Claviceps purpurea (sclerotium)

PLANTS WITH SIMILAR PROPERTIES Similar alkaloids occur in the family Convolvulaceae (*Argyreia, Ipomoea tricolor, Turbina corymbosa*).

CHARACTERS Ergot is a fungus with black-violet spore bodies (sclerotia). It grows on rye (*Secale cereale*) (thus of toxicological importance) but also on more than 600 other species of Poaceae.

OCCURRENCE Cosmopolitan distribution on grasses and wherever rye is cultivated. Alkaloids for medicinal purposes are obtained from field cultivation and saprophytic fermentations.

CLASSIFICATION Neurotoxin; hallucinogen; highly hazardous, Ib

ACTIVE INGREDIENTS The permanent sclerotium is rich in indole alkaloids (up to 1 %), so-called ergot alkaloids. One series are simple amides of lysergic acid, such as ergometrine and ergotamine. The other series has a peptide complex instead of an amide group; examples are ergocristine, ergocornine and α-ergocryptine.

UTILISATION Hippocrates described "Melanthion" (ergot) and its use to stop bleeding after childbirth. The Eleusinian Mysteries of ancient Greece have been ascribed to ergot-induced hallucinations. Isolated alkaloids rather than crude extracts are nowadays used in medicine, such as ergometrine and derivatives (in obstetrics) and ergotamine (to treat migraine).

TOXICITY The toxic effects of ergot ("ergotism"), which has killed tens of thousands of people, have long been known. LD_{50} values in rabbit (i.v.); ergometrine: LD_{50} 3.2 mg/kg; ergotamine 3.0 mg/kg; ergocristine 2 mg/kg; ergocornine 1.1 mg/kg. About 10 mg ergotamine (p.o.) are lethal for humans. Lethal dose of sclerotia in humans (p.o.) 5–10 g (see ergot alkaloids).

SYMPTOMS Acute toxicity: nausea, vomiting, diarrhoea, thirst, pruritus, enhanced but weak pulse, paraesthesia, no feeling in legs and arms, confusion and loss of consciousness. Besides central nervous disturbances (leading to convulsions and twitching, i.e. epileptic seizures), long-term toxicity from contaminated rye flour is caused through painful vasoconstriction. Legs and arms go numb and eventually become gangrenous; miscarriages are also common.

PHARMACOLOGICAL EFFECTS Since ergot alkaloids can interact with several neuroreceptors, especially those of serotonin, but also noradrenaline and dopamine, a wide variety of pharmacological effects have been described.

FIRST AID Administration of sodium sulphate and medicinal charcoal. **Clinical therapy:** Gastric lavage (possibly with potassium permanganate); medicinal charcoal and sodium sulphate; electrolyte substitution; check acidosis. Intubation and oxygen respiration may be needed (see ergot alkaloids).

Claviceps purpurea (Fries) Tul.

family: Clavicipitaceae

ergot de seigle (French); *Mutterkorn* (German); *grano cornuto* (Italian)

Clivia miniata

bush lily • orange lily

Clivia miniata

PLANTS WITH SIMILAR PROPERTIES A genus with 5 species in S Africa: *C. miniata*, *C. nobilis*, *C. gardenii*, *C. mirabilis* and *C. caulescens*.

PLANT CHARACTERS *Clivia* species are shade-loving, perennial, lily-like plants with rosettes of dark green, strap-shaped leaves arising from elongated fleshy rhizomes. *C. miniata* differs in the open, bell-shaped flowers (not tubular as in all other species) that are various shades of orange or less often yellow. The fruits are fleshy, egg-shaped, bright red capsules containing large, short-lived seeds.

OCCURRENCE *C. miniata* is indigenous to the eastern parts of S Africa. The wild form of the species, as well as numerous cultivars (some with variegated leaves), have become popular as garden plants and are also used in many parts of the world as indoor pot plants.

CLASSIFICATION Cell poison; highly hazardous; Ib–II

ACTIVE INGREDIENTS The toxic components are isoquinoline alkaloids, of which lycorine is very well known, since it also occurs in many other Amaryllidaceae. The level of lycorine in *Clivia* was found to be up to 0.4 % dry weight. Several other alkaloids, including cliviamine, clivonine and cliviamartine have been isolated from *Clivia* roots and leaves. It is claimed that the plant also contains hippeastrine, which is structurally related to lycorine.

UTILISATION *Clivia* is mainly used in traditional medicine as a uterotonic to ensure successful childbirth. Other uses include the treatment of pain, fever and snake bite. It is very popular in horticulture, especially in China and Japan.

TOXICITY Children and domestic animals may be poisoned when *C. miniata* is grown in the house or garden. Most cases of poisoning are mild. The traditional medicinal use may lead to accidents through overdose. Lycorine is very poisonous; the LD_{50} in dogs is 41 mg/kg.

SYMPTOMS Symptoms of poisoning are salivation, vomiting, gastrointestinal disturbance with diarrhoea, leading to paralysis and collapse.

PHARMACOLOGICAL EFFECTS The leaves of *C. miniata* have demonstrated uterotonic effects. Lycorine, an inhibitor of protein biosynthesis, has cytotoxic and virustatic properties; it is emetic and diuretic (see lycorine).

FIRST AID Give medicinal charcoal and sodium sulphate. Allow patient to drink plenty of tea; provide shock prophylaxis. **Clinical therapy:** After ingestion of large doses, provide gastric lavage (possibly with potassium permanganate), electrolyte substitution, control acidosis with sodium bicarbonate (urine pH 7.5) and kidney function; in case of spasms give diazepam; in severe cases supply oxygen and intubation (see lycorine).

Clivia miniata (Lindl.) Regel

family: Amaryllidaceae

livia (French); *Zimmer-Clivie, Riemenblatt* (German); *clivia* (Italian); *clivia* (Spanish)

Coffea arabica

coffee tree

Coffea arabica flowers

Coffea arabica fruits

PLANTS WITH SIMILAR PROPERTIES Genus with 90 species in tropical Africa. Of lesser commercial importance are *C. canephora* (robusta or Congo coffee) which is often used in instant coffee and *C. liberica* (Liberian or Abeokuta coffee) which is added to blends for its bitter flavour. Several other plants produce caffeine.

PLANT CHARACTERS Coffee is a shrub or small tree of 4–7 m in height with dark green, glossy leaves and clusters of fragrant, white flowers in the leaf axils. The fruits are rounded berries that turn yellow, red or purple when they ripen.

OCCURRENCE Indigenous to Ethiopia. Cultivated in Brazil, Colombia, Indonesia, Ivory Coast, Mexico and Kenya (6 million tons per year).

CLASSIFICATION Stimulant; slightly hazardous, III

ACTIVE INGREDIENTS The main stimulant in coffee is caffeine, a purine alkaloid. Roasted seeds have 1–2 % caffeine, so that a cup of coffee typically contains about 150 mg. Caffeine for pharmaceutical use is obtained as a by-product of decaffeinated coffee.

UTILISATION Plantations were first started in Yemen in the 9th century. Coffee was introduced to England as "kahveh" in 1601 and became famous throughout Europe. Coffee is a stimulating beverage used to increase alertness, promote concentration and reduce fatigue. Drinks that contain caffeine are widely used by mankind and our literature, music, arts and thus our culture have certainly been positively influenced by caffeine (think of the Viennese coffee houses and their influence on artists and scientists).

TOXICITY Toxic effects occur only at high doses: 150–200 mg/kg appears to be the LD_{50} for caffeine in humans; about 5 g caffeine has caused the death of a child.

SYMPTOMS Excessive amounts of coffee may lead to nervousness, headache, tremor, spasms, palpitations, high blood pressure, insomnia and indigestion. Caffeine is known to be addictive. The topical application of ointments with 30 % caffeine can induce toxic effects.

PHARMACOLOGICAL EFFECTS Caffeine is an inhibitor of cAMP phosphodiesterase and an antagonist at adenosine receptors. As a consequence, dopamine is released and many brain parts become activated. Caffeine is a cortical stimulant that increases the carbon dioxide sensitivity of the brain stem. It affects the cardiovascular system and has a positive inotropic action, resulting in tachycardia and an enhanced output.

FIRST AID In case of overdose: Administration of sodium sulphate and medicinal charcoal.

Clinical therapy: Gastric lavage, instillation of medicinal charcoal and sodium sulphate, and diazepam in case of spasms.

Coffea arabica L.

family: Rubiaceae

caféier d'Arabie (French); *Kaffeestrauch* (German); *caffè* (Italian); *cafeto* (Spanish)

Colchicum autumnale

autumn crocus • meadow saffron

olchicum autumnale flowers

Colchicum autumnale fruits

LANTS WITH SIMILAR PROPERTIES A geus with more than 65 species in Europe, N Afri-
a, C Asia and N India, including *C. pannonicum*,
. speciosum and *C. variegatum*.

LANT CHARACTERS A bulbous plant with
arrow leaves and an unusual flowering and fruit-
g cycle. The young fruit and leaves grow from
e fleshy corm in spring (the fertilised ovary es-
apes winter frosts hidden 20–40 cm below ground
d matures during the following spring and early
mmer). The plant is leafless in autumn, when the
tractive pink, tubular flowers appear. The fruit is
3-locular capsule with many small black seeds.

OCCURRENCE Europe and North Africa; *Col-
hicum* species are popular as ornamental plants.

CLASSIFICATION Cell toxin; extremely hazrdous, Ia

ACTIVE INGREDIENTS The main active com-
ound is colchicine (an unusual phenethylisoqui-
oline alkaloid), at 0.3–1.2 % of dry weight of the
orm, flower or seeds. Also present are demecol-
ine and *N*-desacetyl *N*-formylcolchicine.

UTILISATION The toxicity of *Colchicum* was
entioned in Assyrian medical texts. Dioscorides
arned that the sweet-tasting bulbs of *Colchicum*
re highly toxic and could be mistaken for onion
ulbs. Isolated colchicine in carefully measured
oses is used in modern medicine to treat acute
ttacks of gout and familial Mediterranean fever.

TOXICITY Confusion with *Allium ursinum* but
also suicide and murder attempts have caused serious poisoning (in C Europe, over 16 fatalities in
a period of 30 years). Colchicine is lipophilic and
thus easily absorbed after oral ingestion. The toxic
dose of colchicine in humans is about 10–20 mg;
40 mg is the lethal dose (adults: 5 g seeds; children: 1.2–1.5 g seeds). Livestock poisoning is also
known (LD_{100} for cattle: 1.5–2.5 kg fresh leaves).

SYMPTOMS Poisoning by colchicine and *Colchicum* starts with nausea and vomiting, followed by
watery and bloody diarrhoea, strong abdominal
pain, haematuria, hypotension, convulsions and
paralysis. At lethal doses, colchicine causes fatal
respiratory arrest and cardiovascular collapse
within a few days (see colchicine).

PHARMACOLOGICAL EFFECTS Colchicine
binds to the β-subunit of tubulin dimers and thus
blocks the polymerisation and depolymerisation
of microtubules and in consequence, mitotic cell
division and cellular transport processes (e.g. vesicle transport in neuronal axons) including cell
migration. It acts as a capillary poison.

FIRST AID Induce vomiting; administer sodium
sulphate and medicinal charcoal. Keep patient in
shock position and warm. **Clinical therapy:** Gastric lavage, electrolyte substitution; in cases of severe intoxication, provide intubation and oxygen
respiration.

Colchicum autumnale L.

family: Colchicaceae

olchique d'automne (French); *Herbstzeitlose* (German); *crocus autumnale, colchico* (Italian); *cólquico* (Spanish)

Conium maculatum

poison hemlock

Conium maculatum spotted stem   *Conium maculatum* flowers

PLANTS WITH SIMILAR PROPERTIES A genus with 6 species in Eurasia and Africa.

PLANT CHARACTERS An erect, biennial herb of up to 2 m, with typical purple or yellow spots on the stems and pinnately compound leaves. Small white flowers develop into dry, ribbed fruits that may be confused with anise fruits (*Pimpinella anisum*). Hemlock and its alkaloids are easily recognised by the strong mousy smell on the hands after handling the plant or alkaloids.

OCCURRENCE *C. maculatum* grows in wet and ruderal places in Europe, Asia and N Africa.

CLASSIFICATION Neurotoxin; extremely hazardous, Ia

ACTIVE INGREDIENTS Piperidine alkaloids: coniine and γ-coniceine are the major toxins; minor ones are *N*-methylconiine, conhydrine and pseudoconhydrine. Fruits have concentrations of up to 3.5 %. Roots also contain toxic polyacetylenes, such as falcarinon and falcarinolon.

UTILISATION Extracts were frequently used for murder, suicide and execution in Greek and Roman times. This infamous plant was given to Socrates as a death potion. In Europe, hemlock has been used medicinally as a sedative, antispasmodic, against ulcers and as an antaphrodisiac.

TOXICITY All parts of the plant are highly toxic and fatalities have occurred after accidental ingestion (rare) of roots, leaves or fruits. Al-

kaloids are easily absorbed by diffusion. Livstock poisoning has been recorded (ingestion 2–4 kg leaves will kill a horse or cow). LD_{50} mice: Coniine, 19 mg/kg i.v., 100 mg/kg p.o γ-coniceine, 2.6 mg/kg, i.v., 12 mg/kg, p.o.; methylconiine, 27.5 mg/kg, i.v., 204.5 mg/k p.o. About 0.5–1 g coniine are lethal for huma (p.o.).

SYMPTOMS Paralysis of motor nerve ending leading to a burning sensation in the mouth an throat, salivation, drowsiness, trembling, nause vomiting and diarrhoea, tachycardia, mydriasi breathing difficulty and finally asphyxia followe by mental disturbance and ascending paralysi which starts at the tip of arms and legs and enc with respiratory failure and death.

PHARMACOLOGICAL EFFECTS Coniu alkaloids affect nicotinic acetylcholine recepto and to a lesser degree muscarinergic receptor they block signal transduction at neuromuscula plates and vegetative ganglia. Teratogenic effec occur in cattle and pigs (crooked calf disease).

FIRST AID Instil medicinal charcoal and sodium su phate. **Clinical therapy:** Immediate gastric lavag (possibly with 0.1 % potassium permanganate instillation of medicinal charcoal and sodium su phate, electrolyte substitution; control acidos with sodium bicarbonate; keep patient warm; i tubation and oxygen respiration may be needed

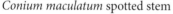

Conium maculatum L.   family: Apiacea

cigué tachée (French); *Gefleckter Schierling* (German); *cicuta maggiore* (Italian)

Convallaria majalis

lily-of-the-valley

Convallaria majalis leaves and flowers

Convallaria majalis fruits

PLANTS WITH SIMILAR PROPERTIES A genus of 3 species in Europe and North America.
PLANT CHARACTERS Lily-of-the-valley is a small leafy perennial (up to 300 mm high) with spreading rhizomes resulting in large clumps. Each plant has a pair of broad leaves with parallel veins. In spring it produces an elegant cluster of bell-shaped, white, fragrant flowers. The fruit are small, red, pea-sized berries with 2–6 seeds.
OCCURRENCE Widely distributed in deciduous forests over most of Europe and Asia; naturalised in N America; cultivated as a garden plant.
CLASSIFICATION Heart poison; highly hazardous, Ib
ACTIVE INGREDIENTS Lily-of-the-valley contains more than 38 cardiac glycosides in yields of 0.1–0.68 % of DW. The level of cardiac glycosides reaches a peak in the flowering period, and the compounds are mainly concentrated in the flowers and seeds. Convallatoxin is the main compound, representing about 40 % of the total heart glycosides. It occurs in related compounds such as convalloside and convallatoxol. Azetidine-2-carboxylic acid (NPAA) and steroidal saponins (convallamaronin) are present in all parts, but especially in the rhizomes and roots.
UTILISATION The herb is used medicinally for the treatment of mild cardiac insufficiency in elderly people. It is also used as a cut flower.

TOXICITY Poisoning may occur through wrong dosage of the drug, or when more than 5–10 of the attractive red fruits are eaten (mostly by children). Lethal dose of cardenolide extract in guinea pigs (i.v.): 0.189 mg/kg; convallatoxin: guinea pigs (i.v.): 0.309 mg/kg; in cats 0.076 mg/kg. In Europe, leaves of *Allium ursinum* are collected in spring in the wild. Since *Convallaria* occurs in similar habitats, inexperienced herb collectors may sample the wrong species (s. p. 29); deaths from eating *Convallaria* leaves have been recorded in Germany.
SYMPTOMS Irritation of skin and eyes, nausea, salivation, diuresis, gastrointestinal disturbance, purging and exhaustion, with the usual respiratory and cardiac symptoms typical of heart glycosides (arrhythmia, tachycardia, hypertension, coma, cardiac arrest) (see cardiac glycosides).
PHARMACOLOGICAL EFFECTS Cardenolides selectively inhibit Na$^+$, K$^+$-ATPase and are therefore general strong poisons (see cardiac glycosides). Azetidine-2-carboxylic acid can be incorporated in proteins in place of proline, thereby increasing the toxicity (see NPAA).
FIRST AID First aid treatment is indicated if plant material from *Convallaria* or isolated cardiac glycosides have been ingested. Treatment involves the induction of vomiting, instillation of medicinal charcoal, and detoxification as outlined under cardiac glycosides.

Convallaria majalis L.

family: Ruscaceae or Convallariaceae

muguet (French); *Maiglöckchen* (German); *mughetto* (Italian); *lirio de los valles* (Spanish)

Cotyledon orbiculata

pig's ears

Cotyledon orbiculata

Cotyledon orbiculata flowers

PLANTS WITH SIMILAR PROPERTIES A genus of 9 species from S and E Africa to Arabia.

PLANT CHARACTERS *C. orbiculata* is a variable succulent shrub with thick, somewhat woody stems and large, fleshy, mostly rounded leaves with a waxy bloom on the surface and reddish margins. The tubular, red to orange flowers are clustered at the tips of erect flowering stalks. They are typically pendulous, with flaring lobes. Several varieties have been described, differing mainly in growth form and leaf shape.

OCCURRENCE *C. orbiculata* is widely distributed in southern Africa. It is commonly cultivated as an ornamental plant in warm regions.

CLASSIFICATION Heart poison; highly poisonous, Ib

ACTIVE INGREDIENTS The toxins in *Cotyledon* species are heart glycosides of the bufadienolide type. Four compounds occur in *C. orbiculata,* namely orbicusides A, B and C and tyledoside C (see also *Tylecodon* species).

UTILISATION Leaves are used in traditional medicine to remove warts and to treat epilepsy, internal parasites, earache and toothache.

TOXICITY The plant is poisonous to small stock and causes a serious chronic disease known as *krimpsiekte* ("shrinking disease"), in the same way as other Crassulaceae such as *Kalanchoe* and *Tylecodon*. Livestock sometimes eat these succulent plants during the dry season when no othe grazing is available. Meat from poisoned anima is known to have caused secondary poisoning c dogs and even humans. *Krimpsiekte* occurs mair ly in sheep and goats, but horses, dogs and chick ens are also susceptible. The toxins have a cumu lative effect. When 14 g of dried leaves were fe daily to a goat over a period of 25 days, it deve oped typical symptoms of *krimpsiekte*. The LD of the major bufadienolides of *C. orbiculata* hav been determined (s.c. in guinea pigs) and varie between 0.1 and 0.25 mg/kg.

SYMPTOMS The effects of *Cotyledon* poisonin are quite unlike the normal symptoms of cardia glycoside poisoning. Typical are a loss of condi tion, convulsions, paralysis of the head and neck respiratory paralysis and finally death. Anima may lie on their sides (fully conscious but para lysed) for several weeks before they die.

PHARMACOLOGICAL EFFECTS Bufadieno lides selectively inhibit Na$^+$, K$^+$-ATPase and ar therefore general heart poisons (see cardiac gly cosides).

FIRST AID First aid treatment is indicated i plant material from *Cotyledon* or isolated cardia glycosides have been ingested. Treatment involve the induction of vomiting, supply of medicina charcoal, and detoxification as outlined unde cardiac glycosides.

Cotyledon orbiculata L.

cotylédon (French); *Schweinsohr* (German)

family: Crassulacea

Crinum bulbispermum

river lily

Crinum bulbispermum flowers

PLANTS WITH SIMILAR PROPERTIES A genus of 120 species in warm and tropical climates, including *C. asiaticum* and *C. flaccidum*.

PLANT CHARACTERS *Crinum* species are large bulbous plants with slender, curving leaves and spectacular pink and white, trumpet-shaped flowers. *C. bulbispermum* is easily distinguished by the blunt leaf tips. The exposed parts of the leaves are killed by frost and veld fires in winter but they grow out again in spring.

OCCURRENCE *C. bulbispermum* is widely distributed in southern Africa, often along rivers. The species and hybrids, e.g. *C.* x *powellii* (*C. bulbispermum* x *C. moorei*) are grown as ornamental plants. *C. asiaticum* and other species are also popular garden plants and cut flowers.

CLASSIFICATION Cell poison; highly hazardous, Ib

ACTIVE INGREDIENTS The river lily contains crinine and powelline, lycorine and crinamine or bulbispermine as main alkaloids (there seems to be regional and seasonal variation), together with acetylcaranine, ambelline, crinasiadine, crinasiaine, galanthamine and hippeastrine. Alkaloids also occur in several other *Crinum* species.

UTILISATION The onion-like bulbs of *C. bulbispermum* are used in traditional medicine to treat colds and scrofula. Bulbs of *C. macowanii* are a Zulu remedy for fever, scrofula, micturition, rheumatic fever, kidney and bladder problems, glandular swellings and skin ailments.

TOXICITY Poisoning through accidental overdose is possible. Stock losses from *Crinum* species have been reported in E Africa. Crinamine and lycorine are highly toxic, with an oral LD_{50} in dogs of 10 mg/kg and 41 mg/kg respectively.

SYMPTOMS Symptoms have not been accurately recorded but probably include nausea, vomiting, dizziness, diarrhoea and kidney trouble, and respiratory arrest.

PHARMACOLOGICAL EFFECTS Crinamine has powerful transient hypotensive and respiratory depressant activity in dogs. Ambelline has analgesic activity similar to that of morphine. Acetylcaranine is a uterine stimulant and is active against some forms of leukaemia, while caranine causes respiratory paralysis and death in animals. Lycorine has cytotoxic properties; it is emetic and diuretic.

FIRST AID Give medicinal charcoal and sodium sulphate. Allow the patient to drink plenty of tea and give shock prophylaxis. **Clinical therapy:** After ingestion, gastric lavage (possibly with potassium permanganate), electrolyte substitution, control acidosis with sodium bicarbonate (urine pH 7.5) and kidney function; in case of spasms give diazepam; in severe cases supply oxygen and intubation.

Crinum bulbispermum (Burm.f.) Milne-Redh. & Schweick.　family: Amaryllidaceae

Crinum (French); *Orangefarbene Hakenlilie* (German)

97

Crocus sativus

saffron

Crocus sativus

Crocus sativus stigmas

PLANTS WITH SIMILAR PROPERTIES A genus of 80 species from the Mediterranean to W China; several are grown as ornamentals.
PLANT CHARACTERS A small bulb bearing firm, grass-like leaves and purple flowers with three bright red style-branches. It is similar to *Colchicum* but has only 3 stamens and not 6 as in *Colchicum*.
OCCURRENCE S Europe and SW Asia. It is a sterile cultigen of uncertain origin that is grown on a commercial scale in Spain. About 130 flowers (dried stigmas) are needed to get 1 g of saffron.
CLASSIFICATION Neurotoxin, mind-altering; moderately hazardous, II
ACTIVE INGREDIENTS A diterpene, crocetin, constitutes the yellow, water-soluble pigment found in the stigmas. A bitter glucoside, picrocrocin, is an important component of the stigmas. The typical saffron aroma is produced when the stigmas are dried; it is due to safranal (the aglycone of picrocrocin).
UTILISATION Saffron was an important dye in ancient Greece and Rome, used to fortify wine and *laudanum* (an opium cocktail); it was thought to be a powerful aphrodisiac. In traditional medicine it has been used as a sedative and spasmolytic to treat numerous ailments (and in large doses to induce abortions). Saffron is a very important spice and natural colouring for food.

TOXICITY Saffron is safe only in the small quantities that are typically used in food. Large quantities (above 1.5 g per day) are dangerous especially during pregnancy: 5 g may cause vomiting and bleeding, 10 g will cause abortion and uterine bleeding, while 20 g may be fatal.
SYMPTOMS Bloody diarrhoea, haematuria and bleeding from the nose and eyelids. CNS effects include excitation with extensive laughing, followed by tachycardia, palpitation, headache, vertigo, loss of appetite, vomiting, delirium, distorted vision, lethargy, paralysis and death. Skin and mucous membranes acquire a yellow colour.
PHARMACOLOGICAL EFFECTS Picrocrocin and its lipophilic derivate safranal have a reactive aldehyde function that can form covalent bonds with amino groups in proteins and nucleic acid. Thus it can affect a multitude of targets and have cytotoxic and mutagenic effects (see aldehydes).
FIRST AID Give medicinal charcoal and sodium sulphate; give liquids to drink. **Clinical therapy** After overdoses: gastric lavage; instillation of medicinal charcoal and sodium sulphate, electrolyte substitution; control acidosis with sodium bicarbonate; in case of excitation, spasms and colic give diazepam and atropine; intubation and oxygen respiration may be required. Check liver, kidney function and blood coagulation.

Crocus sativus L.

family: Iridaceae

safran (French); *Safran* (German); *zafferano vero* (Italian); *azafrán* (Spanish)

Crotalaria burkeana

rattle bush

Crotalaria burkeana

Crotalaria agatiflora

PLANTS WITH SIMILAR PROPERTIES A genus of more than 600 species on all continents (mostly Africa), e.g. *C. juncea, C. anagyroides, C. spectabilis, C. spartioides* and *C. usaramoensis*.

PLANT CHARACTERS Typical for *Crotalaria* species are a pointed keel formed by the lower-most two petals of the flower and the swollen, much-inflated pods. *C. burkeana* is recognised by the spreading red hairs on all parts of the plant. *C. spartioides* has a virgate, broom-like growth form, yellow flowers and narrow leaflets, while the related *C. virgultalis* has wider leaflets and the flowers are partly white and red. *C. dura* is similar to *C. globifera* but has more densely hairy leaves and shorter, more rounded pods.

OCCURRENCE *C. burkeana, C. spartioides, C. globifera* and *C. dura* occur in S Africa, *C. juncea* in India, *C. anagyroides* and *C. spectabilis* in C & S America and *C. usaramoensis* in China.

CLASSIFICATION Cell toxin, mutagen; moderately hazardous, II

ACTIVE INGREDIENTS Numerous macrocyclic pyrrolizidine alkaloids (PAs), especially monocrotaline and senecionine. *C. spartioides* contains retrorsine, *C. dura* and *C. globifera* dicrotaline. Some species with NPAAs.

UTILISATION *C. juncea* is a source of fibres for making ropes, nets and even cigarette paper. Some species are grown for fodder and green manure or as ornamental plants, e.g. *C. agatiflora*.

TOXICITY *Crotalaria* species are the cause of severe poisoning in livestock. Crotalism is an acute or chronic condition resulting from the ingestion of large quantities of *Crotalaria*, leading to serious liver and lung damage. PAs can be transferred into the milk of lactating cows and goats. Cereals contaminated with *Crotalaria* seeds caused serious poisoning of 67 people in India in 1975; 28 of them died. LD_{50} of monocrotaline: rat: 71 mg/kg p.o.; mouse: 166–170 mg/kg p.o.

SYMPTOMS PAs damage the liver (necrosis, fibrosis, megalocytosis), lung (pneumonia, emphysema, oedema), kidneys (megalocytosis), pancreas and GI tract. Liver damage occurs in horses, cattle and sheep; *C. juncea* causes hair loss in sheep.

PHARMACOLOGICAL EFFECTS PAs become metabolically activated in the liver (see senecionine). They can then alkylate proteins and DNA. As a result, they cause damage to liver (veno-occlusive disease), lungs, kidneys, pancreas and the GI tract. After prolonged intake they are mutagenic, teratogenic and carcinogenic.

FIRST AID Give medicinal charcoal to absorb the alkaloids; administer plenty of warm black tea and sodium sulphate. **Clinical therapy:** In case of ingestion of larger doses: gastric lavage and further treatment as described under senecionine.

Crotalaria burkeana Benth.

family: Fabaceae

Crotalaria (French); *Klapperhülse* (German)

Croton tiglium

purging croton

Croton tiglium

Croton tiglium

PLANTS WITH SIMILAR PROPERTIES
A large genus with more than 750 species in warm climates of Africa, Asia, and America, including *C. bonplandianus, C. eluteria, C. flavens, C. laccifer, C. malambo, C. setigerus* and *C. texensis.*

PLANT CHARACTERS A shrub or small tree (4–6 m) with glossy alternate leaves furnished with two glands at the base. Flowers have straw-coloured petals and occur in erect terminal racemes. The seeds are egg-shaped and flattened, 10–13 mm long, 6–9 mm wide, marbled with reddish, yellowish and blackish dots. They resemble castor beans or ticks (*Kroton* means *ticks*).

OCCURRENCE Asia (India, China, Malaysia).

CLASSIFICATION Cell toxin, co-carcinogen, skin irritant; extremely hazardous, Ia

ACTIVE INGREDIENTS Several phorbol esters of the tigliane type, such as TPA (12-O-tetradecanoylphorbol-13-acetate); seeds contain 30–50 % oil; one of the fatty acids is crotonic acid, which is one of the purging compounds. Seeds also have up to 3.8 % crotonide, a purine alkaloid, as well as a toxic lectin, crotin.

UTILISATION In Curacao, where a beverage is produced from *Croton flavens* (Welensali tea), a high incidence of oesophagus cancer has been recorded. Seed oil has been used medicinally as counterirritant to treat rheumatism; together with *Ricinus* oil, it is also used as a purgative.

TOXICITY Seeds and oil are highly toxic because of phorbol esters; crotonic acid is a very strong purgative. About 4 seeds or 0.5–1 ml (20 drops) of the oil can be lethal for humans, 15 seeds for horses. Symptoms set in after 5–10 minutes.

SYMPTOMS Ingestion of *Croton* seeds or oil results in burning and scratching in the mouth and throat (with urticaria and inflammation), salivation, vomiting, impaired consciousness, abdominal pain, watery and painful, bloody diarrhoea, vertigo, arrhythmia, kidney inflammation and delirium. Death may occur after 1–3 days. Externally, the oil leads to reddening, blistering and swollen oedema. Phorbol esters cause temporary blindness after contact with eyes.

PHARMACOLOGICAL EFFECTS. Phorbol esters activate protein kinase C, which is a key enzyme in several signalling pathways. They are important co-carcinogens. Crotin has similar ribosome-inactivating activities as ricin (see abrin).

FIRST AID Give medicinal charcoal and sodium sulphate. Provide plenty of tea. Contact with eye or skin: wash with polyethylene glycol or water in case of skin blisters, cover with sterile cotton and locally with corticoid ointment. **Clinical therapy:** Gastric lavage (possibly with potassium permanganate), instillation of medicinal charcoal and sodium sulphate (see phorbol esters).

Croton tiglium L.

family: Euphorbiaceae

croton revulsif (French); *Krotonölbaum* (German); *crotone* (Italian)

Cucumis africanus

wild cucumber

ucumis africanus

Cucumis myriocarpus

PLANTS WITH SIMILAR PROPERTIES A genus of 32 species in the tropics of the Old World. **PLANT CHARACTERS** A perennial creeper with trailing, hairy stems. The leaves are divided into 5 lobes, with toothed margins and rough hairs. The small yellow male and female flowers occur on the same plant. The characteristic fruits are oblong to ellipsoid, with sparse, stiff bristles. Oblong fruits are said to be non-bitter, while the smaller ellipsoid fruit are usually bitter and poisonous. *C. myriocarpus* is similar to *C. africanus* but it is an annual plant, the female flowers are minutely hairy inside, the leaf stalks have three different hair types and the fruits are smaller, more rounded and have prominent stripes. **OCCURRENCE** *C. africanus* and *C. myriocarpus* are widely distributed in S Africa. The latter occurs as a weed on old lands and disturbed places. **CLASSIFICATION** Cell toxin; highly hazardous, Ib **ACTIVE INGREDIENTS** A series of structurally related triterpenoids known as cucurbitacins occur in bitter fruits of the pumpkin family, including wild cucumbers, pumpkins and calabashes. An example is cucurbitacin B, a widely distributed compound that has been isolated from fruits of *C. africanus*. The free cucurbitacins represent the active principle. **UTILISATION** The use of bitter fruits or bitter

fruit juice of wild cucumbers and wild watermelons as purgative and enemas is potentially lethal. **TOXICITY** *Cucumis* species have caused fatal and near-fatal human poisoning. *C. myriocarpus* and *C. africanus* have been responsible for livestock losses. Cucurbitacins are transferred into milk of lactating women. Cucurbitacin B has an LD_{50} of 0.5 mg/kg (when injected, rabbit) or when given orally to mice, the LD_{50} is 5 mg/kg body weight. **SYMPTOMS** Symptoms after oral intake include vomiting, drastic, even bloody diarrhoea, kidney damage and spasms. Large doses result in respiratory arrest and death. In pregnant women, root extracts cause abortion (see cucurbitacin E). Exposure to skin may cause blister formation and inflammation. Cucurbitacins are amongst the most bitter of all substances known to man. **PHARMACOLOGICAL EFFECTS** Cucurbitacins inhibit mitotic cell division and are therefore cytotoxic (see cucurbitacins); they have antitumour activity but are too toxic to be used in medicine. Cucurbitacins activate intestinal peristaltic movement and cause spasms with colic. **FIRST AID** Administer sodium sulphate and medicinal charcoal; provide plenty of liquid (tea). **Clinical therapy:** Gastric lavage (possibly with 0.1 % potassium permanganate), instillation of medicinal charcoal and sodium sulphate; for further detoxification see cucurbitacin E.

Cucumis africanus L.f.

family: Cucurbitaceae

Cyclamen persicum

cyclamen

Cyclamen persicum (cultivar)

Cyclamen purpurascens (wild type)

PLANTS WITH SIMILAR PROPERTIES
A genus of 19 species in Europe, the Mediterranean, Asia (east to Iran) and E Africa; including *C. hederifolium* and *C. purpurascens*. *Cyclamen* had been a member of the Primulaceae; according to a recent revision it is now included in the Myrsinaceae.

PLANT CHARACTERS This well-known pot plant has simple, heart-shaped leaves arising from a rounded, fleshy tuberous rhizome below the ground. Leaves are dark green but beautifully decorated with pale green or whitish lines, especially along the margins. The attractive flowers are gracefully curved downwards and are usually white, pink or purple, depending on the cultivar.

OCCURRENCE *C. persicum* originates from the eastern Mediterranean region and has been developed into numerous cultivars kept as indoor plants. *C. hederifolium* occurs in S Europe and SW Asia; it can be kept in gardens and has been naturalised in Great Britain. *C. purpurascens* is a wild European species.

CLASSIFICATION Cell poison; moderately hazardous, II

ACTIVE INGREDIENTS Tubers and leaves are rich in a series of triterpene saponins such as cyclamin (having an aldehyde group as additional active moiety).

UTILISATION Saponins from *Cyclamen persicum* and other species have been used as fish poison for fishing. The rhizomes of *C. purpurascens* were formerly used as a drastic purgative.

TOXICITY Cyclamens are popular pot plants so that children or family pets may be at risk. However, the rhizomes are bitter and hidden below the ground. Despite the fact that the plant is quite toxic, very few actual cases of poisoning have been recorded in recent years. About 0.2 g of the tuber causes toxic symptoms, 8 g are lethal for humans.

SYMPTOMS The plant sap causes skin irritation. Symptoms after ingestion include nausea, vomiting, diarrhoea, severe stomach pain, haemolysis, extensive sweat formation, circulatory disturbance, convulsions and respiratory arrest.

PHARMACOLOGICAL EFFECTS Cyclamin is absorbed after oral ingestion and has a very high haemolytic index. Saponins interfere with membrane stability and permeability. They are therefore highly cytotoxic (see saponins).

FIRST AID Administer medicinal charcoal and sodium sulphate. Treat skin blisters with cortisone ointment. **Clinical therapy:** After overdoses: gastric lavage, instillation of medicinal charcoal and sodium sulphate, electrolyte substitution; control acidosis with sodium bicarbonate; in case of spasms and colic give diazepam and atropine; intubation and oxygen respiration may be necessary. Check kidney function.

Cyclamen persicum Mill.

family: Myrsinaceae or Primulaceae

cyclamen de Perse (French); *Alpenveilchen* (German); *ciclamio* (Italian)

Cynoglossum officinale

hound's tongue

Cynoglossum officinale

Borago officinalis

PLANTS WITH SIMILAR PROPERTIES A genus with 75 species in Europe, Africa and Asia, including the Chinese *C. amabile*. *Borago* is a related genus with 3 European species, of which *B. officinalis* (borage) is well known.

PLANT CHARACTERS A biennial plant of up to 0.8 m, covered with soft hairs on all aerial parts and bearing densely hairy, lanceolate leaves. The funnel-shaped flowers have 5 petals that are dark violet when young, turning purple when mature. Fruits are pale brown, with 4 nutlets covered with barbed prickles to facilitate animal dispersal.

OCCURRENCE On sandy and sunny places in Europe and Asia; naturalised in N America. Some species (e.g. *C. amabile*) are occasionally cultivated as ornamental plants.

CLASSIFICATION Liver poison, mutagen; moderately hazardous, II

ACTIVE INGREDIENTS All plant parts contain pyrrolizidine alkaloids (PAs); especially rich in PAs are roots, leaves and flowers (PA content of leaves is about 2.1 %). The main PA is heliosupine (an open diesteralkaloid), followed by echimidine and consolidine; cynoglossine occurs in roots. *Borago* produces lycopsamine, intermedine and their 7-acetyl derivatives.

UTILISATION *C. officinale* has been used in traditional phytomedicine and as a salad (young leaves). It should, however, not be ingested because of its high PA content. *Borago* seed oil, which is rich in gamma linolenic acid and used to treat neurodermitis, has no or very low and uncritical levels of PA. *Borago* leaves are used as fresh herbs; they have sometimes been confused with extremely hazardous *Digitalis* leaves (see *Digitalis*). Generally, aerial parts of Boraginaceae should not be ingested.

TOXICITY Large quantities of *Cynoglossum* material cause serious liver and lung damage. Fatal PA poisoning due to *Cynoglossum* has been reported in livestock (calves and horses).

SYMPTOMS Cynoglossine has a paralytic effect (similar to curare) on frog muscles. Consolidine paralyses the central nervous system. In cattle, large doses cause thirst and palsy of hind legs; see senecionine.

PHARMACOLOGICAL EFFECTS PAs become metabolically activated in the liver (see senecionine). They can then alkylate proteins and DNA. As a result they cause liver damage (venoocclusive disease). After prolonged intake they are mutagenic, teratogenic and carcinogenic.

FIRST AID Give medicinal charcoal to absorb the alkaloids; administer plenty of warm black tea and sodium sulphate. Provide shock prophylaxis (keep patient quiet and warm). **Clinical therapy:** In case of ingestion of large doses: gastric lavage and further treatments described under senecionine.

Cynoglossum officinale L.

family: Boraginaceae

langue de chien (French); *Gewöhnliche Hundszunge* (German); *lingua canina* (Italian)

Cytisus scoparius

common broom • Scotch broom

Cytisus scoparius flowers

Cytisus scoparius

PLANTS WITH SIMILAR PROPERTIES A genus of 33 species in Europe and N Africa. Closely related is *Chamaecytisus* (30 species in Europe).

PLANT CHARACTERS A branched shrub of up to 2 m in height, with ridged photosynthetically active stems, small trifoliolate leaves and bright yellow explosive flowers borne towards the branch ends. The oblong, flat, slightly sickle-shaped pods turn black when they ripen.

OCCURRENCE Central, S and E Europe on light, sandy or acidic soil. The plant is naturalised in N and S America, Australia, and other countries; several broom species are grown as ornamental plants.

CLASSIFICATION Neurotoxin; moderately hazardous, II

ACTIVE INGREDIENTS Tetracyclic quinolizidine alkaloids (QAs), of which sparteine is the main compound in the stems (typical for most members of *Cytisus* and *Chamaecytisus*). Other QAs include lupanine and 13-hydroxylupanine. Also present, especially in the flowers, are amines (tyramine, dopamine and epinine).

UTILISATION The herb has been used medicinally to regulate the circulation and as a diuretic. It has abortive activity and should be avoided during pregnancy. Broom is grown as a sand-binder (e.g. along motorways), as a bee forage and formerly for fibre and dyes. Broom is a source of sparteine that has been used medicinally for uterus contraction and as anti-arrhythmic.

TOXICITY Poisoning with the plant is rare, but more common with extracts or isolated sparteine. Livestock poisoning has been reported. Sparteine LD_{50} mouse: 36 mg/kg i.p., 220 mg/kg p.o.

SYMPTOMS Circulatory collapse, arrhythmia, tachycardia, mydriasis, perspiration, vomiting, diarrhoea, dizziness and even euphoria, convulsions and headache. Ascending paralysis and death from respiratory arrest (see cytisine).

PHARMACOLOGICAL EFFECTS Sparteine is an agonist at mAChR and blocks Na^+ channel (as quinidine); it decreases the overstimulation of the nerve impulses in the heart. Furthermore, it causes bradycardia, a decrease of intracardial pressure, and arterial blood pressure. Tyramine has vasoconstrictive and hypertensive effects. The herb is known to be oxytocic and increases the tone and strength of contraction of the uterus.

FIRST AID Administer medicinal charcoal and sodium sulphate; provide plenty liquids to drink.

Clinical therapy: After overdoses: Gastric lavage (possibly with potassium permanganate), instillation of medicinal charcoal and sodium sulphate; electrolyte substitution; control acidosis with sodium bicarbonate; in case of spasms and colic give diazepam and atropine; intubation and oxygen respiration if necessary (see cytisine).

Cytisus scoparius (L.) Link (= *Sarothamnus scoparius*)   family: Fabaceae

genêt à balai (French); *Besenginster* (German); *ginestra scopareccia* (Italian)

Daphne mezereum

mezereon

Daphne mezereum fruits

Daphne mezereum flowers

PLANTS WITH SIMILAR PROPERTIES A genus of 50 Eurasian species, including *D. odora, D. cneorum, D. gnidium, D. laureola, D. striata* and *D. tangutica*. Several are grown for their fragrant flowers but nearly all are hazardous.

PLANT CHARACTERS Deciduous shrub with a height of up to 1.5 m bearing ovate, pale green leaves. The tubular rosy-pink and strongly scented flowers emerge before the leaves. Fruits (up to 7 mm in diameter) are green when immature and scarlet red upon ripening.

OCCURRENCE Europe and W Asia, often in deciduous forests; cultivated as an ornamental.

CLASSIFICATION Cell poison, co-carcinogen, skin irritant; extremely hazardous, Ia

ACTIVE INGREDIENTS All plant parts, but especially the seeds of red berries and the bark, contain phorbol esters of the daphnane type, such as mezerein and daphnetoxin (up to 0.1 %). Furthermore, several coumarins are present, such as daphnin and umbelliferone.

UTILISATION Fruits have been used as a spice, cosmetic and medicinally as purgative, emetic, and against cancer and pain. *D. cneorum* has been used as a fish poison for fishing. Beggars are known to have inflicted skin wounds with mezereon extracts to induce compassion.

TOXICITY All plant parts except the fruit pulp are toxic. The bright red fruits are very attractive and several instances of human poisoning have occurred when fruits were eaten (10–12 fruits can be lethal for adults, about 2–3 for children). Livestock poisoning have been recorded (3–5 berries kill a pig; 30 g bark a horse or 10 g a dog). Daphnetoxin: LD_{50} mouse: 0.27 mg/kg p.o.

SYMPTOMS Ingestion may lead to a strong burning and scratching in mouth and throat, enhanced salivation, difficulties to swallow, sneezing, small pupils, nausea, vertigo, stomach pain, bloody vomiting, gastrointestinal disturbance, bloody and watery diarrhoea, fever, spasms, paralysis, tachycardia, serious kidney damage, and circulatory collapse with death. External exposure to fresh plant material leads to skin inflammation with reddening and necrotic blister formation.

PHARMACOLOGICAL EFFECTS Mezerein and daphnetoxin are phorbol esters that activate protein kinase C. They are strong cellular poisons, abortifacients and co-carcinogens; see phorbol esters for physiological and toxicological consequences.

FIRST AID Immediately instil 200 ml paraffinum liquidum, large volumes of medicinal charcoal and sodium sulphate. Provide plenty of tea.

Clinical therapy: Gastric lavage (possibly with 0.1 % potassium permanganate), instillation of medicinal charcoal and sodium sulphate; for further treatment see phorbol esters.

Daphne mezereum L.   family: Thymelaeaceae

mézeréon, bois jentil, bois joli (French); *Gemeiner Seidelbast* (German); *mezereo* (Italian)

Datura stramonium

thorn-apple

Datura stramonium flower

Datura stramonium flowering plant

Datura stramonium fruit and seeds

PLANTS WITH SIMILAR PROPERTIES A genus of 9 annuals (all of similar toxicity) in S and C America, e.g. *D. ceratocaula, D. ferox, D. metel* and *D. innoxia*. The related *Brugmansia* includes perennials with hanging flowers (see *Brugmansia*).

PLANT CHARACTERS A robust annual (up to 1.5 m) with large, unpleasantly scented leaves, large, white or purplish, tubular, erect flowers and typical thorny capsules that split open to release numerous small, kidney-shaped, black seeds.

OCCURRENCE Tropical N America; now a cosmopolitan weed and cultivated ornamental.

CLASSIFICATION Neurotoxin, hallucinogen; extremely hazardous, Ia

ACTIVE INGREDIENTS Dried leaves contain 0.3–1 % tropane alkaloids, seeds 0.6 % and roots 0.2 % (mainly hyoscyamine and scopolamine).

UTILISATION *Datura* is medicinally used as an analgesic, as an ingredient of cough syrup and in cigarettes to treat asthma (see hyoscyamine). Both *Datura* and *Brugmansia* were used by the Aztecs and modern Indians as intoxicant and hallucinogen in the same way as tropane plants in the Old World. Plants or extract were smoked, drunk (in fortified beer, chicha or pulque) or placed on the skin. Since the 16th century *Datura* has been used as an intoxicant in Europe. Ointments administered on the skin resulted in the "ability to fly". Extracts were also misused for infanticide,

suicide and murder. The use of these plants for crime, seduction and as narcotic drugs is revealed by some of the common names such as "herbe aux sorciers" or "herbe au diable".

TOXICITY The seeds or less often the leaves or roots are a common cause of accidental poisoning or of deliberate intoxication. Seeds may be accidentally eaten by small children or more often experimentally as a dangerous hallucinogen, sometimes with fatal results. Poisoning may also result from contaminated grain and flour, or from leaves accidentally collected as wild spinach. Cases of livestock poisoning have been reported.

SYMPTOMS At low concentrations, the alkaloids have a depressant and sedative effect, but high doses lead to powerful hallucinations, excitation, reddening of the face, dry mouth, euphoria, mydriasis, confusion, insomnia, respiratory arrest and death. Scopolamine is favoured as a hallucinogen (see hyoscyamine).

PHARMACOLOGICAL EFFECTS Hyoscyamine and scopolamine are antagonists at the muscarinic acetylcholine receptor and therefore function as a parasympatholytic (see hyoscyamine).

FIRST AID Induce vomiting; treat with sodium sulphate and medicinal charcoal. **Clinical therapy:** Immediate gastric lavage; inject 2 mg physostigmine as an antidote. See hyoscyamine for further measures.

Datura stramonium L.   family: Solanaceae

stramoine (French); *Stechapfel* (German); *stramonio* (Italian); *estramonio* (Spanish)

Delphinium elatum

giant larkspur • delphinium

Delphinium elatum plants

Consolida regalis flowers

Delphinium staphisagria seeds

PLANTS WITH SIMILAR PROPERTIES *Delphinium* is a large genus of more than 300 species, including *D. grandiflorum, D. staphisagria* and *D. nuttallianum. Consolida* is a related genus with 43 species in Europe and Asia, such as *C. ajacis* (the true larkspur), *C. regalis* and *C. orientalis. Aconitum* (e.g. *A. napellus*) is also closely related.

PLANT CHARACTERS A tall, usually single-stemmed annual (up to 2.5 m) with toothed leaves and purple, blue or white flowers in spectacular erect racemes. *D. grandiflorum* is smaller, with few-flowered spikes of relatively large flowers.

OCCURRENCE *Delphinium* and related plants are widely distributed in the northern hemisphere, in Europe, Asia, N America, and a few in Africa. These plants are commonly cultivated for the intense blue colours of the flowers (purple, dark blue or pale blue, but also pinkish or white in some hybrids and cultivars).

CLASSIFICATION Neurotoxin; extremely hazardous, Ia

ACTIVE INGREDIENTS *Delphinium* and *Consolida* contain complex mixtures of diterpenoid alkaloids, such as delphinine, delcosine, staphisine, lycoctonine, or nudicauline; especially in seeds (up to 1.3 %), leaves (1 %), and roots (2.4 %).

UTILISATION Larkspur must have been known in ancient Greece since Dioscorides mentions that "Delphinion" is also named "Paralysis". The same author describes the application of extracts from *Delphinium* against ectoparasites (such as lice) and taken with wine as an antidote against scorpion stings. Extracts of *D. staphisagria* have been used against itch mites and other ectoparasites, as well as to eradicate rats and ants.

TOXICITY Young children or domestic animals may be accidentally poisoned. In Europe and North America, poisoning of livestock by *Delphinium* species can be a serious problem. The toxic dose of nudicauline in cattle is about 6 mg/kg. If the alkaloids come into contact with the skin, severe inflammation can result.

SYMPTOMS *Delphinium* alkaloids are similar to albeit weaker than aconitine but can still lead to death by respiratory and cardiac arrest. Toxic symptoms include nausea, excitation, cardiac arrhythmia, spasms, inflammation of the GI tract, paralysis, muscular weakness and shortness of breath. Externally, the alkaloids cause strong inflammation.

PHARMACOLOGICAL EFFECTS The diterpenoid alkaloids are Na^+-channel and nAChR agonists that cause paralysis of heart muscles (with bradycardia, hypotonia) and of neuromuscular plates (see aconitine).

FIRST AID Induce vomiting; treat with sodium sulphate and medicinal charcoal; for clinical therapy with gastric lavage, see aconitine.

Delphinium elatum L.
family: Ranunculaceae

Dauphinette élevée (French); *Hoher Rittersporn* (German); *speronella elevata* (Italian)

Dendrocalamus asper

bamboo

Dendrocalamus asper

Dendrocalamus giganteus stems

PLANTS WITH SIMILAR PROPERTIES Bamboo is a common name for 45 genera of grasses with woody culms, used for many purposes, such as building, pipes, walking-sticks, furniture, mats, blinds, baskets, fans, umbrellas and brushes. Young shoots of several species are edible, including *Dendrocalamus asper, D. giganteus, D. latifolius, Phyllostachys edulis, P. nigra, P. bambusoides, P. dulcis, Bambusa bambos, B. beecheyana, B. vulgaris* and *Arundinaria gigantea*.

PLANT CHARACTERS Bamboo is a large woody grass with pronounced rhizome formation. Stems are hollow between the nodes. Bamboos flower in synchrony, i.e. all plants of an area flower at the same time and die afterwards.

OCCURRENCE. *D. aspera* originates from SE Asia and is grown for its shoots. *D. giganteus* is cultivated in Burma, *Bambusa* and *Phyllostachys* in China and Taiwan.

CLASSIFICATION Cellular poison; moderately hazardous, II

ACTIVE INGREDIENTS Cyanogenic glucosides (CGs) that release toxic HCN (hydrogen cyanide, prussic acid). The main compound is often taxiphyllin; bitter varieties with up to 0.8 % CGs. Young shoots are particularly rich in CGs.

UTILISATION Bamboo shoots are part of a staple diet in Asia. Fresh shoots are sliced in half lengthwise, the outer leaves peeled away and any

fibrous tissue at the base trimmed. It is then thinly sliced into strips and boiled in lightly salted water for 8–10 min. Complete shoots that are poisonous when raw will be safe to eat if boiled or steamed for 20–40 min. The cyanide (HCN) that is liberated in such a process is removed by decanting the cooking water. If products taste bitter, it is a sign that there is still HCN present.

TOXICITY When bamboo shoots are not processed correctly, severe intoxications can occur. HCN toxicity: LD_{50} mouse (p.o.): 3.0 mg/kg; rat (p.o.): 10–15 mg/kg. In humans, 1 mg/kg HCN is lethal (see amygdalin).

SYMPTOMS Poisoning symptoms are typical for cyanogenic glucosides in general (irritation, flushing face, heavy breathing, scratchy throat, headache, and in severe cases, respiratory and cardiac arrest).

PHARMACOLOGICAL EFFECTS Taxiphyllin releases HCN after enzymatic or acid hydrolysis. HCN is rapidly inactivated in the human body at low concentrations. Higher doses stop cellular respiration and lead to general paralysis and respiratory arrest (see amygdalin).

FIRST AID After ingestion of bitter bamboo products, induce vomiting and allow the intake of large volumes of drinks. **Clinical therapy:** In case of severe intoxication, see detoxification, see also amygdalin.

Dendrocalamus asper (Schultes f.) Heyne

bambou (French); Bambus (German); bambú (Italian); bambú (Spanish)

family: Poaceae

Derris elliptica

tuba root • derris root

Derris elliptica compound leaf

PLANTS WITH SIMILAR PROPERTIES A genus of about 40 species from SE Asia to N Australia and E Africa, including *D. trifoliata, D. malaccensis* and *D. uliginosa*. Many of them produce rotenone, which also occurs in plants of the genera *Lonchocarpus, Mundulea* and *Tephrosia*.

PLANT CHARACTERS A woody climber (vine) with tough roots up to 80 mm in diameter, bearing large pinnate leaves, each with 7–15 leaflets. Most parts of the plant are covered in reddish hairs. The white flowers are flushed with pink and are borne in racemes. The fruits are oblong pods with narrow wings along both sides.

OCCURRENCE *Derris elliptica* occurs in tropical Asia. It is cultivated in tropical regions for rotenone production. Some *Derris* species are grown as ornamental plants.

CLASSIFICATION Cell poison; highly hazardous, Ib

ACTIVE INGREDIENTS The typical toxic ingredients of *Derris* are isoflavones – mainly rotenone (up to 16 % in roots). Also present are elliptone, malaccol, deguelin, and tephrosin.

UTILISATION Derris root powder and isolated rotenone are efficacious as natural insecticides, fish and arrow poisons. It has been reported that derris root has been chewed in New Ireland to commit suicide. On Malacca, small pieces of root are used as abortifacients. Roots are aromatic and have a metallic, slightly sour taste. Rotenone is being used as a biopesticide.

TOXICITY Derris is a strong paralytic toxin. The toxicity is enhanced when derris powder is suspended in mineral and fatty oils. It is more toxic to invertebrates than vertebrates. Rotenone is quite toxic when injected (LD_{50} of 2.8 mg per kg in mice); less toxic when ingested.

SYMPTOMS First symptoms include numbness of the tongue that spreads along the digestive tract. A main effect is a paralysis of all muscles, both skeletal and smooth muscles. Higher doses paralyse the brain and the spinal cord. Further symptoms are convulsions and severe vomiting.

PHARMACOLOGICAL EFFECTS Rotenone is an inhibitor of the mitochondrial respiratory chain. This leads rapidly to ATP deficiency and thus blocks most cellular processes, including neuronal and muscular activity.

FIRST AID Administer medicinal charcoal and sodium sulphate; give shock prophylaxis (calmness, keep warm, offer warm drinks). **Clinical therapy:** After overdoses: Gastric lavage (possibly with potassium permanganate), instillation of medicinal charcoal and sodium sulphate, electrolyte substitution; control acidosis with sodium bicarbonate; in case of spasms give diazepam; intubation, oxygen respiration and cardiac massage if necessary.

Derris elliptica (Wall.) Benth.

family: Fabaceae

Derris (French); *Tubawurzel* (German)

Dichapetalum cymosum

poison leaf

Dichapetalum cymosum

PLANTS WITH SIMILAR PROPERTIES A genus of 124 species in tropics of SE Asia, Africa, Madagascar and America. *D. toxicarum* is a toxic liana from W Africa.

PLANT CHARACTERS *D. cymosum* is an enormous woody plant that grows underground except for the numerous branch tips that emerge above the ground. The leaves are oblong, bright green above and below, with the secondary veins forming loops near the leaf margin. Young leaves are hairy but become more or less smooth with age. Clusters of white flowers are produced in early spring. On rare occasions, large, orange fruits are formed.

OCCURRENCE *D. cymosum* has a limited distribution and occurs only in southern Africa.

CLASSIFICATION Cell poison; extremely hazardous, Ia

ACTIVE INGREDIENTS *D. cymosum* contains monofluoroacetic acid. Fluoroacetic acid is relatively harmless but is converted in the body of the animal to highly toxic fluorocitrate. *D. toxicarum* has fluoro-fatty acids in its seed oil.

UTILISATION Extracts have been used in Africa to kill wild boars, monkeys and rats.

TOXICITY Poison leaf is a common cause of serious livestock losses in S Africa, Botswana, Namibia and Zimbabwe. Because it is such a deep-rooted plant, it sprouts at a time when no other green pasture is available. Young, newly emerged shoots are particularly toxic. The lethal oral dose of monofluoroacetate is less than 0.5 mg/kg in humans and in most animals. As little as 20 g of fresh leaves may kill a sheep. Ruminants typically die of heart failure within 24 hours of ingesting the plants, without showing any symptoms.

SYMPTOMS Since cellular respiration is blocked a paralysis of brain, the spinal cord and all muscles, skeletal, cardiac and smooth muscles, is likely consequence.

PHARMACOLOGICAL EFFECTS Fluoroacetic acid interferes with the metabolism of acetic acid in the Krebs cycle and causes a fatal disruption of cellular respiration. It mimics the action of acetic acid, so that fluorocitric acid is formed instead of citric acid. Fluorocitric acid inhibits the activity of aconitase enzymes and a mitochondrial acetate carrier.

FIRST AID Administer medicinal charcoal and sodium sulphate; shock prophylaxis (calmness, keep warm, provide warm drinks). **Clinical therapy:** Gastric lavage (possibly with potassium permanganate), instillation of medicinal charcoal and sodium sulphate, electrolyte substitution, control acidosis with sodium bicarbonate; in case of spasms give diazepam; intubation, oxygen respiration, and cardiac massage if necessary.

Dichapetalum cymosum (Hook.) Engl.

Giftblatt (German)

family: Dichapetalaceae

Dictamnus albus

dittany • burning bush

Dictamnus albus   *Dictamnus albus*

PLANTS WITH SIMILAR PROPERTIES A monotypic genus (1 species but with several morphologically defined varieties).

PLANT CHARACTERS A perennial herb of up to 1.2 m in height. The lower leaves are simple, the upper ones pinnate. Stems, leaves and flowers all bear glandular hairs that are filled with highly scented volatile oil. The oil is inflammable (hence 'burning bush"). Flowers occur in terminal racemes; they are irregular (zygomorphic) and 25 mm long, each with 5 pink, dark-veined petals and 10 upward-pointing stamens.

OCCURRENCE Sunny and dry habitats (often on lime) in Europe and Asia; widely cultivated as an ornamental garden plant.

CLASSIFICATION Cell toxin; moderately hazardous, II

ACTIVE INGREDIENTS Dittany is rich in furanoquinoline alkaloids (dictamnine, γ-fagarine, skimmianine), furanocoumarins (bergapten, xanthotoxin, psoralen). The essential oil is rich in monoterpenes with limonene as main compound (53 %), followed by α- and β-pinene, camphene, sabinen, myrcene, α- and β-phellandrene, α- and γ-terpinene, *cis*- and *trans*-ocimene, *p*-cymene, and several sesquiterpenes (β-bourbonene, β-caryophyllene, α-humulene and germacrene D). Leaves also contain limonoids such as fraxinellon, limonine, obacunone and rutaevin.

UTILISATION Extracts of *Dictamnus* have been used for wound-healing, especially wounds caused by spears or arrows, infections, epilepsy and convulsions. Dioscorides recognised the plant under the name "Beluakos" (remedy against arrow wounds) and "Artemidion" (after Artemis, the goddess of hunters). In addition, extracts from *Dictamnus* were used as abortifacients. In rats, the alkaloids have contraceptive effects.

TOXICITY LD_{50}: Fraxinellon: rats (i.p.) 120 mg/kg and 355 mg/kg for mice. Oral toxicity: mice LD_{50} 430 mg/kg and 275 mg/kg for rats.

SYMPTOMS Both the alkaloids and furocoumarins are strong skin irritants; when exposed to sunlight, they cause photodermatitis and contact dermatitis.

PHARMACOLOGICAL EFFECTS Furanocoumarins and alkaloids can intercalate DNA. Whereas furanocoumarins form diadducts with DNA, the furanoquinolines form monoadducts when activated by sunlight. They are thus phototoxic but also mutagenic and perhaps even carcinogenic after chronic exposure to the skin. Furanoquinoline alkaloids have cytotoxic properties and cause apoptosis.

FIRST AID When skin or eyes have been in contact with the oil, wash with water or polyethylene. Treat skin with cortisone ointments and eyes with antiallergic eyedrops (e.g. cromoglycate).

Dictamnus albus L. (= *D. fraxinella* Pers.)   family: Rutaceae

dictame blanc, fraxinelle (French); *Diptam* (German); *dittamobianco* (Italian)

Dieffenbachia seguine

spotted dumb cane

Dieffenbachia seguine

Dieffenbachia seguine

PLANTS WITH SIMILAR PROPERTIES
A genus of 20 species from tropical America, often grown as ornamental pot plants.

PLANT CHARACTERS Stout perennial herb (up to 2 m) with large, oblong-ovate, heavily spotted leaves. Raphides (needle-shaped crystals) are located in special cells which can shoot oxalate needles into cells when pressure is excerted.

OCCURRENCE Tropical America; many cultivars and hybrids are derived from *D. maculata* and *D. seguine* (they are perhaps the same species).

CLASSIFICATION Cytotoxin; highly hazardous, Ib

ACTIVE INGREDIENTS All parts, especially stems, are poisonous. The toxic principles are not yet fully understood: besides calcium oxalate raphides (up to 0.54 %), also saponins, cyanogenic glucosides, alkaloids and proteases.

UTILISATION Portions of dumb cane stem have been used for torture – when chewed, people (and pets) become silent due to a swollen tongue. The raphides may also cause temporary sterility. Traditional uses: contraceptive, aphrodisiac; treatment of cancer, oedema and skin disorders.

TOXICITY *Dieffenbachia* is a common cause for consultations in poison centres. Children are especially affected. Lethal dose in humans: 5–15 g of oxalic acid (ca. 3–4 g of leaves). LD_{50} in guinea

pigs (p.o.): about 0.6–0.9 g stem sap/animal.

SYMPTOMS First symptoms: burning and severe inflammation of the mouth, throat and oesophagus. The swollen throat and tongue can lead to suffocation and loss of speech. Within 1 hour salivation, difficulty to swallow, nausea, vomiting, severe GI disorder, arrhythmia, tachycardia, dyspnoea, coma and death by cardiac arrest or fatal kidney damage. When sap gets into the eye intensive pain, conjunctivitis and keratitis.

PHARMACOLOGICAL EFFECTS Raphides of oxalate are potent irritants of skin and mucosa (mouth, GI tract and eyes); they can penetrate cells mediate the entry of other toxins and cause further harm (acidity, reduction). These tiny needles can activate mast cells that release histamine, which induces local swelling and pain. If calcium oxalate is deposited in kidney tubules, kidney tissue becomes damaged (see oxalic acid).

FIRST AID Give medicinal charcoal and sodium sulphate. Provide plenty of tea. Contact with eye or skin: wash with polyethylene glycol or water in case of skin blisters, cover with sterile cotton and locally with corticoid ointment (see oxalic acid). **Clinical therapy:** Wash eyes with polyethylene glycol. In severe cases: gastric lavage, possibly with calcium salts, such as calcium gluconate or with 0.1 % potassium permanganate; for further detoxification see oxalic acid.

Dieffenbachia seguine (Jacq.) Schott [= *D. maculata* (Lodd.) Bunting]   family: Araceae
pédiveau vénéneux (French); *Dieffenbachie* (German)

Digitalis purpurea

purple foxglove

Digitalis purpurea

Digitalis grandiflora flowers

PLANTS WITH SIMILAR PROPERTIES A genus of more than 19 toxic species, e.g. *D. lanata* (Austrian foxglove), *D. lutea* (straw foxglove) and *D. grandiflora* (yellow foxglove). *Isoplexis* (3 species, Atlantic Islands) also produces cardenolides.
PLANT CHARACTERS Purple foxglove is a biennial or perennial that forms a dense rosette of leaves in the first year and elegant, 2 m high flowering stalks of purple, spotted flowers in the second year. Various cultivars have been developed.
OCCURRENCE Foxglove is indigenous to the Mediterranean region and Europe, where it often grows on clear-cuts. It is cultivated for medicinal purposes and as garden ornamentals.
CLASSIFICATION Heart poison; extremely hazardous, Ia
ACTIVE INGREDIENTS Purpurea glycosides, digitoxin and several other cardenolides (up to 0.4 %); *D. lanata* contains digoxin, lanatosides, gitoxin and related constituents (about 1.5 %). Leaves are also rich in saponins.
UTILISATION *Digitalis* plants are well-known sources of heart glycosides, used since 1785 to treat the symptoms of heart insufficiency, hypertonia and rhythm abnormalities.
TOXICITY *Digitalis* is highly toxic but cases of human and animal poisoning with wild plants are quite rare. The leaves are similar to those of comfrey (*Symphytum officinale*) or borage (*Borago*

officinalis) and accidental poisoning through herbal teas have been described. However, poisoning with isolated cardiac glycosides, which are used in heart medication, are not uncommon. This may be due to wrong dosages of heart medication (narrow therapeutic window; a 60 % overdose has toxic effects), suicide and even murder. About 0.3 g of dried leaves are toxic, 2.5–5 g (=2 or 3 leaves) are lethal for adults. 25 g dried leaves kill a horse, 150 g cattle, 4–5 g a pig, and 5 g a dog.
SYMPTOMS Concentrated decoctions may cause rapid death due to heart failure. Symptoms include nausea, salivation, gastrointestinal disturbance, purging and exhaustion, with the usual respiratory and cardiac symptoms (visual disturbance, arrhythmia, hypertension, coma, cardiac arrest) associated with heart glycoside poisoning.
PHARMACOLOGICAL EFFECTS Cardiac glycosides inhibit the Na^+, K^+-ATPase, which is essential for neuronal activity and transport processes in all cells (see cardiac glycosides). In heart cells, the inhibition leads to enhanced calcium levels and thus contraction of the heart muscle.
FIRST AID First aid treatment is indicated if plant material of *Digitalis* or isolated cardiac glycosides have been ingested. Treatment involves the induction of vomiting, supply of medicinal charcoal, and detoxification as outlined under cardiac glycosides.

Digitalis purpurea L.

family: Plantaginaceae (formerly Scrophulariaceae)

digitale pourpre (French); *Roter Fingerhut* (German); *digitale rossa* (Italian)

Dimorphotheca cuneata

karoo bietou

Dimorphotheca cuneata

Dimorphotheca cuneata

PLANTS WITH SIMILAR PROPERTIES A genus of 19 species in S & tropical Africa. Poisonous species include *D. nudicaulis*, *D. zeyheri*, *D. spectabilis* and *D. sinuata;* also *Arctotheca* and *Osteospermum* species.

PLANT CHARACTERS *D. cuneata* is a woody shrublet of 0.5 m in height. The leaves are oblong and typically toothed along the margins, and are sticky and aromatic when young. Attractive flower heads are borne on the tips of the branches, usually in early spring. They have yellow disc florets in the middle, surrounded by numerous pure white (or rarely purple) ray florets that fold open in the daytime. Two types of one-seeded fruits are formed (hence the name *Dimorphotheca*): a larger, rounded, flat, winged fruit and a much smaller, narrow, wingless one with a warty surface.

OCCURRENCE *D. cuneata* is one of the most widely distributed species of the genus and occurs over most of the dry interior of S Africa.

CLASSIFICATION Cellular poison; moderately hazardous, II

ACTIVE INGREDIENTS The presence of the toxic cyanogenic glycoside, linamarin, from several *Dimorphotheca* species has been reported.

UTILISATION Several species are cultivated as ornamental plants.

TOXICITY *Dimorphotheca* species and other plants from the family loosely referred to as *bietou* are well known for causing prussic acid poisoning in stock. There may be severe losses in sheep in some years. Human poisoning is rare.

SYMPTOMS Poisoning symptoms are typical for cyanogenic glycosides in general (irritation, flushing face, heavy breathing, scratchy throat, headache, and in severe cases, respiratory and cardiac arrest).

PHARMACOLOGICAL EFFECTS Damage to plant cells through wilting or digestion causes the cyanogenic glycosides to be mixed with hydrolysing enzymes, a beta glucosidase (the glucosides are compartmentalised separately from the enzymes in intact cells), causing a reaction that releases hydrogen cyanide (HCN), a lethal inhibitor of the mitochondrial respiratory chain; see amygdalin for details.

FIRST AID Quick action is required. Induce vomiting and perform gastric lavage (possibly with 300 ml of 0.2 % potassium permanganate).

Clinical therapy: In lite cases (patient still conscious): give sodium thiosulphate (10 %; 100 ml i.v.), which helps the body to detoxify HCN to rhodanide. In severe cases (patient unconscious): immediately inject 3.25 mg/kg 4-dimethylaminophenol (DMAP) i.v. (otherwise i.m.); then inject sodium thiosulphate (10 %; 100 ml i.v.). Control of acidosis with sodium bicarbonate. In severe cases, provide intubation and oxygen respiration.

Dimorphotheca cuneata (Thunb.) Less.   family: Asteraceae

souci de Cap (French); *Kapkörbchen, Kapringelblume* (German)

Dioscorea dregeana

African wild yam

Dioscorea dregeana

Dioscorea dregeana rootstock

PLANTS WITH SIMILAR PROPERTIES A very large cosmopolitan genus of about 850 species in tropical and warm climates; among them *D. batatas, D. pentaphylla, D. alata, D. abyssinica, D. elephantipes, D. esculenta (*potato-yam*), D. trifida, D. floribunda* and *D. daemona.*

PLANT CHARACTERS A perennial climber (vine) with slender, slightly thorny stems growing annually from a fleshy, tuberous rootstock. The large, somewhat hairy leaves have three pointed leaflets. Male and female flowers are borne on separate plants (in slender, drooping clusters), followed by oblong fruit capsules.

OCCURRENCE *D. dregeana* is limited to the moist, eastern parts of southern Africa. *D. elephantipes* is grown as a pot plant.

CLASSIFICATION Cell poison; mind-altering; highly hazardous, Ib

ACTIVE INGREDIENTS The toxic principle of several *Dioscorea* species is the piperidine alkaloid dioscorine. Also present are steroidal saponins, such as glycosides of diosgenin (perhaps partly responsible for the medicinal uses but at the same time implicated in sheep losses). Other compounds include dioscin, deltonin, deltoside, hircinol and demethylbatatstin.

UTILISATION *Dioscorea* species are widely used in traditional African medicine to treat hysteria, convulsions and epilepsy and to pacify psychotic patients. Some are used as a starch food (yam), usually after leaching in water for several days to remove the water-soluble poison. Derivatives of plant-derived steroids are nowadays used as oral contraceptives, anti-inflammatory agents and as androgens, oestrogens and progestins.

TOXICITY Human deaths have been reported after the use of the plant as famine food or as medicine. It has been used in poison bait to destroy monkeys. Dioscorine is highly toxic to humans. It has an LD_{50} of 60 mg/kg in mice when administered by injection.

SYMPTOMS *D. dregeana* is reported to make a person "mad drunk". It is hallucinogenic.

PHARMACOLOGICAL EFFECTS Steroidal saponins interfere with membrane stability and permeability. They are therefore highly cytotoxic and haemolytic. Piperidine alkaloids modulate acetylcholine receptors and can thus affect the sympathetic and parasympathetic nerve system.

FIRST AID Administer medicinal charcoal and sodium sulphate. **Clinical therapy:** After overdoses: gastric lavage (possibly with potassium permanganate), instillation of medicinal charcoal and sodium sulphate, electrolyte substitution; control acidosis with sodium bicarbonate; in case of spasms and colic give diazepam and atropine; intubation and oxygen respiration may be necessary. Check kidney function.

Dioscorea dregeana (Kunth) T.Durand & Schinz

family: Dioscoreaceae

igname sauvage (French); *Yamswurzel* (German); *dioscorea* (Italian)

Duboisia myoporoides

corkwood duboisia

Duboisia myoporoides

Duboisia hopwoodii

PLANTS WITH SIMILAR PROPERTIES A genus of 4 species, including *D. hopwoodii* ("pituri"). *Atropa belladonna* and several other members of the Solanaceae also contain tropane alkaloids.

PLANT CHARACTERS Shrub or small tree (up to 12 m in height); the leaves are simple, alternate, narrowly elliptic and up to 120 x 30 mm. Small bell-shaped white flowers are borne in terminal cymose panicles. The fruits are green and pea-like when unripe; black and fleshy when mature.

OCCURRENCE The genus is restricted to Australia and New Caledonia.

CLASSIFICATION Neurotoxin, hallucinogen; highly hazardous, Ib

ACTIVE INGREDIENTS Leaves contain up to 2 % tropane alkaloids, mainly scopolamine, with lesser quantities of hyoscyamine, tigloidine and other esters; pyridine and piperidine alkaloids are also present, such as nicotine, nornicotine and anabasine.

UTILISATION Dried leaves of pituri have been smoked or chewed by Australian aborigines. It is documented in old stone paintings. *Duboisia* is a commercial source of scopolamine, which is used as premedication in surgery and as anti-emetic. The drug might be useful to treat Parkinson's disease. *Duboisia* was formerly employed as an emu poison and as a natural insecticide (due to the presence of nicotine and anabasine). The timber (known as

corkwood) was used for carving. *D. myoporoides* is a commercial source of alkaloids.

TOXICITY Scopolamine: LD_{50} mouse: 1.7 to 5.9 g/kg s.c.; 163 mg/kg i.v.; lethal dose for humans: 100 mg; (see hyoscyamine).

SYMPTOMS At low concentrations, the alkaloids have a depressant and sedative effect, but higher doses lead to hallucinations, euphoria, confusion, insomnia, vertigo, trembling, difficulties to speak, excitation, nausea, vomiting, palpitation and mydriasis and death. Scopolamine is known for its intoxicating and hallucinogenic activity; nicotine for its stimulating properties (see hyoscyamine).

PHARMACOLOGICAL EFFECTS Scopolamine is an antagonist at the muscarinic acetylcholine receptor and therefore works as a parasympatholytic (see hyoscyamine). Scopolamine shows stronger CNS sedating properties even at low concentrations than hyoscyamine. Nicotine and anabasine are agonists at nAChR and thus activate the parasympathetic nervous system.

FIRST AID. Induce vomiting; administer sodium sulphate and medicinal charcoal. **Clinical therapy:** Parasympathomimetics (i.e. inhibitors of acetylcholine esterase) such as pilocarpine and physostigmine (adults 1–2 mg i.v. or i.m.; children 0.5 mg) can be given as antidotes. For further clinical treatment see hyoscyamine.

Duboisia myoporoides R. Br.

family: Solanaceae

duboisie (French); *Duboisia, Korkholz* (German)

Ecballium elaterium

squirting cucumber

Ecballium elaterium

Ecballium elaterium

PLANTS WITH SIMILAR PROPERTIES A genus with only 1 species; plants with similar properties are found amongst several other genera of the Cucurbitaceae, including *Bryonia, Citrullus, Cucumis, Fevillea* and *Solena*. Species of *Gratiola* (Plantaginaceae) and *Iberis* (Brassicaceae) have a few SMs in common.

PLANT CHARACTERS The squirting cucumber is a perennial herb with trailing, hairy stems and fleshy roots. The somewhat succulent leaves are 5-lobed, 50–150 mm in diameter, pubescent above and densely tomentose below. Flowers are yellow and about 20 mm in diameter. Male flowers are borne in racemes; female flowers are solitary. The characteristic fruits are oblong to ellipsoid (40–60 x 15–25 mm), with sparse, soft bristles on the surface. Fruit fall away from the plant when they ripen and the pericarp contracts. The reduced turgidity causes an explosive release of the seeds and watery fluid through a basal hole.

OCCURRENCE Mediterranean, W Asia; introduced to C America.

CLASSIFICATION Cell toxin; highly hazardous, Ib

ACTIVE INGREDIENTS *Ecballium* contains mixtures of bitter triterpenoids (2.2 % FW), so-called cucurbitacins, which occur in free and glycosidic form. Examples are cucurbitacin E and I.

UTILISATION The plant is extremely poisonous and was formerly employed as drastic purgative (*elaterium album*) and anti-inflammatory drug.

TOXICITY Cucurbitacins are strong cell poisons at high doses. A mere 0.6 ml of the fruit juice can be lethal.

SYMPTOMS Symptoms after oral intake include salivation, vomiting, headache, tachycardia, drastic, even bloody diarrhoea, kidney damage, convulsions and death by respiratory arrest; in pregnant women, root extracts cause abortion. Exposure to skin and mucous membranes may cause blister formation and inflammation. A patient who had aspirated a diluted extract into the nose to treat sinusitis suffered dyspnoea and serious kidney damage, and died after 6 days from cardiac failure. See cucurbitacin E for details.

PHARMACOLOGICAL EFFECTS Free cucurbitacins inhibit mitotic cell division and are therefore cytotoxic (see cucurbitacins); they have antitumour activity but are too toxic to be used in medicine. Cucurbitacins activate intestinal peristaltic movements and cause spasms with colic.

FIRST AID Induce vomiting; administer sodium sulphate and medicinal charcoal; provide plenty of liquid (tea) to drink. **Clinical therapy:** Gastric lavage (possibly with 0.1 % potassium permanganate), instillation of medicinal charcoal and sodium sulphate; for further detoxification see cucurbitacin E.

Ecballium elaterium (L.) A. Rich

family: Cucurbitaceae

concombre sauvage (French); *Spritzgurke* (German); *cocomero asinino* (Italian)

Echium vulgare

viper's bugloss

Echium vulgare   *Echium plantagineum*

PLANTS WITH SIMILAR PROPERTIES A genus of 80 species in Europe, Africa and Asia. It is especially diverse on the Atlantic islands Macaronesia where giant *Echium* species have evolved such as *E. wilpretii*.

PLANT CHARACTERS A biennial plant of up to 1.2 m with erect flowering branches formed in the second year from a rosette of leaves produced in the first year. Characteristic are the bristly hairs on the stems and leaves and the deep blue (not purple) flowers that are borne in one-sided clusters. The flowers are funnel-shaped, with all 5 stamens protruding beyond the rim.

OCCURRENCE On sandy, arid and sunny places in Europe and Asia; naturalised in N America and Australia. Perennial species from the Canary Islands with tall inflorescences are popular ornamentals. Several species have become weeds: the purple-flowered *E. plantagineum* from the Mediterranean was introduced to Australia in 1843; today it is a pernicious weed (Patterson's curse) that causes intoxication of livestock.

CLASSIFICATION Liver poison, neurotoxin, mutagen; moderately hazardous, II

ACTIVE INGREDIENTS All parts of these plants contain pyrrolizidine alkaloids (PAs); especially rich in PAs are the roots, leaves and flowers. The major PA in *E. vulgare* is heliosupine (an open diester alkaloid). *E. plantagineum*

produces echimidine, echiumine and heliotrine.

UTILISATION *Echium* species have been used t some extent in traditional European medicine.

TOXICITY Teas containing *Echium* and othe PA plants are toxic and should not be used, chronic consumption may lead to liver damag Children are more susceptible to *Echium* po soning than adults. Livestock losses ascribed t *E. plantagineum* have occurred in Australia bu the plants are extremely hairy so that anima generally tend to avoid them, except in times food shortage. Liver damage is a serious risk. S vere skin irritation sometimes occurs when pe ple come into contact with the bristly hairs.

SYMPTOMS Symptoms of severe liver distu bance (see senecionine).

PHARMACOLOGICAL EFFECTS PAs be come metabolically activated in the liver (se senecionine). They can then alkylate proteins an DNA. As a result, they cause liver damage (ven occlusive disease). After prolonged intake the are mutagenic, teratogenic and carcinogenic.

FIRST AID Give medicinal charcoal to absor the alkaloids; administer plenty of warm blac tea, and sodium sulphate. Provide shock prophy laxis (keep patient at a quiet place and warm **Clinical therapy:** In case of ingestion of larg doses: gastric lavage and further treatments d scribed under senecionine.

Echium vulgare L.   family: Boraginacea

vipérine (French); *Gemeiner Natternkopf* (German); *echio comune* (Italian)

Encephalartos longifolius

Suurberg cycad

Encephalartos longifolius

Encephalartos longifolius female cone

PLANTS WITH SIMILAR PROPERTIES A genus of almost 50 species in Africa. Other cycad genera with similar toxicity include *Cycas, Macrozamia, Zamia, Stangeria, Bowenia, Lepidozamia, Dioon, Ceratozamia, Chigua* and *Microcycas*.

PLANT CHARACTERS Cycads typically have thick, palm-like stems and very large, pinnate, frond-like leaves. The female cones of *E. longifolius* are enormous (they weigh several kilograms) and contain a large number of bright red, fleshy seeds not unlike dates in size and shape.

OCCURRENCE *E. longifolius* occurs in the Eastern Cape Province of S Africa. The genus is endemic to southern tropical Africa.

CLASSIFICATION Cell poison, mutagen; moderately hazardous, II

ACTIVE INGREDIENTS Cycads contain two toxic alkaloidal glycosides of methylazoxymethanol, known as cycasin and macrozamin. Seed kernels of most cycads contain these compounds (0.2–0.5 %), as well as fresh (0.17–2.5 %) and dried (0.21 %) stems. Stems are said to contain sodium sulphate. Neurotoxic amino acids (3-methylamino-*L*-alanine) are known from *Cycas* species.

UTILISATION Stems decoctions of *Encephalartos* species (and *Stangeria eriopus*) are popular as traditional emetics and this practice may lead to acute or chronic poisoning. Cycad stems are rich in starch and some have been used as staple food

(e.g. *Encephalartos* and *Cycas* stems). The toxins are removed by traditional leaching methods.

TOXICITY Seed kernels are very poisonous and are known to have been the cause of deaths in humans. The pulp around the seeds is considered edible but it is likely to contain some toxins as well. An interesting case of human poisoning occurred during the Anglo-Boer War of 1899–1902, when General Jan Smuts and his commando ate seeds of *E. longifolius*. Only some soldiers were poisoned – perhaps those who also ate the poisonous seed kernels and not only the fleshy parts.

SYMPTOMS Bloody diarrhoea and vomiting. Liver damage and neuronal disturbances have been recorded. NPAAs may cause amyotropic lateral sclerosis ("Guam disease")

PHARMACOLOGICAL EFFECTS The azoxy compounds cycasin and macrozamin release methylazoxymethanol (MAM) after cleavage by glucosidase. MAM is a methylating agent that methylates DNA and proteins, resulting in severe liver damage and cancer of the liver and kidneys.

FIRST AID Give medicinal charcoal to absorb the alkaloids and plenty of warm black tea. **Clinical therapy:** After overdosing: gastric lavage, instillation of medicinal charcoal and sodium sulphate; electrolyte substitution, control of acidosis with sodium bicarbonate. Check kidney and liver function.

Encephalartos longifolius (Jacq.) Lehm.

family: Zamiaceae

Arbre à pain (French); *Suurberg Brotpalmfarn* (German)

Ephedra sinica

ephedra • desert tea

Ephedra sinica

Ephedra americana

Ephedra americana cones

PLANTS WITH SIMILAR PROPERTIES A gymnosperm genus of 40 species in Europe, Asia and America. Alkaloidal species used in traditional medicine include *E. equisetina*, *E. distachya*, *E. gerardiana*, *E. intermedia* and *E. shennungiana*.

PLANT CHARACTERS A multi-stemmed perennial of about 1 m high with wiry, greyish, furrowed stems. Leaves are small and scale-like. Male flowers are inconspicuous, while female flowers develop into small, fleshy, bright red cones.

OCCURRENCE *E. sinica* is a Chinese species; *E. distachya* occurs in the Mediterranean region.

CLASSIFICATION Mind-altering; moderately hazardous, II

ACTIVE INGREDIENTS The main compound in *E. sinica* is mostly ephedrine, but pseudoephedrine and other minor alkaloids are also present. The total alkaloid content is 1–2 %, depending on the species. Tyrosine derivatives, spermine alkaloids and procyanidins occur in roots.

UTILISATION *E. sinica* has recorded medicinal uses in China dating back about 5 000 years. It is traditionally used to treat asthma, bronchitis, rhinitis and sinusitis. *Ephedra* is an ingredient of weight loss products. It is also famous for its presence in performance-enhancing products (prohibited dope) that are used by some athletes. Indians and settlers of N America prepared a stimulatory tea from *E. trifurca* (Mormon tea), a plant rich in norpseudoephedrine. In N America, ephedrin has been used as a "legal" mind-altering drug.

TOXICITY The lethal oral dose in mice is 4.3 g kg; in rats 3.5 g. LD_{50} mouse: 150 mg/kg (i.p.).

SYMPTOMS The drug stimulates the mind an increases mental power and sociability. Feeling of fatigue, hunger and thirst are suppressed. E cessive or prolonged use may lead to insomni hypertension, dependency, aggressive behaviou and personality disorders. Overdoses cause stron perspiration, dilated pupils, contraction of smoot muscles, excitation of the breath centre and uter us, hyperglycaemia, insomnia, constipation an urinary retention.

PHARMACOLOGICAL EFFECTS Ephedrin stimulates α- and β-adrenergic receptors by in creasing the release of noradrenaline from ca echolic synapses and inhibiting its re-uptake. I has central stimulatory effects (similar to am phetamine but weaker) and causes periphera bronchodilation and vasoconstriction and thu increases blood pressure. The central stimulatio results in improved powers of concentration an a suppression of pain, fatigue and hunger. Pr longed use may lead to dependency. Alcohol an caffeine can enhance the toxicity of ephedrine.

FIRST AID Administer medicinal charcoal an plenty of liquids. **Clinical therapy:** After ove dose, see ephedrine.

Ephedra sinica Stapf   family: Ephedracea

ephedra, raisin de mer (French); *Ephedra, Chinesisches Meerträubel* (German); *efedra, uva marina* (Italian)

Equisetum palustre

marsh horsetail

Equisetum palustre sterile and fertile stems

Equisetum telmateia

PLANTS WITH SIMILAR PROPERTIES The horsetail family is an ancient relict of the Carboniferous age, before the time of flowering plants. Presently the genus comprises some 15 species with a cosmopolitan distribution. In addition to *E. palustris*, *E. arvense*, *E. sylvaticum* and *E. telmateia* are also of toxicological interest.

PLANT CHARACTERS A perennial herb with erect, unbranched fertile stems emerging in early spring. These have characteristic oblong cones at their tips that produce spores. The sterile stems are formed in summer – about 0.5 m high, with evenly spaced whorls of side branches. Each node has a circle of small brown scales that represent the highly reduced leaves. *E. palustre* is distinguished from other species by the shorter length of the first internodes of the lateral branches.

OCCURRENCE Wet marshes of the N hemisphere (America, Europe and Asia).

CLASSIFICATION Metabolic poison; moderately hazardous, II

ACTIVE INGREDIENTS Marsh horsetail produces potentially toxic spermidine alkaloids such as palustrine. Also present is a factor (thiaminase) that inactivates the coenzyme thiamine (better known as vitamin B_1) in animals. Furthermore, 5–8 % silicates accumulate in stems (including water-soluble silicates), as well as aluminium salts. Several flavonoids are present, including glycosides and esters of quercetin and kaempherol.

UTILISATION *E. palustre* is not used medicinally but is sometimes an adulterant of *E. arvense*. Dried stems of the latter (*Equiseti herba*) is used mainly as a diuretic for treating inflammation of the lower urinary tract, kidney gravel and post-traumatic and static oedema. Traditional uses include the treatment of enhanced menstrual bleeding, skin disorders and slow healing wounds. The Romans boiled shoots of *E. arvensis* and ate them like asparagus. The cones are eaten in Japan. Due to the high silica content, stems are used for scouring and polishing (hence the common name "scouring rush").

TOXICITY The toxicity of *E. palustre* is mostly known from livestock poisonings, especially in horses which develop "staggering disease" when feeding on *E. palustre*, *E. arvense*, *E. sylvaticum* and *E. telmateia*.

SYMPTOMS Symptoms in horses include excitability, tremor of facial muscles, staggering, falling down and death from total fatigue. Cows show a decreased milk production, weight loss, diarrhoea and paralysis.

PHARMACOLOGICAL EFFECTS Symptoms agree with those of vitamin B_1 deficiency.

FIRST AID In case of animal poisoning, daily inject 250–1 000 mg thiamine or feed the animals with yeast, which is rich in vitamins.

Equisetum palustre L.

prêle de marais (French); *Sumpfschachtelhalm* (German)

family: Equisetaceae

Erysimum cheiri

wallflower

Erysimum cheiri garden form *Erysimum cheiri* *Erysimum cheiri*

PLANTS WITH SIMILAR PROPERTIES More than 250 species are described within *Erysimum*, a genus from Europe, Asia and the Mediterranean area. *Erysimum crepidifolium* and *E. diffusum* are also of special interest, as both have poisonous properties. *Erysimum* species were previously included in the genus *Cheiranthus*.

PLANT CHARACTERS Spring-bedding, scented perennial shrub (0.2–0.6 m in height) with lanceolate leaves and yellow, brown, pink or red flowers. *E. crepidifolium* and *E. diffusum* are biennual or perennial plants of 0.5–0.6 m in height, with light yellow flowers. The brown seeds are formed in linear capsules of up to 70 mm long.

OCCURRENCE *E. cheiri* appears to be a hybrid between 2 or more Aegean species. It is a popular ornamental and various garden forms are known (better known as *Cheiranthus cheiri*). *E. crepidifolium* and *E. diffusum* occur in C, E and S Europe.

CLASSIFICATION Heart poison; moderately hazardous, II

ACTIVE INGREDIENTS Cardenolides, such as cheirotoxin and cheiroside A, with erysimoside especially in seeds. Also present are glucosinolates, such as glucocheirolin (which releases 3-methyl-sulphonyl-propyl-isothiocyanate) and glucoiberin. *E. crepidifolium* and *E. diffusum* accumulate cardenolides in seeds (3.5 %), such as erysimoside, helveticoside and 20 others.

UTILISATION *Erysimum* species were formerly used in medicine to treat spasms, constipation and disturbances of the liver, heart and menstrual cycle. *E. cheiri* was also used for scent production.

TOXICITY *E. crepidifolium* appears to be especially poisonous to geese (German name "Gänse sterbe" = geese killer) and rabbits. Concentrated decoctions may cause rapid death due to heart failure. Lethal dose in cats is 0.12 mg/kg for cheirotoxin and 0.68 mg/kg for cheiroside A. Helveticoside: LD_{50} (i.v.) in pigeons 0.28 mg/kg, in guinea pigs 0.87 mg/kg, in cats 0.11 mg/kg, rat 54 mg/kg; (i.p.) and mice 7.8 mg/kg.

SYMPTOMS Symptoms include nausea, salivation, gastrointestinal disturbance, diuresis, purging and exhaustion, with the usual respiratory and cardiac symptoms (arrhythmia, hypertension, coma, cardiac arrest) associated with heart glycoside poisoning. Isothiocyanates have skin irritating properties (see mustard oil).

PHARMACOLOGICAL EFFECTS In general, cardiac glycosides inhibit the Na^+, K^+-ATPase (see cardiac glycosides).

FIRST AID First aid treatment is indicated if large amounts of seeds of *Erysimum* or isolated cardiac glycosides have been ingested. Treatment involves the induction of vomiting, supply of medicinal charcoal, and in severe cases, detoxification as outlined under cardiac glycosides.

Erysimum cheiri (L.) Crantz (= *Cheiranthus cheiri* L.)   family: Brassicaceae

giroflée jaune (French); *Goldlack* (German); *violaciocca gialla* (Italian)

Erythrina caffra

coast coral tree

Erythrina caffra flowers and pods

Erythrina crista-galli

PLANTS WITH SIMILAR PROPERTIES A genus of 112 species in warm regions of Asia, Africa and America. Several are grown as shade and ornamental trees, among them *E. americana, E. berteroana, E. crista-galli, E. herbacea, E. mitis, E. subumbrans* and *E. variegata*.

PLANT CHARACTERS A medium-sized tree with thick stems that are covered with numerous prickles. The leaves each comprises three pointed, horny leaflets. Attractive orange-red flowers are followed by cylindrical, constricted pods bearing bright red seeds with contrasting dark spots.

OCCURRENCE *E. caffra* is confined to the east coast of S Africa. Several species are cultivated in gardens, e.g. cockspur coral tree, *E. crista-galli*.

CLASSIFICATION Neurotoxin, mind-altering; highly hazardous, Ib

ACTIVE INGREDIENTS Numerous tetracyclic isoquinoline alkaloids ("Erythrina alkaloids") are known from several *Erythrina* species, e.g. erythaline and erysotrine. *E. caffra* seeds contain a potent trypsin inhibitor; toxic proteins and lectins are likely to be present in this and other species.

UTILISATION The bark, whole stems or rarely the leaves or roots of *E. caffra* and other species are commonly used in traditional medicine in Africa, Asia and C & S America. They are mainly applied topically for sores, wounds and arthritis, or to relieve earache or toothache. The Mayas already knew of the intoxicating properties. Mexican Indians have used the alkaloids, which dramatically enhance blood pressure, for revenge since they can induce apoplectic stroke in captured enemies. The attractive seeds have been used as beads; punctured seeds can cause poisoning.

TOXICITY Seeds have a highly resistant seed coat that has to be damaged in order to release the toxic alkaloids. Accidental poisoning may occur from medicinal use. *Erythrina* alkaloids are highly toxic, especially when injected. About 490 mg seed powder per kg BW are lethal for guinea pigs.

SYMPTOMS Curare-like effects, with hypotention, paralysis of motoric nerves and psychic effects, such as euphoria, drunkenness, enhanced libido, fever, substantial reddening of skin, followed by a deep sleep. Death from respiratory arrest may occur after 2–3 days.

PHARMACOLOGICAL EFFECTS Several *Erythrina* alkaloids, such as erysotrine, are known to have curare-like neuromuscular blocking effects, as they affect mAChR as antagonists.

FIRST AID Provide fluids to drink; induce vomiting, give medicinal charcoal and sodium sulphate. **Clinical therapy:** Immediately provide intubation and artificial respiration; gastric lavage, electrolyte substitution, control of acidosis with sodium bicarbonate. Check heart function.

Erythrina caffra Thunb.   family: Fabaceae

arbre-à-corail (French); *Kap-Korallenbaum* (German)

Erythrophleum suaveolens

ordeal tree

Erythrophleum lasianthum

Erythrophleum lasianthum bark

PLANTS WITH SIMILAR PROPERTIES A genus of 9 species in the tropics of Africa, Madagascar, Asia and Australia, including *E. lasianthum*, *E. succirubrum* and *E. africanum*.

PLANT CHARACTERS *E. suaveolens* is a medium-sized tree with pinnate leaves, yellowish flowers in dense elongated clusters, followed by large, flat pods. This species is distinguished from *E. africanum* by the narrow leaflet tips and the many-seeded pods that split mostly along one suture. The pods are up to five-seeded in *E. africanum* and split along both sutures.

OCCURRENCE *E. suaveolens* and *E. africanum* are widely distributed in Africa.

CLASSIFICATION Heart poison; neurotoxin; extremely hazardous, Ia

ACTIVE INGREDIENTS Several diterpenoid alkaloids occur in bark and seeds of *Erythrophleum* species, including cassaine and an alkaloid known as erythrophleine or norcassamidide.

UTILISATION *Erythrophleum* species are notorious for their use in the African trial by ordeal ritual. A tea made from pounded bark is given to the accused. If he or she vomits, it is taken as a sign of innocence. If the poison is not expelled by vomiting, the person will almost certainly die. The person administering the poison can control the process, because very high doses are likely to have an emetic effect. Furthermore, an inno-

cent person is likely to gulp down the poison and therefore vomits, while a guilty person may drink more slowly and hesitantly and is therefore more likely to retain the poison and be killed. Bark extracts were used to prepare arrow and fish poison in W Africa. Bark powder of *E. lasianthum* is traditionally used by Zulu people to treat pain such as fever and headache. It has also been been implicated in murder cases.

TOXICITY Seeds are considered to be more poisonous than bark. The lethal doses are 0.5 g of fruit (rabbit) or 60 g of bark and leaves (sheep).

SYMPTOMS Alkaloids have inotropic properties and lead to cardiac arrest. Furthermore locally anaesthetic effects were recorded. Symptoms of poisoning include spontaneous vomiting, diarrhoea, hypotension, arrhythmia, bradycardia and death from respiratory and cardiac arrest.

PHARMACOLOGICAL EFFECTS *Erythrophleum* alkaloids (including the two main seed alkaloid of *E. lasianthum*) are highly toxic and have powerful cardiotonic activity (positively inotropic similar to that of digitalis and scillaren A.

FIRST AID Provide plenty of fluids to drink; give medicinal charcoal and sodium sulphate. **Clinical therapy:** Gastric lavage, electrolyte substitution, control of acidosis with sodium bicarbonate. Check heart function; intubation and artificial respiration if necessary.

Erythrophleum suaveolens (Guill. & Perrott.) Brenan (= *E. guineense*)   family: Fabaceae

Rotwasserbaum, Gottesurteilsbaum (German)

Erythroxylum coca

coca plant

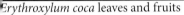
Erythroxylum coca leaves and fruits

Erythroxylum novogranatense flowers

PLANTS WITH SIMILAR PROPERTIES A large pantropical genus with 240 species.

PLANT CHARACTERS A woody shrub of up to 1 m high, with hairless leaves, white flowers and small, red berries. Cultivated forms include *E. coca* var. *coca* (the wild form; with warty stems), var. *ipadu* (a cultigen) and two varieties of *E. novogranatense* (with smooth stems).

OCCURRENCE S America: E Andes – Peru, Bolivia (var. *coca*); Amazon (var. *ipadu*); Colombia and Venezuela (var. *novogranatense*) and N Peru (var. *truxillense*). Cultivated since ancient times.

CLASSIFICATION Neurotoxin; mind-altering; highly hazardous, Ib

ACTIVE INGREDIENTS Tropane alkaloids in leaves (0.1–1.4 %); mainly cocaine, also hygrine and cuscohygrine.

UTILISATION Thousands of Andean people still use 50 g of the leaves, mixed with lime, every day as a masticatory or as tea bags (*mate de coca*) to relieve hunger and fatigue. Cocaine was the first commercial anaesthetic but it is no longer much used in modern medicine. As illegal drug, cocaine is used by millions of people (5 million in the USA alone). Thin lines of cocaine powder (free base) is usually snorted with a straw from a flat surface. It may also be smoked as "crack" (raisin-sized lumps) or dissolved and injected i.v.

TOXICITY Cocaine abuse (with debilitating, addictive results) is a severe problem in many countries of the world. 1–2 g (p.o.) and 0.2–0.3 g s.c. cocaine can be lethal for humans. Sensitive persons have died after snorting 30 mg cocaine.

SYMPTOMS Euphoria, urge to speak, enhanced libido, loss of inhibitions, hallucinations, dilated pupils, vasoconstriction, hypertension, tachycardia and death from respiratory arrest. Schizophrenic episodes may occur. Cocaine induces strong psychic but no physical dependence.

PHARMACOLOGICAL EFFECTS Cocaine is a local anaesthetic and euphoric (powerful and rapid stimulation of the brain). The activity results from blocking Na^+ channels in neurons, enhanced release and inhibition of the re-uptake of dopamine and noradrenaline. Dopamine is normally quickly released from dopamine receptors in the postsynaptic neuron and taken up in the presynaptic neuron by a special carrier mechanism. When the re-uptake is blocked by cocaine, dopamine concentration in the synaptic gap is increased, leading to a prolonged activation of dopamine receptors. Since dopamine is involved with the control of the pleasure response, cocaine stimulates the so-called pleasure centres.

FIRST AID Sedate patient. Administration of medicinal charcoal and sodium sulphate. **Clinical therapy:** Gastric lavage, instillation of sodium sulphate and medicinal charcoal.

Erythroxylum coca Lam.

family: Erythroxylaceae

cocalier (French); *Kokastrauch* (German); *coca* (Italian); *cocal* (Spanish)

Euonymus europaeus

spindle tree

Euonymus europaeus flowers

Euonymus europaeus fruits

PLANTS WITH SIMILAR PROPERTIES A genus of 177 species, including *E. alatus*, *E. atropurpureus*, *E. fortunei*, *E. globularis*, *E.hamiltonianus* and *E. japonicus*. *E. verrucosus* is toxicologically similar to *E. europaeus*.

PLANT CHARACTERS A deciduous shrub or small tree (2–6 m in height) with green, four-angled twigs. Flowers are tiny (8–10 mm in diameter) and greenish white. The fruit is a 3–5-valved, colourful red capsule that dehisces to expose the bird-dispersed, bright orange, arillate seeds.

OCCURRENCE *Euonymus* species occur in the northern temperate zone (Europe, Asia and America) and also Australia. *E. europaeus* is an inhabitant of hedges in Europe and W Asia. Several species are grown as ornamental plants.

CLASSIFICATION Heart poison; highly hazardous, Ib

ACTIVE INGREDIENTS Cardenolides, such as evonoside ($C_{41}H_{64}O_{19}$; MW 860.90), evobioside and evomonoside ($C_{29}H_{44}O_8$; MW 520.70); toxic alkaloids (0.1 %), such as evonine ($C_{36}H_{43}NO_{17}$; MW 761.75) and lectins are also present.

UTILISATION The spindle tree was formerly used in traditional medicine and to make charcoal for gunpowder. Seeds yield a yellow dye used to colour butter. The timber has been used for skewers, spindles, toothpicks and violin bows.

TOXICITY All parts are toxic, especially fruits;

36 fruits are said to kill an adult person; 2 fruit caused a severe poisoning in a 7-year-old child Dried fruit powder is insecticidal. LD_{100} for evc noside in cats: 0.84 mg/kg i.v.

SYMPTOMS Toxic symptoms occur late, ofter 10 to 18 h after intake. These include nausea spasms, shock, increased temperature, blood diarrhoea with colic, liver and kidney damage cardiac arrhythmia, tachycardia (150 heartbeat per minute), kidney and liver damage, paralysi of masticatory muscles, coma and death.

PHARMACOLOGICAL EFFECTS Cardeno lides inhibit Na^+,K^+-ATPase which is essential fo neuronal activity and cellular transport processe (see cardiac glycosides). The alkaloids appear tc have cytoxic properties.

FIRST AID If more than 5 seeds were eaten, in duce vomiting and treat with sodium sulphat and medicinal charcoal. Allow patient to drink plenty of black tea. **Clinical therapy:** In case where large amounts of seeds were ingested gastric lavage (possibly with potassium permar ganate), instillation of medicinal charcoal anc sodium sulphate, electrolyte substitution anc control of acidosis with sodium bicarbonate ma be required. In case of spasms give diazepam; fo shock plasma expander, in case of severe intoxica tion, provide intubation and oxygen respiration See also cardiac glycosides.

Euonymus europaeus L.

family: Celastraceae

fusain d'Europe (French); *Pfaffenhütchen* (German); *fusaria commune, evonimo* (Italian)

Euphorbia cyparissias

cypress spurge

Euphorbia cyparissias

Euphorbia lathyris

PLANTS WITH SIMILAR PROPERTIES A very large genus of about 2 000 species on all continents. They produce large amounts of latex with phorbol esters and are thus potentially toxic. Important herbaceous species include *E. lathyris, E. myrsinitis, E. marginata* and *E. pulcherrima*.

PLANT CHARACTERS Cypress spurge is a typical example of a perennial milkweed. It is 0.4 m high, with lanceolate, hairless leaves along erect stems. Yellow flowers are borne in terminal, 9–15-rayed umbels. The bracts are oblong; the upper ones 3–6 mm long and yellow turning red.

OCCURRENCE Europe and C Asia; common in arid places and dry meadows. Several species are cultivated as ornamental plants; *E. pulcherrima* (with large red bracts) is a favoured ornamental during Christmas time (Christmas Star).

CLASSIFICATION Cell poison, co-carcinogen, skin irritant; highly hazardous, Ib

ACTIVE INGREDIENTS The irritant compounds in the latex are phorbol esters of the ingenane type, called cyparissias factor Cy6, Cy11 and Cy14; other compounds include euphorbon (a mixture of α and β-euphorbol, a C30 steroid).

UTILISATION Latex has been used to promote hair growth and to eliminate freckles, warts and skin ulcers.

TOXICITY *Euphorbia* species contain irritant and toxic latex and several of them have been implicated in human and livestock poisoning. Toxic effects of *E. pulcherrima* are rather rare (only slight GI disturbance). However, the plant is a cause for many consultations at poison centres.

SYMPTOMS The latex causes severe local reactions, including inflammation (erythema formation), oedema and blister formation. Eye: swollen lids, conjunctivitis and keratitis. Ingestion of *Euphorbia* latex or aerial parts leads to lesions in mouth and throat, abdominal pain, bloody diarrhoea, mydriasis, arrhythmia, vertigo, kidney inflammation, coma and even death after 1–3 days. Livestock usually avoid feeding on *Euphorbia* species; however when hay with spurge is ingested, intoxication may occur with heavy gastroenteritis, bloody diarrhoea, liver damage and paralysis.

PHARMACOLOGICAL EFFECTS. Phorbol esters activate the protein kinase C, which is a key enzyme in several signalling pathways. Phorbol esters are important as co-carcinogens (see phorbol esters).

FIRST AID Give medicinal charcoal and sodium sulphate. Provide plenty of tea. Contact with eyes or skin: wash with polyethylene glycol or water (contact ophthalmologist); in case of skin blisters, cover with sterile cotton and locally with a corticoid ointment. **Clinical therapy:** After ingestion, gastric lavage; for further treatment, see phorbol esters.

Euphorbia cyparissias L.   family: Euphorbiaceae

ithymale (French); *Zypressenwolfsmilch* (German); *euforbia cipressina* (Italian)

Euphorbia tirucalli

pencil tree • rubber euphorbia

Euphorbia tirucalli

Euphorbia tirucalli fruits

PLANTS WITH SIMILAR PROPERTIES Of the approximately 2 000 species of *Euphorbia* several are succulent trees or shrubs with toxic milky latex (e.g. *E. tirucalli, E. ingens* and *E. milii*).

PLANT CHARACTERS *E. tirucalli* is a typical example of a succulent *Euphorbia*. It is a tree (5 m in height) with numerous, robust, cylindrical stems. Small, yellow, fleshy flowers are formed on the stems and they mature into three-lobed capsules. Whereas many African *Euphorbia* species are succulents (shrubs and trees), most European milkweeds are annuals and perennial herbs.

OCCURRENCE *E. tirucalli* occurs in Africa, Madagascar and India. Some succulent *Euphorbia* species are used as pot plants.

CLASSIFICATION Cellular poison, co-carcinogen, skin irritant; highly hazardous, Ib

ACTIVE INGREDIENTS The irritant compounds of the white latex are phorbol esters. Several phorbol-related compounds, including esters of 4-deoxyphorbol and ingenol, have been found in *E. tirucalli*. In *E. ingens*, various esters of ingenol are the main irritant principles.

UTILISATION The latex of *E. tirucalli* is a potential source of fuel hydrocarbons. Latex of *E. virosa* is an ingredient of arrow poisons in Namibia and Angola.

TOXICITY *Euphorbia* species contain irritant and toxic latex and several of them have been implicated in both human and livestock poisoning. *E. ingens* is extremely toxic and the latex may cause temporary or even permanent blindness.

SYMPTOMS The latex produces severe local reactions, including inflammation, oedema and blistering of the skin. In animals, severe injuries may be caused to the face, eyes, mouth and tongue. The latex of *E. tirucalli* causes burning and inflammation of the eyes and mucus membranes. *E. mauritanica* causes a fatal nervous disorder in sheep, with muscle tremors, foaming at the mouth, bloat, diarrhoea and fever. Honey made during the flowering periods of species such as *E. ingens* and *E. ledienii* causes severe burning of the mouth and throat and is therefore not suitable for human consumption.

PHARMACOLOGICAL EFFECTS. Phorbol esters activate the protein kinase C, which is a key enzyme in several signalling pathways. Phorbol esters are important as co-carcinogens.

FIRST AID Contact with eyes or skin: wash with polyethylene glycol or water; in case of skin blisters, cover with sterile cotton and locally with corticoid ointment. Give medicinal charcoal and sodium sulphate. Provide plenty of tea. **Clinical therapy:** After oral ingestion gastric lavage, instillation of medicinal charcoal and sodium sulphate; for further treatment see phorbol esters.

Euphorbia tirucalli L.

family: Euphorbiaceae

tirucalli (French); *Bleistiftbaum* (German); *consuelda* (Spanish)

Fabiana imbricata

Chilean false heath • pichi

Fabiana imbricata

PLANTS WITH SIMILAR PROPERTIES The genus *Fabiana* comprises 25 species in warm temperate S. America, including *F. bryoides* and *F. friesii*. Another Solanaceae from Chile is *Latua pubiflora* that is also used as mind-altering plant.

PLANT CHARACTERS An evergreen, ericoid shrub of up to 2 m in height, with small scale-like leaves. White or violet, tubular, small flowers with 5 lobes are borne along the terminal parts of the twigs. Fruits are oval capsules of 5–6 mm long. *Latua pubiflora* is a tree (2–10 m high) with tiny branches bearing pendulous, violet-coloured, bell-shaped flowers (40 mm long). The fruits are small, greenish berries.

OCCURRENCE Originally from S Chile and Argentina, but now grown as a greenhouse and ornamental garden plant in many countries.

CLASSIFICATION Neurotoxin, mind-altering; moderately hazardous, II

ACTIVE INGREDIENTS Aerial parts contain an essential oil. Secondary metabolites include fabianine (a quinoline alkaloid), physcione and other anthraquinones, 3,11-amorphadien and other sesquiterpenes, coumarins and flavonoid glycosides. *Latua pubiflora* accumulates tropane alkaloids (hyoscyamine and scopolamine).

UTILISATION Plants were brought to Europe by early discoverers and grown in the Botanical Garden of Madrid. Dried twigs have been used by S American Indians as an intoxicating drug called pichi-pichi. Twigs were dried and fumigated together with *Latua pubiflora* (Solanaceae) and the smoke was extensively inhaled. As a medicinal plant, pichi was used in Chile and in Europe as diuretic and to treat disturbances of the kidneys and urinary tract. The drug was also used to expel worms in livestock. *Latua* has been widely used as intoxicant and aphrodisiac by the Mapuche Indians in S Chile.

TOXICITY Crude extracts were shown to have a very low toxicity in rats, in doses of up to 5 g of extract per kg body weight.

SYMPTOMS The extracts have diuretic and general tonic properties. The inhaled smoke induces euphoria and intoxication. For *Latua*, see symptoms described under *Atropa* and *Hyoscyamus*.

PHARMACOLOGICAL EFFECTS Fabianine can modulate neuroreceptors, probably of the cholinergic system. Extracts inhibit β-glucuronidase and are antibacterial.

FIRST AID Apply medicinal charcoal. **Clinical therapy:** After overdoses: gastric lavage (possibly with potassium permanganate), instillation of medicinal charcoal and sodium sulphate, electrolyte substitution; control acidosis with sodium bicarbonate; in case of spasms give diazepam; intubation and oxygen respiration may be required.

Fabiana imbricata Ruiz & Pav.   family: Solanaceae

Fabiane imbriquée (French); *Fabiane* (German); *pichi-pichi* (Italian); *pitchi, pichi romero* (Spanish)

Gaultheria procumbens

wintergreen • checkerberry

Gaultheria procumbens fruit

PLANTS WITH SIMILAR PROPERTIES
A cosmopolitan genus with more than 130 species. *G. mucronata* (= *Pernettya mucronata)* and *G. shallon* are cultivated in gardens.
PLANT CHARACTERS A mat-forming woody shrub with smooth, reddish brown, flaking bark, glossy green, leathery leaves, white or pinkish bell-shaped flowers and bright red, edible berries.
OCCURRENCE N America (eastern USA and Canada). The plant is cultivated to a limited extent in gardens.
CLASSIFICATION Cell toxin; moderately hazardous, II
ACTIVE INGREDIENTS An essential oil is present at levels of about 0.8 % of wet weight, of which methylsalicylate represents 99 % of the total monoterpenoids. Methylsalicylate is actually present in the form of gaultherin (also known as monotropitoside), a glycoside with glucose and xylose. Monotropitoside appears to be the pro-drug that is hydrolysed in the intestine and liver to form the active compound, which is methylsalicylate. It has a typical sharp smell and sweetish taste. Synthetic methylsalicylate has largely replaced the natural compound as the active ingredient of modern products. The leaves also contain arbutin (see hydroquinones) and tannins.
UTILISATION Wintergreen is mainly used externally in the form of wintergreen oil, as a coun-

ter-irritant to treat painful muscles and joint including rheumatism. It is also used internal to some extent as a tonic and to treat rheumatis and minor stomach ailments. Canadian India and early settlers traditionally made a refresling tea from the leaves. The oil is still used as a aromatic flavour ingredient for sweets, chewing gum, and in certain beverages (e.g. root beer).
TOXICITY Poisoning was not uncommon i earlier times when wintergreen oil was applie topically or internally. Sometimes, even deat has occurred as a result of the presence of methy salicylate. The lethal dose in humans is 30 g oil.
SYMPTOMS High doses of methylsalicyla cause stomach and kidney damage and gastriti collapse, hallucinations, internal bleeding an death. Ingestion of fruits from *Gaultheria insar* (= *Pernettya furens*) from Chile can cause toxi hallucinogenic and psychic effects.
PHARMACOLOGICAL EFFECTS Salicylat are inhibitors of cyclooxygenase and prevent th formation of prostaglandins, so that analgesi anti-inflammatory and anti-rheumatic effec are plausible. Methylsalicylate and wintergree products are toxic when ingested and should l avoided by people who are allergic to aspirin.
FIRST AID Medicinal charcoal; externally: was skin, mucosa or eye extensively with water c polyethylene glycol.

Gaultheria procumbens L.

family: Ericacea

gaulthérie du Canada, thé des bois (French); *Niederliegende Scheinbeere, Wintergrün* (German); *uva di monte* (Italian)

Gelsemium sempervirens

yellow jasmine

...elsemium sempervirens

Gelsemium sempervirens flowers

...LANTS WITH SIMILAR PROPERTIES A ge-
...us of 3 species, including *G. rankinii* (SE USA)
...d *G. elegans* (SE Asia). Complex indole alka-
...ids also occur in the Rubiaceae.

...LANT CHARACTERS A climber (vine) with
...virling stems and simple, glossy leaves arranged
... opposite pairs. The attractive flowers are bright
...llow and trumpet-shaped.

...CCURRENCE N America (SE USA). *G. sem-
...rvirens* is an ornamental climber grown world-
...ide. It is the state flower of S Carolina.

...LASSIFICATION Neurotoxin; extremely haz-
...dous, Ia

...CTIVE INGREDIENTS *Gelsemium* contains
...onoterpene indole alkaloids at a concentration
...about 0.5 %. The main alkaloids are gelsemine,
...lseverine, gelsemicine, gelsedine, sempervirine
...d various hydroxylated derivates. It also con-
...ins coumarins, iridoid glycosides (gelsemide,
...mperoside) and steroids of the pregnane type.

...TILISATION Tinctures or alkaloidal extracts
...e used to treat facial and oral neuralgia (severe
...in caused by damaged nerves). Extracts have
...so been used as ingredients of cough syrups and
... topical treatment for pain (e.g. pinched nerves
... the spinal cord), neurological conditions and
...eeding of haemorrhoids and the uterus. The
...oisonous *G. elegans* is traditionally used for
...urder and suicide in Indomalaysia.

TOXICITY The plant and its alkaloids are known
to be very poisonous: 12–15 g of the tincture
are said to be potentially lethal to children and
adults. About 180 mg/kg of the total extract is a
fatal dose for rats. Gelsemine: LD_{50} mouse (i.v.)
4 mg/kg; p.o. 1 240 mg/kg. *Gelsemium* is used as
herbal medicine and an overdose may result in
poisoning, sometimes with fatal consequences.
Children and domestic animals may accidentally
be exposed to it in the garden.

SYMPTOMS Mydriasis, convulsions, muscular
weakness, visual disturbance, hypotension and analgesic, sedative and paralytic activities. Death is due
to respiratory arrest.

PHARMACOLOGICAL EFFECTS The alkaloids (especially gelsemine) activate acetylcholine
receptors (especially the muscarinic AChR and
nAChR), cholinesterase and the synaptic re-uptake of dopamine, noradrenaline and serotonin.
It influences the central and peripheral nervous
system.

FIRST AID Provide plenty of fluids to drink;
give medicinal charcoal and sodium sulphate;
shock prophylaxis. **Clinical therapy:** Gastric
lavage, electrolyte substitution, control of acidosis with sodium bicarbonate; in case of spasms,
diazepam; check kidney function. In case of respiratory arrest, provide intubation and artificial
respiration.

...elsemium sempervirens (L.) J. St-Hil.   family: Gelsemiaceae

...smin sauvage (French); *Gelber Jasmin, Giftjasmin* (German); *gelsemio, falso gelsemino* (Italian)

Genista tinctoria

dyer's greenweed

Genista tinctoria

Genista tinctoria flowers

Genista tinctoria fruits

PLANTS WITH SIMILAR PROPERTIES A genus with 87 species in Europe, North Africa, and W Asia, including *G. aetnensis*, the spiny *G. anglica*, *G. germanica*, *G. tridentata* and *G. sagittaria*. A closely related genus, *Teline*, is often included in *Genista*. *Ulex* (gorse) is a genus of 20 species in Europe and the Mediterranean. These spiny plants (e.g. *U. europaeus* and *U. gallii*) accumulate cytisine-type alkaloids in seeds and young shoots. *Spartium junceum* (a monotypic Mediterranean genus, naturalised in many parts of the world and ornamental plant) resembles broom (s. p. 259) but the stems are cylindrical and the flowers fragrant; chemically it is similar to *Genista*.

PLANT CHARACTERS A branched shrub of up to 0.6 m in height, with photosynthetically active stems, narrowly elliptic leaves, small bright yellow flowers borne towards the branch ends and small narrow pods with 6–10 rounded seeds.

OCCURRENCE Europe and W Asia. The plant is naturalised in N America; *Genista* species and allies are grown as ornamental plants.

CLASSIFICATION Neurotoxin; mind-altering; highly hazardous, Ib

ACTIVE INGREDIENTS Cytisine, *N*-methylcytisine, anagyrine, and other quinolizidine alkaloids; up to 0.3 % total alkaloids in aerial parts.

UTILISATION *G. tinctoria* has been used as a diuretic and was formerly grown for yellow dyes.

Flowers of *Genista canariensis* (also known as *Teline canariensis* or *Cytisus canariensis*) have been smoked as stimulant and weak hallucinogen by shamans of the Yaqui Indians in N Mexico. The plant originates from the Canary Islands where it was a ritual plant of the indigenous people (Guanches). It was taken to the New World by the slave transporters who often stopped on the Canary Islands to replenish their water reserves. An artist who drank tea made from flowers of *Spartium junceum*, which are rich in *N*-methylcytisine, experienced vivid and colourful dreams.

TOXICITY Cytisine: LD_{50} mouse: 101 mg/kg p.o.; LD_{50} cat: 3 mg/kg p.o.; dog 4 mg/kg p.o.

SYMPTOMS Small doses cause diarrhoea, large doses induce vomiting, drastic diarrhoea, severe gastroenteritis and heart disorder (see cytisine). *N*-Methylcytisine and cytisine appear to induce slight hallucinogenic activity, as seen in *G. canariensis* and *Spartium junceum*.

PHARMACOLOGICAL EFFECTS Cytisine is a strong agonist at nAChR (similar to nicotine) and thus affects the central and peripheral nervous system (see cytisine).

FIRST AID Administer sodium sulphate, medicinal charcoal, and plenty of warm tea or fruit juice. **Clinical therapy:** After intake of large doses, gastric lavage and further treatments (see cytisine).

Genista tinctoria L.

family: Fabaceae

genêt des teinturiers (French); *Färberginster* (German); *ginestra minore* (Italian)

Gloriosa superba

flame lily • climbing lily

Gloriosa superba

Gloriosa superba

PLANTS WITH SIMILAR PROPERTIES A monotypic genus with 1 species and up to 9 varieties in the tropics of the Old World. *Littonia modesta* (climbing bellflower) and *Sandersonia aurantiaca* (Christmas bells) are ornamentals and cut flowers that are closely related to *Gloriosa*.

PLANT CHARACTERS A geophyte with distinctive branched and finger-like rhizomes giving rise to slender stems. The glossy green leaves have coiled climbing tendrils at their tips. The flowers are unique and spectacular: they have six reflexed and erect, brightly coloured and often wavy petals in various shades of red and yellow (hence flame lily). Also distinctive are the conspicuous arching stamens and style. The oblong fruit capsules split open to release several bright orange seeds. Various colour forms of *G. superba* have been known as *G. rothschildiana*, *G. simplex* and *G. virescens*, but all of them are now considered to belong to a single species.

OCCURRENCE The plant is widely distributed in tropical regions, from S and E Africa to India. It is sometimes cultivated in warm or tropical gardens as an ornamental plant.

CLASSIFICATION Cellular poison; extremely hazardous, Ia

ACTIVE INGREDIENTS The flame lily contains colchicine, a well-known phenethylisoquinoline alkaloid with antimitotic activity. Colchi-

cine occurs as the main compound in rhizomes (0.3 %), leaves (2.4 %) and seeds (0.9 %).

UTILISATION *Gloriosa* has been cultivated on a commercial scale in Sri Lanka and South Africa for the extraction of colchicine from rhizomes or seeds. It has been used in traditional medicine.

TOXICITY Rhizomes and seeds are very toxic and have caused many fatalities in Africa, India and Sri Lanka as a result of accidental ingestion, use as abortifacient, overdosing with traditional medicine, and especially suicide and homicide. More than 40 mg of colchicine cause death within 3 days. Tubers have been mistaken for sweet potatoes (*Ipomoea batatas*). Livestock losses are rare.

SYMPTOMS Intoxication starts with nausea and vomiting, followed by watery and bloody diarrhoea. At toxic doses, colchicine causes fatal respiratory and cardiovascular disruption within a few days (see colchicine).

PHARMACOLOGICAL EFFECTS Colchicine blocks the formation of microtubules and in consequence stops mitotic cell division.

FIRST AID Induce vomiting; administer sodium sulphate and medicinal charcoal. Keep patient in shock position and warm; provide warm tea or coffee. **Clinical therapy:** Gastric lavage, electrolyte substitution; in cases of severe intoxication, provide intubation and oxygen therapy (see colchicine).

Gloriosa superba L.

family: Colchicaceae

Gloire de Malabar, lis de Malabar (French); *Ruhmeskrone, Prachtlilie* (German)

133

Gnidia kraussiana

yellow heads

Gnidia kraussiana

Gnidia burchellii

PLANTS WITH SIMILAR PROPERTIES A genus of 140 species in tropics, from S Africa to Arabia and India, including *G. burchellii* and *G. polycephala*. Some *Gnidia* species were previously known as *Arthrosolen* and *Lasiosiphon*.

PLANT CHARACTERS An erect, hairy perennial shrublet with a tuberous rootstock and several more or less unbranched stems bearing small, elliptic leaves. Numerous tubular yellow flowers are borne in dense, rounded heads on leafless flowering stalks.

OCCURRENCE *G. kraussiana*, *G. polycephala* and *G. burchellii* all occur in C and S Africa.

CLASSIFICATION Cellular poison, co-carcinogen, skin irritant; extremely hazardous, Ia

ACTIVE INGREDIENTS *Gnidia* species produce diterpenoids of the daphnane type (phorbol esters). *G. kraussiana* contains kraussianin, gnidilatin and gnidilatidin as main compounds. *G. burchellii* has 12-hydroxydaphnetoxin as main constituent, while other *Gnidia* species produce gnidicin, gnididin, gniditrin and several other 12-esters of 12-hydroxydaphnetoxin.

UTILISATION *G. kraussiana* is used as fish or arrow poison but is nevertheless commonly used in traditional medicine (often as enema, snuff or topical application). Overdoses or inappropriate applications may be fatal. In E Africa, *G. glauca* is a source of high-quality paper.

TOXICITY *G. kraussiana* has caused numerou human and animal deaths in various parts of A rica. *G. burchellii*, *G. polycephala* and others ar the cause of livestock losses. LD_{50} for *Gnidia* fac tor K6 and K7: mouse, i.p. 0.06 and 0.3 mg/kg.

SYMPTOMS In livestock, *Gnidia* leads to s vere diarrhoea, fever, weakness and a rapid, wea pulse. In cases where large quantities have bee ingested, there may be no obvious symptom prior to sudden death. Powdered roots or leave used as snuff may cause extreme irritation of th nose and throat, coughing and sneezing, followe by headache and nausea.

PHARMACOLOGICAL EFFECTS Diterpene of *Gnidia* belong to the class of phorbol ester that activate protein kinase C, a central regulato ry enzyme in cells; see "phorbol esters" for phys ological and toxicological consequences. *Gnidi* diterpenoids have antitumour and antileukaemi activity.

FIRST AID Give medicinal charcoal and sodiur sulphate. Provide plenty of tea to drink. **Clinica therapy:** Gastric lavage (possibly with potassiu permanganate), instillation of medicinal charcoa and sodium sulphate; electrolyte substitutior control of acidosis with sodium bicarbonate. I case of spasms give diazepam. Provide intubatio and oxygen respiration in case of respiratory a rest. Check kidney function.

Gnidia kraussiana Meisn.   family: Thymelaeacea

Krauss-Gnidia (German)

Gratiola officinalis

hedge hyssop

Gratiola officinalis flowers

Gratiola officinalis

PLANTS WITH SIMILAR PROPERTIES A cosmopolitan genus of 20–25 species in temperate and tropical mountainous regions. *G. neglecta* has similar properties. The genus has been transferred from Scrophulariaceae to Plantaginaceae.

PLANT CHARACTERS Perennial herb (up to 0.6 m in height) with thin creeping rhizomes. The stems are four-angled and bear sessile, linear-lanceolate leaves with dentate tips and minute glands. The flowers are solitary in the axils of the leaves, borne on a short pedicel in an upright orientation, each subtended by a deeply lobed calyx and two bracteoles. The petals are fused into a two-lipped, zygomorphic perianth of 8–10 mm long. The tube is usually yellowish and the spreading lobes are white, often with purple-red dark veins. Fruit capsules contain several yellowish-brown seeds of 0.25 x 0.9 mm.

OCCURRENCE Central Europe, W & E Asia. *G. officinalis* and the N American *G. neglecta* are cultivated in gardens. *G. neglecta* is naturalised in France.

CLASSIFICATION Cytotoxin; highly hazardous, Ib

ACTIVE INGREDIENTS The toxic compounds are tetracyclic triterpenes, such as gratiogenin, gratioside, and cucurbitacins E and I (present in concentrations of up to 0.08 % of fresh weight in leaves). The glycosides of cucurbitacins can be regarded as prodrugs which become activated after enzymatic cleavage. When the tissues are broken, a glucosidase enzyme, elaterase, is released.

UTILISATION Hedge hyssop has been used in traditional medicine to treat constipation and gout, as well as liver and skin disorders. It has also been taken as a diuretic medicine.

TOXICITY Cucurbitacins are strong cell poisons at high doses.

SYMPTOMS Strong purgative and heart weakening properties. Symptoms include nausea, salivation, vomiting, colic, bloody diarrhoea, kidney disturbance, spasms, disturbance of heart and respiration, and death (probably from respiratory arrest). Vision, especially colour vision, becomes distorted. The drug has abortifacient properties. Triterpenoids cause digitalis-like symptoms.

PHARMACOLOGICAL EFFECTS Cucurbitacins inhibit mitotic cell division and are therefore cytotoxic (see cucurbitacin E); they have antitumour activity but are too toxic to be used in medicine. Cucurbitacins activate intestinal peristalsis and cause spasms with colic.

FIRST AID Administration of sodium sulphate and medicinal charcoal; provide plenty of liquid.

Clinical therapy: Gastric lavage (possibly with 0.1 % potassium permanganate), instillation of medicinal charcoal and sodium sulphate; further detoxification (see cucurbitacin E).

Gratiola officinalis L.

family: Plantaginaceae

gratiole, herbe au pauvre (French); *Gottesgnadenkraut* (German); *graziona, fossicaria* (Italian)

Hedera helix

ivy • common ivy

Hedera helix

Hedera helix fruits

PLANTS WITH SIMILAR PROPERTIES A genus with 4–11 species (species boundaries are unclear) in Europe, the Mediterranean and E Asia; many horticultural cultivars have been developed.

PLANT CHARACTERS The common ivy is a woody climber with creeping stems that cling to objects by means of aerial climbing roots. Leaves on vegetative stems are three- to five-lobed and much smaller than the leaves of flowering stems. They are large and ovate, without distinct lobes. Inconspicuous flowers are borne in umbels and are followed by black, bitter-tasting berries.

OCCURRENCE W Asia and Europe. Ivy is widely grown as ornamental climber.

CLASSIFICATION Cell poison; moderately hazardous, II

ACTIVE INGREDIENTS Leaves and fruits contain 5–8 % bidesmosidic saponins with oleanolic acid, hederagenin, and bayogenin as aglycones, with hederasaponin C (= hederacoside C) as the main compound (7 %). Upon wounding and drying, the more toxic monodesmosidic saponins can be formed by hydrolysis. Also present are various phenolics, polyacetylenes (typically including falcarinol, falcarinone and 11-dehydrofalcarinol) and sesquiterpenes.

UTILISATION Extracts from dried leaves are used in phytomedicine to treat cough. They also exhibit antifungal, antiparasitic and molluscicidal activities. Extracts have been used to treat corns and to counteract the effects of alcohol. Extracts of the wood are used as emollients and anti-irritants in skin care products, anticellulitis products and lotions that are applied for relief of skin disorders.

TOXICITY Fruits are considered highly toxic especially for children (2–3 berries cause toxic symptoms, although serious poisoning is rare). LD_{50} of leaf extracts: rat (i.v.) 13 mg/kg; mouse (i.p.) 2.3 mg/kg; (p.o.) > 4 mg/kg.

SYMPTOMS Ivy saponins are irritant to mucosa of the GI tract. Symptoms include nausea, vomiting, headache, enhanced pulse, dizziness, delirium, shock, fever, respiratory arrest and even death. When clipping ivy, gardeners can suffer from contact allergy caused by falcarinol, known from other Araliaceae (see polyacetylenes).

PHARMACOLOGICAL EFFECTS The saponins (especially their monodesmoside derivatives) disturb membrane permeability; they are cytotoxic and haemolytic in higher concentrations. Hederin has vasoconstrictive activity.

FIRST AID Medicinal charcoal. Wash eyes; skin inflammation can be treated with a corticoid ointment. **Clinical therapy:** Gastric lavage, instillation of medicinal charcoal and sodium sulphate for further treatment see saponins.

Hedera helix L.

family: Araliaceae

lierre grimpant, lierre commun (French); *Efeu, Gewöhnlicher Efeu* (German); *edera* (Italian); *hiedra* (Spanish)

Heimia salicifolia

sinicuiche • sun-opener

Heimia salicifolia

Heimia salicifolia flowers

PLANTS WITH SIMILAR PROPERTIES A genus of 3 closely related plants in C & S America.
PLANT CHARACTERS A deciduous perennial shrub of 3 m in height. The stems are square in cross section and bear linear-lanceolate leaves 20–90 mm long. Flowers are yellow (12 mm in diameter) and have 5–7 ovate petals of 17 mm long and 10–18 stamens. The flowers are solitary but two of them emerge from each leaf axil. Fruits are small capsules with tiny seeds.
OCCURRENCE Highland of Mexico, but also S USA and C & S America.
CLASSIFICATION Neurotoxin; mind-altering; moderately hazardous, II
ACTIVE INGREDIENTS The plant accumulates a special type of quinolizidine alkaloids, such as lythrine, cryogenine, heimine, vertine, lyfoline, and nesodine. Tannins are also present (up to 15 % in leaves).
UTILISATION Sinicuiche has been used by Mexican Indians for intoxication. Leaves were crushed and fermented; the potion was said to have mind-altering properties. Traditional medicine in Mexico employs the drug against syphilis, as narcotic, diuretic and febrifuge. It is also used to treat wounds caused by thorns.
TOXICITY High doses are probably toxic. People have reported extremely unpleasant experiences after ingesting extracts from 20–30 g of leaves. The symptoms are those of general poisoning and may last up to three days.
SYMPTOMS Extracts have weak mind-altering properties. It is possible that the plant was traditionally used in mixtures with others or that special varieties with high levels of particular alkaloids were responsible for the claimed hallucinogenic effects. The plant is also known as "sun-opener" and some people have reported hallucinations where "everything seems yellow". Other symptoms reported include euphoria, chills, dizziness, deafness, acoustic hallucinations (although this claim has been questioned), disorientation, loss of coordination, difficulty to move, urge to urinate, unpleasant dreams, pain in the limbs and internal organs and headache.
PHARMACOLOGICAL EFFECTS Cryogenine has anticholinergic and spasmolytic activities. Lythrine has proven diuretic properties. Because of structural similarities with lupin alkaloids, which affect cholinergic neuroreceptors, such an interaction would be also plausible for *Heimia* alkaloids (see cytisine).
FIRST AID Induce vomiting; administer sodium sulphate, medicinal charcoal and give plenty of warm tea or fruit juice to drink. **Clinical therapy:** After intake of larger doses, gastric lavage and further treatments used for quinolizidine alkaloids (see cytisine).

Heimia salicifolia (H.B.K.) Link et Otto

family: Lythraceae

sinicuiche (French); *Sinicuiche* (German)

137

Heliotropium arborescens

heliotrope • cherry pie

Heliotropium arborescens

Heliotropium europaeum

PLANTS WITH SIMILAR PROPERTIES A genus with 250 species in warm regions, mainly Eurasia and America; only 10 in Europe, e.g. *H. europaeum*, *H. porovii*, *H. lasiocarpum* and *H. amplexicaule*. A related genus with pyrrolizidine alkaloids is *Lithospermum* with 45 species in Asia, Europe and America, including *L. officinale*, *L. purpureo-coeruleum*. *Lithodora* (with 7 European species) is a close relative and has sometimes been merged with *Lithospermum*.

PLANT CHARACTERS Perennial shrub of up to 2 m in height with pubescent, ovate to elliptic-oblong leaves. The small, purple or violet to white flowers are borne in scorpioid racemes and emit a pronounced, attractive vanilla scent (therefore popular as a garden plant).

OCCURRENCE Sandy and sunny places in Peru and Ecuador, but widely cultivated as an ornamental garden shrub.

CLASSIFICATION Liver poison, neurotoxin, mutagen; moderately hazardous, II

ACTIVE INGREDIENTS All parts of *Heliotropium* plants contain pyrrolizidine alkaloids (PAs); especially rich in PAs are roots, leaves, and flowers. The PA content of aerial parts of *H. europaeum* is about 0.9 %, of which the main compounds are indicine, acetylindicine, heliotrine and cynoglossine (open monoester alkaloids).

UTILISATION *H. arborescens* has been grown in S Europe for its scent. Some species are used in traditional medicine. *Lithospermum* species are also used medicinally, as a contraceptive (the N American *L. ruderale*) and as natural dyes (*L. arvense*). *L. canescens* is the source of a N American Indian red face-paint.

TOXICITY Contaminated flour in Afghanistan (*H. porovii*) and Tajikistan (*H. lasiocarpum*) has resulted in lethal human poisoning. PA poisoning occurs in livestock when feed is contaminated with *Heliotropium* seeds or straw.

SYMPTOMS Cynoglossine is known to cause tachycardia, dilated pupils, paralysis of the respiratory centre and cardiac arrest. PAs are cumulative poisons and can cause veno-occlusive liver disorder after long-term ingestion.

PHARMACOLOGICAL EFFECTS PAs become metabolically activated in the liver (see senecionine). They can then alkylate proteins and DNA. As a result, they cause liver damage, and after prolonged intake they are mutagenic, teratogenic and carcinogenic.

FIRST AID Give medicinal charcoal to absorb the alkaloids; administer plenty of warm black tea and sodium sulphate. Provide shock prophylaxis (keep patient at a quiet place and warm).

Clinical therapy: In case of ingestion of larger doses: gastric lavage and further treatments described under senecionine.

Heliotropium arborescens L. (= *H. peruvianum* L.)   family: Boraginaceae

héliotrope, fleur des dames (French); *Vanilleblume, Heliotrop* (German); *eliotropio, girasole del Peru* (Italian)

138

Helleborus viridis

green hellebore

Helleborus viridis

Helleborus niger

PLANTS WITH SIMILAR PROPERTIES A genus with 21 species in Europe and Asia, such as *H. niger, H. foetidus, H. lividus, H. orientalis, H. vesicarius, H. dumetorum* and *H. purpurascens*.

PLANT CHARACTERS Perennial herb of 0.2–4 m in height. Typical for *Helleborus* are rhizomes with aerial shoots that take several years to flower. Leaves are deeply lobed and the flowers are nodding, flat, with 5 green petals. Seeds are arilate to attract ants for dispersal.

OCCURRENCE W and C Europe; several species are widely cultivated as ornamental plants.

CLASSIFICATION Heart poison; extremely hazardous, Ia

ACTIVE INGREDIENTS Extremely bitter cardiac glycosides in all parts of the plant (up to 1.5 %); hellebrin as main bufadienolide, with saponins and alkaloids such as celliamine and sprintilline. Similar compounds occur in *H. niger*, which additionally accumulates the irritating ranunculin (converted after hydrolysis to protoanemonin).

UTILISATION Powders from *H. viridis* and *H. niger* were formerly used in C Europe as sneezing powder (since they are mucosal irritants) but are banned today. The drug has been used in traditional medicine to treat constipation, nausea, intestinal worms and nephritis, but was also employed as an abortifacient.

TOXICITY All parts are poisonous, with a burning taste. LD_{50} for hellebrin; cat (i.v.) 0.1 mg/kg; rat (i.v.) 21 mg/kg; mouse (i.p.) 8.4 mg/kg.

SYMPTOMS Concentrated decoctions may cause rapid death due to heart failure. Symptoms include nausea, salivation, scratching in mouth and throat, dilated pupils, burning thirst, abdominal pain and colic, gastrointestinal disturbance, drastic diarrhoea and exhaustion, with the usual respiratory and cardiac symptoms (bradycardia, arrhythmia, fibrillation, hypertension, coma, death from cardiac and respiratory arrest) associated with heart glycoside poisoning. The alkaloids cause excitation of motoric CNS centres, unrest, spasms, paralysis, bradycardia, negative inotropic effects, and death by respiratory arrest (see aconitine). The glycosides cause bleeding and blister formation on contact with mucosal tissues.

PHARMACOLOGICAL EFFECTS In general, cardiac glycosides inhibit the Na^+, K^+-ATPase, which is important for neuronal signal transduction and cellular transport processes (see cardiac glycosides). The alkaloids have similar properties as aconitine.

FIRST AID Immediate first aid treatment is indicated if plant material from *Helleborus* or isolated cardiac glycosides have been ingested. It involves the induction of vomiting; the instillation of sodium sulphate and medicinal charcoal; for clinical treatment, see aconitine and cardiac glycosides.

Helleborus viridis L.   family: Ranunculaceae

hellébore vert (French); *Grüne Nieswurz* (German); *elleboro verde* (Italian)

Heracleum mantegazzianum

giant hogweed

Heracleum mantegazzianum

Heracleum mantegazzianum with large umbels

PLANTS WITH SIMILAR PROPERTIES The genus *Heracleum* comprises 65 species in northern temperate climates. A related umbellifer with similar properties is *Peucedanum galbanum* (blister bush).

PLANT CHARACTERS Giant hogweed is an impressive perennial of up to 3.5 m tall. The red-spotted stems can be 100 mm in diameter at the base. Numerous white flowers are borne in large umbels (0.5 m across). *Peucedanum galbanum* is a robust, erect shrub of 2 m in height with a resinous smell. The large leaves are divided into diamond-shaped leaflets with serrated edges. Small, yellow flowers are arranged in large umbels and are followed by dry, flat, winged fruit.

OCCURRENCE Originally endemic to the Caucasian mountains; introduced as garden plant (since 1890), now widely spread over Europe and the USA as a neophytic weed. *P. galbanum* occurs only in fynbos areas of S Africa.

CLASSIFICATION Cytotoxin; mutagen; moderately hazardous, II

ACTIVE INGREDIENTS Toxicologically important are furanocoumarins, such as 5-methoxypsoralen and imperatorin: fruits contain 3.28 %, leaves 0.28 % and stems 0.28 % furanocoumarins. *P. galbanum* has similar furanocoumarins.

UTILISATION Extracts from *H. sphondylium* have been used medicinally and in liqueurs and other alcoholic drinks in E Europe. Tinctures of *P. galbanum* have been employed in the Cape to treat bladder and kidney problems, as well as obesity in men. Furanocoumarins are employed medicinally to treat psoriasis (see furanocoumarins).

TOXICITY Contact with the plant or plant sap results in severe blistering of the skin. Contact with bare skin should be avoided. *p*-Cymene is toxic to mammals. The oral LD_{50} in rats is reported to be 4.75 mg/kg.

SYMPTOMS Accidental skin contact with toxic furanocoumarins (e.g. when gardeners cut hogweed on a sunny day) results in itching and blistering, about 40–50 h after exposure. Sunlight is required for the blistering effect to develop, so that affected parts should be kept out of the sun. The blisters (similar to third degree burns) heal slowly and leave dark patches on the skin for several weeks (see furanocoumarins).

PHARMACOLOGICAL EFFECTS Furanocoumarins are lipophilic and can enter the skin and cells by free diffusion. They can intercalate DNA. When activated by sunlight (UV 310–340 nm), they form bivalent crosslinks with both DNA strands. This effect leads to apoptotic cell death and also to mutations.

FIRST AID Apply sterile tissue on inflamed blisters; external treatment with corticoids. After ingestion, instil medicinal charcoal; see furanocoumarins.

Heracleum mantegazzianum Sommier & Levier

family: Apiaceae

berse géante du Caucase, grand berce (French); *Riesen-Bärenklau* (German)

Hippomane mancinella

manchineel

Hippomane mancinella flowers

Hippomane mancinella fruit

PLANTS WITH SIMILAR PROPERTIES A small genus of 5 species in the New World. The famous upas tree or *ipoh* (*Antiaris toxicaria*) of the family Moraceae also has a poisonous milky latex used as arrow and ordeal poisons. This Indonesian tree is thought to poison its surroundings, so that it might be fatal to approach. The similarity is superficial, as the latex contains cardiac glycosides.

PLANT CHARACTERS A tree of 3–12 m in height. The leaves are elliptic in shape, about 25–100 x 25–50 mm, with a shiny surface. Small yellow flowers are borne in clusters. The fruits are drupes that resemble small apples; they are 30 mm in diameter, yellow when unripe and reddish and aromatic when mature. They are eaten and dispersed by fruit bats. All plant parts produce a very poisonous white latex.

OCCURRENCE *H. mancinella* originates from the West Indies (Carribean, Bahamas) and C and S America. It often forms dense groves on the coast and has been used as a windbreak; trees have been cultivated in Brazil.

CLASSIFICATION Cell poison, co-carcinogen; highly hazardous, Ib

ACTIVE INGREDIENTS The latex of manchineel contains complex mixtures of phorbol esters of the daphnane and tigliane types, such as huratoxin and mancinellin. Also present are toxic ellagitannins, such as hippomanine A and B.

UTILISATION The tree is used in traditional medicine as a strong purgative, diuretic and vermifuge, and also to treat tumours, skin disturbances, syphilis, oedema, malaria, warts and corns. The latex has been used as arrow poison.

TOXICITY If latex comes into contact with the eye, it can cause blindness. Ingestion of 30–35 g of fruits or about 20 drops of the latex can cause death in humans. The lethal dose of hippomanin A in the mouse (i.p.) is 40–60 mg/kg.

SYMPTOMS The latex has strong skin irritating properties (burning, inflammation, blister formation). Internally, extracts of *H. mancinella* cause vomiting, severe gastrointestinal inflammation, bloody diarrhoea, fever, diuresis, cystitis, paralysis and death from respiratory arrest.

PHARMACOLOGICAL EFFECTS. Phorbol esters activate protein kinase C, a key enzyme in several signalling pathways. Phorbol esters are important as co-carcinogens (see phorbol esters).

FIRST AID Give medicinal charcoal and sodium sulphate. Provide plenty of tea. Contact with eyes or skin: wash with polyethylene glycol or water; in case of skin blisters, cover with sterile cotton and locally with corticoid ointment. **Clinical therapy:** After ingestion, gastric lavage (possibly with potassium permanganate), instillation of medicinal charcoal and sodium sulphate; for further treatment see phorbol esters.

Hippomane mancinella L.   family: Euphorbiaceae

figuier vénéneux (French); *Manzinellenbaum, Manzanilla* (German)

Homeria pallida

homeria

Homeria pallida

Homeria miniata

PLANTS WITH SIMILAR PROPERTIES
A South African genus with 32 species. It is closely related to the genus *Moraea* with 120 species in tropical and southern Africa.

PLANT CHARACTERS *H. pallida* has a fibrous corm below the ground bearing a single, narrow, leathery leaf and a cluster of 1–10 star-shaped, yellow or rarely orange flowers. *Moraea polystachya* has broad leaves and large blue flowers.

OCCURRENCE *H. pallida* is widely distributed in the central parts of southern Africa. *Moraea polystachya* is widely distributed in the central interior of South Africa and northwards into Namibia. Several other poisonous species occur in tropical and southern Africa.

CLASSIFICATION Heart poison; highly hazardous, Ib

ACTIVE INGREDIENTS *Homeria* species produce bufadienolides such as 1α,2α-epoxyscillirosidin. Three cardiac glycosides (among them 16β-formyloxybovogenin A) have been isolated from *M. polystachya* and *M. graminicola*.

UTILISATION *Moraea* species have been cultivated to a limited extent. Some are non-toxic and have been used by the Khoisan people of southern Africa as important food sources.

TOXICITY *Homeria* and *Moraea* species are extremely toxic. Poisoning sporadically causes severe losses of large and small stock. The toxins are stable, so that dried leaves that are left in hay remain poisonous. Animals usually avoid these plants but may be poisoned during periods of drought or scarcity. Human poisoning with fatal outcome are on record, where children accidentally ate the corms of toxic species instead of *M. fugax* and other edible species. A dose of 700 g of flowering plants of *H. pallida* has killed a sheep within 2 hours. The LD_{50} of 1α,2α-epoxyscillirosidin (by injection) was less than 0.2 mg/kg for guinea pigs and 3.6 mg/kg for mice. The lethal dose of dried flowering *Moraea* in sheep and cattle is less than 2 g per kg, with death occurring within 1 or 2 days.

SYMPTOMS In livestock, the symptoms are typical of heart glycoside poisoning. In addition to an abnormal heartbeat, nervousness and apathetic behaviour, the animals may hang their heads and show paralysis of the hindquarters. Other symptoms include bloat and diarrhoea.

PHARMACOLOGICAL PROPERTIES Bufadienolides inhibit the Na^+, K^+-ATPase, which is one of the most important ion pumps of humans and animals (see cardiac glycosides).

FIRST AID First aid treatment is indicated if plant material from *Homeria* or *Moraea* or isolated cardiac glycosides have been ingested. Treatment involves the induction of vomiting and detoxification as outlined under cardiac glycosides.

Homeria pallida Bak.   family: Iridaceae
Homerie (German)

Hura crepitans

hura • sandbox tree

Hura crepitans female

Hura crepitans male

PLANTS WITH SIMILAR PROPERTIES A genus with 2 species in tropical America (*Hura crepitans* and *H. polyandra*). *Hippomane mancinella*, also of the Euphorbiaceae, is chemically similar.

PLANT CHARACTERS A tree with a grey, spiny stem and large, heart-shaped leaves of up to 0.3 m long. Male and female flowers are quite different (but both dark red) and are borne on the same tree. The large fruits are explosively dehiscent, expelling the seeds (7 g each) to a distance of 40 m. The plant produces a white toxic latex.

OCCURRENCE Tropical America, including the West Indies. It is widely cultivated in the tropics (India, Florida, California and the Bahamas).

CLASSIFICATION Cellular poison, co-carcinogen, skin irritant; highly hazardous, Ib

ACTIVE INGREDIENTS Several phorbol esters of the tigliane and daphnane types, such as huratoxin ($C_{34}H_{48}O_8$, MW 488), a monoester. A toxic ribosome inactivating lectin, crepitin or hurin, is also present in the seeds and latex.

UTILISATION Fruits have been used, wired together as sandboxes before blotting paper became available. As a fish poison it is 10x more potent than rotenone and was even used to kill snakes (anacondas). Leaves and latex are used topically in traditional medicine to treat skin disorders and infections, rheumatism, boils and neuralgic pains. Oral applications have been used to induce vomiting and for constipation and intestinal worms.

TOXICITY Hura latex is highly irritant, especially to mucosal tissues (mouth, throat and nose) and the eye. Only 2–3 seeds cause toxic effects.

SYMPTOMS Ingestion of latex or any other part of the plants causes severe and bloody diarrhoea and vomiting. Externally, latex causes serious skin inflammation with oedema and blister formation. If latex gets into contact with eyes (e.g. when cutting sandbox trees) it can lead to inflammation of the cornea and blindness.

PHARMACOLOGICAL EFFECTS. Phorbol esters activate protein kinase C, an important key enzyme in several signalling pathways. Phorbol esters are important as co-carcinogens (see phorbol esters). The lectin is similar to ricin and abrin and can block ribosomal protein synthesis in cells (see abrin).

FIRST AID Give medicinal charcoal and sodium sulphate. Provide plenty of tea. Contact with eyes or skin: wash with polyethylene glycol or water; in case of skin blisters, cover with sterile cotton and locally with corticoid ointment. **Clinical therapy:** Gastric lavage (possibly with potassium permanganate), instillation of medicinal charcoal and sodium sulphate. In case of contact with eyes, consult an ophthalmologist. For further treatment see phorbol esters. Check liver and kidney function after poisoning.

Hura crepitans L.   family: Euphorbiaceae

sablier blanc (French); *Sandbüchsenbaum* (German)

Hyaenanche globosa

hyena poison

Hyaenanche globosa

Hyaenanche globosa

PLANTS WITH SIMILAR PROPERTIES A monotypic genus with only one species and no close relatives.

PLANT CHARACTERS *Hyaenanche globosa* is a large shrub or small, rounded tree (up to 5 m) with leathery, dark green leaves arranged in whorls of four along the stems. Male and female flowers occur on separate trees. They are reddish in colour and inconspicuous, since there are no petals. The male flowers occur in many-flowered clusters, while the female flowers are single or in groups of up to three. The large fruit capsules have three or four valves that split open when ripe. Each fruit has six shiny black seeds.

OCCURRENCE *H. globosa* has an exceptionally small natural distribution area and is endemic to the *Gifberg* ("Poison Mountain"), an isolated, flat-topped mountain in the western part of S Africa.

CLASSIFICATION Neurotoxin; highly hazardous, Ib

ACTIVE INGREDIENTS The poisonous compounds in Hyaenanche are toxic sesquiterpene lactones. Several compounds have been isolated and described, of which tutin (also known as hyaenanchine) appears to be the main constituent. Other lactones present include mellitoxin, urushiol III and isodihydrohyaenanchine.

UTILISATION Fruits of *Hyaenanche globosa* were used in former times as a source of a potent poison to destroy hyenas and other vermin. The fruits were placed in sheep carcasses. The name of the genus is well chosen, as *Hyaenanche* is the Greek word for hyena poison. Powdered seeds were used by the San people as arrow poison.

TOXICITY There are no recent cases of human or livestock poisoning. The fruit and seeds are known to be extremely toxic. The dried rind of an immature fruit was fed to a rabbit; 10 g caused convulsions and death within a few minutes.

SYMPTOMS Tutin, the main toxic compound, causes convulsions, delirium and coma in humans (see helenalin).

PHARMACOLOGICAL EFFECTS The sesquiterpene lactones are reactive substances that can form covalent bonds with proteins and DNA. They appear to affect important protein targets in the brain (see helenalib).

FIRST AID Treat with sodium sulphate and medicinal charcoal. Allow patient to drink plenty of fluids, such as black tea. **Clinical therapy:** Gastric lavage (possibly with potassium permanganate), instillation of medicinal charcoal and sodium sulphate, electrolyte substitution, control of acidosis with sodium bicarbonate. In case of spasms give diazepam; for shock plasma expander; in case of severe poisoning, intubation and oxygen respiration may be necessary.

Hyaenanche globosa (Gaertn.) Lamb & Vahl (= *Toxicodendron capense*)   family: Picrodendraceae or Euphorbiaceae

Hyaenanche (German)

Hyoscyamus niger

henbane

Hyoscyamus niger

Hyoscyamus aureus

Hyoscyamus albus

PLANTS WITH SIMILAR PROPERTIES A genus of 15 species, including *H. albus*, *H. muticus* and *H. aureus*. Members of the genera *Atropa*, *Brugmansia*, *Datura*, *Duboisia*, *Mandragora* and *Scopolia* are toxicologically similar.

PLANT CHARACTERS An annual or biennial herb (up to 0.8 m) with soft, hairy stems and soft, lobed and sparsely hairy leaves. The petals are greyish yellow with dark purple veins. Henbane is characterised by a disgusting smell.

OCCURRENCE *Hyoscyamus* species occur in W Europe, N Africa and in SW and C Asia. Several species are common weeds of disturbed places; *H. niger* is naturalised in N America and Australia.

CLASSIFICATION Neurotoxin, mind-altering; extremely hazardous, Ia

ACTIVE INGREDIENTS The tropane alkaloids hyoscyamine (the major compound), scopolamine (up to 40–60 % of total alkaloids) and several minor alkaloids. Alkaloids concentrations: 0.04–0.17 % DW in leaves, up to 0.3 % in the seeds.

UTILISATION Henbane is a legendary plant with an interesting history of applications, including narcotic uses during surgery and hallucinogenic uses in witchcraft. It was known to induce mania. Since ancient times, henbane was administered for relief of pain, toothache and nervous disorders. Leaves were smoked to treat asthma. Henbane extracts were once used in beer brewing: the bitter tropane alkaloids were a substitute for hops and they also enhanced the effects of the ethanol. In modern medicine, henbane leaf or henbane alkaloids (isolated hyoscyamine and atropine, its racemate) are used in ophthalmology and to treat spasms of the gastrointestinal tract.

TOXICITY Cases of accidental or deliberate poisoning are rare today, except when henbane is taken as a mind-altering drug ("dream tea"). In antiquity and during medieval times, henbane was a common means of murder and suicide. About 15 seeds can be lethal in children.

SYMPTOMS At low concentrations, the alkaloids have depressant and sedative effects, but high doses lead to hallucinations, euphoria, confusion, insomnia, tantrum, unconsciousness and death from respiratory arrest. Typical symptoms include reddening of the face, dry mouth, dilated pupils and enhanced pulse. Scopolamine is favoured as a hallucinogen (see hyoscyamine).

PHARMACOLOGICAL EFFECTS Hyoscyamine and scopolamine are antagonists at the muscarinic acetylcholine receptor and therefore work as parasympatholytics (see hyoscyamine).

FIRST AID Induce vomiting; treat with sodium sulphate and medicinal charcoal. **Clinical therapy:** Immediate gastric lavage. Physostigmine can be used as an antidote; for further detoxification see hyoscyamine.

Hyoscyamus niger L.

family: Solanaceae

jusquiame noire (French); *Schwarzes Bilsenkraut* (German); *giusquiamo nero* (Italian); *veleño nero* (Spanish)

Hypericum perforatum

St John's wort

Hypericum perforatum flowers

Hypericum perforatum

PLANTS WITH SIMILAR PROPERTIES A large genus with 370 species in Europe and the Old World tropics.

PLANT CHARACTERS St John's wort is a perennial herb or shrublet (up to 0.6 m), with erect, longitudinally ridged branches bearing small, gland-dotted leaves. Oil glands are visible as translucent dots when holding a leaf against the light, hence the name "*perforatum*". The flowers are bright yellow and are dotted with small blackish glands. They have five somewhat reflexed petals and numerous erect stamens.

OCCURRENCE Europe and Asia (a naturalised weed in N America, S Africa and Australia). Several species are grown as garden plants.

CLASSIFICATION Cell poison; phototoxin; antidepressant; slightly hazardous, III

ACTIVE INGREDIENTS The main compounds in *Hypericum* are a phloroglucinol derivative known as hyperforin and a naphthodianthrone called hypericin (which gives a red colour to isolated oil). Also present are various phenolic compounds and several terpenoids.

UTILISATION St John's wort has been used for healing since ancient times. In modern phytomedicine it is mainly recommended for mild depression, mood disturbances and anxiety (supported by clinical studies). A red oil produced from fresh plant is popular to treat wounds and burns. *Hypericum* is taken internally as vermifuge and to treat inflammation of the GI tract.

TOXICITY *Hypericum* species cause photosensitivity in animals when more than 3–5 g/kg have been consumed. The effect is ascribed to naphthodianthranones (mainly hypericin). *Hypericum* poisoning (hypericism) in livestock has been reported in several parts of the world.

SYMPTOMS After exposure of skin to hypericin-containing extracts, erythema, sores and necrosis can occur, so that the skin turns leathery and parchment-like. In addition, the lips and eyelids become immobilised. The condition may also be caused by ingestion of buckwheat (*Fagopyrum esculentum*). Further symptoms of *Hypericum* poisoning include loss of appetite, peeling of the skin, hair loss, convulsions and coma.

PHARMACOLOGICAL EFFECTS The toxic principle is hypericin, a photodynamic agent which is activated by sunlight so that it can bind to proteins and inactivate them. Black or brown animals are not affected, only white or partially white ones (non-pigmented skin). An inhibition of neurotransmitter re-uptake provides a plausible hypothesis for the antidepressant activity. Extracts can induce CYP enzymes in the liver.

FIRST AID After overdosing, provide medicinal charcoal and sodium sulphate; keep patients away from sunlight for 1–2 weeks.

Hypericum perforatum L.

family: Hypericaceae or Clusiaceae

millepertius perforé (French); *Echtes Johanniskraut* (German); *iperico, erba di San Giovanni* (Italian)

Ilex aquifolium

holly

Ilex aquifolium leaves and fruits

PLANTS WITH SIMILAR PROPERTIES A large genus with over 400 species on all continents, among them *I. vomitoria*, *I. cassine*, *I. guayusa* and *I. paraguariensis* (maté or Paraguay tea).

PLANT CHARACTERS A shrub or small tree (3–15 m) with dark green, leathery leaves having spine-pointed teeth along the margins. The white flowers are 6 mm in diameter: male and female flowers occur on separate trees. The fruits are bright scarlet-red berries, 7–12 mm in diameter.

OCCURRENCE *I. aquifolium* is indigenous to Europe, W Asia and N Africa but naturalised in several countries. Numerous cultivars, including variegated forms, are cultivated in gardens.

CLASSIFICATION Cell poison; moderately to highly hazardous, Ib–II

ACTIVE INGREDIENTS Triterpenes and their esters: α-amyrin, β-amyrin, baurenol, erythrodiol, oleanolic acid, ursolic acid, uvaol, ursolaldehyde and oleanolaldehyde. Also present are several steroids and a nitrile (menisdaurin; 0.28 % in fruits). *I. cassine* and *I. vomitoria* accumulate up to 0.3 % caffeine, *I. paraguariensis* 1.6 % caffeine and 0.45 % theobromine in their leaves. *I. guayusa* accumulates 3–4 % caffeine.

UTILISATION Holly is used in traditional medicine to treat fever, rheumatism, gout, chronic bronchitis and constipation. In Europe, twigs with berries are famous as Christmas decorations. The N American *I. vomitoria* and *I. cassine* have been used by Indians to make "black drink", which had stimulating properties. Together with smoking tobacco, this black tea induced bravery for going into battle. The S American *I. paraguariensis* is a famous source of maté tea and contains stimulating purine alkaloids and tannins.

TOXICITY Holly berries are commonly found in gardens and used as decorations, so that children or domestic animals may be poisoned. Most cases are harmless. About 20–30 fruits can be lethal for humans, while more than 5 berries can cause poisoning symptoms in children.

SYMPTOMS Symptoms include sleepiness, vomiting, gastrointestinal disturbance, gastritis, abdominal pain, diarrhoea, arrhythmia, paralysis and kidney damage, and even death.

PHARMACOLOGICAL EFFECTS The triterpenes have cytotoxic properties but the mode of action of holly toxins has not been elucidated. For caffeine effects, see caffeine.

FIRST AID Provide plenty of fluids to drink; give medicinal charcoal and sodium sulphate; shock prophylaxis. **Clinical therapy:** Gastric lavage (>10 berries), electrolyte substitution, control of acidosis with sodium bicarbonate; in case of spasms, diazepam; check kidney function. In case of respiratory arrest, provide intubation and artificial respiration.

Ilex aquifolium L.   family: Aquifoliaceae

houx (French); *Stechpalme, Ilex* (German); *agrifolio* (Italian)

Illicium anisatum

Japanese anise • shikimi

Illicium anisatum leaves

Illicium anisatum fruit

Illicium anisatum flowers

Illicium verum dry fruits

PLANTS WITH SIMILAR PROPERTIES A genus with 42 species in E Asia and N America, including *I. floridanum* (purple anise) and *I. verum* (Chinese anise).

PLANT CHARACTERS A highly aromatic shrub or small tree (up to 8 m) with ovate to lanceolate, bright green leaves, 40–120 mm long. The attractive flowers are yellowish green to white, with about 30 slender, petaloid segments. The characteristic star-shaped fruit is made up of several free carpels that each form a small, single-seeded capsule.

OCCURRENCE Japan, Korea and Taiwan, where it is commonly cultivated. *I. verum* occurs naturally in China and Vietnam, while *I. floridanum* is indigenous to SE USA.

CLASSIFICATION Neurotoxin; highly hazardous, Ib

ACTIVE INGREDIENTS Fruits contain sesquiterpene dilactones known as anisatin and neoanisatin, and an essential oil (1 %) with monoterpenes and traces of anethol. Shikimic acid and the shikimic acid pathway is named after *shikimi*, the Japanese common name for the plant. Shikimic acid is an important biochemical intermediate in plants and microorganisms.

UTILISATION The star-shaped fruits of *I. verum* are used as a spice and to flavour liqueurs. It is a source of anethol. In the pharmaceutical industry,

shikimic acid from star anise is used as a precursor for the production of Tamiflu (an inhibitor of neuraminidase in influenza viruses). A first insecticidal fumigant has been made from *Illicium* oil in ancient China. The toxic seeds of *I. anisatum* have been used to kill fish, and topically as a therapeutic agent. Shikimi is traditionally planted at Buddhist shrines and still today on graves and in graveyards in Japan.

TOXICITY *I. anisatum* has been used as an adulterant of *I. verum* and has caused serious poisoning and even death. 1.5 g fruits can cause toxic effects. LD_{50} anisatin: mouse 0.7–1.0 mg/kg. Livestock poisoning has been reported.

SYMPTOMS Ingestion of shikimi fruits or anisatin cause vomiting, diarrhoea, salivation and even death. Typical are tonic and clonic spasms, tremor, epileptic seizures with respiratory arrest and arrhythmia.

PHARMACOLOGICAL EFFECTS Anisatin has an activity similar to that of picrotoxinin from *Anamirta paniculata* (Menispermaceae). Both compounds act as non-competitive antagonist of GABA and bind to the $GABA_A$ receptor.

FIRST AID After ingestion, instil medicinal charcoal, sodium sulphate. **Clinical therapy** After overdoses: gastric lavage with 0.1 % potassium permanganate. Provide artificial respiration if necessary.

Illicium anisatum L. (= *I. japonicum* Sieb., *I. religiosum* Sieb. et Zucc.)   family: Illiciaceae or Schisandraceae

badiane du Japon (French); *Shikimibaum, Japanischer Sternanis* (German)

Inocybe fastigiata

split fibrecap

ıocybe fastigiata

Inocybe fastigiata

FUNGI WITH SIMILAR PROPERTIES Mushrooms with muscarine include *Inocybe patouillar-ii, I. napiges, I. geophylla, Clitocybe rivulosa* and specially the deadly *C. dealbata*. The original source of muscarine, *Amanita muscaria*, accumulates this substance only as a minor alkaloid.

CHARACTERS The cap is up to 70 mm in diameter; the shape is conical or nearly so, but becomes mbonate; it splits radially when expanding and straw-coloured or pale brown with long silky adial fibres. The stem is up to 90 mm long and p to 12 mm in diameter, pale brown at the base nd whitish at the apex. There is no basal swelling ulb). The spore colour is light to dark brown.

OCCURRENCE Europe and North America. he typical habitat is deciduous woodlands, especially beech woods, where the split fibrecap can e found growing on the ground or on wood. It is very common fungus and may grow alone, scattered or gregariously. The season is early summer late autumn.

CLASSIFICATION Mind-altering; highly hazrdous, Ib

ACTIVE INGREDIENTS Muscarine is a main netabolite and responsible for the mind-altering nd toxic effects. It is found mainly in *Inocybe* nd *Clitocybe* species.

UTILISATION These fungi are small and not onsidered to be edible – some are indeed poisonous. The split fibrecap has been used as a recreational drug and people have experimented with this and other muscarine-containing mushrooms.

TOXICITY LD_{50} mouse : 0.23 mg/kg i.v.; death after 3–10 minutes. Muscarine does not pass the blood-brain barrier.

SYMPTOMS Symptoms start 15 to 20 minutes after ingestion of muscarine-rich mushrooms. Symptoms include extensive perspiration, salivation, tear production, vomiting, abdominal colic, diarrhoea, severe nausea, small pupils and visual disturbance, decrease of blood pressure, bradycardia, bronchial spasms and dyspnoea, with death from cardiac arrest.

PHARMACOLOGICAL EFFECTS Muscarine is an agonist at muscarinic acetylcholine receptor and thus functions as a direct parasympathomimetic. It activates postganglionary parasympathic receptors in the same way as acetylcholine. Since muscarine cannot be cleaved by cholinesterase, it provokes a much longer excitation.

FIRST AID Immediately inject atropine as an antidote, since it can replace muscarine from mAChR. Initial dose: for adults 1–2 mg i.v., for children 0.05 mg/kg. If the secretions stop, then the atropine was effective. Untreated muscarine poisoning can be deadly but with medication, chances of recovery are good.

Inocybe fastigiata (Schaeff. ex Fr.) Quél. [= *I. rimosa* (Bull.) P.Kumm.]

family: Cortinariaceae

Inocybe fastigié (French); *Kegeliger Risspilz* (German)

Ipomoea tricolor

morning glory

Ipomoea tricolor

Ipomoea tricolor seeds

PLANTS WITH SIMILAR PROPERTIES
A very large genus of more than 650 species in warm and temperate climates, including species with ergot alkaloids such as *Ipomoea tricolor, I. carnea, I. hederacea, I. muricata, I. nil* and *I. pes-caprae*.
PLANT CHARACTERS *Ipomoea* species are common climbers (vines) with twining stems and large purple, pink or white flowers that last only for one morning. Those of *I. tricolor* are distinctive in their attractive blue, white and yellow colour pattern. *I. purpurea* is commonly cultivated and has sparsely hairy leaves and large flowers that are usually dark blue or purple. *Ipomoea* seeds are oblong and dark-green or black.
OCCURRENCE Mexico and C. America. The plant is widely cultivated in gardens. Cultivars such as "Heavenly Blue", "Pearly Gates", "Summer Skies", "Blue Star" and "Wedding Bells" all produce ergot alkaloids in their seeds.
CLASSIFICATION Neurotoxin, mind-altering; highly hazardous, Ib
ACTIVE INGREDIENTS Ergot alkaloids, such as ergometrine; chanoclavine, elymoclavine, lysergic acid amide and other derivatives of lysergic acid. *I. purpurea* does not sequester ergot alkaloids in its seeds. The alkaloids are apparently produced by a fungal symbiont.
UTILISATION Mexican Indians used ololiuqui (the origin is unresolved; it might have been *Ipo-*

moea or *Turbina corymbosa*) for ritual and oracl purposes. Seeds were collected, squashed and in bibed in pulque and tepache.
TOXICITY *Ipomoea* species are not particular poisonous but their seeds contain toxic alkaloid and cases of fatal poisoning have been recorded Ingestion of about 164–214 g/kg seeds were leth: in rats, 1–2 g/kg seeds induce hallucinations i humans.
SYMPTOMS Seeds and seed extracts are ha lucinogenic, similar to LSD but weaker. In th hypnotic state, Indians are said to have talke to dead relatives and gave away secrets. Furthe symptoms from seed ingestion include nause: vomiting and weakness; high doses cause deat from respiratory arrest (see ergot alkaloids).
PHARMACOLOGICAL EFFECTS Like othe indole alkaloids of the Convolvulaceae, ergin and lysergic acid derivates are powerful halluc nogens, since they modulate the activity of sero onin, dopamine and noradrenaline receptors (se ergot alkaloids).
FIRST AID Administration of sodium sulphate and give medicinal charcoal. **Clinical therapy** Gastric lavage (possibly with potassium perman ganate), instillation of medicinal charcoal an sodium sulphate, electrolyte substitution; chec acidosis. Intubation and oxygen respiration ma be necessary (see ergot alkaloids).

Ipomoea tricolor Cav. (= *I. violacea* auct. non L.)   family: Convolvulacea

ipomée (French); *Prachtwinde* (German)

Jatropha curcas

physic nut

Jatropha curcas

Jatropha curcas fruits

PLANTS WITH SIMILAR PROPERTIES A genus of 175 species in warm parts of Africa and America, including *J. aconitifolia, J. cuneata, multifida, J. podagrica* and *J. spathulata*.

PLANT CHARACTERS A small tree (up to 6 m) with large, lobed leaves on long stalks and small, hairy, yellowish flowers. Male and female flowers occur on the same plant. The fruits are large capsules that split open to release three dark brown and black seeds or nuts of about 20 mm long.

OCCURRENCE Tropical America. It is grown in India and Africa as a source of biodiesel.

CLASSIFICATION Cellular poison, co-carcinogen, skin irritant; highly hazardous, Ib

ACTIVE INGREDIENTS The seeds of *J. curcas* contain toxic lectins known as curcin I and II (see ricin). Curcin is water-soluble and is therefore not present in the extracted seed oil. It can be deactivated by heat. The laxative properties of the seed oil are due to various diterpenoids of the tigliane (phorbol) type. They include curcuson A and C, 12-desoxy-16-hydroxyphorbol-13,16-diester and the fatty acid curcanoleic acid. The ricinoleic acid found in castor oil and the crotonoleic acid of croton oil are similar to curcanoleic acid.

UTILISATION *J. curcas* is grown in C America, Africa and India as a boundary hedge and for soil melioration. The seed oil is used for making candles and soap and especially for the production of biodiesel. A non-toxic variety (free from phorbol esters) is used in Mexico for cooking oil.

TOXICITY The tasty nuts are sometimes eaten by children, causing serious and even lethal poisoning. About 5 to 20 seeds are sufficient to cause toxic effects. The lethal dose of seed oil in rats (p.o.) is 9 ml/kg.

SYMPTOMS The latex is highly irritant and causes blistering of skin and destruction of mucous membranes. Ingestion of two seeds results in purgation, while larger amounts cause vomiting, abdominal pain, severe diarrhoea, delirium, vertigo, impairment of vision and memory, as well as muscular twitching. Death is caused by respiratory arrest.

PHARMACOLOGICAL EFFECTS. The toxicity of seeds is due to the concerted action of both phorbol esters and the lectin. Phorbol esters activate protein kinase C, which is a key enzyme in several signalling pathways. Curcin has similar ribosome-inactivating activities to ricin from *Ricinus communis* (see abrin).

FIRST AID Give medicinal charcoal and sodium sulphate. Provide plenty of tea. Contact with eyes or skin: wash with polyethylene glycol or water; in case of skin blisters, cover with sterile cotton and locally with a corticoid ointment. **Clinical therapy:** After ingestion, gastric lavage; for further treatment see phorbol esters.

Jatropha curcas L. family: Euphorbiaceae

pignon d'Inde (French); *Purgiernuss* (German); *giatrofa catartica* (Italian)

Juniperus sabina

savin

Juniperus sabina cones

Juniperus virginiana

PLANTS WITH SIMILAR PROPERTIES A genus of 50 species in the N hemispere and Africa. *J. virginiana* (Virginia cedar) and *J. horizontalis* from N America have similar toxic properties.

PLANT CHARACTERS Evergreen gymnosperm shrub or tree (4 m high) with adult leaves modified to scales. The fleshy cones are 3–7 mm in diameter and dark blue, bloomed white. Aerial parts produce an unpleasant smell when crushed.

OCCURRENCE *J. sabina* is native to S and C Europe and W Asia; *J. virginiana* is from N America. Both species are widely cultivated.

CLASSIFICATION Cell poison; extremely hazardous, Ia

ACTIVE INGREDIENTS All plant parts contain an essential oil (3–5 %), consisting of sabinene (20 %), sabinylacetate (40 %), thujone and several other monoterpenes. Plants produce the cytotoxic podophyllotoxin, a lignan (see *Podophyllum*). The oil of *J. virginiana* is also rich in sabinene.

UTILISATION Savin has a long history as an abortifacient plant (and as an insecticide). Topically, the oil was used to treat warts and condylomata. Smoke from *J. recurva* has been inhaled by shamans in E Asia to achieve intoxication.

TOXICITY Ingestion of 6 drops of the oil or 5–20 g of the twigs can be fatal in adults.

SYMPTOMS Ingestion causes nausea, vomiting, excitation, arrhythmia, tachycardia, convulsions, abdominal pain, respiratory arrest, serious kidney and liver damage, severe inner bleeding and haematuria, uterus contractions and bleeding resulting in abortion. Death occurs from central respiratory paralysis and deep unconsciousness after 10 h or several days. On the skin, the oil causes blister formation and deep necrotic scars.

PHARMACOLOGICAL EFFECTS The terpenoids, such as sabinene, probably interfere with membrane permeability. Sabinene is a reactive monoterpene with a highly reactive cyclopropane ring and an exocyclic methylene group which can form covalent bonds with SH-groups of proteins. The observed cytotoxic and neurotoxic effects are therefore plausible.

FIRST AID Give medicinal charcoal and sodium sulphate. Provide plenty of tea. Contact with eyes or skin: wash with polyethylene glycol or water; in case of skin blisters, cover with sterile cotton and locally with a corticoid ointment. **Clinical therapy:** After ingestion, instil 200 ml liquid paraffin to absorb the toxin, then gastric lavage (possibly with potassium permanganate), instillation of medicinal charcoal and sodium sulphate; control of acidosis with sodium bicarbonate; in case of spasms, and colic give diazepam and atropine; monitor kidney and liver function. In case of respiratory arrest, provide intubation and artificial respiration.

Juniperus sabina L.

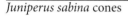

sabine, genévrier sabine (French); *Sadebaum* (German); *Sabina* (Italian)

family: Cupressaceae

Kalanchoe lanceolata

lance-leaved kalanchoe

Kalanchoe lanceolata

Kalanchoe delagoensis flowers

PLANTS WITH SIMILAR PROPERTIES A genus of 125 species in the Old World tropics, mainly Africa and Madagascar, e.g. *K. blossfeldiana, K. flammea, K. daigremontiana* (= *Bryophyllum d.*) *K. delagoensis* (= *Bryophyllum delagoense*) and *K. pinnata*.

PLANT CHARACTERS *K. lanceolata* is an erect annual or short-lived perennial with bright green, succulent leaves and yellow to orange flowers. *Kalanchoe* species have only four petals and four fruit segments; the similar-looking *Cotyledon* and *Tylecodon* have five.

OCCURRENCE *K. lanceolata* occurs in southern Africa. Many *Kalanchoe* species are grown in gardens.

CLASSIFICATION Heart poison; moderately hazardous, II

ACTIVE INGREDIENTS The main bufadienolides in *K. lanceolata* are lanceotoxin A, lanceotoxin B and hellebrigenin. No bufadienolides are present in *K. blossfeldiana*.

UTILISATION *K. blossfeldiana* and *K. daigremontiana* are only two examples of the numerous species grown world-wide as ornamental plants.

TOXICITY Several *Kalanchoe* species, as well as other members of the Crassulaceae, particularly *Tylecodon* and *Cotyledon*, are well known for causing a poisoning syndrome known in S Africa as *krimpsiekte*. Under field conditions, goats are most commonly poisoned, but sheep, cattle, and sometimes even horses are affected. An interesting phenomenon is secondary *krimpsiekte*, where people and dogs are poisoned by eating meat (even cooked meat) of an animal that has died from *krimpsiekte*. Animals are usually poisoned during droughts when grazing deteriorates. *K. rotundifolia* is very poisonous – only 300 g of flowers can kill a sheep within 36 hours. The lanceotoxins A and B have LD_{50} values of respectively 0.2 and 0.1 mg/kg when injected into guinea pigs.

SYMPTOMS Symptoms of *Kalanchoe* poisoning are very different from those normally seen with cardiac glycoside poisoning: exhaustion, paralysis of the head and neck, convulsions, respiratory paralysis and finally death. Poisoned animals may sometimes lie paralysed but fully conscious on their sides for several weeks before they die. Acute poisoning results in symptoms similar to those of other bufadienolides, while chronic and cumulative poisoning produce symptoms of *krimpsiekte* (see *Cotyledon orbiculata*).

PHARMACOLOGICAL EFFECTS Bufadienolides selectively inhibit Na+, K+-ATPase and are therefore general poisons (see cardiac glycosides).

FIRST AID First aid treatment is indicated if plant material from *Kalanchoe* or isolated cardiac glycosides have been ingested. Treatment involves supply of medicinal charcoal, and detoxification as outlined under cardiac glycosides.

Kalanchoe lanceolata (Forssk.) Pers.

Kalanchoe (French); *Kalanchoe* (German)

family: Crassulaceae

Laburnum anagyroides

golden chain • golden rain

Laburnum anagyroides flowers

Laburnum anagyroides pods and seeds

PLANTS WITH SIMILAR PROPERTIES A genus with 2 species in Europe, including *L. alpinum* and a hybrid *L.* x *watereri*. *Petteria ramentacea* (a tree from the Balkans) is chemically similar.

PLANT CHARACTERS Long-lived trees of up to 7 (17) m in height, with photosynthetically active stems, large trifoliolate clover-like leaves, bright yellow flowers in hanging racemes and oblong, flat, slightly sickle-shaped pods that turn brown when they ripen. Seeds are shiny black.

OCCURRENCE Central and SE Europe; widely grown as ornamental plant in parks and gardens.

CLASSIFICATION Neurotoxin; highly hazardous, Ib

ACTIVE INGREDIENTS Cytisine, *N*-methylcytisine, anagyrine, and other quinolizidine alkaloids in all parts of the plant; seeds up to 3 % DW, leaves up to 0.3 % DW. *L. alpinum* has high levels of ammodendrine in leaves. Also present are traces of simple pyrrolizidine alkaloids.

UTILISATION The timber is hard and has been used as ebony substitute, e.g. in inlay work.

TOXICITY Golden rain is often grown in parks, sometimes close to children's playgrounds. Children can ingest the attractive seeds, or may chew on twigs and sweet-tasting roots. In Germany, golden rain is a common cause for consultations in poison centres and has been reported more than 4 000 times in 15 years. 3–4 pods (equivalent to 15–20 seeds) are lethal for children. 0.5 g seeds/kg kill a horse, 6 g/kg chickens and doves Cytisine: LD_{50} mouse: 101 mg/kg p.o.; cat: 3 mg/kg p.o.; dog 4 mg/kg p.o. (see cytisine).

SYMPTOMS The first effects of cytisine poisoning set in after 0.25–1 h and include burning of the mouth and throat, intensive salivation, thirst and nausea, followed by long hours of sometimes bloody vomiting, cold sweat, mydriasis, visual disturbance, vertigo, excitation and confusion (with hallucinations, delirium), anxiety, tachycardia, trembling muscles, spasms and collapse. Symptoms are weaker in heavy tobacco consumers. Overdosing can lead to a strychnine-type of paralysis, convulsions and death from respiratory or circulatory arrest after 1–2 days while the heart is still beating. Death is a rare event when first aid is carried out.

PHARMACOLOGICAL EFFECTS Cytisine is a strong agonist at nicotinic acetylcholine receptor and thus affects the parasympathetic and sympathetic nervous system (see cytisine). Anagyrine and ammodendrine are teratogenic and cause malformation in animals.

FIRST AID Induce vomiting; administration of sodium sulphate, medicinal charcoal, and plenty of warm tea or fruit juice. **Clinical therapy:** Gastric lavage with potassium permanganate; for further treatments see cytisine.

Laburnum anagyroides (L.) Medikus

family: Fabaceae

aubour, cytise faux ébénier (French); *Goldregen* (German); *avorniello* (Italian)

Lactuca virosa

bitter lettuce • opium lettuce

Lactuca virosa

Lactuca virosa flower head

PLANTS WITH SIMILAR PROPERTIES A cosmopolitan genus of 75 species, among them lettuce (*L. sativa*) and prickly lettuce (*L. serriola*) subfamily Cichorioideae).

PLANT CHARACTERS An erect and hairless biennial, up to 2 m tall. Stems and leaves are purple-flushed and rich in milky latex. Leaves are partly pinnatifid with fine spines along the edges. The inflorescence is a loose panicle with branches at acute angles to main stem. Flower heads are 1–13 mm in diameter, with 7–12 yellow florets. The achenes are purple-black without bristles but with a white pappus of equal simple hairs.

OCCURRENCE Central, southern Europe and N Africa. Plants typically grow in dry places.

CLASSIFICATION Cellular poison, mind-altering; highly hazardous, Ib

ACTIVE INGREDIENTS Bitter sesquiterpene lactones, especially lactucin and lactucopicrin, in the latex of aerial parts (3.5 %). Also present are lacquinelin and 8-desoxy-11,13-dihydrolactucin in root latex and the triterpenes α-lactucerol (= taraxasterol) and β-lactucerol in leaves.

UTILISATION Lactucarium (the dried latex) has sedating, narcotic and hypnotic properties. It has been used in traditional medicine to treat various forms of cough and bronchitis, pains, nymphomania, insomnia, unrest and excitation. When lactucarium is smoked, it produces euphoria and happiness, accompanied by sedating and analgesic effects (hence the name "opium lettuce").

TOXICITY Poisoning may result from confusion with *L. sativa* or from overdosing with lactucarium. Lactucin and lactupicrin also occur in small amounts in other members of the genus *Lactuca,* including lettuce and dandelion (*Taraxacum officinale*), where they contribute to the bitter taste and to some allergic reactions. The lethal dose in dogs: 2 g; 0.5–06 g/kg s.c. in mice. Livestock poisoning has been reported.

SYMPTOMS Symptoms include perspiration, enhanced breathing and heart palpitations, dilated pupils, itching, vertigo, headache, distorted vision, diuresis, sleepiness and even death. The sesquiterpene lactones have allergenic properties.

PHARMACOLOGICAL EFFECTS The sesquiterpene lactones are lipophilic compounds that can pass the blood-brain barrier; many sesquiterpene lactones are chemically highly reactive compounds that can form covalent bonds with proteins, such as ion channels, receptors and enzymes (see helenalin). Root extracts have analgesic properties in mice.

FIRST AID Give medicinal charcoal and sodium sulphate. **Clinical therapy:** Gastric lavage, medicinal charcoal and sodium sulphate, in case of spasms provide diazepam or physostigmine to reduce them; oxygen respiration may be needed (see helenalin).

Lactuca virosa L.

family: Asteraceae

laitue vireuse (French); *Gift-Lattich* (German); *lattuga venenosa* (Italian)

Lantana camara

lantana

Lantana camara fruits

Lantana camara flowers

PLANTS WITH SIMILAR PROPERTIES A genus of 150 species in tropical America and Africa. Cultivated species include *L. montevidensis*, *L. rugulosa*, *L. tiliifolia* and *L. trifolia*. Several other members of the Verbenaceae are grown in gardens for their conspicuous fruits, including *Callicarpa bodinieri* and *Clerodendron trichotomum*.

PLANT CHARACTERS A perennial shrub of up to 2 m height with square, sometimes thorny stems. Small, tubular florets are borne in hemispherical heads; they usually open yellow and turn to orange when fertilised, but several colour forms have been developed as garden plants, including shades of yellow, salmon, orange, pink and red. The fruits are clusters of small drupes; green when unripe, blue-black when mature.

OCCURRENCE Lantana originates from tropical America. It grows as an aggressive weed (being dispersed by birds) and is presently widely distributed around the world (including S Europe, India, SE Asia, Australia, USA, Hawaii, New Caledonia and parts of S and E Africa).

CLASSIFICATION Cell poison; moderately hazardous, II

ACTIVE INGREDIENTS Toxic pentacyclic triterpenoid esters, including lantadene A (angelate ester) and lantadene B (dimethylacrylate), whose concentrations vary with the cultivars. Also present are verbascoside (an iridoid glycoside), an essential oil with citral and sesquiterpenes.

UTILISATION Modern cultivars are grown as ornamental garden shrubs and pot plants.

TOXICITY Lantana is a common cause of livestock poisoning in tropical countries, especially India and Australia. Human poisoning can occur when children ingest green fruits. A case is known where several children were poisoned and one of them (who was not treated with gastric lavage) died from pulmonary oedema.

SYMPTOMS In children, mydriasis as in *Atropa* as well as vomiting, diarrhoea, ataxia, cyanosis, dyspnoea. Symptoms of lantana poisoning in animals include vomiting, constipation, photosensitivity, anorexia and icterus. Kidney, gall bladder and liver damage are typical.

PHARMACOLOGICAL EFFECTS Inhibition of protein kinase C has been reported. Lantadene is structurally similar to cholesterol and there is evidence that it can penetrate biomembranes in a similar fashion. The lipophilic part will orientate parallel to the fatty acid residues of phospholipids and to cholesterol, whereas its polar moiety will be exposed to the extracellular space. This arrangement will influence membrane fluidity.

FIRST AID Instil medicinal charcoal and sodium sulphate. In case of higher doses, carry out gastric lavage with 0.1 % potassium permanganate and give medicinal charcoal.

Lantana camara L.

family: Verbenaceae

lantanier (French); *Wandelröschen* (German)

Letharia vulpina

wolf lichen

Letharia vulpina

PLANTS WITH SIMILAR PROPERTIES A genus with two species of filamentous lichens that mostly grow on trees; the second species is *L. columbiana* from N America. Lichens are unusual because they are actually a combination of two separate organisms that live in close symbiosis: algal cells (usually green algae) living amongst fungal filaments.

PLANT CHARACTERS A filamentous, heavily branched lichen that varies from 10 mm to 120 mm in length. The branches are usually hanging and a brilliant yellow green to duller yellow ochre colour. The soredia (clonal symbiotic propagules, made from a weft of fungal hyphae surrounding a hollow with algal cells) are easily rubbed off. *L. columbaria* is less heavily branched and does not have soredia.

OCCURRENCE Coniferous woods in mountains of N America and C and N Europe, especially the Alps and Scandinavia. It grows in dry sunny places, often on dead trees and epiphytic on the bark of conifers, especially pine, larch and cedar. *L. columbiana* occurs in N America and has similar properties.

CLASSIFICATION Neurotoxin; highly hazardous, Ib

ACTIVE INGREDIENTS The main component in *L. vulpina* is vulpic acid ($C_{19}H_{14}O_5$; MW 322.33), a furanolactone.

UTILISATION Wolf lichen was formerly used to kill wolves and foxes (hence the scientific and other names). Extracts have been used as a yellow dye for baskets, wool and textiles. Weak infusions were used to treat stomach complaints and externally applied to wounds and sores. The Achomawi Indians are said to have used the lichen to make poison arrow heads.

TOXICITY About 30 mg/kg vulpic acid are lethal.

SYMPTOMS Vulpic acid is a neurotoxin that attacks the nervous system and causes death from respiratory and circulatory arrest.

PHARMACOLOGICAL EFFECTS Vulpic acid possibly reacts with proteins of the nervous system, partly by forming covalent bonds.

FIRST AID If vulpic acid or *Letharia* extracts come into contact with eyes or skin, wash them immediately with water or polyethylene glycol. After ingestion give medicinal charcoal and sodium sulphate; in case of respiratory arrest, provide artificial respiration and cardiac massage.

Clinical therapy: After overdosing immediate gastric lavage (possibly with potassium permanganate); in severe cases supply oxygen and intubation. Electrolyte substitution may be necessary. Control acidosis with sodium bicarbonate (urine pH 7.5) and monitor kidney and liver function for up to 8 days after poisoning.

Letharia vulpina (L.) Vain.   family: Parmeliaceae

lichen tue-loup (French); *Fuchsflechte, Wolfsflechte* (German)

Leucaena leucocephala

lead tree • giant wattle

Leucaena leucocephala

Leucaena leucocephala leaves

PLANTS WITH SIMILAR PROPERTIES A legume genus with 22 species distributed from Texas to Peru, including *L. esculenta* and *L. leucocephala* (= *L. glauca*).

PLANT CHARACTERS Giant wattle is a multi-stemmed tree with a wide crown, hairy twigs, compound leaves and globose, white or cream-coloured flower heads in small groups of up to three. The fruits are flat, oblong, brown pods with numerous seeds.

OCCURRENCE *L. leucocephala* occurs naturally in the extreme southern parts of N America but is widely used in agroforestry and as garden ornamental. The species has become naturalised in parts of Africa and Asia.

CLASSIFICATION Cell poison; moderately hazardous, II

ACTIVE INGREDIENTS The toxic compound in *Leucaena* is *L*-mimosine, a non-protein amino acid (NPAA) that occurs in concentrations of 2–5 % in leaves or up to 9 % in seeds. Mimosine is modified to 3-hydroxy-4(1*H*)-pyridone when it is detoxified in ruminants, but the latter is also considered to be poisonous.

UTILISATION *L. leucocephala* is popular as a multi-purpose tree in agroforestry. In addition to its fodder value, it is also used for green manure, pulp for paper-making, poles and fire wood. It is cultivated as biodiesel plant, producing about 1 million barrels of oil per year from 12 000 ha. The brown and flat seeds have been used for beads (jumbie beans). Salted, ripe seeds of *L. esculenta* are said to be edible.

TOXICITY *Leucaena leucocephala* becomes toxic to ruminants when it exceeds 25 % of the diet. Human poisoning has occurred when young shoots, green pods and seeds were mistakenly used as food (see NPAA).

SYMPTOMS Mimosine typically causes complete or partial hair loss (alopecia) in animals and humans. The accumulation of 3-hydroxy-4(1*H*)-pyridone (in both humans and animals) may lead to thyroid impairment and goiter (see NPAA).

PHARMACOLOGICAL EFFECTS Non-protein amino acids mimic endogenous protein amino acids. They can interfere with amino acid synthesis and transport. When incorporated into proteins, they lead to a different 3D-structure and activity. Mimosine is an analogue of tyrosine and inhibits tyrosine decarboxylase, tyrosinase and collagen biosynthesis (see NPAA).

FIRST AID Give medicinal charcoal to absorb the NPAA; give plenty of warm black tea to drink. **Clinical therapy:** After overdosing: gastric lavage, instillation of medicinal charcoal and sodium sulphate; electrolyte substitution, control of acidosis with sodium bicarbonate. Check kidney and liver function.

Leucaena leucocephala (Lam.) De Wit   family: Fabaceae

tamarinier sauvage (French); *Weißfaden, Pferdetamarinde* (German)

Ligustrum vulgare

common privet

Ligustrum vulgare flowers   Ligustrum ovalifolium fruit

PLANTS WITH SIMILAR PROPERTIES A genus of 40 species in Europe, N Africa, E & SE Asia to Australia, including *L. japonicum*, *L. lucidum*, *L. ovalifolium*, *L. sinense* and *L. indicum*.

PLANT CHARACTERS *L. vulgare* is a semi-evergreen shrub (up to 5 m tall) with lanceolate leaves (30–60 mm long), small, white flowers and black berries. *L. lucidum* is a small tree with large, leathery leaves, while *L. ovalifolium* (California privet) is a shrub with much smaller, pale green, oval leaves. Variegated forms of both species are grown in gardens. Fruits are black shiny berries, 5–8 mm in diameter. All the species have white-dotted stems and opposite leaves.

OCCURRENCE *L. vulgare* originates from Europe and the Mediterranean; it is widely planted as hedges in gardens and parks (naturalised in N America). Chinese privet (*L. lucidum*) originates from China and Korea, while California privet (*L. ovalifolium*) occurs naturally only in Japan. Both are popular garden and hedge plants.

CLASSIFICATION Cell poison; moderately hazardous, II

ACTIVE INGREDIENTS The berries (but also the bark and leaves) of *L. vulgare* contain high concentrations of secoiridoid glucosides (up to 3.85 %). The main compounds are ligustroside and oleuropein. The same or similar secoiridoids probably also occur in other species.

UTILISATION Privets are tolerant of city pollution, hence their popularity as garden shrubs, hedge plants and street trees.

TOXICITY The berries of *Ligustrum* are attractive to children and the cause for many consultations in poison centres. They are only poisonous when large numbers (more than 15) are eaten.

SYMPTOMS When only a few berries have been eaten, there may be no symptoms at all. Depending on the dose, nausea, vomiting, diarrhoea, dizziness, headache and stomach pain may occur. Large quantities are more dangerous, and can lead to severe gastroenteritis, shock, convulsions, irregular heartbeat and respiratory problems.

PHARMACOLOGICAL EFFECTS The iridoid glucosides become reactive metabolites when the glucose is cleaved by β-glucosidase. Then the lactol ring opens and reactive aldehyde groups are formed. These can form covalent bonds with important proteins (see aldehydes).

FIRST AID Provide plenty of fluids to drink; after 5–10 berries: give medicinal charcoal and sodium sulphate. **Clinical therapy:** Only in cases where more than 20 fruits have been ingested: gastric lavage, electrolyte substitution, control of acidosis with sodium bicarbonate; in case of spasms, diazepam; check kidney function. In case of respiratory arrest, provide intubation and artificial respiration.

Ligustrum vulgare L.   family: Oleaceae

troène commun (French); *Gewöhnlicher Liguster* (German); *ligustro commune* (Italian)

Linum usitatissimum

flax • linseed

Linum usitatissimum

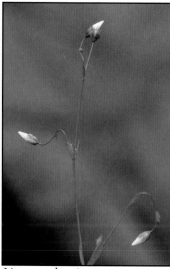

Linum catharticum

PLANTS WITH SIMILAR PROPERTIES *Linum* is a large genus of 180 species from temperate and subtropical regions. Among them are *L. marginale*, *L. catharticum*, *L. flavum*, *L. perenne*, *L. narbonense* and *L. grandiflorum*.

PLANT CHARACTERS Flax is an erect annual that grow to a height of 1 m. It has thin stems, small, oblong leaves and pale blue flowers (20–30 m in diameter). The typical flax seeds (linseeds) are borne in rounded, segmented capsules.

OCCURRENCE The natural distribution includes the Mediterranean region and western Europe. Flax is an ancient cultigen (probably derived from *L. angustifolium*), selected for indehiscent capsules and large seeds (linseed cultivars) or fibrous stems (cultivars grown for flax fibre).

CLASSIFICATION Cellular poison; moderately hazardous, II

ACTIVE INGREDIENTS Seeds contain up to 1 % cyanogenic di- and monoglycosides (linustatin, neolinustatin, and linamarin, lotaustralin, respectively). Linamarase splits the glycoside into acetone, glucose and cyanide (HCN). Intact seeds release 2 mg HCN/100 g; crushed seeds: 7 mg/100 g and milled seeds 25 mg/100 g. Also present are sterols and triterpenes (cholesterol, campesterol, stigmasterol, sitosterol and others). Several *Linum* species contain lignanes, such as methylpodophyllotoxins, especially in their roots.

UTILISATION Seeds have become popular a bulk-forming laxatives to treat chronic constipation and the symptoms of irritable colon, spasmodic colitis and diverticulitis. Linseed oil ha been used for nutrition for 10 000 years and ha been included as a drying oil in paints, varnishe and printing inks. Stems are used to produce fla (for textiles, carpets, twine, ropes and bags) bu has largely been replaced by cotton. The seed cake is a good livestock feed and fertiliser. Som *Linum* species with podophyllotoxins have bee used as purgatives (such as *L. catharticum*).

TOXICITY Human poisoning from flax cyano genic glycosides is very rare, as seeds are normall ingested intact.

SYMPTOMS Animals fed on large amounts o linseed or seed-cake exhibited spasms, paralysi and even death. Death is caused by respiratory arrest (see amygdalin).

PHARMACOLOGICAL EFFECTS The cyano genic glycosides release HCN, which blocks cel lular respiration (see amygdalin). The podophyl lotoxins cause severe diarrhoea and block cel division (see *Podophyllum*).

FIRST AID Only in case of poisoning, after in gestion of large amounts of milled seeds, induc vomiting; administer medicinal charcoal. **Clini cal therapy:** In case of severe intoxication, se amygdalin.

Linum usitatissimum L.

family: Linacea

lin (French); *Lein, Flachs* (German); *lino* (Italian); *lino* (Spanish)

Lobelia inflata

Indian tobacco

obelia tupa   Lobelia inflata   Lobelia cardinalis

PLANTS WITH SIMILAR PROPERTIES large genus of more than 365 species in warm nd tropical climates. Several are important ornaental plants, such as *L. cardinalis, L. chinensis, inflata, L. siphilitica* and *L. tupa.*

PLANT CHARACTERS *L. inflata* is an annual f up to 0.6 m high with slender stems bearing ssile leaves and and small, pale blue flowers. The alyx is persistent and becomes swollen or inflat- d in the fruiting stage (hence "*inflata*").

OCCURRENCE N America; widely cultivated. thers are from China (*L. chinensis*), N America . *cardinalis, L. siphilitica*) and Chile (*L. tupa*).

CLASSIFICATION Neurotoxin, mind-altering; ghly hazardous, Ib

ACTIVE INGREDIENTS Several piperidine al- aloids (up to 0.5 % of dry weight), mainly lobe- ne. Alkaloids are also present in the latex and rm complexes with chelidonic acid.

UTILISATION *L. inflata* is a traditional medi- ne for symptomatic relief of asthma, bronchitis nd pertussis. Isolated lobeline is used in oral anti- moking preparations to lessen the symptoms of icotine withdrawal. It was once injected as re- scitation treatment after ethanol, morphine or eeping pill poisoning, in case of problems with naesthetics, and against asphyxia and apnoea in abies. It has been used as a tobacco substitute nd emetic by N American Indians (hence the

common name). Chinese lobelia (*L. chinensis*) is a diuretic, blue cardinal flower (*L. siphilitica*) is a putative syphilis cure, while the Chilean cardinal flower (*L. tupa*) is a traditional narcotic and possibly a mind-altering, hallucinogenic plant.

TOXICITY Lobeline and the other piperidine alkaloids are paralytic and lethal in high doses.

SYMPTOMS Low doses stimulate, high doses block respiration. Toxic symptoms include bradycardia, hypotension, arrhythmia, dilated pupils, anxiety, nausea, vomiting, diarrhoea with colic, convulsions and death from respiratory arrest.

PHARMACOLOGICAL EFFECTS Lobeline activates nicotinic acetylcholine receptors (similar to nicotine). It is a respiratory stimulant, which accelerates respiratory movements by a direct action on the central nervous system. It also acts as a bronchodilator and ganglionic stimulant.

FIRST AID Provide plenty of tea to drink; give medicinal charcoal and sodium sulphate. Contact with eyes or skin: wash with polyethylene glycol or water. Keep patient warm and calm. **Clinical therapy:** After overdosing: Gastric lavage (possibly with 0.1 % potassium permanganate), instillation of medicinal charcoal and sodium sulphate; in case of spasms give diazepam or chloralhydrate (for children). In case of respiratory arrest, provide intubation, cardiac massage and artificial respiration.

Lonicera xylosteum

fly honeysuckle

Lonicera xylosteum fruits

Lonicera periclymenum flower

PLANTS WITH SIMILAR PROPERTIES A large genus of 180 species from temperate and subtropical parts of the N hemisphere. Among them are *L. caprifolium*, *L. etrusca*, *L. fragrantissima*, *L. hildebrandiana*, *L. japonica*, *L. nigra*, *L. nitida* and *L. periclymenum*. Other Caprifoliaceae often grown in gardens for their conspicuous fruits include the snowberry, *Symphorocarpos albus* (white fruits) and various species of *Viburnum opulus*, *V. lantana* (red fruits).

PLANT CHARACTERS Fly honeysuckle is a deciduous, non-climbing shrub (2–3 m tall) with opposite, broadly elliptic leaves. White to yellow flowers occur in pairs on 10–20 mm long petioles. The bright scarlet, paired fruits are attractive.

OCCURRENCE Europe and N Asia; common in C Europe at the edge of woods or in hedges. *L. caprifolium*, its hybrids and cultivars are common garden plants.

CLASSIFICATION Cellular poison; slightly hazardous, III

ACTIVE INGREDIENTS Fruits contain "xylostein", a cyanogenic glucoside, triterpene saponins, secoiridoids and traces of glycoalkaloids.

UTILISATION Several *Lonicera* and other species of Caprifoliaceae are grown as ornamental or garden plants, such as *L. periclymenum*, *L. caprifolium* and *L. fragrantissima* or various *Symphorocarpos* and *Viburnum* species. Fruits of these plants are often eaten by birds during winter.

TOXICITY Fruits of *Lonicera*, *Viburnum* an Symphoricarpos are frequently a cause of consu tations in poison centres, although their actu toxicity is quite low. Ingestion of more than 2 *Lonicera* fruits have caused toxic symptom 2.5 g/kg DW of fruits causes diarrhoea in rabbit Intraperitoneal injections of 20–40 g/kg extra were deadly to mice.

SYMPTOMS Symptoms include fever, brea pain and vomiting. High doses lead to nause abdominal pain, vomiting, tremors, cold swea dizziness, bloody diarrhoea, cyanosis, tachyca dia, arrhythmia, respiratory arrest and kidne disturbance.

PHARMACOLOGICAL EFFECTS The cyan genic glucosides release cyanide (HCN) whe fruits are squashed or eaten. HCN blocks cell lar respiration (see amygdalin). Saponins disru biomembranes and appear to enhance the HC toxicity. They also cause gastrointestinal and ki ney problems.

FIRST AID If only few fruits were ingeste keep calm and wait. When 5–10 and more frui have been eaten administer sodium sulphate an medicinal charcoal. Provide plenty of fluids t drink. **Clinical therapy:** In case of ingestion o large amounts of fruits, provide gastric lavag (possibly with potassium permanganate).

Lonicera xylosteum L.

family: Caprifoliace

chèvrefeuille des buissons, lonicéra camérisier (French); *Rote Heckenkirsche* (German); *caprifoglio peloso* (Italian)

Lophophora williamsii

mescal • peyote

Lophophora williamsii

Lophophora williamsii

PLANTS WITH SIMILAR PROPERTIES A genus of 1 or 2 species in S Texas to N & E Mexico. Mescaline also occurs in *Trichocereus* species and phenethylamines in *Ariocarpus fissuratus, Gymnocalycium gibbosum, Pachycereus pecten-aboriginum* and *Pelecyphora aselliformis.*

PLANT CHARACTERS A spineless, dark green, lobular cactus up to 100 mm in diameter. It has 8–14 ribs with evenly spaced, woolly areoles. The pink flowers are up to 25 mm in diameter.

OCCURRENCE Arid regions of Mexico and S Texas. It is also grown as an ornamental cactus.

CLASSIFICATION Neurotoxic, hallucinogen; highly hazardous, Ib

ACTIVE INGREDIENTS The main compound is mescaline (4.5–7 % DW); other simple alkaloids include *N*-formylmescalin, *N*-acetylmescaline, anhalonine, anhalonidine and lophophorine.

UTILISATION Peyotl (the Aztec name) has substantial psychedelic properties and was widely used as a hallucinogen in C America. The isolated mescaline even gained literary celebrity through A. Huxley (1954) in his book "The Doors of Perception". Mescaline has been used by many artists. Mescaline and new designer drugs such as ecstasy (synthetic amphetamines resembling mescaline in chemical structure) have found their way into the modern narcotic drug culture.

TOXICITY Chewing of 10–12 dried stems (called "mescal buttons") provoke intoxication after 1–2 hours. Drug users apply mescaline parenterally. LD_{50} mouse: 212 mg/kg i.p.; rat: 132 mg/kg i.p., 157 mg/kg i.v., 330–410 mg/kg i.m.

SYMPTOMS Mescaline paralyses the CNS in doses of >400 mg. Symptoms: decreased blood pressure, pulse of 50–150, dilated pupils, pains, respiratory depression, vasodilatation, weakness, liver damage and vomiting. Doses above 0.2 % have progressive paralytic effects. Mind-altering symptoms include euphoria, colourful hallucinations, distorted vision and thinking, aphrodisiac, strong even schizophrenic emotions, loss of awareness of time and space, and sleeplessness.

PHARMACOLOGICAL EFFECTS Mescaline is structurally similar to the catecholamine-type neurotransmitters, although its activities resemble those of LSD (which mainly interacts with serotonin receptors, such as $5-HT_2$). It has been speculated that a metabolite of mescaline causes the psychedelic effects.

FIRST AID Try to sedate the patient by talking. In case of strong excitation, apply doxepin (better not diazepam). After oral overdosing keep the respiratory tract free and the patient warm.
Clinical therapy: In severe cases, gastric lavage, instillation of medicinal charcoal, sodium sulphate, plasma substitution; provide intubation and artificial respiration.

Lophophora williamsii (Lem. ex Salm-Dyck) J.M.Coult.

family: Cactaceae

peyote, peyoti (French); *Meskalkaktus, Peyotekaktus* (German)

Lotononis laxa

wild lucern

Lotononis laxa

PLANTS WITH SIMILAR PROPERTIES A genus of 150 species, mostly in southern Africa.
PLANT CHARACTERS *L. laxa* is a prostrate perennial with small, densely hairy trifoliate leaves and a single stipule at the base of each leaf stalk. The small, yellow flowers are usually single at each node, not clustered. They turn orange after pollination. The fruits are oblong and hairy. *L. involucrata* is a perennial herb with a strong taproot and prostrate, radiating, hairy branches and clusters of yellow flowers at the tips. *L. fruticoides* is an erect annual that may be mistaken for a perennial. It has yellowish twigs growing in a zig-zag pattern which bears very small yellow flowers and flat, oblong pods characterised by sharp tips.
OCCURRENCE *L. laxa* occurs in southern and eastern Africa, while the other toxic species are restricted to S Africa.
CLASSIFICATION Cell poison, neurotoxin, mutagen; moderately hazardous, II
ACTIVE INGREDIENTS About 60 of the 150 species of *Lotononis* accumulate prunasin, a well-known cyanogenic glucoside, as well as malonyl esters of prunasin. The toxic species listed above are known to be strongly cyanogenic. Pyrrolizidine alkaloids (PAs) such as senecionine and integerrimine (both hepatotoxic macrocyclic esters), are also found in *L. involucrata*. Some *Lotononis*

species produce the neurotoxic quinolizidine alkaloids lupanine, cytisine and *N*-methylcytisine.
UTILISATION *L. bainesii* is a well-known cultivated pasture legume. It is not poisonous.
TOXICITY *L. laxa*, *L. carnosa* and *L. involucrata* have been responsible for several incidences of fatal poisoning of livestock. The very toxic *L. fruticoides* has been implicated in sheep losses but the species was not yet described and named at that time. It contains about 468 mg of cyanide per 100 g dry weight (plants with more than 20 mg per 100 g are usually considered poisonous).
SYMPTOMS The symptoms of fatal *Lotononis* poisoning agree with the typical symptoms associated with hydrogen cyanide poisoning, and may also include severe diarrhoea (see amygdalin). PA-containing plants can produce veno-occlusive disease in humans and animals.
PHARMACOLOGICAL EFFECTS Prunasin releases HCN after cleavage by β-glucosidase. HCN is rapidly inactivated in the body at low concentrations. Higher doses stop cellular respiration and lead to general paralysis and respiratory arrest (see amygdalin). PAs are liver toxins and mutagens that can cause cancer (see senecionine).
FIRST AID After ingestion, allow the intake of large volumes of drinks. **Clinical therapy:** In case of severe intoxication: for further detoxification see amygdalin.

Lotononis laxa Eckl. & Zeyh.   family: Fabaceae
Lotononis (German)

164

Lupinus polyphyllus

garden lupin

Lupinus polyphyllus

Russell hybrid

PLANTS WITH SIMILAR PROPERTIES A large genus of about 200 species in the New World (e.g. *L. mutabilis, L. texensis, L. perennis, L. arboreus*) and a few in Europe and Africa (e.g. *L. angustifolius, L. albus* and *L. luteus*).

PLANT CHARACTERS *L. polyphyllus* is an erect perennial herb with digitately compound leaves and attractive, elongated clusters of blue flowers (or various shades of white, yellow, red or purple in garden forms).

OCCURRENCE *L. polyphyllus* is indigenous to N. America, but is widely naturalised in Europe, New Zealand, C and SE Europe. It is widely grown as a garden plant. Russell hybrids (*L. polyphyllus* x *L. arboreus*) are especially popular.

CLASSIFICATION Neurotoxin; moderately hazardous, II

ACTIVE INGREDIENTS Lupanine, 13-hydroxylupanine, sparteine, and other quinolizidine alkaloids (QAs) occur in all parts of the plant; seeds have up to 3 % and leaves up to 0.6 % alkaloids.

UTILISATION Lupins are grown as green manure and for land restoration. Annual sweet lupins (low-alkaloid cultivars of *L. angustifolius, L. albus* and *L. luteus*) are cultivated as a protein crop for human and animal consumption in Australia, S Africa, S America and Europe. Even in antiquity, alkaloid-rich lupin seeds were debittered by cooking and leaching before consumption.

TOXICITY Lupins are responsible for different diseases in animals: lupinosis is a poisoning with liver and muscle damage caused by mycotoxins from a fungus (*Phomopsis leptostromiformis*) that grows on lupins, while "crooked calf disease" occurs when cows graze on lupins containing anagyrine and/or ammodendrine (*L. sericeus, L. caudatus, L. formosus* or *L. argenteus*) during gestation. Lupin poisoning occurs when dried plant material or seeds of bitter lupins (mainly *L. angustifolius*) are fed to animals. Lupanine: LD_{50} rat: 177 mg/kg i.p. (death after 6–24 min), 1.6 g/kg (death after 2 min to 4 h); 13-hydroxylupanine: LD_{50} rat: 199 mg/kg i.p.; sparteine: LD_{50} mouse: 36 mg/kg i.p., 220 mg/kg p.o.

SYMPTOMS Lupin (alkaloidal) poisoning may result in salivation, vomiting, difficulties to swallow, arrhythmia, convulsions, paralysis and death from respiratory arrest. Plants with quinolizidine alkaloids can be abortifacient because of uterus-contracting properties of the alkaloids.

PHARMACOLOGICAL EFFECTS Lupanine is an agonist at nAChR; other QAs are agonistic at mAChR or modulate ion channels and thus affect the central and peripheral nervous system.

FIRST AID Administer sodium sulphate and medicinal charcoal. **Clinical therapy:** In case of larger doses: gastric lavage and further treatments (see cytisine).

Lupinus polyphyllus Lindl.

family: Fabaceae

lupin (French); *Vielblättrige Lupine* (German); *lupino* (Italian)

Lycopodium clavatum

common clubmoss

Lycopodium clavatum

Lycopodium squarrosum

PLANTS WITH SIMILAR PROPERTIES The clubmoss family was abundant in the Carboniferous period and represents a relict of the early evolution of land plants. It is a genus with 40 species that grow in both tropical and temperate climates. In Europe, the best known species are *Lycopodium clavatum* and *L. annotinum*; *Huperzia selago* is related and has similar toxic properties.

PLANT CHARACTERS The clubmoss is a member of the order Lycopodiales (Pteridophyta) with creeping and branched stems (length up to 1 m). The small, lance-shaped leaves occur in several rows along the stems. Some erect stems produce spores at their tips. In *Huperzia selago*, the spores are borne along the stems and not at the tips.

OCCURRENCE In moist woodland in Europe, Asia and America.

CLASSIFICATION Neurotoxin; moderately hazardous, II

ACTIVE INGREDIENTS The plant is rich in alkaloids (more than 0.2 % dry weight) such as lycopodine, clavatine, annotidine and huperzine A, which can be formally regarded as quinolizidine alkaloids.

UTILISATION *Lycopodium* was one of the holy plants (besides mistletoe) of the Druids (the priests of the ancient Celts). Names such as the German ("Drudenkraut") and the Welsh ("gras duw" = donation of god) reflect this old connection with history. The clubmoss produces fine yellow spore in high numbers, which were used as "sulphu vegetabile" as a conspergent to manufacture pills hair powder and dusting agent on condoms. Sinc the spores are easily inflammable, they were use for flashlights. Another application was finger print detection. Some clubmosses have been use for stuffing upholstery, basket- and bag-making *L. alpinum* was used as a yellow dye for wool. I recent times, *Lycopodium* species have also becom popular as pot plants and as florist greens for us in mixed flower arrangements.

TOXICITY Allergies with granulomatosis wa observed when condoms were used that had bee dusted with *Lycopodium* spores. Lethal dose i mice and frogs: 0.2 g.

SYMPTOMS Taken orally, strong perspiration nausea, vertigo, lalopathy, spasms, vomiting an diarrhoea are observed.

PHARMACOLOGICAL EFFECTS *Lycopod um* alkaloids are a antagonist at nAChR and in hibitors of cholinesterase (especially huperzine A and thus affect the central nervous system. The show toxic properties similar to those of curar (see cytisine, curarine).

FIRST AID Induce vomiting; administration o sodium sulphate, medicinal charcoal. **Clinica therapy:** Gastric lavage and further treatment (see cytisine).

Lycopodium clavatum L.

family: Lycopodiacea

jalousie (French); *Keulenbärlapp* (German); *licopodio clavato* (Italian)

Mandragora officinarum

mandrake

Mandragora officinarum flowers

Mandragora officinarum fruit

PLANTS WITH SIMILAR PROPERTIES A genus of 6 species (Mediterranean to the Himalayas), including *M. autumnalis* and *M. turcomanica*.
PLANT CHARACTERS A perennial with thick, tuberous roots of 0.6 m long that are traditionally thought to resemble the human body and a rosette of dark green leaves. Greenish white flowers appear in spring, which turn into ball-shaped fruits of up to 40 mm in diameter. The fruits turn from green to yellow when they ripen.
OCCURRENCE Southern Europe (arid places).
CLASSIFICATION Neuropoison, hallucinogen; extremely hazardous, Ib
ACTIVE INGREDIENTS Tropane alkaloids; up to 0.6 % dry wt in roots; mature fruits with low alkaloid levels; mainly scopolamine, accompanied by hyoscyamine and other minor alkaloids.
UTILISATION Since antiquity, mandrake has quelled the fantasies of humans and it features in many old stories. The Ebers Papyrus (1500 BC) described a recipe against intestinal worms, which included fruits from mandrake. Dioscorides noted its use as a sleep-inducing drug, as a narcotic for surgery, against eye-disease and as an abortifacient. Mandrake was widely known as an aphrodisiac in ancient Greece (also known as Kirkaia, reflecting its amorous use by Circe). The military strategist Maharbal (200 BC) left his enemy several amphoras of wine mixed with mandrake extract. When the enemy was drunk and immobilised, Maharbal and his warriors came back and killed them. During medieval times, *Mandragora* extracts were used by witches and sorcerers for hallucinogenic purposes (together with *Atropa*, *Datura* and *Hyoscyamus*). *M. turcomanica* (white mandrake) might represent the "soma" of ancient Iran and India.
TOXICITY *Mandragora* poisoning was common until medieval times, but is rare today. A salad of wild herbs that included mandrake leaves caused toxication in Italy. Scopolamine: LD_{50} mouse: 1.7 to 5.9 g/kg s.c.; 163 mg/kg i.v.; lethal dose for humans: adults 100 mg; children 2–10 mg.
SYMPTOMS At low concentrations, the alkaloids have a depressant and sedative effect, but higher doses lead to hallucinations, euphoria, confusion, insomnia, vertigo, trembling, difficulties to speak, excitation, nausea, vomiting, mydriasis and death from central paralysis.
PHARMACOLOGICAL EFFECTS Scopolamine is an antagonist at the muscarinic acetylcholine receptor and therefore works as a parasympatholytic (see hyoscyamine).
FIRST AID After ingestion of roots or leaves, induce vomiting; treat with sodium sulphate and medicinal charcoal. **Clinical therapy:** Physostigmine can be used as an antidote (see hyoscyamine).

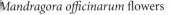

Mandragora officinarum L.

family: Solanaceae

mandragore (French); *Alraune* (German); *mandragora primaverile* (Italian)

Manihot esculenta

cassava • tapioca • manioc

Manihot esculenta

Manihot esculenta roots

PLANTS WITH SIMILAR PROPERTIES *Manihot* comprises a genus with more than 98 species in warm and tropical America.

PLANT CHARACTERS Cassava is a shrub of up to 5 m in height, with green digitate leaves and small, dioecious, inconspicuous yellow flowers and ribbed fruits. Important for use are the large tuberous roots, which are brown outside and white inside. Wild forms are bitter and poisonous; cultivated forms are sweet and non-toxic.

OCCURRENCE. Cassava probably originates from Brazil or Mexico, where the roots had been used as a food source since ancient times. Later it was introduced to tropical Africa and Asia as a starch-rich food plant.

CLASSIFICATION Cellular poison; moderately hazardous, Ib–II

ACTIVE INGREDIENTS Cyanogenic glucosides (CGs), such as linamarin occur in all parts of the plant; bitter varieties with CG contents up to 0.1 % of fresh weight; sweet forms with less than 50 mg/kg linamarin.

UTILISATION Cassava roots are widely used as a good starch source (30 % dry weight) and as an alternative to potatoes; it is part of a staple diet in S America and in Africa (fufu). Several products (e.g. tapioca) are made from cassava. The toxicity of cyanogenic glucosides was known and special care was taken to remove it. Cassava roots were ground and left standing for hours in water. The volatile HCN that was liberated in such a process was removed by extended periods of squeezing and by evaporation. HCN extracts have been used in America as natural pesticides.

TOXICITY When bitter cassava is not processed correctly, severe intoxications occur (as happened to early explorers into Africa). Bitter cassava is resistant to herbivores because of its high CG content. Livestock poisonings are quite common.

SYMPTOMS Poisoning symptoms are typical for HCN in general (irritation, flushing face, heavy breathing, scratchy throat, headache, and in severe cases, respiratory and cardiac arrest). Chronic intake cause neuropathies (deafness, paralysis, myelopathy), hyperglycaemia or cretinism. In parts of Africa, thousands of women and children suffer from this syndrome.

PHARMACOLOGICAL EFFECTS Linamarin releases HCN after cleavage by linamarase or in the stomach; HCN is rapidly inactivated in the human body at low concentrations. Higher doses stop cellular respiration and lead to general paralysis and respiratory arrest (see amygdalin).

FIRST AID After ingestion of bitter cassava products, induce vomiting and allow the intake of large volumes of liquids. **Clinical therapy:** In case of severe intoxication, see amygdalin.

Manihot esculenta Crantz   family: Euphorbiaceae

manioc, tapioca (French); *Cassava, Maniok* (German); *manioco* (Italian); *yuca* (Spanish)

Melia azedarach

Persian lilac • Chinaberry • syringa tree

Melia azedarach

Melia azedarach flowers and mature fruits

PLANTS WITH SIMILAR PROPERTIES A genus of 3 species in tropical Asia to Australia.

PLANT CHARACTERS Persian lilac is usually a single-stemmed, medium-sized tree with a spreading crown and glossy green, pinnately compound leaves. The clusters of small, lilac-coloured, sweetly scented flowers are followed by globose, smooth green berries that become pale yellowish brown and wrinkled at maturity.

OCCURRENCE Persian lilac or syringa tree is indigenous to Asia, but has been introduced into many parts of the world as ornamental tree.

CLASSIFICATION Cell poison; highly hazardous, Ib

ACTIVE INGREDIENTS *M. azedarach* produces complex mixtures of limonoids, with melinoon and melianol as the two major compounds in ripe fruits. The toxicity of the fruits is very variable and some trees are known to be more toxic than others. Four tetranortriterpenoids called meliatoxins A_1, A_2, B_1 and B_2 are considered to be the toxic compounds and are present at about 0.5 % of dry weight. They appear to be unstable in acid medium.

UTILISATION Wood of *M. azedarach* has been used for construction. Bark and leaves have been employed in traditional medicine. Triterpenoids have antifeedant and insecticidal properties and seeds and leaves have been used in China as pesticides.

TOXICITY The ripe fruits are a very common cause of human poisoning, as infants and children tend to be attracted to the berries. Fatalities are rare, however, as the toxicity appears to be very variable. Ruminants are usually able to ingest large amounts of leaves without showing any symptoms but pigs are very sensitive to the ripe berries (which are more poisonous than green berries or leaves). The meliatoxins vary from tree to tree and were found to have an LD_{50} in pigs (p.o.) of about 6.4 mg/kg body weight. The LD_{50} in mice (per injection) was 16 mg/kg.

SYMPTOMS Symptoms of poisoning include perspiration, vomiting, respiratory distress, convulsions, rapid and weak heartbeat, below normal body temperature and dilated pupils.

PHARMACOLOGICAL EFFECTS The triterpenoids probably affect membrane stability and permeability. High doses can cause cell death.

FIRST AID Administer medicinal charcoal. Wash eyes; treat skin inflammation with corticoid ointment. **Clinical therapy:** Gastric lavage (possibly with potassium permanganate), instillation of 10 g medicinal charcoal and sodium sulphate, electrolyte substitution, control of acidosis with sodium bicarbonate. In case of spasms and colic, give diazepam and atropine; in severe cases, provide intubation and oxygen respiration.

Melia azedarach L.   family: Meliaceae

lilas des Indes (French); *Paternosterbaum, Indischer Zederachbaum* (German); *albero da rosari* (Italian)

Melianthus comosus

honeyflower

Melianthus comosus

Melianthus comosus flowers with black nectar

PLANTS WITH SIMILAR PROPERTIES A southern African genus with 8 species, including *M. major.*

PLANT CHARACTERS *M. comosus* is a rounded shrub of about 2.5 m with a characteristic unpleasant smell. The leaves are pinnately compound and have coarsely toothed leaflet margins and two filiform stipules. Also typical are the short racemes of green and red flowers producing black nectar. The fruits are four-angled, bladdery capsules that become papery when they mature. *M. major* is a taller plant with hairless, greyish green leaves and dark purple flowers borne on purple flowering stalks.

OCCURRENCE *M. comosus* is widely distributed in the dry interior region of South Africa; *M. major* has a more restricted distribution in the Western Cape Province. Both species are grown as ornamental plants in Europe. *M. major* has become naturalised in parts of Africa, India, Australia, New Zealand and America.

CLASSIFICATION Heart poison; highly hazardous, Ib

ACTIVE INGREDIENTS The roots contain several bufadienolides, including melianthugenin, melianthusigenin, and hellebrigenin-3-acetate.

UTILISATION *M. comosus, M. major* and other species are commonly used in rural areas as topical treatments for wounds, sores, burns and rheu-matic pains. Although weak infusions have been taken in traditional medicine, this practice should be discouraged, as it is potentially lethal.

TOXICITY There are several records of human and animal deaths caused by *Melianthus* species. The unpleasant smell of the leaves acts as a deterrent but animals sometimes ingest the leaves when grazing becomes scarce. As little as 500 g of fresh leaves proved fatal to a sheep. The roots are known to be very toxic and even parasitic plants such as mistletoes (*Viscum* and *Tapinanthus* species) growing on the above-ground parts are said to become poisonous.

SYMPTOMS Symptoms include nausea, salivation, foaming vomit, colic, gastrointestinal disturbance, purging and exhaustion, with the usual respiratory and cardiac symptoms (arrhythmia, hypertension, coma, cardiac arrest) associated with heart glycoside poisoning.

PHARMACOLOGICAL EFFECTS In general, cardiac glycosides inhibit the Na^+, K^+-ATPase (see cardiac glycosides). Concentrated decoctions may cause rapid death due to heart failure.

FIRST AID First aid treatment is indicated if plant material from *Melianthus* or isolated cardiac glycosides have been ingested. Treatment involves the induction of vomiting, instillation of medicinal charcoal, and detoxification as outlined under cardiac glycosides.

Melianthus comosus L.

family: Melianthaceae

buisson-à-miel, mélianthe (French); *Honigblume, Honigstrauch* (German)

Mesembryanthemum tortuosum

sceletium

Mesembryanthemum tortuosum

Mesembryanthemum tortuosum flower

PLANTS WITH SIMILAR PROPERTIES There are 8 species of *Sceletium*, but only *S. tortuosum* (now *Mesembryanthemum tortuosum*) is well known and used on a commercial scale. Mind-altering alkaloids have also been reported from *Mesembryanthemum crystallinum* and other *Mesembryanthemum* species, *Drosanthemum hispidum* and several *Trichodiadema* species. These reports need to be verified, however. It has recently been proposed that *Sceletium* and several other genera should be included in an enlarged genus *Mesembryanthemum*.

PLANT CHARACTERS A short-lived, creeping perennial with overlapping pairs of succulent leaves. Characteristic are the glistening water cells and persistent leaf veins (the leaves become "skeletonised" in summer, hence the name *Sceletium*). Single yellow to orange flowers are followed by pale brown, papery and spongy seed capsules.

OCCURRENCE *Mesembryanthemum* species are indigenous to S Africa. Forms of *M. tortuosum* with a high alkaloid content are being grown commercially.

CLASSIFICATION Mind-altering; slightly hazardous, III

ACTIVE INGREDIENTS Sceletium contains mesembrine-type alkaloids, of which mesembrine is usually the main alkaloid in leaves. Several structuraly related alkaloids have been found in various species and plant parts, including mesembrenone, mesembrenol and tortuosamine. The alkaloid concentrations vary from 0.05–2.3 % of dry weight. Oxalates are also present.

UTILISATION It has been speculated that even the Assyrians knew of a mesemb used for medicinal purposes. Fermented and dried whole plants are traditionally chewed by Khoisan people (like chewing tobacco) to elevate mood, to relieve hunger and thirst and to treat colic in infants. Modern uses include replacement therapy for alcoholics and relief of stress and anxiety. It has also been suggested that S African rock paintings, with their visionary pictures, have been made under the influence of sceletium. *Mesembryanthemum crystallinum* has been used as an ingredient of an intoxicating beer.

TOXICITY No adverse side effects have been recorded. Overdosing leads to nausea.

SYMPTOMS Extracts with mesembrine have sedating properties. High doses may cause euphoria but the plant is not hallucinogenic.

PHARMACOLOGICAL EFFECTS Mesembrine is a strong inhibitor of serotonin-uptake; other alkaloids may be active at other neuroreceptors. Even after years of habitual use, there is no physical and psychological dependency.

FIRST AID After serious overdosing, instil medicinal charcoal, sodium sulphate and provide plenty of tea to drink.

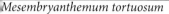

Mesembryanthemum tortuosum L. (= *Sceletium tortuosum* (L.) N.E. Br.)   family: Aizoaceae

sceletium (French); *Sceletium* (German); *sceletium* (Italian)

Mundulea sericea

cork bush • silver bush

Mundulea sericea

PLANTS WITH SIMILAR PROPERTIES A genus with 15 species on Madagascar and one extending into the Old World tropics. Other species with rotenone include *Derris elliptica, Lonchocarpus violaceus, Neorautanenia amboensis,* and *Tephrosia vogelii.*

PLANT CHARACTERS A multi-stemmed, erect shrub or small tree with pinnate, silvery leaves and thick, corky bark. The flowers are purple or rarely white and are followed by oblong, flat, velvety pods.

OCCURRENCE *M. sericea* has a wide natural distribution area in Africa, Madagascar, Sri Lanka and India.

CLASSIFICATION Cell poison; highly hazardous, Ib

ACTIVE INGREDIENTS Rotenoids such as deguelin and tephrosin occur in the bark. Early studies suggested that rotenone is present, but these reports need confirmation.

UTILISATION The plant has been used as an arrow poison and for committing suicide. The bark, leaves, roots and seeds have been used to kill insects, fish and even crocodiles. Cork bush is also used as a traditional emetic for treating human poisoning.

TOXICITY Ingestion of bark infusions or decoctions is often not thought of as being particularly dangerous to humans, but some reports state that it is very toxic. Rotenone is poisonous when injected (LD_{50} of 2.8 mg per kg body weight in mice). Rotenone and tephrosin appear to be relatively harmless when ingested, as cattle and game regularly feed on the leaves without ill effects. However, the compounds are considered dangerous and even lethal when inhaled or injected.

SYMPTOMS First symptoms include numbness of the tongue, which spreads along the digestive tract. A main effect is a paralysis of all muscles, both skeletal and smooth muscles. Higher doses paralyse the brain and the spinal cord. Further symptoms have been recorded, including convulsions and strong vomiting.

PHARMACOLOGICAL EFFECTS Rotenone and related isoflavonoids inhibit the mitochondrial respiratory chain. This leads rapidly to ATP depletion and thus blocks most cellular processes, including neuronal and muscular activity.

FIRST AID Administer medicinal charcoal and sodium sulphate, induce vomiting; shock prophylaxis (calmness, keep warm, provide warm drinks).

Clinical therapy: After overdoses: Gastric lavage (possibly with potassium permanganate), instillation of medicinal charcoal and sodium sulphate; electrolyte substitution; control acidosis with sodium bicarbonate; in case of spasms give diazepam; if necessary, provide intubation, oxygen respiration and cardiac massage.

Mundulea sericea (Willd.) A. Chev.   family: Fabaceae
Silberbusch (German); *famano* (Malagasy)

Myristica fragrans

nutmeg tree

Myristica fragrans opened fruit

Myristica fragrans seeds and mace

PLANTS WITH SIMILAR PROPERTIES A genus with 72 species from tropical Asia to Australia, several of which produce myristicin.

PLANT CHARACTERS Nutmeg is an evergreen tree of up to 20 m in height. Female trees produce a pale brownish yellow, fleshy fruit that splits into two at maturity. Inside is a large, hard seed, surrounded by a peculiar bright red aril (mace) that forms a thin, net-like fleshy layer. Trees are grown from seeds, which germinate during a short period of 8–10 days.

OCCURRENCE SE Asia. The nutmeg tree is indigenous to Amboine Island, one of the Malaccan Islands. It was introduced to Brazil, Mauritius, Malaysia, India, Sri Lanka, Sumatra and the Caribbean Islands (especially Grenada).

CLASSIFICATION Neurotoxin, mind-altering; moderately hazardous, II

ACTIVE INGREDIENTS Nutmeg nuts and mace contain an essential oil (up to 15 % of dry weight). The main compounds are sabinene, α-pinene and β-pinene (80 %) and myristicin (10 %), together with other phenylpropanoids, elemicin, eugenol, isoeugenol, methyleugenol, safrol and others.

UTILISATION Nutmeg and mace are used as a spice to flavour milk puddings, biscuits and other food items. The essential oil is known to have a spasmolytic activity and relieves stomach cramps, flatulence and catarrh of the respiratory tract.

Nutmeg oil is traditionally applied as a counter-irritant for relief of aches and pains. The oil has been used as abortifacient and as a marijuana substitute. It is also used to flavour toothpaste and cigarettes. Nutmeg has widely been employed as a stimulating and intoxicating drug that can be addictive. It has a reputation as aphrodisiac.

TOXICITY High doses can be hazardous. An 8-year-old boy died after eating 2 nuts.

SYMPTOMS Large doses may cause headache and dizziness. Doses of 5 g or more are dangerous and may cause psychic disturbances, hallucinations, heart palpitations, headache, vertigo, mydriasis, vomiting, tachycardia and abortion.

PHARMACOLOGICAL EFFECTS The drug has pronounced antimicrobial and anti-inflammatory activity. The addictive and hallucinogenic effects are ascribed to myristicin and elemicin that are probably converted to amphetamine-like compounds. Phenylpropanoids with terminal methylene groups can be activated in the liver to a reactive epoxide that can bind to proteins and DNA. They are potentially cytotoxic and mutagenic (see myristicin).

FIRST AID After overdose, administration of medicinal charcoal. **Clinical therapy:** In severe cases: gastric lavage (possibly with potassium permanganate), instillation of medicinal charcoal and sodium sulphate (see myristicin).

Myristica fragrans Houtt.

family: Myristicaceae

muscadier (French); *Muskatnussbaum* (German); *noce moscata* (Italian); *nuez moscada* (Spanish)

Narcissus pseudonarcissus

wild daffodil • lent lily

Narcissus pseudonarcissus wild type

PLANTS WITH SIMILAR PROPERTIES A genus with 27 species in the Mediterranean region, including *N. triandrus*, *N. asturiensis*, *N. confusus*, *N. poeticus*, *N. papyraceus*, *N. tazetta* and *N. viridiflorus*. Many species are grown in horticulture and there are numerous hybrids and cultivars.

PLANT CHARACTERS *Narcissus* species are attractive perennials with fleshy bulbs and dark green, strap-shaped leaves. The flowers of *N. pseudonarcissus* are medium-sized, usually dark to pale yellow. The corona (the characteristic bell-shaped outgrowth in the centre of the flower) is large and free from the stamens.

OCCURRENCE *N. pseudonarcissus* occurs naturally in western parts of Europe and has become naturalised in several places. The plant has been a very popular garden ornamental since 500 years ago and is grown as a spring garden plant and cut flower in many parts of the world.

CLASSIFICATION Cell poison; highly hazardous; Ib–II

ACTIVE INGREDIENTS *N. pseudonarcissus* produces several isoquinoline alkaloids. The best-known of these are lycorine and galanthamine, toxic substances also found in other members of the family. *N. confusus* is especially rich in galanthamine.

UTILISATION The plant has been used in traditional medicine. Many species of the genus are grown as ornamental plants and a few are cultivated specifically for their scent.

TOXICITY *N. pseudonarcissus* bulbs and leaves are toxic and there is a danger that children and domestic animals may be poisoned in gardens. Since the bulbs superficially resemble onions, poisoning has sometimes resulted from confusion between the two. Lycorine is highly toxic with an LD_{50} in dogs of 41 mg/kg.

SYMPTOMS Symptoms of poisoning are salivation, vomiting and diarrhoea, leading to paralysis, liver damage, collapse and even death. Oxalate crystals (raphides) of the outer bulb scales or leaves can cause dermatitis in gardeners, when handling bulbs (known as daffodil itch).

PHARMACOLOGICAL EFFECTS Lycorine inhibits protein biosynthesis and has cytotoxic and virustatic properties. It is emetic and diuretic. Galanthamine blocks cholinesterase and therefore has parasympathomimetic effects.

FIRST AID Give medicinal charcoal and sodium sulphate. The patient should drink plenty of tea; apply shock prophylaxis. **Clinical therapy:** After ingestion, gastric lavage (possibly with potassium permanganate), electrolyte substitution, control acidosis with sodium bicarbonate (urine pH 7.5) and kidney function; in case of spasms give diazepam; in severe cases supply intubation and oxygen.

Narcissus pseudonarcissus L.   family: Amaryllidaceae

narcisse jaune, bonhomme, chaudron (French); *Osterglocke, Gelbe Narzisse* (German); *narciso trombone* (Italian)

Nerium oleander

oleander

Nerium oleander cultivar

Nerium oleander

PLANTS WITH SIMILAR PROPERTIES A monotypic genus (only one species).

PLANT CHARACTERS A multi-stemmed, woody shrub of up to 6 m high. The leathery leaves are oblong in shape, with a distinct white midrib and parallel lateral veins. Attractive, single or double flowers in a range of colours (shades of white, yellow, pink or red) are produced in summer, followed by narrow capsules containing hairy seeds.

OCCURRENCE *N. oleander* has a very wide area of distribution, from the Mediterranean region to W China. It is tolerant of drought and has become a popular garden shrub in many warm parts of the world.

CLASSIFICATION Heart poison; extremely hazardous, Ia

ACTIVE INGREDIENTS All parts, especially leaves and seeds, contain more than 30 cardenolides (1–2 % of dry weight). The main toxin is oleandrin; aglycones include digitoxigenin, gitoxigenin, oleandrigenin, adynerigenin, neriagenin, oleagenin and uzarigenin. Toxins can be transferred into honey via the pollen and nectar.

UTILISATION Oleander leaves or tea made thereof are traditionally used to treat functional disorders of the heart. It is extremely poisonous however, so that the herb is no longer much used except within the safeguards of traditional prac-

tices. The herb has also been used topically to treat skin rashes and scabies, as abortifacient and for suicide.

TOXICITY All parts of the plant are extremely poisonous and are of regular concern in poison centres. Poisonings may result from eating meat skewered on the stems or inhaling the smoke from burning wood or leaves. Human fatalities can occur after drinking extracts or ingesting less than 3–4 leaves (estimated at 4 g). Poisoning of livestock and pets occurs sporadically, when animals are exposed to plants in gardens, hedge clippings or stagnant water into which flowers or leaves had fallen. The lethal dose of green leaves is 3–15 g in sheep and 10–20 g in horses and cattle.

SYMPTOMS Poisoning leads to numbness of the tongue and throat, nausea, vomiting, weakness, spasms, confusion and visual disturbances. A slowing and weakening of the heartbeat to 30 or 40 beats per minute may occur and finally also bradycardia, atrio-ventricular block and fibrillation, followed by death after 2 to 5 h.

PHARMACOLOGICAL EFFECTS Cardenolides inhibit Na^+, K^+-ATPase, thereby blocking the signal transduction from nerves to muscles (see cardiac glycosides).

FIRST AID Induce vomiting; administration of sodium sulphate and medicinal charcoal. **Clinical therapy:** see cardiac glycosides.

Nerium oleander L.

family: Apocynaceae

Laurier rose (French); *Oleander, Rosenlorbeer* (German); *oleandro* (Italian)

Nicotiana glauca

tree tobacco • tobacco tree

Nicotiana glauca flowers　　Nicotiana glauca

PLANTS WITH SIMILAR PROPERTIES
A genus of 66 species in the New World (including *N. tabacum*, *N. sylvestris*, *N. alata*, *N. rustica* and *N. paniculata*) and one in Namibia (*N. africana*).

PLANT CHARACTERS An erect, sparsely branched shrub or small tree with arching branches bearing glaucous, hairless, fleshy leaves and clusters of tubular yellow flowers. The fruits are small capsules that develop within the persistent calyces.

OCCURRENCE Tree tobacco is indigenous to Argentina and was accidentally introduced to the Mediterranean, Africa and Asia. It is drought tolerant and herbivore resistant (mostly avoided by sheep and goats) and has therefore become a common invasive weed of disturbed places in many subtropical regions of the world.

CLASSIFICATION Neurotoxin, mutagen; highly hazardous, Ib

ACTIVE INGREDIENTS All above-ground parts of the plant contain high levels of the piperidine alkaloid anabasine as practically the only alkaloid.

UTILISATION Tree tobacco has been used by different Indian tribes in traditional medicine and as insecticide. It is found in archaeological sites of the Nazca culture.

TOXICITY Several tragic cases of fatal human poisoning have been recorded. These accidents resulted from young plants that were accidentally mistaken for other herbs that are cooked as traditional spinaches. Fatal livestock poisonings and teratogenic defects in sheep have been recorded but the plant is rarely grazed by animals.

SYMPTOMS Anabasine poisoning is similar to nicotine poisoning and the symptoms appear rapidly. Typical are headache, weakness, nausea, salivation, convulsions, confusion, difficulty in breathing and hypertension. In case of large doses, rapid death may result from paralysis of the respiratory muscles.

PHARMACOLOGICAL EFFECTS Anabasine (and nicotine) are powerful agonists at nicotinic acetylcholine receptors and thus affect neuromuscular and neuronal activity. Lower doses activate the central and vegetative nervous system, whereas higher doses are paralysing.

FIRST AID Evoke vomiting; give medicinal charcoal and sodium sulphate. Drink plenty of tea or water and keep warm. In case of contact of the alkaloids with skin, wash with water and soap.

Clinical therapy: After ingestion of large doses, gastric lavage (possibly with 0.1 % potassium permanganate), electrolyte substitution, control acidosis with sodium bicarbonate (urine pH 7.5); in case of spasms give diazepam; in severe cases supply oxygen intubation and respiration. Atropine, orciprenaline, and phentolamine can be given as antidotes.

Nicotiana glauca Graham　　　　　　　　　　　　　　　family: Solanaceae

tabac en arbre (French); *Bauerntabak* (German); *tabacco glauco, tabacco orecchiuto* (Italian)

Nicotiana tabacum

tobacco

Nicotiana tabacum cultivation

Nicotiana tabacum

PLANTS WITH SIMILAR PROPERTIES A genus of 67 species, mainly in the New World.

PLANT CHARACTERS An erect, leafy and single-stemmed annual herb with large, glandular leaves and pink tubular flowers.

OCCURRENCE *N. tabacum* is a cultigen that was domesticated in pre-Columbian times from *N. sylvestris*, *N. tomentosiformis* and *N. otophora*.

CLASSIFICATION Neurotoxin, mind-altering; highly hazardous, Ib

ACTIVE INGREDIENTS Nicotine is the major alkaloid, while nornicotine, anabasine and nicotyrine are minor components. The leaf alkaloid content is up to 9 % (in *N. rustica* up to 18 %).

UTILISATION Indians have been smoking tobacco leaves (also *N. rustica*) since pre-Columbian times for ritual purposes or just for pleasure. Smoking rapidly became fashionable in W Europe (more so when it was said to have aphrodisiac properties). Despite the high risk of heart disease and cancer, smoking remains popular in most countries. Nicotine (or tobacco leaf powder) is very poisonous to most insects and has been used (since 1664) as an insecticide.

TOXICITY Accidental poisoning due to ingestion of cigarettes (with 10 mg nicotine) or cigars (90 mg nicotine) by small children is quite common but also rarely fatal. Tobacco is frequently used for suicide. Tobacco harvesters may suffer from "green tobacco sickness" if no protective clothing is used. Taken by mouth, 40–60 mg nicotine is lethal for adults; chronic smokers are less susceptible. 5–10 g of dry leaves will kill a dog or cat, 300–500 g a horse or cow.

SYMPTOMS The symptoms of nicotine poisoning appear rapidly since the lipophilic free base is readily absorbed (also through the skin). They include nausea, weakness, headache, salivation, tremor, convulsions, tachycardia, cold perspiration, confusion, diarrhoea, difficulty in breathing and hypertension. High doses cause convulsions, unconsciousness and rapid death through respiratory failure. Nicotine is the addictive component of tobacco, and has tranquillising and stimulating properties. Chronic smoking causes serious circulatory and heart disorders. Tobacco smoke contains tar, nitrosamines and benzopyrenes with mutagenic properties; they can cause cancer (especially lung carcinoma).

PHARMACOLOGICAL EFFECTS Nicotine is a powerful agonist at nicotinic acetylcholine receptors and thus affects neuromuscular and neuronal activity, and vegetative ganglia. Lower doses activate the central and vegetative nervous system, whereas higher doses are paralytic. Nicotine induces the release of dopamine, which produces a feeling of happiness and well-being.

FIRST AID See *Nicotiana glauca*.

Nicotiana tabacum L.   family: Solanaceae

tabac de Virginie (French); *Tabak, Virginischer Tabak* (German); *tabacco Virginia* (Italian); *tabaco de Virginia* (Span.)

Ornithogalum thyrsoides

star-of-Bethlehem • chincherinchee

Ornithogalum thyrsoides

Ornithogalum umbellatum

PLANTS WITH SIMILAR PROPERTIES A large genus with more than 250 species in Africa and Eurasia, including *O. toxicarum*, *O. boucheanum*, *O. prasinum*, *O. longibracteatum*, *O. angustifolium*, *O. umbellatum* and *O. pyrenaicum*. Several species are known to be toxic to animals and humans. These include *O. saundersiae*, *O. conicum*, *O. ornithogaloides*, *O. prasinum*, *O. tenellum*, and a recently described miniature species, *O. toxicarium*. A related genus that accumulates cardiac glycosides is *Scilla* (40 species in Europe, Asia and Africa).

PLANT CHARACTERS The plant is an erect perennial herb with a rosette of strap-shaped leaves and a single, fleshy flowering stem growing from a small bulb. The attractive flowers are star-shaped, with a dark centre and white petals.

OCCURRENCE *O. thyrsoides* is restricted to S Africa; introduced to S Australia. As cut flower it is exported to Europe. Several species are grown as ornamental plants.

CLASSIFICATION Heart poison; moderately to highly poisonous, Ib–II

ACTIVE INGREDIENTS The toxic compound in *O. thyrsoides* is a steroidal glycoside known as prasinoside G. Similar compounds have been isolated from *O. prasinum* and *O. saundersiae*. However, *O. toxicarium*, *O. longibracteatum* and *O. umbellatum* contain cardiac glycosides (carde-

nolides), such as rhodexin A, B, ornithogalin and convallatoxin.

UTILISATION Some species have been used in traditional medicine but today they are mainly grown as ornamental plants. In Europe, bulbs of *O. umbellatum* and green shoots ("asparagus") of *O. pyrenaicum* are sometimes eaten.

TOXICITY People and especially children are regularly exposed to the plants in gardens and the home (when used as cut flowers). Animals may be poisoned by garden waste or contaminated fodder. *Ornithogalum* species are extremely poisonous, with fatal doses in animals often measured in mg/kg rather than the usual g/kg body weight. *O. toxicarium* killed nearly 3 000 small livestock in one season.

SYMPTOMS Symptoms include nausea, salivation, colic, gastrointestinal disturbance, purging and exhaustion, with the usual respiratory and cardiac symptoms (arrhythmia, hypertension, coma, cardiac arrest) associated with heart glycoside poisoning.

PHARMACOLOGICAL EFFECTS In general cardiac glycosides inhibit the Na$^+$, K$^+$-ATPase (see cardiac glycosides).

FIRST AID Induce vomiting; administer sodium sulphate and medicinal charcoal. **Clinical therapy:** In case of severe poisoning: see cardiac glycosides.

Ornithogalum thyrsoides Jacq.   family: Hyacinthaceae or Asparagaceae

étoile de Bethléem, ornithogale (French); *Milchstern* (German)

Oxalis acetosella

wood sorrel

Oxalis acetosella

Oxalis pes-caprae

PLANTS WITH SIMILAR PROPERTIES
A very large genus of more than 700 species with a cosmopolitan distribution. Many species occur in S America and S Africa, including *O. pes-caprae*, *O. magellanica* and *O. tuberosa*. Some species of *Rumex* (such as *R. acetosa*) are also known as sorrel and together with rhubarb (*Rheum rhabarbarum = R.* x *hybridum*) are regularly eaten as vegetables because of their sour taste. *O. corniculata* is a cosmopolitan weed of unknown origin that is found in practically all gardens and greenhouses around the world.

PLANT CHARACTERS Perennial herbs with clover-like leaves that display sleeping motions – they characteristically droop or fold at night. The attractive, funnel-shaped flowers have five white petals. *O. pes-caprae* is somewhat larger and has conspicuous yellow flowers. It has large, fleshy and edible tubers.

OCCURRENCE *O. acetosella* occurs in woods of central and southern Europe, Asia and N America. *O. pes-caprae* originated from S Africa but has become a troublesome weed in areas with a Mediterranean climate (Europe, Africa, Australia and S America).

CLASSIFICATION Cytotoxin; slightly hazardous, III

ACTIVE INGREDIENTS Leaves are rich in oxalates (0.3–1.25 %) and have a sour taste.

UTILISATION Leaves may be eaten as a vegetable. Leaves of *O. pes-caprae* have been used in cooking as a replacement for vinegar.

TOXICITY Poisoning has been reported in children and livestock, resulting in kidney damage. Excessive consumption of sorrel or rhubarb leaves has resulted in some human fatalities in Europe. Oxalate poisoning of livestock occurs quite frequently in pastures infested with *Oxalis* species, especially *O. pes-caprae* in Australia and New Zealand.

SYMPTOMS Irregular pulse, hypotonia, spasms, collapse of circulation, coma and death by central paralysis or fatal kidney damage (see oxalic acid). Patients with reduced kidney function are especially vulnerable (see oxalic acid). About 5 to 15 g of oxalic acid are lethal for adults.

PHARMACOLOGICAL EFFECTS Oxalic acid forms insoluble salts with calcium. If calcium oxalate is deposited in kidney tubules, kidney tissue becomes damaged and other consequences follow (see oxalic acid).

FIRST AID Give calcium to precipitate oxalic acid; milk is a good source of calcium (see oxalic acid). **Clinical therapy:** In severe cases: gastric lavage (possibly with calcium salts, such as calcium gluconate; or with 0.1 % potassium permanganate); for further detoxification see oxalic acid.

Oxalis acetosella L.   family: Oxalidaceae

surelle petite oxalide, pain de concon (French); *Wald-Sauerklee* (German); *acetosella del boschi* (Italian)

Papaver somniferum

opium poppy

Papaver somniferum

Papaver somniferum capsule with latex

PLANTS WITH SIMILAR PROPERTIES A genus of 80 species in Europe, Africa, Asia and N America, including *P. glaucum, P. bracteatum, P. setigerum, P. orientale, P. rhoeas* and *P. nudicaule.*

PLANT CHARACTERS A leafy annual herb of up to 1.5 m high with glaucous, hairless and markedly toothed leaves. The attractive purple flowers and characteristic fruit capsules are borne individually on erect, slender stalks.

OCCURRENCE SW Asia. The cultivated form of poppy is an ancient cultigen that has been grown in Europe, N Africa and Asia (India, Afghanistan, Turkey) for centuries. Poppy and other *Papaver* species are grown as ornamentals.

CLASSIFICATION Neurotoxin; mind-altering; highly hazardous, Ib

ACTIVE INGREDIENTS Opium poppy and opium contain several isoquinoline alkaloids, especially morphine (10–12 %), noscapine (= narcotine, 2–10 %) and codeine (2.5–10 %).

UTILISATION Opium is the latex that oozes out of shallow cuts made in unripe poppy fruits. Its sedating and mind-altering properties have been known since antiquity. Pure alkaloids are nowadays used in standardised medicines to treat intense pain (e.g. in cancer patients) and cough. Morphine can be chemically converted into heroin, a major intoxicant. Afghanistan produces (mostly illegally) about 90 % of opium (about 8 200 tons) in the world (with an increasing trend).

TOXICITY Opium poisoning results from overdosing for medical, intoxication, or suicidal purposes. A sedating tea, made from dried capsules, has caused several cases of serious poisoning. About 200–400 mg morphine, given orally, are lethal for humans; 100–200 mg after parenteral application. Heroin: 50–75 mg, papaverine: >100 mg/kg are lethal (see morphine).

SYMPTOMS Deep sleep, irregular slow respiration, narcotic muscle relaxation, vomiting, bradycardia, reddening of face, dizziness, vertigo and contracted pupils. Death occurs after 6–12 hours from respiratory arrest. Opium and morphine cause constipation (used as anti-diarrhoeals).

PHARMACOLOGICAL EFFECTS Morphine modulates endorphine receptors and promotes powerful sleep-inducing, analgesic and hallucinogenic effects but is highly addictive. Codeine is an effective painkiller (weaker than morphine and less addictive), and sedates the cough centre (antitussive activity). Noscapine is a specific antitussive but is cytotoxic, as it binds to microtubules and inhibits spindle formation. Papaverine inhibits phosphodiesterase and is used as spasmolytic.

FIRST AID After oral poisoning: Use medicinal charcoal and sodium sulphate; induce vomiting. **Clinical therapy:** Gastric lavage within 1–2 h after ingestion (see morphine).

Papaver somniferum L.

family: Papaveraceae

pavot officinal, pavot somnifère (French); *Schlafmohn* (German); *papavero domestico, papavero sonnolente* (Italian)

Pausinystalia yohimbe

yohimbe tree

Pausinystalia yohimbe

Pausinystalia yohimbe bark

PLANTS WITH SIMILAR PROPERTIES A genus of 13 species in tropical W Africa. Other sources of yohimbine include *Aspidiosperma quebracho-blanco, Alstonia angustifolia* (Apocynaceae) and *Corynanthe pachyceras* (Rubiaceae).

PLANT CHARACTERS A large forest tree (up to 30 m) with opposite pairs of simple leaves, small tubular, white or yellowish flowers and small fruit capsules containing winged seeds.

OCCURRENCE W & C Africa (Cameroon to Congo). Most of the commercial bark material is collected from natural stands and there is concern about the sustainability of wild-harvesting.

CLASSIFICATION Neurotoxin, mind-altering; highly hazardous, Ib

ACTIVE INGREDIENTS Yohimbe bark contains monoterpene indole alkaloids of the yohimbine type at levels of 3–15 % dry wt. The main compound, (+)-yohimbine, occurs with several minor alkaloids (mostly isomers of yohimbine).

UTILISATION Bark extracts are traditionally used in West and Central Africa as male tonics and stimulants. The drug was used together with *Tabernanthe iboga* in initiation rites. It has nowadays become a popular phytomedicine for treating impotence and frigidity. There is scientific support for these uses, as yohimbine was shown to increase copulation frequency in rats. In addition, the product is reported to alleviate hyper-tension and fatigue.

TOXICITY No fatalities resulting from acute overdosage have yet been reported but 1.8 g of yohimbine (100 times the average daily dose) is said to have resulted in unconsciousness for some hours but complete recovery within a few days. Yohimbine is moderately toxic when injected (LD_{50} of 20 mg/kg in mice).

SYMPTOMS Yohimbine is a central stimulant and can enhance general anxiety. Side effects include over-stimulation, irritability and anxiety attacks. Due to the MAO inhibitory action, yohimbine could cause serious side effects when used with tyramine-rich foods. The excitability of the lower abdomen is also increased, so that there is a plausible explanation for the use in urinary incontinence (weak bladder) and impotence.

PHARMACOLOGICAL EFFECTS The alkaloid yohimbine inhibits α-adrenergic receptors, especially α_2-receptors; it thus affects the sympathic nervous system, leading to vasodilatation and a decreased blood pressure. Yohimbine is a local anaesthetic and is said to produce mild hallucinations when smoked.

FIRST AID After oral poisoning: Instil medicinal charcoal and sodium sulphate, induce vomiting and provide plenty of tea to drink. **Clinical therapy:** Gastric lavage within 1–2 h after ingestion (possibly with potassium permanganate).

Pausinystalia yohimbe (K. Schum.) Beille (= *Corynanthe yohimbe* K. Schum.)   family: Rubiaceae
yohimbe, yohimbéhé (French); *Yohimbe-Baum, Potenzholz* (German); *yohimbe* (Italian)

Peddiea africana

poison olive

Peddiea africana

Peddiea africana fruits

PLANTS WITH SIMILAR PROPERTIES A genus of 12 species in E and S Africa. *P. volkensii* is chemically similar to *P. africana*. Several members of the family Thymelaeaceae are known to produce daphnane-type diterpenes (phorbol esters). Similar compounds are found in the irritant latex of many *Euphorbia* species and other members of the Euphorbiaceae.

PLANT CHARACTERS The poison olive is a shrub or small tree with tough, fibrous bark and glossy green, hairless leaves on very short petioles. Small, green to yellowish green, tubular flowers are borne in rounded clusters. The fruits are egg-shaped, olive-like berries that turn black and shiny when they ripen.

OCCURRENCE *P. africana* grows on the margins of evergreen forests and is distributed along the South African coastal region and further north to Zimbabwe. Many of the species occur in the mountains of E Africa (especially Tanzania).

CLASSIFICATION Cellular poison, cocarcinogen, skin irritant; extremely hazardous, Ia

ACTIVE INGREDIENTS Roots of *P. africana* contain a diterpene of the daphnane type, known as *Peddiea* factor A_1. Two structurally similar diterpenes have also been found in *P. volkensii*. The presence of several quinones and coumarins have been reported from *P. fischeri*.

UTILISATION Infusions of the whole plant have been used for human poisoning in Zimbabwe. Poison olive is used to some extent in traditional medicine but few details have been recorded.

TOXICITY Despite the fact that *Peddiea* species are used in traditional medicine and also for murder, almost nothing appears to be known about their toxicity.

SYMPTOMS The plant or its extracts are known to be lethally poisonous but the exact symptoms are as yet unknown. However, it is likely to be similar to those observed with phorbol ester poisoning. The phorbol esters are also known to be skin irritant compounds that are responsible for severe allergic reactions in some people. Exposure to fresh plant material leads to skin inflammation with reddening and blister formation.

PHARMACOLOGICAL EFFECTS The daphnane type phorbol esters activate protein kinase C see "phorbol esters" for physiological and toxicological consequences.

FIRST AID Induce vomiting, give medicinal charcoal and sodium sulphate. Provide plenty of tea. **Clinical therapy:** Gastric lavage, instillation of medicinal charcoal and sodium sulphate; electrolyte substitution, control of acidosis with sodium bicarbonate. In case of spasms give diazepam. Provide intubation and oxygen therapy in case of respiratory arrest. Check kidney function (see phorbol esters).

Peddiea africana Harv.

Giftolive, Grüner Blütenbaum (German)

family: Thymelaeaceae

Peganum harmala

African rue • harmala

Peganum harmala

Peganum harmala flower

PLANTS WITH SIMILAR PROPERTIES A genus with 4 or 5 species in the N hemisphere. *Peganum* is chemically similar to *Banisteriopsis caapi* (Malpighiaceae).

PLANT CHARACTERS Harmala is a much-branched small shrub (about 0.5 m high) bearing pinnately compound leaves with linear segments. The attractive cream-coloured, star-shaped flowers are followed by dehiscent seed capsules.

OCCURRENCE Mediterranean region (Europe and N Africa), Arabia and western Asia. Harmala is typical of dry and sunny habitats.

CLASSIFICATION Neurotoxin, hallucinogen; highly hazardous, Ib

ACTIVE INGREDIENTS Seeds are toxic and hallucinogenic because of β-carboline type indole alkaloids (so-called harman alkaloids; 2–6 % dry weight, mainly harmine, with harman, harmol and harmatol). Leaves and stems produce vasicine, vasicinone and other quinazoline alkaloids.

UTILISATION Seeds are traditionally used as analgesic and antispasmodic to treat wounds and rashes, eye diseases, stomach pain, nervous disorders, rheumatism, impotence and even Parkinson's disease. Rue had been introduced to Iran and India from N Africa by Muslims. Smoke generated by burning the seeds is inhaled as a tradiional intoxicant and sexual stimulant in central Asia and Arabian countries. It is claimed that the hallucinogenic effects of the seed alkaloids have given rise to the intricate designs on oriental carpets, as well as the concept of the flying carpet. Seeds thrown into the fire may cause the sensation of flying, similar to the effects of atropine and witches that fly on brooms.

TOXICITY (see *Banisteriopsis*).

SYMPTOMS About 4 mg of harmine and harmaline induce hallucinogenic and euphoric effects (see *Banisteriopsis*).

PHARMACOLOGICAL EFFECTS The alkaloids interact with the central nervous system. They can activate serotonin receptors because of structural similarity between serotonine and β-carboline alkaloids. They also inhibit monoamine oxidase (MAO), an enzyme that breaks down catechol amines. As a consequence, serotoninergic and dopaminergic synapses are stimulated, leading to euphoric and hallucinogenic effects. The alkaloids intercalate with DNA and are cytotoxic, especially after exposure to light.

FIRST AID Provide plenty of fluids to drink; induce vomiting, give medicinal charcoal and sodium sulphate; apply shock prophylaxis. **Clinical therapy:** Gastric lavage, electrolyte substitution, control of acidosis with sodium bicarbonate; in case of spasms, diazepam; check kidney function. In case of respiratory arrest, provide intubation and artificial respiration.

Peganum harmala L.   family: Nitrariaceae or Zygophyllaceae

harmel (French); *Steppenraute* (German); *péganum* (Italian)

Petroselinum crispum

parsley

Petroselinum crispum (curly cultivar)   *Petroselinum crispum* (Italian parsley)

PLANTS WITH SIMILAR PROPERTIES A genus of 2 European species.

PLANT CHARACTERS Parsley is a small, biennial herb with a fleshy root and compound leaves borne in a basal rosette. Inconspicuous yellow flowers and small dry fruits (schizocarps) are produced in characteristic umbels in the second year. The wild form of parsley, as well as the cultivated Italian parsley (var. *neapolitanum*) have flat leaf segments, while modern cultivars have finely divided and curly, moss-like leaves. Turnip-rooted parsley (var. *tuberosum*) has edible roots.

OCCURRENCE Parsley originated in Europe and western Asia but is widely cultivated for culinary and medicinal uses. It is perhaps the most popular of all garnishes and salad ingredients.

CLASSIFICATION Cell poison; slightly hazardous, III

ACTIVE INGREDIENTS An essential oil rich in the phenylpropanoids apiol, myristicin or allyltetramethoxybenzol occur in parsley fruits (2–6 %) and in leaves (0.5 %). Also present are biologically active flavonoids (including apiin) and traces of furanocoumarins (e.g. bergapten, oxypeucedanin, psoralen). Of toxicological importance are chemotypes that are rich in myristicin, apiol or allyltetramethoxybenzol.

UTILISATION Leaves and roots are traditionally used to treat stomach and urinary tract disorders (including kidney gravel). The leaves are also used topically against skin ailments and to relieve itching. The fruits are much more active and have been used to treat painful periods but are no longer considered safe because of possible abortifacient effects. Parsley extracts are popular ingredients in diuretic and laxative products used for their "slimming" effects.

TOXICITY Apiol-rich extracts or oil increase contraction of smooth muscles of the gut, bladder and uterus. Parsley has therefore been used traditionally as an abortifacient. High doses are toxic and may cause liver damage.

SYMPTOMS Ingestion of large amounts of leaves can induce the mind-altering effects known from myristicin (see *Myristica fragrans*).

PHARMACOLOGICAL EFFECTS Myristicin and apiol have a terminal methylene group that is chemically highly reactive. They can bond to proteins or DNA. The diuretic activity is ascribed to the irritant and stimulant effects the phenylpropanoids and flavonoids have on the kidneys. Pure apiol is known to be abortifacient in high doses. Myristicin is a stimulant and hallucinogen that is also present in nutmeg oil from *Myristica fragrans* (see myristicin and furanocoumarins).

FIRST AID In case of overdosing, instil medicinal charcoal and sodium sulphate, and provide ample amounts of liquids to drink.

Petroselinum crispum (Mill.) A.W. Hill   family: Apiaceae

persil (French); *Petersilie* (German); *prezzemolo* (Italian); *perejil* (Spanish)

Phaseolus lunatus

lima bean

Phaseolus lunatus

Phaseolus lunatus

PLANTS WITH SIMILAR PROPERTIES *Phaseolus* comprises a genus of more than 36 species in warm and tropical America. Several species have been domesticated and cultivated, such as wild bean or tepary bean (*P. acutifolius*), scarlet runner bean (*P. coccineus*), Andes bean (*P. flavescens*) and the common bean or French bean (*P. vulgaris*).

PLANT CHARACTERS Lima bean is a perennial plant with typical pinnately trifoliate leaves, white flowers and broad, flattened pods. Each pod contains several white or pale green seeds. Commercial cultivars differ in seed size (large- or small-seeded) and in growth form (climbing or bushy).

OCCURRENCE. Lima bean originates from C & S America, including Mexico, Ecuador, Peru and Argentina. Large-sized lima beans were domesticated about 8 000 years ago in Peru. Small beans have been cultivated in Mexico and C America from about 500 years ago. The lima bean was later introduced into Africa and Madagascar, as well as N America, where it has become an important commercial food crop. It requires warm, tropical conditions and more than 100 000 tons of dry beans are produced each year.

CLASSIFICATION Cellular poison; moderately hazardous, II

ACTIVE INGREDIENTS Lima beans contain cyanogenic glucosides (CGs) such as linamarin.

Bitter cultivars produce up to 0.1–0.4 % dry wt.

UTILISATION The seeds are used mainly as dry beans. It is an excellent protein source and is part of the staple diet in S America and in parts of Africa. Lima beans are soaked in water for several hours, then cooked for at least 10 minutes. CGs and HCN that are liberated during this process are removed by decanting the cooking water.

TOXICITY High doses can have lethal effects. When lima beans are not processed correctly, severe intoxications can occur.

SYMPTOMS Poisoning symptoms are typical for HCN in general (irritation, flushing face, heavy breathing, scratchy throat, headache, and in severe cases, respiratory and cardiac arrest).

PHARMACOLOGICAL EFFECTS Linamarin releases HCN after cleavage by the enzyme linamarase or by acid; HCN is rapidly inactivated in the human body at low concentrations. Higher doses stop cellular respiration and thus the production of the essential ATP which drives all energy dependent processes; this leads to general paralysis and respiratory arrest (see amygdalin).

FIRST AID After ingestion of bitter lima bean products, induce vomiting and allow the intake of large volumes of drinks. **Clinical therapy:** In case of severe intoxication, follow the general procedure for HCN poisoning (see amygdalin).

Phaseolus lunatus L.   family: Fabaceae

haricot de Lima (French); *Limabohne, Mondbohne* (German); *fagiolo de Lima* (Italian); *frijol de Lima* (Spanish)

Phaseolus vulgaris

common bean • French bean • kidney bean

Phaseolus vulgaris pods

Phaseolus coccineus flowers

PLANTS WITH SIMILAR PROPERTIES There are more than 36 species of *Phaseolus*, all from the warm and tropical parts of N and S America. Several species have been domesticated and cultivated, such as *P. acutifolius, P. coccineus, P. flavescens* and *P. lunatus*. Other well-known plants with poisonous lectins include *Abrus precatorius, Robinia pseudoacacia, Wisteria sinensis* (all Fabaceae) and also *Ricinus communis* (Euphorbiaceae).

PLANT CHARACTERS The common bean is an annual plant with two basic growth habits: climbers, with long twining stems and bushy types, with short erect stems. The leaves are pinnately trifoliate, with two small stipules where they join the stem. Typical white to violet, pea-like flowers are borne in small clusters, followed by long, hanging pods. Each pod contains several kidney-shaped seeds (beans). A large number of cultivars have been developed, differing in the size, shape and colour of the seeds.

OCCURRENCE The common bean originates from tropical and subtropical C America, from Mexico to Peru, Bolivia and Argentina. It is a popular crop plant and staple diet in many parts of the world.

CLASSIFICATION Cell poison; moderately hazardous, II–III (raw)

ACTIVE INGREDIENTS The main toxic substance in bean seeds is the agglutinating lectin mixture known as phasin, which can be destroyed by heat (e.g. during cooking).

UTILISATION Beans are an important protein source and are widely produced and consumed. Dried bean pods (the pericarp without seeds) are used in traditional medicine to induce diuresis.

TOXICITY Although it is known that beans should be cooked and not eaten raw, the ingestion of raw beans by children has caused toxic effects.

SYMPTOMS Symptoms after eating raw seeds or pods include serious haemorrhagic gastrointestinal problems, bloody vomiting, perspiration, chill, tachycardia, weakness and shock, abdominal spasms. Phasin is known to be less toxic than abrin (see *Abrus*) and the symptoms are usually temporary. In rats, ingestion of raw beans causes atrophy of the thymus and spleen. Some people (especially those working in bean processing companies) develop allergic skin reactions.

PHARMACOLOGICAL EFFECTS Phasin is a haemagglutinating and mitogenic substance with the typical properties of ribosome inactivating lectins (see abrin).

FIRST AID When less than 6 seeds were ingested provide medicinal charcoal and plenty to drink. First aid treatment is indicated if several damaged or chewed uncooked seeds have been ingested. Treatment involves the induction of vomiting and detoxification as outlined under abrin.

Phaseolus vulgaris L.   family: Fabaceae

haricot (French); *Gartenbohne* (German), *fagiuolo comune* (Italian); *judia* (Spanish)

Physostigma venenosum

calabar bean • ordeal bean

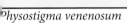
Physostigma venenosum

Physostigma venenosum seeds

PLANTS WITH SIMILAR PROPERTIES A genus of 4 species in tropical Africa.

PLANT CHARACTERS A woody climber that can reach up to 15 m in height. It has large, pinnately trifoliate leaves, pale purple flowers produced in clusters and large pods that each contain about three reddish brown seeds (beans) (15 by 8 mm).

OCCURRENCE Mainly in W Africa, but introduced to warm parts of America and Asia. Since it is common on the African Calabar Coast, it became widely known as the calabar bean.

CLASSIFICATION Neurotoxin; extremely hazardous, Ia

ACTIVE INGREDIENTS The seeds contain the indole alkaloid physostigmine (or eserine) and several derivates.

UTILISATION In W Africa, the calabar bean was used as ordeal poison. For example, the Efik people (according to a report from the missionary W. Daniell in 1840) would force the condemned person to drink a potion made of the calabar bean ("esere"). The person either died or was able to vomit and get rid of the poison. In the latter case, the person was considered innocent and allowed to depart unmolested. The kings of Calabar were usually buried together with hundreds of their people, who were decapitated, poisoned with esere or just buried alive. Physostigmine is used medicinally to treat glaucoma, since it reduces intraocular pressure. Another treatment is relevant for patients who produce antibodies against the endogenous acetylcholine receptor, resulting in myasthenia gravis. For treatment of Alzheimer's disease, physostigmine shows positive effects similar to those of galanthamine.

TOXICITY Poisoning can occur from overdosing the drug or isolated physostigmine. Neostigmine: LD_{50} mouse: 0.6 mg/kg s.c, 0.5 mg/kg i.v., 8 mg/kg p.o.

SYMPTOMS Physostigmine intoxication leads to numbness in the extremities, followed by a prolonged giddiness. Death is caused by paralysis of the heart and by respiratory failure.

PHARMACOLOGICAL EFFECTS Physostigmine is a potent inhibitor of cholinesterase, acting as an indirect parasympathomimetic. Physostigmine and its derivative neostigmine extend the biological half-life of acetylcholine and the elevated acetylcholine concentrations increase the activity of the cholinergic system.

FIRST AID Evoke vomiting; give medicinal charcoal and sodium sulphate. The patient should drink plenty of tea or water. **Clinical therapy:** Gastric lavage, electrolyte substitution, control acidosis with sodium bicarbonate (urine pH 7.5); in case of spasms give diazepam; in severe cases supply oxygen respiration and intubation. Atropine and vasoconstrictors can be given as antidotes.

Physostigma venenosum Balf.

family: Fabaceae

fève de Calabar (French); *Calabarbohne* (German); *fava del Calabar* (Italian); *haba de Calabar, nuez esere* (Spanish)

Phytolacca americana

pokeweed • inkberry

Phytolacca americana

Phytolacca americana

Phytolacca esculenta

PLANTS WITH SIMILAR PROPERTIES A genus of 25 species in warm and tropical climates, including *P. acinosa* (= *P. esculenta*) from East Asia, *P. polyandra* from China and the soapberry tree, *P. dodecandra* from Africa. Also well known is the bella umbra or elephant tree (*P. dioica*).

PLANT CHARACTERS Pokeweed is a leafy perennial herb of up to 3 m in height, bearing soft fleshy leaves, small greenish-white flowers and pendulous clusters of dark purple berries. Each berry comprises several fleshy segments. The plant was previously known as *P. decandra* and is still widely known by this name.

OCCURRENCE N America. Pokeweed is sometimes found as a weed of disturbed places, especially in the Mediterranean region and in C Europe.

CLASSIFICATION Cell toxin; moderately hazardous, II

ACTIVE INGREDIENTS Pokeweed fruits contain coloured alkaloids (betacyanins), such as betanidine, betanine, phytolaccine and others. Important are saponins, phytolaccosides and several different triterpenoid aglycones (e.g. phytolaccagenin) in all plant parts. Oleanolic acid and phytolaccagenic acid are the main aglycones in *P. dodecandra*. Roots are chemically diverse and contain a lectin (known as pokeweed mitogen, PWM), an antiviral protein (known as pokeweed antiviral protein, PAP) and neolignans (isoameri-

canin A and others).

UTILISATION Roots or berries are used in traditional medicine as anti-rheumatic, emetic, analgesic, anti-inflammatory, anti-catarrhal, purgative, immune stimulant and lymphatic system stimulants. The fruit sap has been used to improve the colour of red wines. In parts of Africa, the berries of *P. dodecandra*, whose saponins are molluscicidal, are a cheap and effective measure to control the water snails that transfer bilharzia. The antiviral proteins are used against leukaemia.

TOXICITY Overdosing of medicinal preparations and eating the leaves as salad have caused poisoning in N America. *P. acinosa* contains proteins (probably lectins) that can cause abortion.

SYMPTOMS Symptoms include disturbance of the GI tract, such as nausea, vomiting, abdominal pain and colic, furthermore headache and circulatory problems. The sap has irritant properties when it comes into contact with the eye.

PHARMACOLOGICAL EFFECTS Saponins, lignans and lectins are probably responsible for the observed activities, including anti-inflammatory, hypotensive, diuretic, molluscicidal, spermicidal and abortive effects. The lectin and saponins are quite toxic, especially when injected.

FIRST AID Instil medicinal charcoal and sodium sulphate.

Phytolacca americana L. (= *P. decandra* L.)   family: Phytolaccaceae

phytolaque (French); *Amerikanische Kermesbeere* (German); *fitolacca, uva turca* (Italian)

Pilocarpus jaborandi

jaborandi

Pilocarpus jaborandi

Pilocarpus pennatifolius

PLANTS WITH SIMILAR PROPERTIES A genus of 22 species in tropical America. *P. microphyllus*, *P. pennatifolius* and *P. racemosus* are also known as "jaborandi" and are acceptable as alternative sources of commercial raw materials.

PLANT CHARACTERS A shrub or small tree (up to 3 m) with large, hairless, compound leaves and small pinkish flowers in pendulous clusters.

OCCURRENCE *Pilocarpus* species are indignous to S and C America. *P. jaborandi* occurs mainly in Brazil.

CLASSIFICATION Neurotoxin; highly hazardous, Ib

ACTIVE INGREDIENTS Leaves contain pilocarpine and numerous other imidazole alkaloids (e.g. isopilocarpine, pilocarpidine and isopilocarpidine) at levels of up to 1 % dry weight.

UTILISATION Jaborandi leaves are an important traditional medicine used against fever, influenza, bronchitis and tonsillitis. Extracts are used topically to reduce the symptoms of psoriasis and to reverse baldness. The alkaloids are nowadays used in ophthalmology to treat glaucoma (they lower intraocular pressure) and after surgery (to stimulate the intestines or bladder).

TOXICITY Poisoning results mostly from pilocarpine used in ophthalmology or in cosmetics. 60 mg are a lethal oral dose for humans. It is easily absorbed after oral ingestion.

SYMPTOMS Pilocarpine stimulates secretion of tears, saliva and sweat. Further symptoms: miosis, inhibition of accommodation, nausea, vomiting, initial tachycardia with a transition to bradycardia, vasodilatation and circulatory collapse, enhanced gut and bladder motility with colic. Enhanced bronchial secretion and bronchial spasms can cause cyanosis, dyspnoea and oedema. Stimulation of smooth uterus muscles can induce abortion. Death results from cardiac arrest.

PHARMACOLOGICAL EFFECTS Pilocarpine is a direct parasympathomimetic and an agonist at muscarinergic acetylcholine receptors. It acts as a miotic and reduces intraocular pressure. It is highly active as a diaphoretic (increases perspiration) and salivation. The anti-baldness activity is ascribed to an increase in capillary blood circulation, the opening of skin pores and the promotion of transdermal absorption of other pharmaceutical compounds.

FIRST AID Evoke vomiting; give medicinal charcoal and sodium sulphate. Patient should drink plenty of tea or water. **Clinical therapy:** Gastric lavage (possibly with 0.1 % potassium permanganate), electrolyte substitution, control acidosis with sodium bicarbonate (urine pH 7.5); in case of spasms give diazepam; in severe cases supply intubation and oxygen respiration. Atropine and vasoconstrictors can be given as antidote.

Pilocarpus jaborandi Holmes

family: Rutaceae

jaborandi (French); *Gewöhnlicher Jaborandistrauch* (German); *iaborandi* (Italian); *jaborandi* (Spanish)

Piper methysticum

kava kava • kava

Piper methysticum

Piper betle

PLANTS WITH SIMILAR PROPERTIES
A very large genus of more than 2 000 species, including several which are used as spices, such as ordinary black pepper (*P. nigrum*), betel (*P. betle*) (see *Areca catechu* for its utilisation), cubebs or Java pepper (*P. cubeba*), Guinea pepper (*P. guineense*) and Indian long pepper (*P. longum*).

PLANT CHARACTERS Kava is an evergreen, dioecious scrambler with woody, purple-spotted stems arising from enormous branched, fleshy rhizomes that can weigh up to 10 kg. The large leaves are heart-shaped, dark green above and paler below, with several veins radiating from the base. Inconspicuous flowers are borne in oblong clusters.

OCCURRENCE Polynesia. The plant appears to be an ancient cultigen derived from *P. wichmannii* and is grown in the western Pacific (mainly the islands of Fiji, Samoa, Tonga and Vanuatu).

CLASSIFICATION Mind-altering; slightly hazardous, III

ACTIVE INGREDIENTS Roots contain styrylpyrones (also called kavapyrones or kavalactones) at levels of more than 3.5 % dry weight. The main styrylpyrones are kawain, methysticin, dihydrokavain, dihydromethysticin, yangonin and desmethoxy-yangonin.

UTILISATION Traditional kava drink, which is stimulant and lightly intoxicant, is part of the cultural tradition of Polynesian people. It wa traditionally prepared by women and childre chewing the roots, spitting out the pulp and the mixing it with water to ferment. In modern time kava roots or root extracts have become popula to treat anxiety, stress and sleep disturbance Kava has been proposed as the herbal alternativ to synthetic anxiolytics such as benzodiazepine with the added advantage that there is no evidenc of physical or psychological dependency. The eff cacy of kava in treating patients with anxiety, ter sion and agitation or psychosomatic menopaus complaints is supported by several controlled clin cal studies. Because of liver damage in a few case the registration of the drug has been withdrawn.

TOXICITY Chronic abuse of kava may lead t serious toxic effects, such as liver damage.

SYMPTOMS Kava has sedative, muscle relax ing, anti-convulsive, tranquillising and analgesi properties.

PHARMACOLOGICAL EFFECTS Kavalac tones bind to various neuroreceptors, especiall GABA and dopamine receptors, and inhibit Na channels. These interactions explain the observe sedative, muscle relaxing, anti-convulsive, trar quillising and analgesic effects.

FIRST AID In case of overdosing, instil medic nal charcoal and sodium sulphate. Encourage th patient to drink plenty of water.

Piper methysticum Forster f.　　　　　　　　　　　　　　　　　　family: Piperacea

kava, kava kava (French); *Kava, Kava-Pfeffer, Rauschpfeffer* (German); *kava-kava* (Italian)

Podophyllum peltatum

may apple • American mandrake

Podophyllum peltatum

Podophyllum peltatum flower

Podophyllum hexandrum fruit

PLANTS WITH SIMILAR PROPERTIES A genus of 5 species from NE America and Asia.

PLANT CHARACTERS The may apple is a perennial herb with a single pair of soft, deeply lobed leaves arising from a branched rhizome. The solitary white flowers turn into oblong, fleshy fruits (toxic when green, edible when ripe). Himalayan may apple (*P. hexandrum*) forms dense clumps.

OCCURRENCE *P. peltatum*: eastern N America; *P. hexandrum* (the most important of the four Asian species): the Himalayas.

CLASSIFICATION Cell poison; highly hazardous, Ib

ACTIVE INGREDIENTS Rhizomes and roots contain up to 6 % of podophyllum resin or podophyllin. This resin contains up to 50 % of podophyllotoxin, a lignan (1-aryltetrahydro-naphthalene). Related compounds are α- and β-peltatin, neosoxypodophyllotoxin and their glycosides.

UTILISATION May apple is traditionally used as a strong purgative and also to remove condylomata. The lignans are chemically converted into chemotherapeutic agents (etoposide, mitoposide, teniposide) for treating testicular carcinomas and lymphomas.

TOXICITY Poisoning results mostly from medical preparations, but also from eating leaves (in salads) or the attractive red fruits. Lignans are mutagenic and can lead to malformations when taken during pregnancy. About 300–600 mg podophyllin is a lethal dose in humans. Podophyllotoxin: LD_{50} rats: 8.7 mg/kg i.v., 15 mg/kg i.p.

SYMPTOMS Ingestion leads to strong irritation of mucosa and eyes, serious gastroenteritis with abdominal colics, vomiting, long-lasting and sometimes bloody diarrhoea, cough, tachycardia, arrhythmia, paraesthesia in legs, polyneuritis, damage of central and peripheral nervous system, ileus, vaginal bleeding, nephritis, spasms, delirium, coma and death from respiratory arrest.

PHARMACOLOGICAL EFFECTS Podophyllotoxin and peltatin (both highly toxic when ingested) inhibit the growth of cells by acting as mitotic spindle poisons — they stop cell division at the beginning of metaphase. Semisynthetic derivatives with similar anticancer activity (inhibition of DNA topoisomerase I) but fewer side effects are available. They cause necrosis when applied to external carcinomas and warts of less than four cubic centimetres.

FIRST AID Give medicinal charcoal and sodium sulphate. **Clinical therapy:** (after overdosing): Gastric lavage, electrolyte substitution, control acidosis with sodium bicarbonate (urine pH 7.5); in case of spasms give diazepam; in severe cases supply intubation and oxygen respiration. Check kidney function; in severe cases perform haemoperfusion.

Podophyllum peltatum L.

family: Berberidaceae

podophylle pelté, pomme de mai (French); *Gewöhnlicher Maiapfel* (German); *podofillo* (Italian)

Prunus dulcis var. *amara*

bitter almond

Prunus dulcis

Prunus dulcis flowers and old fruits

PLANTS WITH SIMILAR PROPERTIES
A large genus of more than 200 species in temperate and subtropical regions (especially the northern hemisphere), including *P. armeniaca* (apricot), *P. avium* (cherry), *P. cerasus* (sour cherry), *P. domestica* (plum), *P. persica* (peach) and *P. spinosa* (blackthorn). Other Rosaceae with seeds rich in cyanogenic glucosides (CGs) include *Malus domestica* (apple), *Pyrus communis* (pear), *Cydonia oblonga* (quince) and *Pyracantha coccineus* (firethorn).

PLANT CHARACTERS A deciduous shrub or tree (up to 10 m in height) with elliptical leaves and white or pale pink flowers, each with 5 petals. The fruits are covered by a velvety outer layer and have a woody or bony inner layer surrounding the single seed (nut).

OCCURRENCE. C & W Asia; the var. *dulcis* has been cultivated for the last 3 000 years in the Near East, N Africa and the Mediterranean. Commercial cultivation nowadays occurs in many parts of the world, especially USA, Spain and Italy.

CLASSIFICATION Cellular poison; highly hazardous, Ib

ACTIVE INGREDIENTS Seeds contain up to 8.5 % cyanogenic diglucosides (amygdalin); seeds of "sweet" almonds (*P. dulcis* var. *dulcis*) have only trace amounts. Cyanogenic glucosides also occur in seeds of *P. armeniaca* (up to 8 %), *P. domestica* (up to 2.5 %), *P. persica* (up to 6 %), *P. cerasus* (up

to 6 %); *Malus sylvestris* has up to 1.5 % CGs.

UTILISATION Almonds (especially sweet almonds) are widely used as snacks, sweets and in confectionery. Small amounts of bitter almond are used for aromatisation, e.g. in marzipan, or in liqueurs such as amaretto. Cold pressed almond oil is used in aromatherapy and in cosmetics. Isolated amygdalin has been used as laetrile in alternative cancer therapy especially in the USA. This application was largely ineffective but there is still a lively debate and controversy about this therapy.

TOXICITY HCN is a strong toxin; 5–12 chewed seeds of bitter almonds, apricots or peaches can be lethal for children, 20–60 for adults. HCN is lethal for humans at a dose of 1 mg/kg. Poisoning has been recorded from eating seeds of cyanogenic species or by overdosing with laetrile.

SYMPTOMS High doses of HCN stop cellular respiration and lead to general paralysis and respiratory arrest (see amygdalin).

PHARMACOLOGICAL EFFECTS Amygdalin releases HCN after cleavage by emulsin; HCN is rapidly inactivated in the human body at low concentrations. HCN is a respiratory poison.

FIRST AID When several bitter almond seeds have been ingested, induce vomiting and allow the intake of large volumes of drinks. **Clinical therapy:** In case of severe intoxication, see amygdalin.

Prunus dulcis (Mill.) D.A. Webb var. *amara* (DC.) Buchheim

family: Rosaceae

amandier amère (French); *Bittermandel* (German); *mandorla amara* (Italian); *almendro amargo* (Spanish)

Prunus laurocerasus

cherry laurel

Prunus laurocerasus flowers

Prunus laurocerasus fruits

PLANTS WITH SIMILAR PROPERTIES The genus *Prunus* is a large group of more than 200 species in the northern hemisphere (especially Europe); species with similar toxicological properties are *P. padus* (bird cherry) and *P. serotina* (black cherry).

PLANT CHARACTERS A branched evergreen shrub of up to 4 m in height, with simple, glossy green leaves (length up to 150 mm) and elongated racemes of small white flowers, borne in the upper leaf axils. The fruits are small black berries.

OCCURRENCE. Cherry laurel is indigenous to SE Europe and Asia Minor; it is widely cultivated in gardens and parks, often as a hedge or screen planting.

CLASSIFICATION Cell poison; moderately hazardous, II

ACTIVE INGREDIENTS The leaves of cherry laurel contain prunasin (1–1.5 % fresh weight), while the seeds contain amygdalin (0.16–0.21 %). Fruit pulp releases 100–200 ppm HCN.

UTILISATION Cherry-laurel water was formerly used as a flavouring agent and medicinally as a respiratory stimulant. The fruits (drupes) of cherry laurel look like tiny black cherries and may therefore be attractive to infants and small children.

TOXICITY The use of cherry-laurel water with 0.1 % HCN for suicide and murder is well recorded. No harm is usually done if only a few fruits have been eaten, as the toxic seeds are spat out or pass through the stomach undamaged. Although ingestion of material from *P. laurocerasus* is quite common in humans (many consultations in poison centres concern ingestion of laurel fruits by children), severe intoxications are rare. Animals may be fatally poisoned however (1 kg of fresh leaves will kill a cow). Goats died after feeding on *P. serotina* leaves. As with all cyanogenic plants, the glycoside is only broken down enzymatically to the toxic HCN once the plant material is damaged (by chewing, for example).

SYMPTOMS Poisoning symptoms are typical for cyanogenic glycosides in general (with irritation, flushing face, heavy breathing, scratchy throat, headache, and in severe cases, respiratory and cardiac arrest).

PHARMACOLOGICAL EFFECTS Amygdalin and prunasin release HCN after cleavage by the enzymes emulsin or β-glucosidase; HCN is rapidly inactivated in the human body at low concentrations. Higher doses stop cellular respiration and ATP production. The loss of ATP as an energy source leads to general paralysis and respiratory arrest (see amygdalin).

FIRST AID After ingestion of several cherry laurel seeds or leaves induce vomiting and allow the intake of large volumes of drinks. **Clinical therapy:** In case of severe toxication, see amygdalin.

Prunus laurocerasus L.   family: Rosaceae

laurier-cerise (French); *Kirschlorbeer, Lorbeerkirsche* (German); *lauroceraso* (Italian); *lauroceraso, laurel-cerezo* (Spanish)

Psilocybe mexicana

teonanacatl • magic mushroom

Psilocybe mexicana

Psilocybe semilanceata

FUNGI WITH SIMILAR PROPERTIES Fungi with psilocybin include *Conocybe cyanopus, Copelandia bispora, Inocybe aeruginascens, I. coelestium, Psilocybe azurescens, Galerina steglichii, Panaeolus cyanescens, Gymnopilus aeruginosus, Pluteus cyanopus, Stropharia cubensis, Pholiotina cyanopus* and *Panaeolina foenisecii*.

CHARACTERS A small saprophytic mushroom, growing singly or in groups, with a pileus (cap) of up to 20 mm in diameter, borne on a slender, 100 mm long bare stipe (stalk).

OCCURRENCE *Psilocybe mexicana* occurs in Mexico; related species occur worldwide. *P. semilanceata, P. cubensis, Pluteus salicinus* and *Inocybe aeruginascens* are psychotropic species of Europe.

CLASSIFICATION Hallucinogen; moderately hazardous, II

ACTIVE INGREDIENTS Up to 2 % alkaloids such as psilocybin and psilocin, which are analogues of the neurotransmitter serotonin.

UTILISATION This small mushroom was known to the Aztecs as *teonanacatl*. It has a long history, as mushroom-shaped stones (approx. 3 000 years old) have been discovered in Guatemala. Even today, the Psilocybe cult is still important in parts of Mexico.

TOXICITY Poisoning is mainly due to deliberate ingestion of mushroom for intoxication. 4–8 mg psilocybin are already mind-altering, whereas

6–20 mg cause serious hallucinogenic effect LD_{50} mouse: 275 mg/kg i.v., 420 mg/kg i.p.

SYMPTOMS Psychedelic effects last 4–12 hou and are both pleasant and unpleasant: happiness coloured pictures, anxiety, laughing, loss of inh bitions, tantrum, aggressiveness, erotic feeling hallucinations, loss of personality, delirium an unconsciousness. Toxic symptoms appear aft ½ to 2 h and include headache, dizziness, dilate pupils, coordination problems, nausea, musc weakness, slow pulse, low blood pressure, parae thesia and restlessness. Overdoses can be letha especially in children or adolescents.

PHARMACOLOGICAL EFFECTS Psilocybi is the phosphate ester of psilocin. The phosphat group increases the stability of the alkaloid in th gastrointestinal tract. It is cleaved in the body an the lipophilic psilocin, which can pass the blooc brain barrier by diffusion, is the neuroactiv compound. Psilocin is a mimic of serotonin (the a $5HT_2$ agonist). It has psychedelic properties sin ilar to those of LSD but is 100 times less active.

FIRST AID Keep patient warm and calm in quiet and dark room; often no further therapy necessary. In case of strong excitation: give do: epin (50 mg p.o. or i.m.), or diazepam. After ove doses: induce vomiting, administer medicin charcoal and sodium sulphate; supply artifici respiration and oxygen if necessary.

Psilocybe mexicana Heim

family: Strophariace

Teonanacatl, Mexikanischer Zauberpilz (German)

Psychotria ipecacuanha

ipecac

sychotria ipecacuanha

Psychotria emetica

PLANTS WITH SIMILAR PROPERTIES A very large genus of 800–1500 species in the tropics. *Psychotria viridis*, *P. poeppigiana* and *P. psychotriaefolia* produce *N,N*-DMT and have strong mind-altering properties.

PLANT CHARACTERS A woody shrublet (up to 0.4 m) with pairs of opposite leaves, characteristic interpetiolar stipules, clusters of small white flowers and small, ovoid, fleshy fruits that turn dark red to purple when they ripen.

OCCURRENCE The species is indigenous to Brazil.

CLASSIFICATION Cell poison; mind-altering, highly poisonous, Ib

ACTIVE INGREDIENTS Roots contain emetine, cephaeline and numerous other isoquinoline alkaloids at a level of up to 4 % of dry weight.

UTILISATION Rhizomes are collected for commercial use in the Matto Grosso region of Brazil. Ipecac is an ingredient of emetic syrups ("syrup of ipecac") that are used to induce vomiting, especially in cases where children have been accidentally poisoned (see Introduction). At low doses it is a traditional medicine used to treat chronic bronchitis (the alkaloids are expectorant).

TOXICITY Poisoning results mainly from overdosing of medicines containing emetine or from syrup of ipecac. Lethal oral dose: emetine 57 mg/kg, cephaeline 32 mg/kg.

SYMPTOMS Although the alkaloids are strongly emetic, dangerous overdosing can occur. Symptoms include rhinitis, serious irritation of skin and mucosa, cough, bronchial oedema, nausea, strong vomiting, gastroenteritis, bloody diarrhoea, abdominal pain, bradycardia, dyspnoea, weak muscle activity, enhanced salivation and perspiration. Death may result from cardiac arrest.

PHARMACOLOGICAL EFFECTS Emetine and cephaeline are very similar in their biological activities. They are irritants and stimulants of the nervus vagus in the stomach, resulting in a reflex reaction that leads to secretion of water in the lungs (similar to the secretolytic activity of saponins). Ipecac therefore has expectorant and secretolytic activities. Emetine strongly intercalates DNA and blocks protein biosynthesis; it causes cell death by apoptosis. This cytotoxicity is the basis for the use of emetine as a cytotoxic to treat tumours and intestinal parasites.

FIRST AID Evoke vomiting; give medicinal charcoal and sodium sulphate. Allow the patient to drink plenty of tea. **Clinical therapy:** After overdosing: gastric lavage (possibly with potassium permanganate), electrolyte substitution, control acidosis with sodium bicarbonate (urine pH 7.5) and kidney function; in case of spasms give diazepam; in severe cases supply intubation and oxygen.

Psychotria ipecacuanha (Brot.) Stokes (= *Cephaelis ipecacuanha* Brot.)   family: Rubiaceae

Ipécacuanha (French); *Brechwurzel, Ruhrwurzel, Ipecacuanha* (German), *ipecacuana* (Italian); *ipecacuana* (Spanish)

Pteridium aquilinum

bracken fern

Pteridium aquilinum

Pteridium aquilinum young leaves

PLANTS WITH SIMILAR PROPERTIES A single species with a cosmopolitan distribution.

PLANT CHARACTERS Bracken is a fern with branched rhizomes below the ground giving rise to leathery, bipinnate leaves. When emerging, the young leaves are coiled like fiddle heads before they unfold. The brown, powdery spores are produced along the lower edges of the leaf.

OCCURRENCE The bracken fern is often regarded as one of the most cosmopolitan of all plants. It is common in moist regions and at high altitudes on all continents, where it often forms dense stands, particularly after fire.

CLASSIFICATION Cell poison, mutagen; highly hazardous, Ib

ACTIVE INGREDIENTS The main toxin is a sesquiterpenoid called ptaquiloside. It occurs with the enzyme thiaminase and prunasin, a cyanogenic glycoside.

UTILISATION Bracken has been used as bed straw in former times and was thus distributed to many places by humans. Young bracken fern leaves are processed by cooking and are then commonly used as food in Japan and other parts of the world. The processing is considered to remove most but not all of the poisonous and carcinogenic compounds.

TOXICITY The carcinogenic and toxic compounds can be passed to humans through milk and other dairy products when cattle are expose[d] to pastures with bracken fern.

SYMPTOMS Bracken causes thiamine deficie[n]cy in non-ruminant animals, with the resultan[t] neuromuscular disturbances, as well as acu[te] haemorrhage syndrome, bovine haematuria an[d] "bright blindness". Serious stock losses are rar[e] and occur only when grazing becomes scarce du[r]ing dry periods.

PHARMACOLOGICAL EFFECTS Ptaqu[i]loside has a reactive cyclopropane ring that ca[n] alkylate DNA and proteins. It is a carcinogen[ic] and mutagenic norsesquiterpene that has bee[n] implicated in the high incidence of bladde[r] oesophagal and stomach cancer amongst peopl[e] who regularly consume young bracken frond[s]. Ptaquiloside destroys bone marrow and thereb[y] reduces the production of blood platelets (resul[t]ing in internal bleeding) and white blood cel[ls] (reducing resistance to infectious diseases). Th[e] enzyme thiaminase destroys thiamine (vit B_1) resulting in nervous symptoms, lack of coordina[tion] tion and internal bleeding in non-ruminants suc[h] as pigs and horses. These are typical symptoms [of] vitamin B_1 deficiency.

FIRST AID After eating large amounts of brack[en], en, evoke vomiting; instil medicinal charcoal an[d] sodium sulphate. Allow the patient to drink plen[ty] ty of tea and apply shock prophylaxis.

Pteridium aquilinum (L.) Kuhn

family: Dennstaedtiacea[e]

fougère aigle (French); *Adlerfarn* (German); *felce aquilina* (Italian)

Pulsatilla vulgaris

pasque flower

Pulsatilla vulgaris

PLANTS WITH SIMILAR PROPERTIES A genus of 38 species in the N hemisphere. The meadow pasque flower (*Pulsatilla pratensis*) and several other *Pulsatilla* species are an alternate source of raw material for the medicinal drug. *P. chinensis* are used as triditional medicine in China. Many other members of the family Ranunculaceae are used in traditional medicine for their topical counter-irritant effects.

PLANT CHARACTERS A small perennial herb with hairy stems bearing compound, finely dissected, feathery leaves. The attractive, anemone-like, purplish-blue flowers are borne individually at the tips of slender stalks. The compound fruits are formed by several carpels, each ending in a hairlike tip.

OCCURRENCE The plant is indigenous to Europe.

CLASSIFICATION Cell poison, mutagenic; moderately hazardous, Ib–II

ACTIVE INGREDIENTS Fresh plant material contains a terpenoid glucoside known as ranunculin that is broken down to protoanemonin when the plant is dried – the compound dimerises to form less toxic anemonin, if it has not alkylated proteins before.

UTILISATION Preparations are used in C and E European traditional medicine and in homeopathy to treat a wide range of ailments, including pain, neurotonic disorders and sleeplessness. In China, the roots of *P. chinensis* (*bai tou weng*) are part of traditional medicine and are used to treat dysentery.

TOXICITY *Pulsatilla* is rarely of toxicological importance; see protoanemonin. In rare cases, large overdoses of medicinal products may lead to intoxication.

SYMPTOMS Protoanemonin is well known as an irritant lactone that causes blistering of human skin. It causes severe allergic dermatitis when it comes in contact with the skin, resulting in itching, blisters and rashes. Internally, protoanemonine causes paralysis of the central nervous system. Gastrointestinal disturbance, including nausea, vomiting and diarrhoea, as well as nephritis are the usual symptoms.

PHARMACOLOGICAL EFFECTS Protoanemonin is a reactive compound with an exocyclic methylene group; it can bond with SH-groups of proteins and DNA (and is therefore mutagenic). The compound is also strongly antibacterial and antifungal (see protoanemonin).

FIRST AID If large amounts were consumed, instil medicinal charcoal, sodium sulphate and induce vomiting. Allergic symptoms can be treated with local application of corticoid ointment. In rare cases of breathing difficulties, intubation and oxygen may be necessary.

Pulsatilla vulgaris Mill. (= *Anemone pratensis* L.)   family: Ranunculaceae

anémone pulsatille (French); *Gemeine Küchenschelle* (German); *pulsatilla* (Italian)

Punica granatum

pomegranate

Punica granatum flower   *Punica granatum* fruit

PLANTS WITH SIMILAR PROPERTIES A genus of 2 species, with *P. protopunica* (on Socotra).
PLANT CHARACTERS A spiny, deciduous shrub or small tree (up to 5 m), with small, glossy leaves and attractive orange flowers. The large fruit has a persistent, crown-like calyx and a tough rind surrounding numerous bright red, edible seeds.
OCCURRENCE SE Europe (probably Turkey and the Caspian region). Pomegranate is believed to be an ancient cultivar and is still an important crop in the Middle East and Mediterranean region. Cultivars with double flowers have been developed for use as ornamental garden plants.
CLASSIFICATION Neurotoxin; moderately hazardous, II
ACTIVE INGREDIENTS Several piperidine alkaloids occur in the roots, bark, leaves and young fruit but not in the rind and ripe fruit. The major active alkaloids are pelletierine, *N*-methylisopelletierine, and pseudopelletierine. The fruit rind contains gallo/ellagitannins – mainly punicalin and punicalagin, at very high levels (up to 28 %).
UTILISATION Pomegranate was mentioned in the Ebers Papyrus and was traded by Punic traders (hence the Roman name "Malum punicum"). It was dedicated to the Phoenician mother goddess Astarte; later to Hera and Aphrodite in ancient Greece. It symbolises "fertility" and "life" to this day. In modern Greece, fruits are for this reason opened at the entrance of a house on New Year's day. It has been speculated that Eve seduced Adam with a pomegranate and not a real apple. Persephone was tempted by Hades with a pomegranate. Dried fruit rind or fruit pulp is a remedy for upset stomachs and especially to treat diarrhoea. Decoctions of root bark were widely used as tapeworm remedy up to the 1950s. Fermented fruit pulp is used to produce grenadine.
TOXICITY Extracts of the bark can be toxic and about 80 g of fruit rind causes toxic effects. The fleshy seeds, however, are not poisonous.
SYMPTOMS The alkaloids are absorbed through the intestine and cause undesirable CNS effects. Symptoms include hypertension, disturbance of vision, spasms, nausea, vomiting, gastroenteritis, vertigo, bradycardia, collapse, convulsions and even death (by respiratory failure).
PHARMACOLOGICAL EFFECTS The alkaloids interact strongly with nicotinic acetylcholine receptors, similar to nicotine.
FIRST AID Give medicinal charcoal, sodium sulphate and plenty of tea to drink. **Clinical therapy:** After ingestion of larger doses, gastric lavage (possibly with potassium permanganate), electrolyte substitution, control acidosis with sodium bicarbonate (urine pH 7.5) and kidney function in case of spasms give diazepam; in severe cases supply intubation and oxygen.

Punica granatum L.   family: Lythraceae (formerly Punicaceae)

grenadier (French); *Granatapfelbaum* (German); *melograno* (Italian); *granado* (Spanish)

Ranunculus acris

meadow buttercup

Ranunculus acris   *Ranunculus acris*   *Ranunculus multifidus*

PLANTS WITH SIMILAR PROPERTIES There are about 250 species of *Ranunculus* all over the world and heavy livestock losses have been reported from Europe, Africa and N America. Especially toxic are *R. sceleratus, R. thora, R. flammula, R. illyricus, R. multifidus* and *R. bulbosus.*

PLANT CHARACTERS Meadow buttercup is a much-branched perennial herb of up 1 m in height, with palmately compound leaves borne on slender stalks. The leaves are usually divided into 5 or 7 wedge-shaped segments. The flowers have 5 bright yellow, glossy petals and are borne at the tips of slender pedicels. The fruit structure is characteristic for the family. It comprises numerous free carpels that each form a small, one-seeded nut.

OCCURRENCE Europe to C Asia; naturalised in N America. Cultivars with double flowers ("Flore Pleno") are known as bachelor's buttons and are popular garden plants.

CLASSIFICATION Cell poison, mutagenic; highly hazardous, Ib–II

ACTIVE INGREDIENTS The toxic compound in several members of the Ranunculaceae is protoanemonin, an acrid oil. It does not occur in fresh material but is produced by enzymatic action (from a bitter-tasting glycoside, ranunculin) when fresh plant material is eaten or bruised (see protoanemonin).

UTILISATION Leaf poultices are used in traditional medicine to treat wounds, external cancers and rheumatism. Beggars are said to have treated their skin with sap of *Ranunculus* to produce horrible blisters in order to induce sympathy.

TOXICITY Buttercup is known to be poisonous to humans and animals. Farm animals are occationally killed even though the plant is not palatable. Ranunculin-containing plants are highly toxic – the lethal dose in sheep was shown to be 11 g/kg fresh weight.

SYMPTOMS Protoanemonin is a highly irritant and vesicant lactone that causes blistering of human skin. It causes severe allergic dermatitis, resulting in itching, blisters and rashes. When ingested, protoanemonin causes paralysis of the central nervous system; gastrointestinal disturbance (nausea, vomiting, diarrhoea) and nephritis are the usual symptoms.

PHARMACOLOGICAL EFFECTS Protoanemonin is a reactive compound with an exocyclic methylene group; it can bond with SH-groups of proteins and with DNA bases (thus mutagenic). The compound is also strongly antibacterial and antifungal (see protoanemonin).

FIRST AID If large amounts were consumed, instil medicinal charcoal, sodium sulphate and induce vomiting. Allergic symptoms can be treated with local application of corticoid ointment.

Ranunculis acris L.   family: Ranunculaceae

bouton d'or, rénoncule àcre (French); *Scharfer Hahnenfuß* (German); *piè corvino, ranuncalo comune* (Italian)

Rauvolfia serpentina

Indian snakeroot

Rauvolfia serpentina   *Rauvolfia vomitoria*

PLANTS WITH SIMILAR PROPERTIES
A genus of 60 species in the tropics. Other species of medicinal and toxicological interest include *R. vomitoria* from W Africa, *R. tetraphylla* from C & S America, and *R. caffra* from S and E Africa.

PLANT CHARACTERS A shrublet of 0.5 m high. The leaves are hairless, pointed and arranged in whorls of 3 to 5. The tiny white or pink flowers are borne in elongated clusters. The attractive red berries turn black when they ripen.

OCCURRENCE Widely distributed in SE Asia, from Pakistan and India to Indonesia. *R. vomitoria* is harvested in Africa as an alternative source of reserpine. *R. caffra* occurs naturally from S to E Africa.

CLASSIFICATION Neurotoxin; highly hazardous, Ib

ACTIVE INGREDIENTS Monoterpene indole alkaloids (1–3 %) of the yohimbane-, heteroyohimbane-, sarpagane- and ajmalane type accumulate in roots. The main hypertensive alkaloids are reserpine and rescinnamine (=reserpinine). Also of medicinal and toxicological interest are the alkaloids serpentine, ajmalicine, and ajmaline.

UTILISATION The drug is used against mild hypertension, anxiety and mental disorders. In traditional medicine, snakeroot is used to treat snake and insect bites and various other ailments such as insomnia and rheumatism.

TOXICITY *R. caffra* is said to cause 70 deaths per year in Tanzania. Overdosing (with the drug or isolated alkaloids) is a main cause of poisoning.

SYMPTOMS Alkaloids affect the central and peripheral nervous systems. Extracts relax smooth muscles, dilatate blood vessels, induce arrhythmia, acidosis, shock (death from cardiac arrest).

PHARMACOLOGICAL EFFECTS Reserpine inhibits the re-uptake of noradrenaline in synaptic vesicles of noradrenergic neurons. As a result, the neurotransmitter is degraded by MAO and COMT and becomes unavailable. These properties explain the efficacy of the drug against high blood pressure: it depletes the tissue stores of catecholamines from peripheral areas of the body. Sedative effects are ascribed to the depletion of noradrenaline and serotonin from the brain. These activities result in vasodilatation, hypotension and central sedation. *Rauvolfia* alkaloids are known to cause long-lasting depression (persisting long after the treatment has stopped), so that persons with depression or suicidal tendencies should avoid them. Ajmaline blocks Na$^+$ channels and is used to treat tachycard arrhythmia.

FIRST AID Induce vomiting; administer sodium sulphate and medicinal charcoal. Gastric lavage, instillation of 10 g medicinal coal and sodium sul-

Rauvolfia serpentina (L.) Benth. ex Kurz

family: Apocynaceae

rauwolfia, arbre aux serpents (French); *Rauwolfia, Schlangenholz* (German); *rauwolfia, serpentina indiana* (Italian)

Rhamnus cathartica

buckthorn

Rhamnus cathartica fruits

Rhamnus frangula fruits

PLANTS WITH SIMILAR PROPERTIES A genus of 135 species (N & S hemispheres), including *R. frangula* (= *Frangula alnus*) and *R. purshiana*. The latin name *cathartica* derives from the Greek "kathartikos", meaning "cleaning" and purging.

PLANT CHARACTERS A woody shrub of 3 m in height, with spiny branches bearing clusters of prominently veined and finely toothed leaves at their tips. The tiny yellow flowers occur in clusters. The fruits are small rounded drupes ("berries") that change from green to black as they ripen.

OCCURRENCE Buckthorn is indigenous to Europe, Asia and North Africa. It is also cultivated as an ornamental shrub.

CLASSIFICATION Cell toxin; moderately hazardous, II

ACTIVE INGREDIENTS The active compounds are anthraquinone glycosides (2–5 %; up to 1.4 mg/ fruit). Other compounds present include emodin, emodianthrone, glucofrangulins and frangulins.

UTILISATION The dried berries have been used in traditional medicine as blood purifier and diuretic. Nowadays it is still used as a laxative medicine.

TOXICITY Pregnant women, children under 12 and patients with obstruction or inflammation of the intestines or with appendicitis, colitis and Crohn's disease should not use buckthorn. Chil-

dren who had ingested *R. cathartica* berries and horses that were fed on *R. frangula* leaves died from serious diarrhoea. Ingestion of a few berries rarely has serious consequences.

SYMPTOMS Anthrones are irritating on skin and mucosa and cause serious disturbance of the GI tract, sometimes with bloody diarrhoea, and irritation of kidneys. Anthraquinones may be mutagenic when used on a long-term basis.

PHARMACOLOGICAL EFFECTS Emodin-type anthranoids (1,8-dihydroxyanthracene derivatives) are present in buckthorn as β-glycosides. These compounds are not active and are merely pro-drugs. Only when they reach the colon, are they activated by being broken down by the microorganisms into anthrones. These anthrones act on the colon motility; they enhance the secretion of water and inhibit its absorption in the colon (see anthraquinones). The result is a soft stool and faster bowel movements. The main molecular target is a chloride channel; of lesser importance are the Na^+ and K^+-ATPase.

FIRST AID Give medicinal charcoal and sodium sulphate. Allow the patient to drink plenty of tea or water. **Clinical therapy:** After overdosing: Gastric lavage, electrolyte substitution, control acidosis with sodium bicarbonate (urine pH 7.5); in case of spasms give diazepam; in severe cases supply intubation and oxygen.

Rhamnus cathartica L.

family: Rhamnaceae

nerprun purgatif (French); *Echter Kreuzdorn* (German); *spina cervina* (Italian); *espino cerval* (Spanish)

Rheum rhabarbarum

garden rhubarb

Rheum rhabarbarum plant

Rheum rhabarbarum (flowering)

PLANTS WITH SIMILAR PROPERTIES The genus *Rheum* comprises more than 30 species in Europe and Asia, including several that are used in traditional medicine: *R. australe* (Himalaya rhubarb), *R. officinale* (Chinese rhubarb) and *R. palmatum* (Chinese rhubarb, Turkish rhubarb) and *R. rhaponticum* (rhapontic rhubarb). Plants of the genus *Rumex* are related; more than 200 species are centred in Eurasia and N America.

PLANT CHARACTERS A robust perennial herb with thick stems below the ground and very large leaves on sturdy, bright red petioles. Many small, green-red flowers are borne in tall clusters.

OCCURRENCE Garden rhubarb is indigenous to Mongolia. The plant is sometimes considered to be a hybrid, with *R. rhabarbarum* as the one parent (it is then called *R.* x *hybridum* or *R.* x *cultorum*). It is widely cultivated as a vegetable.

CLASSIFICATION Cytotoxin; slightly hazardous, III

ACTIVE INGREDIENTS Petioles and leaves are rich in oxalates (0.3–0.7 %). Roots of *Rheum* and *Rumex* species also contain anthraquinones and have strong purgative properties.

UTILISATION Petioles have an attractive sour taste and are often eaten as a vegetable. They may also be stewed as pudding or used in pie fillings, chutney, jam and wine. The rhizomes and roots of species such as *R. officinale* and *R. rhaponticum*

are commercial laxative medicines.

TOXICITY High doses of rhubarb (especially leaves) and long-term ingestion can cause toxic effects. A few people have died in Europe as a result of excessive consumption of rhubarb or sorrel leaves (as dessert or as soup).

SYMPTOMS Symptoms after ingestion of large amounts of petioles include irregular pulse, hypotension, spasms, collapse of circulation, coma and death by central paralysis or fatal kidney damage (see oxalic acid). Patients with reduced kidney function are especially vulnerable (see oxalic acid). About 5 g of oxalic acid can be lethal for humans. Small children can develop kidney problems after eating small amounts of rhubarb petioles.

PHARMACOLOGICAL EFFECTS Oxalic acid forms insoluble salts with calcium. If calcium oxalate is deposited in the kidney tubules, kidney tissue becomes damaged and other consequences follow (see oxalic acid).

FIRST AID Only necessary after large overdoses: give calcium or chalk to precipitate oxalic acid; milk is a good source of calcium (see oxalic acid). **Clinical therapy:** In severe cases: gastric lavage (possibly with calcium salts, such as calcium gluconate; or with 0.1 % potassium permanganate; for further methods of detoxification see oxalic acid.

Rheum rhabarbarum L. (= *R.* x *hybridum* Murray)   family: Polygonaceae

rhubarbe (French); *Rhabarber, Speiserhabarber* (German); *rabarbaro* (Italian); *ruibarbo* (Spanish)

Rhododendron ponticum

purple rhododendron • ponticum

Rhododendron ponticum

Rhododendron canadense

PLANTS WITH SIMILAR PROPERTIES A very large genus with more than 850 species. Several are cultivated as ornamental plants. Genera with similar toxicity include *Kalmia* and *Pieris*.

PLANT CHARACTERS A shrub with oval-lanceolate leaves and attractive, funnel-shaped, pink or purple flowers borne in umbel-like clusters.

OCCURRENCE Originally from Europe (Iberian peninsula, Balkans, Turkey), but presently cultivated worldwide.

CLASSIFICATION Cell toxin; highly hazardous, Ib–II

ACTIVE INGREDIENTS Mainly diterpenes such as andromedotoxin (= grayanotoxin I or acetylandromedol) and several iridoid glucosides. Andromedotoxin is present in the nectar of some but not all species. The hydroquinone arbutin occurs in some species (see hydroquinone).

UTILISATION Some species have been used in traditional medicine as teas. The main importance, however, is as ornamental plants.

TOXICITY If bees feed on rhododendron, the honey may be enriched with andromedotoxin and can thus become toxic (see andromedotoxin). Livestock poisoning (sheep, goats, cattle) has been reported. Animals feed on evergreen *Rhododendron* when other food is no longer available. Problems may occur in zoos where visitors feed zoo animals with leaves from ornamental *Rhododendron* plants. Human poisoning has been reported from children eating leaves, adults drinking tea or ingesting *Rhododendron* honey from Turkey, Oregon or Lesbos (Greece). Gardeners may suffer from contact allergies. LD_{50} for acetylandromedol in mice: 1.28 mg/kg i.p., 5.1 mg/kg p.o.; guinea pig: 1.3 mg/kg i.p.; rat: 2–3 mg/kg p.o.

SYMPTOMS Burning sensation in mouth and throat, salivation, cold sweat, nausea, difficulties to swallow, vertigo, diarrhoea, abdominal pain, intoxication, spasms, peripheral paralysis, irregular and weak pulse, blue coloration of the skin (cyanosis), arrhythmia, arterial hypotension, apnoea and death from respiratory arrest.

PHARMACOLOGICAL EFFECTS Andromedotoxins are inhibitors of sodium channels (similar to aconitine), which bind to receptor site II of the channel. Sensitive membranes remain depolarised and can no longer transmit action potentials. After initial excitation nerve cells soon become paralysed. The depolarisation leads to an enhanced level of intracellular calcium ions, which cause a positive inotropic reaction in heart muscles.

FIRST AID Induce vomiting, give medicinal charcoal and sodium sulphate. Provide plenty of warm tea. Blistering of skin or mucous membranes: apply sterile tissue and hydrocortisone.
Clinical therapy: Gastric lavage, charcoal; for further treatment see andromedotoxin.

Rhododendron tomentosum

marsh tea • wild rosemary

Rhododendron tomentosum

Rhododendron tomentosum flower buds

PLANTS WITH SIMILAR PROPERTIES The genus *Rhododendron* includes more than 850 species, mainly in N temperate regions. Marsh tea is well known in traditional medicine and in toxicology by its old name, *Ledum palustre*.

PLANT CHARACTERS A woody shrub (up to 1.5 m) bearing narrowly oblong leaves with rolled in margins. They are glossy green above, silvery hairy below and resemble rosemary leaves, hence the common name. Attractive clusters of white or pink flowers are borne on the branch tips.

OCCURRENCE Marshy places and bogs in N Europe, N Asia and N America.

CLASSIFICATION Neurotoxin, mind-altering; moderately hazardous, II

ACTIVE INGREDIENTS Marsh tea contains an essential oil (up to 2.5 % of dry weight) with two sesquiterpene alcohols (palustrol and ledol) as the main components. Also present are carvacrol, thymol and coumarins such as fraxin and esculin. In other *Rhododendron* species, diterpenes are the toxic compounds.

UTILISATION Extracts are mainly used as ingredients of cough mixtures (whooping cough), anti-rheumatic preparations and also as emetic, diuretic and diaphoretic medicines. Shamans in Asia have inhaled the smoke from burning leaves of this species and *Juniperus recurva* as intoxicant. The Vikings and later people in northern Ger-

many used the plant to increase the potency of their beer. This use was banned in 1723 by the duke of Hanover. Marsh tea is traditionally used as insecticide against moths and bedbugs. The oil has been used as abortifacient.

TOXICITY Toxic effects have been recorded when high doses of oil were used as abortifacient.

SYMPTOMS The essential oil causes strong irritation of the GI tract with vomiting, diarrhoea and damage to the kidneys and the urinary tract. Further symptoms include perspiration, muscle and joint pain, central excitation and euphoria followed by paralysis.

PHARMACOLOGICAL EFFECTS. Ledol and palustrol have highly reactive cyclopropane in their structure, which can alkylate a diversity of proteins, ion channels and neuroreceptors. The oil irritates skin and mucous membranes. Ledol exhibits intoxicant and narcotic properties.

FIRST AID Give medicinal charcoal and sodium sulphate. Provide plenty of tea to drink. Contact with eyes or skin: wash with polyethylene glycol or water; in case of skin blisters, cover with sterile cotton and locally with a corticoid ointment.

Clinical therapy: After ingestion, instil 200 ml liquid paraffin to absorb the toxin, then gastric lavage (possibly with potassium permanganate). Instillation of medicinal charcoal and sodium sulphate is recommended (see thujone).

Rhododendron tomentosum Harmaja (= *Ledum palustre* L.)   family: Ericaceae
lède, léde des marais (French); *Sumpfporst* (German); *ledum palustre* (Italian)

Rhus toxicodendron

poison ivy • poison oak

Rhus toxicodendron (autumn leaves)

PLANTS WITH SIMILAR PROPERTIES
A cosmopolitan genus of 200 species, including *R. typhina* (stag's horn sumac), *R. vernix* (poison sumac), *R. radicans* (poison ivy) and *R. succedanea* (wax tree); most of them have strong irritant properties. Other genera of the Anacardiaceae also produce irritating alkyl- and alkenylphenols.

PLANT CHARACTERS A climber, shrub or tree with large trifoliate leaves on long petioles, greenish-white flowers and white fruits. All parts produce a white latex that rapidly turns brown.

OCCURRENCE America (S Canada to Guatemala) and E Asia (C China and Japan).

CLASSIFICATION Cell poison, skin irritant; extremely hazardous, Ia

ACTIVE INGREDIENTS Latex contains toxic 3-alkylbrenzcatechins, called urushiols, toxicodendron acid, or toxicodendrin. The alkyl side chain with 15 to 17 carbons can vary in length and number of double bonds. Leaves and fruits accumulate more than 3 % toxins.

UTILISATION Several *Rhus* species are grown in gardens and parks. Some are used for the production of dyes, lacquers, and tanning material.

TOXICITY Extremely allergenic, responsible for over 2 million poisonings (*Rhus* dermatitis) in the USA.

SYMPTOMS Contact with skin or mucosa causes serious allergic reactions in sensitive persons. Inflammations rapidly develop into dermatitis with swelling, reddening, fever, pain and unpleasant itching. *Rhus* dermatitis can persist for weeks and months. Even years later, parts of the affected skin can develop a relapse of dermatitis. Contact with the eye is extremely hazardous and can lead to blindness. Oral ingestion causes severe irritation of the mucosa of mouth, throat and GI tract. Symptoms include nausea, vomiting, gastroenteritis with bloody diarrhoea and colic, vertigo, dizziness, excitation and serious kidney damage.

PHARMACOLOGICAL EFFECTS Urushiols can penetrate biomembranes with their lipophilic side chain, exposing the catechol moiety to the external side. They are extremely allergenic.

FIRST AID Contact with eyes or skin: wash with water, 0.1 % potassium permanganate or Roticlean (polyethylene glycol); in case of skin blisters, cover with sterile cotton and treat with corticoid or ichthyol ointment. Itching can be treated with antihistaminic agents. After ingestion: Induce vomiting or perform gastric lavage (possibly with 0.1 % potassium permanganate), instillation of medicinal charcoal, sodium sulphate; electrolyte substitution, plasma expander, control of acidosis with sodium bicarbonate. In case of spasms inject diazepam. Provide intubation and oxygen respiration in case of respiratory arrest. Check kidney function.

Rhus toxicodendron L. (= *Toxicodendron pubescens* Mill.)   family: Anacardiaceae

sumac vénéneux, sumac grimpant (French); *Giftsumach* (German); *edera velenosa* (Italian); *hiedra venenosa* (Spanish)

Ricinus communis

castor oil plant

Ricinus communis

Ricinus communis flowers

PLANTS WITH SIMILAR PROPERTIES A monotypic genus (only 1 species). Poisonous lectins are also found in *Abrus precatorius*, *Robinia pseudoacacia* and *Phaseolus vulgaris*.

PLANT CHARACTERS A large shrub (up to 4 m), with very large, hand-shaped leaves. Female and male flowers appear separately near the tips of the branches. The spiny capsules contain 3 shiny seeds, each about 10–22 mm long, and mottled with silver, brown and black (s. p. 29).

OCCURRENCE Origin uncertain (probably NE Africa and India). It has become a naturalised weed in many subtropics and tropic countries. Plants are often grown as garden ornamentals.

CLASSIFICATION Cell poison; extremely hazardous, Ia

ACTIVE INGREDIENTS Two different toxins occur in the seed – ricinine (an alkaloid) and ricin (a mixture of four lectins). These compounds do not occur in seed oil because they are not soluble in oil. A single seed contains about 1 mg ricin.

UTILISATION Castor oil has been known since antiquity (4000 BC) as an effective purgative medicine. About 800 000 tons per year of the oil is nowadays used as biodiesel, lubricant and as raw material for polymer manufacturing. Press cake is used as animal feed or fertiliser.

TOXICITY Excessive doses of the oil may be toxic. Punctured seeds used in rosaries may cause intoxication if ingested by infants or small children. Inhaled dust from seeds can cause severe allergic reactions and anaphylactic shock. Animals are usually not poisoned, except from seed cake that has not been heat-inactivated. The alkaloid ricinine is highly toxic and is suspected in rare cases when leaves are browsed by livestock. The seeds are extremely poisonous; 5–6 will kill a child, 10–20 an adult. About 0.18 g/kg milled seeds or 1 mg/kg p.o. or 1 µg/kg i.p. of the lectin are fatal. 0.2 g/kg of seeds are lethal for horses and geese, 1–2 g/kg for other domestic animals.

SYMPTOMS Symptoms always appear after a latent period of several hours: nausea, bloody diarrhoea and vomiting, followed by intestinal inflammation (acute gastroenteritis), severe liver and kidney damage, weak pulse. Death is caused by heart and circulatory failure or uraemia.

PHARMACOLOGICAL EFFECTS Ricinoleic acid reduces the net absorption of fluids and electrolytes and stimulates peristalsis. Ricin is one of the most toxic substances known to man; it is a strong cell poison and inhibits ribosomal protein biosynthesis (see abrin).

FIRST AID First aid treatment is indicated if damaged or chewed seeds, e.g. from necklaces have been ingested. Treatment involves the induction of vomiting and detoxification as outlined under ricin.

Ricinus communis L.

family: Euphorbiaceae

ricin (French); *Rizinus, Christuspalme* (German); *ricino* (Italian); *ricino* (Spanish)

Robinia pseudoacacia

false acacia • black locust

Robinia pseudoacacia fruits

Robinia pseudoacacia flowers

PLANTS WITH SIMILAR PROPERTIES A North American genus with 4 species, including *R. hispida*. Other plants with toxic lectins include *Abrus precatorius, Ricinus communis, Phaseolus vulgaris* and *Viscum album*.

PLANT CHARACTERS The false acacia is a tree of 10–25 m in height with a rough bark and thorny branches bearing pinnate leaves with 9–19 leaflets. Small white flowers (typically blotched with yellowish green) are borne in clusters. The fruits are oblong, flat pods, each with several small black seeds.

OCCURRENCE *Robinia pseudoacacia* is indigenous to the eastern and central parts of the USA but has become naturalised in many parts of the world (especially Europe and Asia). It is also a popular garden tree.

CLASSIFICATION Cell poison; highly hazardous, Ib

ACTIVE INGREDIENTS The main toxic substance in seeds is the lectin mixture known as robin (or phasin in bark). The bark lectin consists of 2 subunits of 31.5 and 29 kDa. Seeds have 2 lectins, RPA1 and RPA2, with a MW of 100 kDa, consisting of 3 subunits. Lectins can be destroyed by heat. Seeds also contain the non-protein amino acid (NPAA) known as *L*-canavanine.

UTILISATION *Robinia* is an important bee plant and produces the so-called "Acacia honey" in Europe, which has a typical flavour. Flowers, leaves and bark have been employed as aromatics in traditional medicine.

TOXICITY The inhalation of *Robinia* dust and the ingestion of seeds have rarely caused deadly poisoning. Toxic symptoms (gastrointestinal disturbance) may occur even when a mere 4–5 seeds are eaten and chewed to break the tough seed coat so that the toxins are released. Horses feeding on *Robinia* bark died after 4 h.

SYMPTOMS Robin and phasin are haemagglutinating and mitogenic substances that cause gastrointestinal problems, abdominal pain, vomiting, mydriasis, weakness, vertigo, fever, sleepiness, spasms, visual problems and headache. These symptoms were observed in children poisoned with black locust bark.

PHARMACOLOGICAL EFFECTS Robin is a cell poison but is known to be less toxic than abrin (see *Abrus*). The symptoms often do not last very long. In experiments with mice, robin (30 mg/kg i.p.) decreased the glycogen level in liver but enhanced it in muscles.

FIRST AID If less than 5 seeds were ingested, provide medicinal charcoal and plenty to drink. First aid treatment is indicated if large amounts of damaged or chewed seeds have been ingested. Treatment involves the induction of vomiting and detoxification as outlined under abrin.

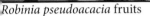

Robinia pseudoacacia L.

family: Fabaceae

faux acacia, robinier (French); *Gewöhnliche Scheinakazie, Robinie* (German); *robinia* (Italian)

Ruta graveolens

rue • herb of grace

Ruta graveolens

Ruta graveolens flowers and fruits

PLANTS WITH SIMILAR PROPERTIES A genus of 7 species in Europe, N Africa and W Asia.
PLANT CHARACTERS A strongly aromatic shrub of about 1 m high bearing compound leaves with minute, translucent glands. The bright yellow flowers occur in flat-topped clusters at the branch ends. The fruits are gland-dotted capsules that release small black seeds when they ripen.
OCCURRENCE S Europe. The plant is commonly cultivated in many parts of the world.
CLASSIFICATION Cell poison, mutagen; moderately hazardous, II
ACTIVE INGREDIENTS *Ruta graveolens* is chemically very complex. Of interest are the numerous coumarins (including coumarin, herniarin, gravelliferon, rutaretin), furanocoumarins (bergapten, psoralen, rutamarin), furanoquinoline alkaloids (dictamnine, skimmianine, rutacridone and various derivatives) and the flavonoid rutin (5 %).
UTILISATION Known as "Peganon" in Greece and "Ruta" in Rome, this plant had a special role in ancient mythology. *Ruta* was dedicated to the goddess Hecate who was recognised as the "daughter of the night" and the "brewer of toxins". The magic touch remained with *Ruta* until medieval times, when the plants were used for exorcism and against witches and the evil eye. It was laid into baby carriages and coffins. *Ruta* was commonly cultivated in monastery gardens and country gardens for treating a very wide range of medical problems, including menstrual disorders, dyspeptic complaints, heart palpitations, arthritis, sprains, injuries and skin diseases. In former times is was also used to improve and stabilise poor quality wine. It is still widely grown in herb gardens and as an ornamental plant.
TOXICITY Rue caused many fatalities in former times when it was used for abortion. Large doses of extracts or the pure essential oil are toxic when ingested.
SYMPTOMS The coumarins in rue can cause severely itching and burning photodermatitis (see furanocoumarins). Ingestion can cause swelling of the tongue, salivation, serious gastroenteritis, diuresis, miosis, visual disturbance, haematuria, and even death.
PHARMACOLOGICAL EFFECTS Reported antimicrobial, antispasmodic, anti-exudative, analgesic and ion channel inhibiting activities are probably due to furanocoumarins and furanoquinoline alkaloids. Several furanocoumarins and furoquinoline alkaloids are mutagenic and show strong phototoxicity. Cytotoxic and antimicrobial activities are known for acridone alkaloids.
FIRST AID When skin or eyes have been in contact with the oil, wash with water or Roticlean. Treat skin with cortisone ointments and eyes with antiallergic eyedrops (opticrom). Avoid sunlight.

Ruta graveolens L.

family: Rutaceae

rue fétide, herbe-de-grâce (French); *Weinraute* (German); *ruta* (Italian); *ruda común* (Spanish)

Salvia divinorum

magic mint • diviner's sage

Salvia divinorum

Salvia fruticosa

PLANTS WITH SIMILAR PROPERTIES *Salvia* is a very large and polyphyletic group of over 900 species. Only a few are known as poisonous or mind-altering. The well-known garden sage (*S. officinalis*) and other species are used in traditional medicine and may contain toxic components (e.g. thujone) in their volatile oil. Another member of the Lamiaceae, *Plectranthus scutellarioides* (= *Coleus blumei*), produces symptoms (when smoked) that resemble those of *S. divinorum*.

PLANT CHARACTERS A perennial plant that can reach a height of more than 2 m. Stems are winged and square in transverse section as in most Lamiaceae; it can have a diameter of up to 20 mm. The leaves are oval-lanceolate. Flowers occur in terminal clusters – they have the typical two-lipped structure of *Salvia* flowers and are pale blue. Plants do not produce viable seeds and can only be propagated vegetatively from cuttings.

OCCURRENCE Originally indigenous to Mexico (Oaxaca); nowadays available as garden plant.

CLASSIFICATION Mind-altering; moderately hazardous, II–III

ACTIVE INGREDIENTS The mind-altering components are diterpenes of the neoclerodan type known as salvinorin A and B (3.7 mg/g dry wt), accompanied by further bicyclic diterpenes. *Plectranthus scutellarioides* contains several diterpenes, which have some similarties with salvino-

rin A. *Plectranthus barbatus* (= *Coleus forskohlii*) accumulates forskolin, an activator of adenylate cyclase, an enzyme that could enhance the activity of those neurotransmitters which use the enzyme in signal transduction.

UTILISATION Aztecs have used a plant in their rituals whose properties resembled those of magic mushroom (*Psilocybe*); it was called *pipiltzintzintl*. It has been speculated that this drug was *S. divinorum*. Leaves of *S. divinorum* have been chewed or smoked as intoxicating drug. Shamans used the plant for soothsaying in divinatory and healing rituals. Mexican Indians have used the drug to treat disorders of the GI and urinary tracts, and for headache and rheumatism.

TOXICITY An intoxicating dose is said to be around 25 leaves. The active principle is absorbed through the mucosa of the mouth.

SYMPTOMS Doses of 150–500 μg salvinorin A induce strong psychedelic effects with strange visions and feelings. The onset is fast, but the symptoms do not last long.

PHARMACOLOGICAL EFFECTS Salvinorin A binds to the κ-opiate receptor; it is the first non-alkaloidal compound that is known to affect this neuroreceptor.

FIRST AID In case of overdosing, induce vomiting; then administration of sodium sulphate and medicinal charcoal.

Salvia divinorum Epling & Játiva

family: Lamiaceae

sauge des devins (French); *Wahrsagesalbei, Zaubersalbei* (German); *salvia dei veggenti* (Italian); *salvia de los adivinos, ska pastora* (Spanish)

Sanguinaria canadensis

bloodroot

Sanguinaria canadensis

Sanguinaria canadensis garden form

PLANTS WITH SIMILAR PROPERTIES A monotypic genus. Sanguinarine is also present in Asian plume poppy, *Macleaya cordata*, *Chelidonium majus*, *Dicentra spectabilis*, *Eschscholzia californica*, *Glaucium flavum*, *Papaver rhoeas* and *Argemone mexicana*.
PLANT CHARACTERS A stemless perennial herb (up to 0.4 m) with lobed leaves and solitary white flowers growing from a fleshy underground rhizome. The fruit is a capsule with numerous small seeds. All parts produce a red latex.
OCCURRENCE N America (Canada and the USA). Rhizomes are mainly wild-harvested in the eastern parts of the USA.
CLASSIFICATION Cell poison, mutagenic; moderately hazardous, Ib–II
ACTIVE INGREDIENTS Rhizomes contain isoquinoline alkaloids (up to 9 %), with 50 % sanguinarine (a benzophenanthridine) as the main compound. Several related alkaloids are present, including chelerythrine, sanguilutine, allocryptopine, protopine, berberine and coptisine.
UTILISATION The rhizome is traditionally used as an emetic, expectorant and spasmolytic to treat various ailments of the respiratory tract. Topically it was used to treat ulcers, warts and cancer. Bloodroot is still used as an ingredient of commercial expectorants, cough syrups and stomachics despite its known toxicity. Sanguinarine has anti-plaque activity and is therefore some-

times included in dental products (mouth rinses and toothpaste). It is effective in counteracting gingivitis but commercial preparations nowadays rarely contain sanguinarine.
TOXICITY Sanguinarine: LD_{50} mouse: 19.4 mg/kg i.v., 102 mg/kg s.c.
SYMPTOMS Extracts with sanguinarine cause narcotic effects, convulsions which resemble those of strychnine, central sedation, activation of gut motility and enhanced salivation. Sensitive nerve terminals are first activated and later paralysed. Death results from respiratory arrest.
PHARMACOLOGICAL EFFECTS The biological activity is ascribed mainly to sanguinarine. It binds to various proteins and shows anti-inflammatory, antifungal and antimicrobial activity. Sanguinarine inhibits several transaminases and Na^+,K^+-ATPase. The expectorant and emetic effects are probably due to irritation of the nervus vagus by the alkaloids in the stomach (see *Psychotria ipecacuanha*). Since some of the alkaloids can intercalate DNA, they are potential mutagens. Prolonged use of sanguinarine in oral hygiene products is no longer considered safe.
FIRST AID Induce vomiting; administration of medicinal charcoal and sodium sulphate. **Clinical therapy:** Gastric lavage (possibly with 0.1 % potassium permanganate), instillation of medicinal charcoal (see isoquinoline alkaloids).

Sanguinaria canadensis L.

family: Papaveraceae

sanguinaire du Canada (French); *Kanadische Blutwurzel* (German); *sanguinaria* (Italian)

Scadoxus puniceus

red paintbrush

Scadoxus puniceus   *Scadoxus multiflorus*   *Scadoxus multiflorus*

PLANTS WITH SIMILAR PROPERTIES A genus with 9 species in tropical Africa. It is closely related to the genus *Haemanthus* and other known sources of Amaryllidaceae alkaloids. *Scadoxus* species were formerly included in the related genus *Haemanthus* (which has distichous leaves without midribs).

PLANT CHARACTERS An attractive bulbous plant bearing long, strap-shaped leaves with wavy margins and distinct midribs. The thick, purple-spotted flowering stems bear dense clusters of orange-red flowers resembling powder puffs. The related *S. multiflorus* has a much larger inflorescence with stalked flowers in a globose arrangement. *H. coccineus* is known as April's fool because, like many African Amaryllidaceae, it flowers in the southern hemisphere autumn.

OCCURRENCE *S. puniceus* occurs in shady places from tropical S Africa northwards to the Ethiopean highlands. *Haemanthus* species occur naturally in South Africa. *Scadoxus* and *Haemanthus* species have become popular pot plants and are also widely grown as garden ornamentals.

CLASSIFICATION Cell poison; moderately hazardous, Ib–II

ACTIVE INGREDIENTS Isoquinoline alkaloids (so-called Amaryllidaceae alkaloids) are known to be the main toxic compounds in both *Scadoxus* and *Haemanthus* species. *Scadoxus puniceus* contains haemanthamine and haemanthidine as main compounds, while *H. coccineus* accumulates coccinine and an isomer known as montanine.

UTILISATION *Scadoxus* species are popular medicinal plants and overdoses may cause poisoning or even fatalities. *Haemanthus coccineus* is traditionally used as an antiseptic and diuretic and overdosing may likewise cause intoxications.

TOXICITY Children and domestic animals may be accidentally poisoned by plants growing in the house or garden. Amaryllidaceae alkaloids are very toxic. Montanine, for example, has an LD_{50} of 42 mg/kg (per injection) in dogs.

SYMPTOMS The alkaloids have hypotensive and convulsive properties.

PHARMACOLOGICAL PROPERTIES The toxicity of Amaryllidaceae alkaloids are linked to their interference with enzyme systems. Haemanthamine has demonstrated antimalarial activity and a high antiretroviral activity but also a low therapeutic index. Its selective apoptosis-inducing activity and marked antiproliferative effects result from complex formation with RNA.

FIRST AID Provide plenty of fluids to drink; give medicinal charcoal and sodium sulphate; apply shock prophylaxis. **Clinical therapy:** After overdosing: gastric lavage; in case of spasms, diazepam; check kidney function. In case of respiratory arrest, provide intubation and artificial respiration.

Scadoxus puniceus (L.) Fris & Nordal (= *Haemanthus puniceus* L.)   family: Amaryllidaceae

ail rouge, hémanthe (French); *Gefleckte frühe Blutblume* (German); *scadoxus puniceus* (Italian, Spanish)

Scopolia carniolica

scopolia

Scopolia carniolica

PLANTS WITH SIMILAR PROPERTIES A genus with 5 species, occurring from the Mediterranean to the Himalayas. Chemically similar are *Atropa, Hyoscyamus, Mandragora* and *Datura*.

PLANT CHARACTERS *Scopolia* is a perennial herb (up to 0.5 m) with persistent underground rhizomes and annual stems bearing soft leaves. The characteristic pendulous flowers are borne individually on slender, arching pedicels. They are bell-shaped, purplish brown outside and yellowish green inside.

OCCURRENCE Grow in low-density forests of SE Europe (Eastern Alps, Carpathian mountains). The plant is sometimes grown in gardens and has been cultivated commercially for harvesting both the rhizomes and the leaves.

CLASSIFICATION Neurotoxin, hallucinogen; highly hazardous, Ib

ACTIVE INGREDIENTS Rhizomes contain tropane alkaloids at 0.3–0.8 % dry weight. The main compound is *L*-hyoscyamine, which co-occurs with small amounts of scopolamine. Leaves have been used commercially for alkaloid extraction. They contain hyoscyamine, scopolamine, cuscohygrine and 3α-tigloyloxytropane in yields of up to 0.5 %.

UTILISATION Scopolia was known as a hallucinogenic plant and was used as aphrodisiac, narcotic, stimulant and hallucinogen in C Europe. It had similar uses to *Atropa* and was also employed in witchcraft. In modern medicine, carefully controlled doses are used to treat intestinal spasm (including the bile duct and urinary tract). Extracts have been applied to treat rheumatic pain.

TOXICITY Poisoning with scopolia is quite rare; sometimes leaves and roots are confused with those of edible plants. Scopolia has been employed as a mind-altering drug in Europe (see hyoscyamine).

SYMPTOMS At low concentrations, the alkaloids have a depressant and sedative effect, but high doses lead to hallucinations, euphoria, confusion, insomnia and death from respiratory arrest. Main symptoms are reddening of the face, dry mucosa, enhanced pulse and dilated pupils.

PHARMACOLOGICAL EFFECTS Hyoscyamine and scopolamine are antagonists at the muscarinic acetylcholine receptor and therefore work as parasympatholytics (see hyoscyamine).

FIRST AID Induce vomiting; administer sodium sulphate and medicinal charcoal. **Clinical therapy:** Parasympathomimetics (i.e. inhibitors of choline esterase) such as physostigmine (adults 1–2 mg i.v. or i.m.; children 0.5 mg) can be given as antidote. Provide gastric lavage (possibly with water or 0.1 % potassium permanganate), instillation of medicinal charcoal and sodium sulphate and even forced diuresis (see hyoscyamine).

Scopolia carniolica Jacq.   family: Solanaceae

scopolie du Caucase (French); *Glockenbilsenkraut, Tollkraut* (German); *scopolia* (Italian)

Senecio jacobaea

common ragwort

Senecio jacobaea flower heads

PLANTS WITH SIMILAR PROPERTIES A large genus with more than 1 500 species on all continents, including *S. inaequidens, S. cineraria, S. squalidus, S. vulgaris, S. ovatus, S. vernalis, S. latifolius, S. retrorsus, S. isatideus* and *S. burchellii.* Related Asteraceae that produce pyrrolizidine alkaloids (PAs) include *Petasites* (19 species), *Adenostyles* (4 species) and *Eupatorium* (45 species); *Tussilago farfara* (1 species) has a low PA content.

PLANT CHARACTERS An erect perennial herb (up to 1 m), with pinnately divided, finely toothed leaves and large numbers of attractive dark yellow flower heads.

OCCURRENCE Europe, Asia and N Africa, and naturalised in North America.

CLASSIFICATION Cell poison, neurotoxin, mutagen; moderately hazardous, II

ACTIVE INGREDIENTS All plant parts store unsaturated pyrrolizidine alkaloids (PAs); dried aerial parts of flowering plants contain 0.3 % PAs. The major PAs (cyclic diesters) are senecionine, seneciphylline, jacobine and erucifoline.

UTILISATION *Senecio* species have been used to treat capillary and arterial bleeding, especially in gynaecological conditions (mainly to induce menstruation and to reduce excessive bleeding), and also in hypertrophic gingivitis. Furthermore, for the improvement of circulation (*S. vulgaris*), to treat diabetes (*S. ovatus*) and externally as a poultice for pain and inflammation (*S. jacobaea*).

TOXICITY Poisoning can arise from medical treatment and from contamination of wheat (bread poisoning) and other foods. *Senecio* is the cause of the most serious of all livestock poisonings in Europe, N America and Africa, known as seneciosis (but animals usually avoid *Senecio*). Pyrrolizidine alkaloids cause acute poisoning of humans and livestock or more often chronic poisoning that may go undetected for a long period. LD_{50} senecionine: mouse 64 mg/kg i.v.; seneciphylline: rat 77 mg/kg i.p. (see senecionine).

SYMPTOMS PAs damage the liver (veno-occlusive disease), lungs, kidneys, pancreas and GI tract. For further symptoms, see senecionine.

PHARMACOLOGICAL EFFECTS PAs become metabolically activated in the liver (see senecionine). They can then alkylate proteins and DNA. As a result they cause liver damage; after prolonged intake they are mutagenic, teratogenic and carcinogenic. PAs can also modulate neuroreceptors and CNS activity.

FIRST AID Give medicinal charcoal to absorb the alkaloids; administration of plenty of warm black tea, and of sodium sulphate. Provide shock prophylaxis (keep patient quiet and warm). **Clinical therapy:** In case of ingestion of larger doses: gastric lavage and further treatments described under senecionine.

Senecio jacobaea L.

family: Asteraceae

herbe St. Jacques (French); *Jakobskreuzkraut* (German); *senecione di San Giacomo* (Italian)

Sesbania punicea

sesbania • rattlebox

Sesbania punicea

PLANTS WITH SIMILAR PROPERTIES A genus of 50 species in warm and wet climates.

PLANT CHARACTERS *Sesbania punicea* is a spindly shrub or small tree with pinnately compound leaves congested at the branch tips. Small clusters of decorative orange flowers are borne in drooping clusters. The pods are oblong and pointed, with wings along the four margins. They contain several greyish brown seeds.

OCCURRENCE *S. punicea* occurs naturally in C and S America (Argentina, Brazil and Mexico). It was introduced to various parts of the world as an ornamental garden plant, but has become an aggressive invader, especially in SE USA and S Africa. Several other species are planted as ornamentals, including the tropical Asian bakphul (*S. grandiflora*), which has edible flowers but inedible fruits.

CLASSIFICATION Neurotoxin, moderately hazardous, II

ACTIVE INGREDIENTS Sesbania seeds contain an alkaloid, sesbanimide A, as the main active compound. It co-occurs with other, related alkaloids such as sesbanimides B and C.

UTILISATION *S. bispinosa* and *S. cannabina* are sources of guar and fibres that can be used to make sails and fishing nets. *S. grandiflora* flowers are eaten as salad and its bark and leaves are used for medicinal purposes. *S. sesban* produces molluscicidal saponins and is a traditional source of soap in Africa. It is also grown for fodder, fibre and green manure.

TOXICITY All parts of the plant, including the leaves, flowers and seeds are known to be poisonous. The toxic seeds are produced in large numbers and are attractive to birds and poultry. Fowls and pigeons are known to have been poisoned but sheep and other animals are very rarely affected. The lethal dose of seeds in fowls and turkeys was found to be about 2.5 g/kg per body weight (which is about 20 to 70 seeds). Seeds are lethal to sheep and show cumulative effects if fed in small amounts over long periods. A mere 0.1 mg/kg/ day of ground seeds caused death in sheep within six days.

SYMPTOMS Amongst the known symptoms are nausea, vomiting, abdominal pain, diarrhoea, weakness, depression and respiratory failure that may prove fatal.

PHARMACOLOGICAL EFFECTS Sesbanimide A is highly active and has antitumour properties.

FIRST AID Administer medicinal charcoal to absorb the alkaloids (and sodium sulphate); give plenty of warm black tea. **Clinical therapy:** In case of ingestion of large doses: gastric lavage and further treatments described under non-protein amino acids.

Sesbania punicea (Cav.) Benth.

family: Fabaceae

flamboyant de Hyères (French); *Sesbanie* (German); *sesbania* (Italian); *sesbania* (Spanish)

Solandra maxima

chalice vine • cup-of-gold

olandra maxima

Solandra maxima

PLANTS WITH SIMILAR PROPERTIES A genus of 10 species, occurring naturally in tropical America. The differences between species are rather subtle and only *S. maxima* is well known. Chemical properties are similar to that *of Atropa, Hyoscyamus, Mandragora* and *Datura*.

PLANT CHARACTERS *Solandra maxima* is a woody, perennial climber (vine) with sturdy stems bearing large (150 mm long), glossy green, elliptical leaves that have prominent white midribs. All parts of the plant are hairless. The enormous bell-shaped flowers (up to 0.25 m long) are borne on the branch ends. They are golden yellow with a prominent purplish brown line on the inside of each petal. A scent reminiscent of coconut is produced at night.

OCCURRENCE Indigenous to Mexico. The plant is sometimes grown in gardens and parks.

CLASSIFICATION Neuropoison, hallucinogen; highly hazardous, Ib

ACTIVE INGREDIENTS The dry roots contain 0.64 % alkaloids; the aerial parts and fruits are also alkaloid-rich. *L*-hyoscyamine is the main compound, together with small amounts of scopolamine. The leaves have been used commercially for alkaloid extraction. They contain hyoscyamine, tigloidine, scopolamine, cuscohygrine and 3α-tigloyloxytropane in yields of up to 0.5 % dry weight.

UTILISATION *Solandra maxima* has been used in traditional medicine, mainly as aphrodisiac and to treat cough. It was known as a hallucinogenic plant and was used as narcotic, stimulant and hallucinogen in Mexico. The Aztecs called it tecomaxochitl.

TOXICITY Poisoning with *Solandra* is quite rare. The plant has been employed as an intoxicating drug (see hyoscyamine).

SYMPTOMS At low concentrations, the alkaloids have a depressant and sedative effect, but high doses lead to hallucinations, euphoria, confusion, insomnia and death from respiratory arrest. The main symptoms are reddening of the face, dry mucosa, enhanced pulse and dilated pupils.

PHARMACOLOGICAL EFFECTS Hyoscyamine and scopolamine are antagonists at the muscarinic acetylcholine receptor and therefore work as parasympatholytics (see hyoscyamine).

FIRST AID Induce vomiting; administer sodium sulphate and medicinal charcoal.

Clinical therapy: Parasympathomimetics (i.e. inhibitors of choline esterase) such as physostigmine (adults 1–2 mg i.v. or i.m.; children 0.5 mg) can be given as antidote. Provide gastric lavage (possibly with water or 0.1 % potassium permanganate), instillation of medicinal charcoal and sodium sulphate and even forced diuresis (see hyoscyamine).

Solandra maxima (Sessé & Moç.) P.S. Green (= *S. hartwegii*)

family: Solanaceae

olandre (French); *Goldkelchwein* (German); *solandra* (Italian); *copa de oro, copa dorada* (Spanish)

Solanum tuberosum

potato

Solanum tuberosum

Solanum tuberosum (green potatoes)

PLANTS WITH SIMILAR PROPERTIES A large, mostly New World genus of 1 700 species. Eurasian *S. dulcamara* has toxicological relevance.

PLANT CHARACTERS The common potato is a perennial with annual stems arising from a fleshy tuber. White or purple flowers (depending on the cultivar) are borne on the branch ends.

OCCURRENCE Indigenous to S America but cultivated as a crop in most parts of the world.

CLASSIFICATION Cell poison; moderately poisonous, Ib–II

ACTIVE INGREDIENTS Potatoes contain α-solanine and chaconine as the main toxins (up to 5 % dry wt in sprouts). Black nightshade (*S. nigrum*) has solasodine and solamargin; tomato (*Lycopersicon esculentum*) has tomatidine. *S. dulcamara* has either tomatidenol, soladulcidine or solasodine (up to 0.6 % dry wt in green fruits). Tubers of domestic potato normally contain only low levels of alkaloids (less than 10 mg per 100 g in most cultivars). The safe limit is 20 mg/100 g fresh wt. Peeling and cooking reduce the alkaloid level, so that healthy, non-diseased potatoes, which are processed properly, rarely cause problems. Green tubers should be avoided and early potatos eaten with caution, since alkaloid levels might be high. Tubers should be stored in the dark!

UTILISATION Potatoes are widely grown for their starch-rich tubers. Species used in tradition-

al medicine include *S. tuberosum* (gastric ulcers) aubergine or *S. melongena* (skin conditions) an *S. xanthocarpum* (seeds used as expectorants).

TOXICITY Potato plants are poisonous and hav been responsible for numerous human and anima deaths. Fatalities usually occur when green po tatoes, potato sprouts or green fruits are eaten c fed to farm animals. The minimum toxic dose c solanine in humans is 2–5 mg/kg; lethal oral dose about 400–500 mg. LD_{50} 75 mg/kg i.p. Teratogen effects are also known in some animals. *S. dulcam ra:* 10 unripe fruits are hazardous, 200 lethal.

SYMPTOMS Potato poisoning results in ston ach pain, prolonged vomiting, non-bloody dia rhoea, fever, confusion, weak and rapid pulse hallucinations, headache and coma.

PHARMACOLOGICAL EFFECTS. Solanin and other steroidal alkaloids can affect neurc receptors and inhibit cholinesterase. Because c their amphiphilic structure, they disturb men brane fluidity and permeability – similar cytc toxic properties as in saponins (see solanine).

FIRST AID If 2–3 fruits were eaten: give medic nal charcoal to absorb the alkaloids and plent of warm black tea to drink. After >4 berrie: sprouts > 5 cm, induce vomiting, instil medic nal charcoal and sodium sulphate (see solanine

Clinical therapy: Larger doses: gastric lavage (possibly with 0.1 % potassium permanganate).

Solanum tuberosum L.

family: Solanacea

pomme de terre (French); *Kartoffel* (German); *patate* (Italian); *papa* (Spanish)

Sophora secundiflora

mescal bean

Sophora secundiflora

Sophora microphylla

PLANTS WITH SIMILAR PROPERTIES A genus of 47 species in tropical and northern temperate Asia, Europe and America, including the New Zealand kowhai (*S. tetraptera*), the Chinese *. davidii* and *S. tonkinensis* (the *shan dou gen* of Chinese medicine) and the Polynesian *S. toromiro* extinct in nature on the Easter Islands but saved by cultivation in botanical gardens). The well-known *kushen* or Japanese pagoda tree is now called *Styphnolobium japonicum* but was previously known as *Sophora japonica*.

PLANT CHARACTERS Evergreen trees, usually about 7 m high (but rarely up to 17 m), with large pinnate leaves, blue flowers in hanging racemes and thick, bean-like pods that turn brown when they ripen. Seeds are bright red and have been used in necklaces.

OCCURRENCE Southwestern parts of North America; commonly cultivated as ornamental plant in many parts of the world.

CLASSIFICATION Neurotoxin, mind-altering; highly hazardous, Ib

ACTIVE INGREDIENTS *S. secundiflora* accumulates cytisine, *N*-methylcytisine, anagyrine, and other quinolizidine alkaloids. Some species (e.g. *S. alopecuroides* and *S. tonkinensis*) produce quinolizidine alkaloids of the matrine type.

UTILISATION Seeds of mescal beans have been found in archaeological sites in Texas that are more than 8 000 years old. The seeds have been used in C America before peyotl as the principal stimulant and hallucinogen. For intoxication, seeds are roasted on a fire and a quarter of a seed is chewed and swallowed. Mescalero Indians have added mescal beans to beer to fortify it. Kickapoo Indians have mixed mescal beans with tobacco and used the decoction to treat earache.

TOXICITY Intoxication with fatal outcome is possible, because of the high cytisine content (see *Laburnum*). Half a bean is a dose sufficient to induce delirium that lasts for 2–3 days.

SYMPTOMS Ingestion of the seeds causes burning of mouth and throat, salivation, and prolonged, even bloody vomiting, diarrhoea, severe gastroenteritis and paralysis. Central excitation with hallucinations, red colour vision, headache, and delirium is typical. Death can set in after 1–9 h or sometimes after a few days.

PHARMACOLOGICAL EFFECTS Cytisine is a strong agonist at nAChR and thus affects the central nervous system and neuromuscular plates (see cytisine).

FIRST AID Induce vomiting; administration of sodium sulphate and medicinal charcoal. Provide plenty of warm tea or fruit juice. **Clinical therapy:** After overdoses: gastric lavage and further treatments, see cytisine.

Sophora secundiflora (Ortega) DC.   family: Fabaceae

haricot mescal (French); *Meskalbohne* (German); *sophora secundiflora* (Italian); *mescal* (Spanish)

Sorghum bicolor

sorghum • great millet

Sorghum bicolor

Sorghum bicolor

PLANTS WITH SIMILAR PROPERTIES A genus of 24 species in the Old World and 1 species in Mexico.

PLANT CHARACTERS S. bicolor is a robust grass with broad, sheathing leaves that have prominent white midribs. The inflorescence is a much-branched, rounded panicle in S. bicolor subsp. bicolor (grain sorghum), a loose, open panicle in S. bicolor subsp. arundinaceum (common wild sorghum), and a compact panicle in S. bicolor subsp. drummondii (Sudan grass, previously known as S. sudanense).

OCCURRENCE Sorghum species are common in many parts of Africa. Grain sorghum was domesticated on the Ethiopean highlands at least 5 000 years ago, from where it spread to India, Southeast Asia and China. It became an important staple crop.

CLASSIFICATION Cell poison; moderately hazardous, II

ACTIVE INGREDIENTS Sorghum species all contain dhurrin, a cyanogenic glucoside. The dhurrin content is known to vary from plant to plant, depending on the species, genetic variation and especially environmental factors. Dhurrin levels are high in actively growing plants (240 mg/ 100 g), while dry seeds are free of glycosides.

UTILISATION Sorghum is widely cultivated for its starch-rich "seeds" or caryopses (small, dry, one-seeded fruits) that are eaten like rice or couscous but more often used for making porridge, unleavened bread or sorghum beer.

TOXICITY When animals rapidly ingest wilted plants, they may be killed by excessive fermentation and bloat. Despite their value as cultivated fodders, Sorghum species often cause prussic acid (HCN) poisoning. Problems usually occur when the plants were damaged by hail, frost, drought or herbicides. HCN toxicity: LD_{50} mouse (p.o.) 3.0 mg/kg ; rat (p.o.): 10–15 mg/kg. For humans 1 mg/kg HCN is lethal (see amygdalin).

SYMPTOMS Poisoning symptoms are typical for cyanogenic glycosides in general and include irritation, a flushing face, heavy breathing, scratchy throat, headache, and in severe cases, respiratory and cardiac arrest (see amygdalin).

PHARMACOLOGICAL EFFECTS The cyanogenic glucosides and hydrolyzing enzymes are located in different parts of intact leaf cells but become mixed when the tissue is damaged through wilting or ingestion. The hydrolysis reaction results in the release of hydrogen cyanide (HCN), a potent inhibitor of cellular respiration.

FIRST AID After ingestion of Sorghum shoots, induce vomiting and allow the intake of large volumes of drinks. **Clinical therapy:** In case of severe intoxication, detoxification is required (see amygdalin).

Sorghum bicolor (L.) Moench (= S. vulgare Pers.)   family: Poaceae

gros millet, sorgho (French); *Mohrenhirse* (German); *sorgo coltivato* (Italian); *millo, sorgo común* (Spanish)

Spirostachys africana

tamboti

Spirostachys africana bark

Spirostachys africana leaves

PLANTS WITH SIMILAR PROPERTIES A genus with 2 species in tropical Africa.

PLANT CHARACTERS A medium-sized, deciduous tree with a rounded crown and rough, blackish bark. The leaves are oblong-elliptic and have two minute glands where the petiole joins the leaf blade. They turn bright yellow or red in autumn. Inconspicuous male and female flowers are borne in small, elongated spikes. Most of the florets are male, but a few near the base of the spikes are female. The fruits are small, three-lobed capsules. Of toxicological interest is the poisonous milky latex that occurs in all parts of the plant, including the wood.

OCCURRENCE Tamboti occurs naturally in the tropical regions of Africa. Trees are gregarious and often form groups of several individuals.

CLASSIFICATION Cellular poison, co-carcinogen, skin irritant; extremely hazardous, Ia

ACTIVE INGREDIENTS The latex contains complex mixtures of diterpenes (phorbol esters), of which stachenone and stachenol are amongst the main compounds. The timber is very poisonous but it is not yet clear to which compound or compounds the toxicity should be ascribed.

UTILISATION Tamboti is important in traditional medicine and stem bark, root bark and latex are used as emetics, purgatives, painkillers, and vermifuges. The timber is much sought after for furniture. It is heavy, hard and very durable, with a beautiful natural lustre. Both the sawdust and the smoke are highly irritant and may cause headache and a burning sensation in the lungs if inhaled. The wood is not only irritant but it is claimed to be unsuitable for use as firewood also because it may poison the meat or food that is cooked on an open fire.

TOXICITY Deaths have been recorded after overdosing with traditional medicine.

SYMPTOMS The irritant latex may cause inflammation and blistering of the skin as well as the destruction of mucous membranes. Accidental contact with eyes may result in severe damage or even blindness.

PHARMACOLOGICAL EFFECTS The phorbol esters activate protein kinase C, which is an important regulatory enzyme; see phorbol esters for physiological and toxicological consequences.

FIRST AID Induce vomiting, give medicinal charcoal and sodium sulfate. Provide plenty of tea or other fluids to drink. **Clinical therapy:** Gastric lavage (possibly with potassium permanganate), instillation of medicinal charcoal and sodium sulphate; electrolyte substitution, control of acidosis with sodium bicarbonate. In case of spasms give diazepam. Provide intubation and oxygen therapy in case of respiratory arrest. Check kidney function.

Spirostachys africana Sond.   family: Euphorbiaceae

tamboti, tambuti (French); *Tamboti* (German); *sandalo africano* (Italian); *sándalo africano* (Spanish)

Strophanthus gratus

strophanthus

Strophantus gratus

PLANTS WITH SIMILAR PROPERTIES A genus of 38 species in the Old World tropics (Africa and Asia), usually growing as lianoid shrubs, including *S. kombe, S. sarmentosus, S. caudatus, S. hispidus* and *S. speciosus.*

PLANT CHARACTERS A woody climber that may extend to heights of more than 10 m. It has robust stems bearing simple, glossy leaves in opposite pairs. The attractive flowers are large, pink or purple in colour and occur in clusters, followed by pairs of narrow capsules. *S. kombe* has hairy leaves and yellow and white flowers.

OCCURRENCE Tropical West Africa (*S. gratus*) and tropical South and East Africa (*S. kombe*); both are cultivated on a small scale.

CLASSIFICATION Heart poison; extremely hazardous, Ia

ACTIVE INGREDIENTS *S. gratus* seeds contain more than 30 cardiac glycosides; of which ouabain (g-strophanthin) is the main compound (3–8 %). *S. kombe* seeds also has a complex mixture of cardiac glycosides (up to 10 %), with k-strophanthin and cymarin as main compounds. *S. speciosus* contains christyoside as a major glycoside.

UTILISATION Purified cardiac glycosides are used to treat heart failure – they slow down the heart rate but increase the force and efficiency of the contractions. *Strophanthus* species are very poisonous and have been used traditionally as ar-row and spear poisons in Africa, both for hunting and for criminal purposes. Arrows with poison made from *S. kombe* seeds are said to bring down large game within 100 metres.

TOXICITY Despite the fact that *Strophanthus* species are very poisonous to humans and animals, only a few cases of poisoning have been recorded. The glycosides impart a very bitter taste to the plants, are usually poorly absorbed when taken orally and induce vomiting, so that lethal quantities are rarely ingested. Concentrated decoctions may cause rapid death due to heart failure. About 1 mg ouabain can be lethal when injected (see cardiac glycosides).

SYMPTOMS Symptoms include nausea, hyperthermia, headache, salivation, gastrointestinal disturbance, purging and exhaustion, with the usual respiratory and cardiac symptoms (arrhythmia, hypertension, coma, cardiac arrest) associated with heart glycoside poisoning.

PHARMACOLOGICAL EFFECTS In general, cardiac glycosides inhibit Na$^+$, K$^+$-ATPase, a very important ion pump (see cardiac glycosides).

FIRST AID First aid treatment is indicated if plant material from *Strophanthus* or isolated cardiac glycosides have been ingested. Treatment involves the induction of vomiting, supply of 10 g medicinal charcoal, and detoxification as outlined under cardiac glycosides.

Strophanthus gratus (Wallich & Hook.) Baillon   family: Apocynaceae
strophanthus (French); Strophanthus (German); strofanto (Italian)

Strychnos nux-vomica

nux vomica

trychnos nux-vomica　　　　　　　　　　*Strychnos nux-vomica* seeds

PLANTS WITH SIMILAR PROPERTIES A genus of 190 species in the tropics and warm areas. *S. ignatii* of SE Asia is also a source of strychnine.

PLANT CHARACTERS An evergreen tree (up to 25 m) with spiny branches and characteristic three-veined leaves. The small white, tubular flowers are followed by large yellow fruits containing about 8 disc-shaped seeds.

OCCURRENCE S Asia to Australia. Wild-harvested and cultivated in Asia and W Africa.

CLASSIFICATION Neurotoxin, mind-altering; extremely hazardous, Ia

ACTIVE INGREDIENTS Seeds contain monoterpene indole alkaloids (1–3 % dry wt), with strychnine and brucine as major compounds. Minor alkaloids include pseudostrychnine, colurine, vomicine and novacine.

UTILISATION Nux vomica has been used as a potent stimulant and bitter tonic, and as analgesic in ointments. Seeds were formerly used mainly to kill rodents, but also as source of arrow and ordeal poisons. South American species such as *S. toxifera* are a source of curare (quaternary bisindole alkaloids – C-toxiferine I and others) that block nicotinic acetylcholine receptors, resulting in a paralysis of muscles. These alkaloids have been used as arrow poison by S American Indians.

TOXICITY Strychnine and related alkaloids are extremely poisonous. Accidental poisoning can result from cocaine adulterated with strychnine, or from tonics or rodenticides containing strychnine. The alkaloid has also been used for suicide and murder. LD_{50} rat: 5 mg/kg p.o., 1.6–3.0 mg/kg i.p. The toxic dose of strychnine in adults is 0.2 mg/kg, lethal are 5 mg (children) and 100–300 mg (adults). Brucine has a minimal lethal dose in humans of 30 mg.

SYMPTOMS Strychnine causes strong convulsions. Poisoning symptoms include anxiety, pain, extreme sensitivity to light, smell and noise, severe spasms and convulsions with tetanus and tremor; death by asphyxia, due to irreversible contraction of bronchial muscles or cardiac arrest. Depending on the dose, death can occur within minutes or a few hours.

PHARMACOLOGICAL EFFECTS Strychnine inhibits the glycine receptor (an inhibiting neuroreceptor) and thus increases excitability of all parts of the CNS. Strychnine is especially active in the spinal cord that has a high density of glycine receptors. Even small sensory signals can cause symmetric muscular spasms.

FIRST AID The main aim is to stop spasms by injecting curare, and hypoxia by providing intubation and artificial respiration. Other measures include instillation of medicinal charcoal, injection of diazepam, barbital to reduce spasms. Keep patient in a quiet, dark and warm room.

Strychnos nux-vomica L. (= *S. colubrina* Wight; = *S. lucida* R.Br.)　　　　　　family: Loganiaceae

noix vomique (French); *Brechnussbaum* (German); *noce vomica* (Italian); *nuez vómica* (Spanish)

Symphytum officinale

comfrey

Symphytum officinale

Symphytum asperum

PLANTS WITH SIMILAR PROPERTIES A genus with 35 species in Europe, Africa and Asia, including *S.* x *uplandicum* (a hybrid between *S. officinale* and *S. asperum*), *S. asperum*, *S. caucasicum* and *S. tuberosum*.

PLANT CHARACTERS Comfrey is a perennial herb with a basal cluster of large, hairy leaves growing from a fleshy rhizome. The annual flowering stems (up to 1 m) produce tubular blue or yellowish white flowers in one-sided clusters.

OCCURRENCE Europe and W Asia. The plant has become naturalised in N America and is widely grown as an ornamental in many parts of the world.

CLASSIFICATION Liver poison, neurotoxin, mutagen; slightly hazardous, II–III

ACTIVE INGREDIENTS All parts of the plants contain pyrrolizidine alkaloids (PAs), especially rich in PAs are roots (0.3 % DW), leaves (0.2 % DW), and flowers. PAs include open diester PAs such as intermedine, symphytine, and echimidine. Other secondary metabolites are allantoin (0.8 %), large amounts of mucilage (fructans), 4–6 % tannins, triterpenes, rosmarinic acid and other organic acids.

UTILISATION Comfrey (with allantoin and rosmarinic acid as the active components) is mainly used externally for the treatment of inflammation, bruises, dislocations, glandular swellings,

inflammation, pulled ligaments and muscle sprains, arthritis, slow healing wounds and boil In traditional medicine, infusions were take against gastritis, lung disorders, stomach ulce: and bleeding. Comfrey (young shoots and roo of several species) is sometimes used as a salac dietetic or traditional medicine; the internal u should be strongly discouraged because of th high content of PAs.

TOXICITY Symphytin LD_{50} mouse: 300 mg/k i.p.; see senecionine. Poisoning has been reporte after ingestion of comfrey.

SYMPTOMS Ingestion of large amounts ca cause CNS disturbance; chronic use may be de rimental to the liver and other organs.

PHARMACOLOGICAL EFFECTS PAs be come metabolically activated in the liver (se senecionine). They can then alkylate proteins an DNA. As a result they cause liver damage (veno occlusive disease), and after prolonged intak they are mutagenic, teratogenic and carcinogen: (especially causing liver tumours).

FIRST AID Give medicinal charcoal to absor the alkaloids; administer plenty of warm blac tea, and sodium sulphate. Provide shock prophy laxis (keep patient warm and at a quiet place **Clinical therapy:** In case of ingestion of larg doses: gastric lavage and further treatments de scribed under senecionine.

Symphytum officinale L. (= *S. consolida* Gueldenst. ex Ledeb.)   family: Boraginacea

grande consoude (French); *Gemeiner Beinwell* (German); *consolida maggiore* (Italian)

Synadenium cupulare

dead-man's tree

ynadenium cupulare   *Synadenium grantii*

PLANTS WITH SIMILAR PROPERTIES A genus of 30 species in tropical E & S Africa and America. *S. grantii* is a close relative (it is usually a much larger plant with larger leaves that are often purple or purple-spotted). *Synadenium* should not be confused with the closely related *Monadenium*, an African genus of more than 50 species with similar properties.

PLANT CHARACTERS A rounded, large succulent shrub of up to 2 m in height, with thick, fleshy stems bearing large, smooth succulent leaves with prominent midribs on the lower surfaces. The flowers are borne in short clusters on the branch tips. They are very small and characteristically surrounded by a cup-shaped involucre. The leaves and stems exude copious amounts of milky latex when broken.

OCCURRENCE *S. cupulare* is indigenous to the eastern tropical parts of southern Africa. *S. grantii* occurs from Mozambique northwards to tropical East Africa. *S. grantii* is commonly grown as a house plant in Europe and N America. It is better known than *S. cupulare*, which is less frequently cultivated.

CLASSIFICATION Cellular poison, co-carcinogen, skin irritant; extremely hazardous, Ia

ACTIVE INGREDIENTS Latex is a rich source of complex mixtures of diterpenes: *S. grantii* contains tigliane-type phorbol esters with a 4-de-

oxyphorbol skeleton. The main compound with skin-irritant properties is 12-*O*-tigloyl-4-deoxy-phorbol-13-isobutyrate. *S. cupulare* probably produces the same or similar compounds.

UTILISATION Both species are used in African traditional medicine and are also popular ornamentals and pot plants.

TOXICITY *Synadenium* causes serious irritation of the skin and mucosa. Domestic animals, adults and especially children may be at risk if plants are cultivated in the house or garden.

SYMPTOMS Latex of *S. cupulare* and *S. grantii* is highly irritant and may cause severe burning and itching of the skin (including inflammation, erythema, oedema and blistering), eyelids and lips that may last for several hours. It is claimed that a burning sensation may be felt even without direct skin contact due to the irritant vapours. The latex may cause permanent blindness by completely destroying the cornea of the eye.

PHARMACOLOGICAL EFFECTS Phorbol esters activate protein kinase C; see "phorbol esters" for physiological and toxicological consequences.

FIRST AID Induce vomiting, give medicinal charcoal, sodium sulphate and plenty of tea. **Clinical therapy:** Gastric lavage (possibly with potassium permanganate), instillation of medicinal charcoal and sodium sulphate (for further treatment see phorbol esters).

Synadenium cupulare (Boiss.) L.C. Wheeler   family: Euphorbiaceae

Euphorbe arborescente (French); *Milchbusch* (German); *synadenium* (Italian)

Tabernanthe iboga

iboga

Tabernanthe iboga   initiation rites

PLANTS WITH SIMILAR PROPERTIES A genus of only 2 species in tropical Africa.

PLANT CHARACTERS A shrub of 1–2 m in height with large yellow roots and lanceolate leaves of about 140 mm long. Flowers are small, white to yellow and arranged in sparse clusters. The fruits are pointed, fleshy berries that turn orange-yellow when they ripen. Leaves and stems exude white latex with a pungent smell.

OCCURRENCE Tropical rainforests of Congo, Cameroon to Angola; cultivated in W Africa.

CLASSIFICATION Neurotoxin, mind-altering; highly hazardous, Ib

ACTIVE INGREDIENTS Seeds are rich in monoterpene indole alkaloids, such as catharanthine, voaphylline and coronaridine. Roots contain 1 % alkaloids with ibogaine (the main compound), ibogamine, tabernanthine and ibogaline.

UTILISATION Iboga is one of the most important psychedelic and hallucinogenic intoxicants used in many African rituals and initiation rites. The root drug has been chewed by indigenous people in tropical rainforests of W Africa to suppress hunger and fatigue. It is strongly associated with the Bwini religion. In Congo, roots are leached in palm wine and drunk as aphrodisiac. The main interest in iboga and ibogaine results from their use in the treatment of drug addiction (especially heroin and other opiates). It is claimed that the anti-addiction effects were accidentally discover ed by heroine addicts who were looking for a new "high". In the sixties, an iboga product was sold in France under the trade name Lambarene. Clinical studies in the USA were discontinued when a patient died from an overdose.

TOXICITY Large doses of ibogaine are said to cause convulsions, paralysis and death from respiratory failure. Several deaths have been recorded. Doses of 100 mg/kg ibogaine caused serious neuro degeneration in rats. A single dose of 500 to 800 mg ibogaine was used in clinical studies for the treatment of opioid addiction.

SYMPTOMS Root extracts have intoxicant properties. About 2–10 mg/kg ibogaine cause hallucinations, visions (meeting and seeing dead ancestors), and euphoria. Higher doses cause paralysis, dizziness and even death.

PHARMACOLOGICAL EFFECTS Ibogaine is an agonist at serotonin receptors and antagonistic at NMDA receptors. Tabernanthine is antagonist at $GABA_A$ receptors.

FIRST AID Induce vomiting; administer sodium sulphate and medicinal charcoal. Provide shock treatment. **Clinical therapy:** Provide gastric lavage (possibly with water or 0.1 % potassium permanganate), instillation of medicinal charcoal and sodium sulphate. In serious cases, intubation and artificial respiration may be necessary.

Tabernanthe iboga Baill.   family: Apocynaceae

iboga (French); *Iboga* (German); *iboga* (Italian, Spanish)

Taxus baccata

yew

axus baccata   *Taxus baccata* cones

LANTS WITH SIMILAR PROPERTIES A genus with 7 species in Eurasia and America.

LANT CHARACTERS A shrub or small tree ith dark green, linear leaves arranged in two ows, inconspicuous male cones and ovoid female ones surrounded by bright red, fleshy arils.

CCURRENCE The common yew occurs naturally in Europe and Asia Minor. *T. brevifolia* is adigenous to the west coast of N America.

LASSIFICATION Cell poison; extremely hazardous, Ia

CTIVE INGREDIENTS *Taxus* species produce iterpenes with a taxane nucleus. The main toxic ompound in *T. baccata* is taxine A ($C_{35}H_{47}NO_{10}$; IW 641.77; 1.79 %). Taxol was first extracted om *T. brevifolia* bark. Nowadays, leaf diterpees from *T. baccata* (such as 10-deacetylbaccatin I) are converted into structural analogues of xol, such as docetaxel (better known as Taxore®). They are used for the treatment of cancer.

TILISATION The toxicity of the yew tree was ell known in antiquity (used for murder, suicide nd abortion): Nikander (200 BC) described the ainful death (the throat becomes contracted). he Celtic chieftain Catuvolcus (53 BC) preferred agesting yew to becoming a Roman slave. The ncient Celts (and the goddess Artemis) used arows and spears poisoned with yew extracts.

OXICITY Leaves and stems (but not the fleshy arils) are very poisonous. Human and animal deaths have been recorded. 50–100 g of leaves can be fatal to humans, pigs, dogs and poultry. 100–200 g kill horses and sheep, 500 g cows. Taxine: LD_{50} rat 20.16 mg/kg i.p.; mouse 19.7 mg/kg p.o., 21.9 mg/kg i.p., 12.96 mg/kg s.c.

SYMPTOMS Symptoms appear after 0.5–1.5 h: mydriasis, nausea, vomiting, dizziness, painful diarrhoea with colic, kidney damage and red lips. Respiration is first activated and then suppressed. Pulmonary spasms, respiratory and circulatory failure lead to death in coma after 1.5–24 h.

PHARMACOLOGICAL EFFECTS Taxol and docetaxel are spindle poisons that inhibit cell division by preventing the depolymerisation of the microtubules. Taxines inhibit Na^+ and Ca^{++} channels in myocardial cells.

FIRST AID Instil medicinal charcoal and then induce vomiting; administration of plenty of warm black tea, and of sodium sulphate. **Clinical therapy:** Gastric lavage (with water or 0.1 % potassium permanganate), instillation of medicinal charcoal and sodium sulphate. Provide intubation and oxygen respiration in case of respiratory arrest and plasma expander when shock occurs. Electrolyte substitution, control of acidosis with sodium bicarbonate. In case of spasms inject diazepam and barbiturates; for arrhythmia use lidocain.

axus baccata L.   family: Taxaceae

commun (French); *Eibe* (German); *tasso* (Italian); *tejo* (Spanish)

Theobroma cacao

cacao tree

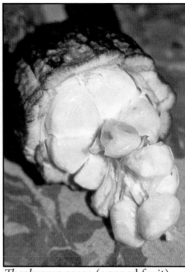

Theobroma cacao flowers   *Theobroma cacao* (opened fruit)   *Theobroma cacao* fruit

PLANTS WITH SIMILAR PROPERTIES A genus of 20 species in tropical America. Other species used to a limited extent in cacao production include *T. bicolor* (tiger chocolate), *T. angustifolium* (monkey chocolate) and *T. grandiflorum* (cupuaçu). *Theobroma* has been placed in the Malvaceae by APGII.

PLANT CHARACTERS Evergreen tree of 15 m with long, leathery leaves. Small white flowers emerge from the stem (cauliflory). The large fruits contain 20–70 seeds (cocoa beans). They are green and turn to yellow and red when mature.

OCCURRENCE The tree is thought to have originated in the Andean foothills but has been cultivated in S Mexico since ancient times. A very large industry is based on cultivation in tropical rainforest areas of Africa, Asia and America.

CLASSIFICATION Stimulant; slightly hazardous, III

ACTIVE INGREDIENTS Seeds are rich in theobromine and caffeine; seed coat with 2.9 %, cocoa powder 3 % alkaloids; tannins of the catechin type are abundant (6–7 %) and useful as antioxidants.

UTILISATION The Mayas and Aztecs recognised "chocolatl" as a food of the gods (therefore named *Theobroma*) with aphrodisiac properties. When Cortés invaded America, he conquered the Aztec kingdom of Montezuma and got hold of cacao. Cocoa was introduced into Europe as a refreshing and reputedly aphrodisiac drink and b[y] 1700 there were 2 000 chocolate houses in Lo[n]don alone. Slab chocolate is a more recent inven[tion] and dates from the early 18th century.

TOXICITY Toxic effects only occur at very hig[h] doses. 150–200 mg/kg appears to be the LD_{50} fo[r] caffeine in humans; about 5 g caffeine has cause[d] the death of a child. Livestock poisoning (and eve[n] fatalities) have resulted from cacao products.

SYMPTOMS Excessive amounts of cocoa ma[y] lead to nervousness, headache, tremor, spasm[s] palpitations, high blood pressure, insomnia an[d] indigestion. The topical application of ointmen[ts] with 30 % caffeine can induce toxic effects.

PHARMACOLOGICAL EFFECTS Caffein[e] is an inhibitor of cAMP phosphodiesterase an[d] an antagonist at adenosine receptors. As a conse[-]quence, dopamine is released and many parts o[f] the brain become activated. Caffeine is a cortic[al] stimulant that increases the carbon dioxide sens[i-]tivity of the brain stem. It affects the cardiova[s-]cular system and has a positive inotropic actio[n] resulting in tachycardia and an enhanced outpu[t]

FIRST AID In case of overdose: administratio[n] of sodium sulphate; induce vomiting, give m[e-]dicinal charcoal. **Clinical therapy:** Gastric lava[ge] (possibly with potassium permanganate), instill[a-]tion of medicinal charcoal and sodium sulphat[e;] diazepam in case of spasms.

Theobroma cacao L.   family: Malvaceae (formerly Sterculiacea[e)]

cacaotier, cacaoyer (French); *Kakaobaum* (German); *cacao* (Italian); *cacao real* (Spanish)

Thesium lineatum

thesium • witstorm

Thesium lineatum

Thesium lineatum fruits

PLANTS WITH SIMILAR PROPERTIES A large genus of 325 species with an Old World distribution. The plants are herbaceous or woody, almost leafless root-parasites that can absorb secondary metabolites from their host plants. In addition to *T. lineatum*, one other species is also known to be toxic to livestock: *T. namaquense* (poison bush). It is a much smaller shrub than *T. lineatum* and has dark brown and persistent (but minute) leaves.

PLANT CHARACTERS *T. lineatum* is a semi-parasitic, woody, much-branched shrub of about 1 m in height, with thorny green stems and small, caducous leaves. The tiny white flowers occur individually along the stems and turn into rounded berries crowned by the remains of the flower. They turn snow white when they ripen.

OCCURRENCE *T. lineatum* occurs over a large part of the dry western region of southern Africa. *T. namaquense* has a more restricted distribution in the central interior of South Africa.

CLASSIFICATION Heart poison; highly hazardous, Ib

ACTIVE INGREDIENTS The active principle of *T. lineatum* was found to be a bufadienolide, thesiuside. The plant grows as a root parasite on other plants but the cardiac glycoside appears to be produced by the *Thesium* plants and not by the host plants.

UTILISATION Species such as *T. hystrix* are used to a limited extent in traditional medicine as "blood purifiers" and to treat chest ailments.

TOXICITY *T. lineatum* and *T. namaquense* have a history of causing livestock losses under certain conditions. Poisoned sheep, goats or cattle usually die suddenly but sometimes show symptoms such as rapid and laboured breathing. About 500 g of stems are fatal to a sheep and death may occur a few hours after ingestion.

SYMPTOMS Symptoms of human poisoning are poorly recorded but ingestion of stem or root infusions are known to cause foaming vomit and sometimes death. Symptoms in animals include nausea, salivation, retching, gastrointestinal disturbance, diuresis, purging and exhaustion, with the usual respiratory and cardiac symptoms (arrhythmia, hypertension, coma and cardiac arrest) associated with heart glycoside poisoning.

PHARMACOLOGICAL EFFECTS In general, cardiac glycosides inhibit the Na^+, K^+-ATPase (see cardiac glycosides). Concentrated decoctions may cause rapid death due to heart failure.

FIRST AID First aid treatment is indicated if plant material from *Thesium* or isolated cardiac glycosides have been ingested. Treatment involves the induction of vomiting, supply of medicinal charcoal, and detoxification as outlined under cardiac glycosides.

Thesium lineatum L.f.

family: Santalaceae

bésion, thésium (French); *Leinblatt* (German)

Thevetia peruviana

yellow oleander

Thevetia peruviana

Thevetia peruviana fruits

PLANTS WITH SIMILAR PROPERTIES A genus of 8 species in tropical America. *Thevetia peruviana* was previously known as *T. neriifolia*, in reference to the leaves that are similar to those of *Nerium oleander* (the common oleander). The latter is a related poisonous plant that also contains cardiac glycosides.

PLANT CHARACTERS A leafy, latex-containing shrub of up to 10 m in height, with glossy, bright green, linear leaves and highly decorative yellow flowers. The distinctive triangular fruit capsules turn black when they ripen. Each fruit contains 2–4 flat seeds.

OCCURRENCE *Thevetia peruviana* is indigenous to Mexico and the northern parts of South America. It is a popular garden shrub in many parts of the world with tropical and Mediterranean climates.

CLASSIFICATION Heart poison; highly hazardous, Ib

ACTIVE INGREDIENTS A large number of cardiac glycosides occur in yellow oleander (up to 5 % in seeds; 0.07 % in leaves). Thevetin A and B (= cerebroside) are the main compounds. A heart stimulant called thevetin (containing a mixture of thevetin A and B) was formerly available as a commercial product.

UTILISATION Leaves and seeds are often used in suicide and murder attempts (especially in S India and Sri Lanka). Powdered seeds have been used a rat poison. Yellow oleander is a common garde shrub and accidental poisonings of humans an livestock are therefor often reported.

TOXICITY All parts of the plant (leaves, flower milky latex and especially the seeds) are extreme ly poisonous (4 seeds can be fatal in children, 8–1 seeds in adults). Two leaves are said to be letha for infants. If first aid treatment starts only afte 4 h or longer, the prognosis is unfavourable. Fo tunately, the leaves and seeds have an intense bi ter taste that acts as a deterrent (and often induc vomiting). Nevertheless, the ingestion of larg quantities may lead to rapid death.

SYMPTOMS Symptoms include numbness c the tongue and throat, nausea, vomiting, weal ness, spasms, confusion and visual disturbance followed by a slowing and weakening of th heartbeat to 30 or 40 beats per minute. In fata cases, bradycardia, atrio-ventricular block an fibrillation are typical. Death can occur 2 to 3 after ingestion.

PHARMACOLOGICAL EFFECTS Cardeno lides inhibit Na$^+$, K$^+$-ATPase, and thereby block ing the signal transduction from nerves to muscle (see cardiac glycosides).

FIRST AID Induce vomiting; administer sodiur sulphate and medicinal charcoal. **Clinical thera py:** see cardiac glycosides.

Thevetia peruviana (Pers.) K. Schum. (= *Cerbera thevetia* L.)　　　　　　　　　family: Apocynacea

arbre à lait, laurier jaune (French); *Gelber Oleander, Gelber Schellenbaum* (German); *oleandro giallo* (Italian)

Thuja occidentalis

American arbor-vitae • white cedar

Thuja occidentalis

Thuja orientalis cones

PLANTS WITH SIMILAR PROPERTIES A genus of 5 species in E Asia and N America. *T. plicata, T. orientalis* and *T. standishii* are also toxic.

PLANT CHARACTERS An evergreen conifer (up to 20 m in the original wild form) with flattened branches, scale-like leaves, inconspicuous male cones and small, oblong female cones with imbricate cone scales.

OCCURRENCE Eastern parts of N America. Cultivars with yellow leaves ("golden cypress") and dwarf forms are popular garden plants.

CLASSIFICATION Cell poison; neurotoxin, mind-altering; highly hazardous, Ib

ACTIVE INGREDIENTS The volatile oil (0.4–1 %) contains the toxic principle. The main compound is thujone (both α-thujone and β-thujone), which is present at levels of up to 65 % of the total oil. Also present are α-pinene, borneol, camphor, fenchone, and some sesquiterpenes. Plicatic acid (a lignan) can cause allergic reactions.

UTILISATION Extracts are used in traditional medicine to treat colds, fever, bronchitis, headache, rheumatism and cystitis. *Thuja* extracts or thuja oil have been used to cause abortions, which sometimes led to fatalities. The essential oil is used commercially as an ingredient of various cosmetic products, disinfectants and insecticides.

TOXICITY Thujone is a pronounced neurotoxin that irritates skin and mucosa and shows cytotoxic effects on liver and kidney cells. It has abortifacient properties. LD_{50} α-thujone mouse: 87.5 mg/kg s.c.; β-thujone mouse 442.4 mg/kg s.c.

SYMPTOMS Mydriasis, fever, visual disturbance, headache, gastroenteritis with severe vomiting, diarrhoea, tachycardia, strong convulsions, degenerative damage of kidney, heart and liver. Death is caused by circulatory and respiratory arrest. Chronic ingestion of thujone initially leads to CNS stimulation, optical and acoustical hallucinations, later to deep depression, epileptic seizures, and loss of personality. Internal use is therefor not recommended. Ingestion over a long period may be particularly harmful due to cumulative effects (see thujone).

PHARMACOLOGICAL EFFECTS Thujone is a reactive monoterpene (with cyclopropane ring), known to be neurotoxic (see *Artemisia absinthium*).

FIRST AID Induce vomiting; instillation of sodium sulphate and medicinal coal. **Clinical therapy:** Provide gastric lavage (possibly with 0.1 % potassium permanganate), medicinal charcoal, sodium sulphate and polyethylene glycol. In case of excitation, inject diazepam. Provide intubation and oxygen respiration in case of respiratory arrest and plasma expander when shock occurs. Treat thujone overdose with atropine against colic and diazepam against spasms; control liver, blood and kidney functions.

Thuja occidentalis L.

family: Cupressaceae

thuya d'occident, thuya américain (French); *Abendländischer Lebensbaum* (German); *thuja* (Italian); *tuya* (Spanish)

Turbina corymbosa

ololiuqui

Turbina corymbosa

PLANTS WITH SIMILAR PROPERTIES A genus of 5 species in tropical America. A related and large genus is *Ipomoea* (morning glory) with over 650 species in warm and temperate climates, including species with ergot alkaloids such as *Ipomoea tricolor* (heavenly blue morning glory), *I. carnea* (morning glory bush), *I. hederacea, I. muricata, I. nil* (blue morning glory) and *I. pes-caprae* (beach morning glory). Well-known cultivated ornamentals are *I. purpurea* (common morning glory) and *I. violacea* (= *I. macrantha, I. tuba*).

PLANT CHARACTERS *Turbina* is a herbaceous, twining creeper with attractive white flowers and heart-shaped leaves. Inflorescences are borne on long peduncles that emerge from the leaf axils. The fruits each contain one light brown seed.

OCCURRENCE Mexico (around the Gulf of Mexico) and also the West Indies and Cuba. It has become naturalised in the Philippines.

CLASSIFICATION Neurotoxin, mind-altering; highly hazardous, Ib

ACTIVE INGREDIENTS Ergot alkaloids, such as ergine, erginine, chanoclavine, elymoclavine, lysergic acid amide and other derivatives of lysergic acid, occur in leaves and seeds but not in the rhizome and roots. The alkaloids are probably produced by symbiotic fungi.

UTILISATION Mexican Indians used ololiuqui (origin unresolved; it might have been *Ipomoea* or *Turbina corymbosa*) for oracle purposes, for divin atory activities, in sacrificial ceremonies and eve as an aphrodisiac. Seeds were collected, squashe and imbibed in pulque and balche.

TOXICITY *Ipomoea* species are not particularl poisonous but their seeds contain toxic alkaloid and cases of fatal poisoning have been recorded 15–20 seeds are an intoxicating dose. More tha 100 seeds caused unpleasant apathy and sever gastrointestinal disturbances.

SYMPTOMS Seeds and seed extracts are hall cinogenic and induce trance or deep sleep wit dreams. In the hypnotic state, Indians are said t have talked to their dead relatives, and gave awa secrets. Further symptoms from seed ingestio include nausea, vomiting and weakness (see ergc alkaloids).

PHARMACOLOGICAL EFFECTS Like othe indole alkaloids of the Convolvulaceae, ergir and lysergic acid derivates are powerful hallucinc gens, since they modulate the activity of serotonir dopamine and noradrenaline (see ergot alkaloids).

FIRST AID Administration of sodium sulphat give medicinal charcoal. **Clinical therapy:** Ga tric lavage (possibly with potassium permar ganate), instillation of medicinal charcoal an sodium sulphate, electrolyte substitution; chec acidosis. If necessary, intubation and oxygen re piration (see ergot alkaloids).

Turbina corymbosa (L.) Raf. (= *Rivea corymbosa*)   family: Convolvulacea

ololiuqui (French); *Ololiuqui* (German); *ololiuqui* (Italian); *yerba de la virgen, yerba de las serpientes* (Spanish)

Turnera diffusa

damiana

Turnera diffusa

Turnera ulmifolia

PLANTS WITH SIMILAR PROPERTIES A genus of 50 species in the tropical and warm parts of America (1 species in Africa). The commercial species is also known as *Damiana diffusa* var. *aphrodisiaca*. Other species also contain alkaloids but only *D. diffusa* is used to any extent.

PLANT CHARACTERS Damiana is an aromatic shrub of up to 2 m in height. The simple leaves have toothed margins and can easily be recognised by their sunken lateral veins. Yellow, solitary, broadly funnel-shaped flowers are borne at the branch tips.

OCCURRENCE Damiana is indigenous to tropical America (Carribean Islands, Mexico and southern California). Commercial material is harvested mainly from wild plants but crop development has been initiated.

CLASSIFICATION Stimulant, mind-altering; slightly hazardous, III

ACTIVE INGREDIENTS Damiana contains a cyanogenic glycoside (tetraphyllin B) and low levels (up to 0.7 %) of arbutin. Also present is an essential oil that contains various monoterpenoids and sesquiterpenoids. Among the main constituents are α-pinene, β-pinene, calamene, α-copaene, δ-cadinene, thymol and perhaps 1,8-cineole and p-cymene. The plant is said to produce gums, tannins, resins, bitter principles and phytosterols. The active substance is sometimes claimed to be damianin, an amorphous bitter substance that has not been chemically characterised.

UTILISATION Damiana has an old reputation (since prehistoric times of the Maya) as a stimulant and aphrodisiac. It has been used both for preventing and treating impotence, loss of libido and other sexual disturbances (and also menstrual problems and prostate problems). The herb (and the essential oil) have a reputation for alleviating the symptoms of fatigue, depression and stress. In Mexico, an infusion is drunk like ordinary black tea. Damiana herb may be included in palm wine or wine and taken as an aphrodisiac.

TOXICITY Toxic symptoms may occur after ingestion of large amounts of the plant or its essential oil.

SYMPTOMS When smoked, daminiana leaves cause euphoria similar to marijuana. Taken orally, extracts are sedating and stimulating.

PHARMACOLOGICAL EFFECTS The traditional use of damiana herb as aphrodisiac is not yet supported by scientific or clinical studies. However, there have been reports of hypoglycaemic and uterotonic activities. The chemical compounds and their toxicological and pharmacological significance are poorly known.

FIRST AID In case of overdosing, instil medicinal charcoal, sodium sulphate and provide plenty of water or tea to drink.

Turnera diffusa Willd. (= *Damiana diffusa* var. *aphrodisiaca*)   family: Turneraceae or Passifloraceae

damiana (French); *Damiana* (German); *damiana* (Italian, Spanish)

Tylecodon wallichii

nenta

Tylecodon wallichii   *Tylecodon cacalioides*

PLANTS WITH SIMILAR PROPERTIES
A southern African genus of 31 species. It is
closely related to *Cotyledon* but can easily be distinguished by the deciduous leaves of all the species (the leaves are shed during the dry summer
months). Bufadienolides also occur in species of
Kalanchoe and *Cotyledon*, but not in *Crassula* and
other members of the family Crassulaceae. *T. wallichii* is very closely related to *T. cacalioides* and
can hardly be distinguished from it except when
in flower (in the latter, the flowers are 17–25 mm
long and sulphur yellow).
PLANT CHARACTERS *T. wallichii* is a perennial shrub with thick, tuberculate stems and
clusters of terete, slender, succulent leaves at the
branch tips. The flowers are small (7–12 mm long)
and usually lemon yellow.
OCCURRENCE *T. wallichii* occurs in the dry
western parts of southern Africa.
CLASSIFICATION Heart poison; highly hazardous, Ib
ACTIVE INGREDIENTS The main toxic compound in *T. wallichii* is a heart glycoside (bufadienolide) called cotyledoside. Other *Tylecodon*
species contain similar bufadienolides.
UTILISATION There are no recorded uses but
some species are grown as ornamental plants.
TOXICITY *Tylecodon* species are known to be
extremely poisonous and are a major cause of

stock losses in southern Africa. They cause an
acute or chronic condition known as *krimpsiekte*
("shrinking disease"). Acute *krimpsiekte* results in
bloating and sudden death, while chronic *krimpsiekte* may develop gradually (see *Cotyledon orbiculata*). As little as 14 g of dried leaves given over
a period of 25 days have produced *krimpsiekte* in
a goat and it has been shown that the toxins have
a cumulative effect. When meat from a poisoned
animal is eaten, the toxins present may cause sec
ondary poisoning of dogs and even humans. LD_{50}
of cotyledoside in guinea pigs: 0.116 mg/kg (sub
cutaneous injection) or 0.173 mg/ kg (p.o.).
SYMPTOMS Acute poisoning results in symp
toms similar to those of other bufadienolides,
while chronic and cumulative poisoning produce
symptoms of *krimpsiekte* (see *Cotyledon orbiculata*). Krimpsiekte is an intoxication of livestock
that affects the nervous and muscular systems (see
Cotyledon orbiculata), usually with fatal results.
PHARMACOLOGICAL EFFECTS Bufadieno-
lides selectively inhibit Na^+, K^+-ATPase and are
therefore general poisons (see cardiac glycosides).
FIRST AID First aid treatment is indicated if
plant material from *Tylecodon* or isolated cardiac
glycosides have been ingested. Treatment involves
the induction of vomiting, supply of medicinal
charcoal, and detoxification as outlined under
cardiac glycosides.

Tylecodon wallichii (Harv.) Tölken   family: Crassulaceae

Urginea maritima

sea squill • sea onion

Urginea maritima   *Urginea maritima*   *Urginea maritima*

PLANTS WITH SIMILAR PROPERTIES The genus *Urginea* is sometimes included in *Drimia*, and comprises 120 species in S Europe, Africa and Asia. *U. maritima* is usually considered to be a species complex comprising several close relatives that are all Mediterranean in distribution. *U. indica* is an alternative source of raw material.

PLANT CHARACTERS Squill is a bulbous perennial with a large, white, onion-like bulb, a cluster of oblong leaves and a robust flowering stalk of up to 1 m high with numerous small white flowers. Some species (e.g. *U. sanguinea*) have red or reddish-purple bulbs.

OCCURRENCE The real *U. maritima* is found only on the Iberian peninsula; the closely related species occur in other parts of the Mediterranean region. Squill was introduced to S California for the production of rodenticides. *Drimia indica* is an Indian and African plant. *U. sanguinea* occurs in tropical southern Africa.

CLASSIFICATION Heart poison; highly hazardous, Ib

ACTIVE INGREDIENTS *Urginea* bulbs contain bufadienolides (1.8 %). The main glycosides in *U. maritima* are glucoscillaren A and scillaren A that occur with 11β-hydroxylated derivatives (scillaphaeoside and glucoscillaphaeoside). Enzymatic hydrolysis of scillaren A (loss of one glucose molecule) gives proscillaridin A.

UTILISATION Squill is traditionally used as a heart tonic and diuretic but also as an expectorant in cough mixtures, for chronic bronchitis, asthmatic bronchitis and whooping cough. Externally it is employed to disinfect wounds. Species with red bulbs are traditionally used as rat poisons.

TOXICITY Species such as *U. sanguinea* are used in traditional medicine and have caused more than 40 human fatalities in one year alone in South Africa (1.5 g of bulbs were lethal). Livestock losses are also common (so-called *slangkop* poisoning). Dried, powdered bulbs are lethal to sheep at a single dose of 2.5 g/kg.

SYMPTOMS Typical are nausea, vomiting, salivation, gastrointestinal disturbance, purging and exhaustion, with the usual respiratory and cardiac symptoms (arrhythmia, hypertension, coma, cardiac arrest) associated with heart glycoside poisoning.

PHARMACOLOGICAL EFFECTS Cardiac glycosides inhibit the Na^+, K^+-ATPase (see cardiac glycosides). Concentrated decoctions (very bitter) may cause rapid death due to heart failure.

FIRST AID First aid treatment is indicated if plant material from *Urginea* or isolated cardiac glycosides have been ingested. Treatment involves the induction of vomiting, supply of medicinal charcoal and detoxification as outlined under cardiac glycosides.

Urginea maritima L. (= *Drimia maritima* (L.) Stearn)   family: Hyacinthaceae or Asparagaceae

scille maritime (French); *Meerzwiebel* (German); *cipolla marina, squilla* (Italian); *cebolla albarrana, escila* (Spanish)

Urtica dioica

stinging nettle

Urtica dioica   *Urtica urens*

PLANTS WITH SIMILAR PROPERTIES An almost cosmopolitan genus of 80 species, most of which have stinging hairs. Stinging hairs are also typical of other Urticaceae (including the genera *Girardinia, Laportea, Obetia, Soleirolia, Urera*). *Soleirolia soleirolii* is a creeping member of the family that is commonly grown as a cultivated ornamental and pot plant. It has tiny, pale green, rounded leaves and is known by the common names helxine, peace-in-the-home, mind-your-own-business and mother-of-thousands.

PLANT CHARACTERS *U. dioica* is a perennial herb of up to 1.5 m high with typical drooping, dull, greyish-green leaves and slender, pendulous flower clusters that are longer than the leaf stalks. In contrast, *U. urens* is an annual herb and a much smaller plant with glossy, bright green leaves and flower clusters that are shorter than the leaf stalks. The flowers of both species are inconspicuous.

OCCURRENCE *U. dioica* is indigenous to Europe and Asia, while *U. urens* occurs naturally over the entire northern hemisphere. Both species are naturalised weed in many countries.

CLASSIFICATION Neurotoxin; slightly hazardous, III

ACTIVE INGREDIENTS Nettle leaf accumulates a wide range of secondary metabolites (amines, flavonol glycosides, phenolic acids, scopoletin, β-sitosterol and tannins). Stinging hairs are rela-

tively stiff due to silica incrustation. They work like minute injection needles and are filled with acetylcholine, histamine, serotonin and further (unknown) compounds (possibly peptides). The ball-like top of the hair breaks when in contact with skin. Stinging hairs of the Australian stinging tree *Laportea moroides* contain a tricyclic octapeptide that causes irritation lasting for weeks.

UTILISATION The leaves and roots are used in supportive treatment of rheumatic complaints and symptoms of an enlarged prostate, and to a lesser extent against inflammatory conditions of the urinary tract. Nettle fibres of *U. cannabina* have been used to produce nettle linen. The leaves are a raw material for industrial chlorophyll production.

TOXICITY Serious reactions are rare, but symptoms are unpleasant.

SYMPTOMS Skin contact with nettles causes a painful itching with welt formation at the places that were pricked by stinging hairs. In sensitive persons, inflammation can last for 36 h.

PHARMACOLOGICAL EFFECTS The stinging hairs act as miniature syringes that inject histamine and acetylcholine, causing extreme irritation. Nettle roots interact with testosterone-binding proteins, 5α-reductase and aromatase. Nettle lectins appear to be immune stimulants.

FIRST AID In serious cases, apply ointment with antihistaminic compounds or corticoids.

Urtica dioica L.   family: Urticaceae

ortie brûlante (French); *Große Brennnessel* (German); *ortica maschio* (Italian); *ortiga mayor* (Spanish)

Veratrum album

white hellebore • false hellebore

Veratrum album

Veratrum album flowers

PLANTS WITH SIMILAR PROPERTIES A genus of 15–20 species in northern hemisphere, including *V. nigrum, V. californicum* and *V. viride*.
PLANT CHARACTERS A robust perennial herb with broad leaves that are pleated lengthwise and arranged in threes. Flowers are yellowish-green, often white inside and star-like.
OCCURRENCE *Veratrum album* occurs in mountainous regions of Europe and Asia.
CLASSIFICATION Cell toxin, neurotoxin; extremely hazardous, Ia
ACTIVE INGREDIENTS Plants are rich in steroidal alkaloids, such as protoveratrine A and B (see aconitine), germerine, jervine or cyclopamine, which are conjugated with chelidonic or veratric acid. Alkaloids are abundant in all plant parts; roots (1.3 %), rhizome (1.6 %), leaves (up to 1.5 %).
UTILISATION Known as "Somphia" to the ancient Egyptians, "Veratrum" to the Romans and "Laginon" to the Celts, *Veratrum* was recognised as a toxic and medicinal plant. In antiquity it was used (according to Dioscorides) to initiate emesis, for abortion, and to kill rodents. In folk medicine, rhizome extracts were used to treat gout, rheumatism, and trigeminal neuralgia and was employed against ectoparasites (such as lice). Powders induce sneezing and were still an ingredient of sneezing powder at the beginning of the 20th century (but was later banned because of its toxicity).

TOXICITY *Veratrum* has been confused with gentian (*Gentiana lutea*) causing serious accidental poisoning. Poisoning of goats and sheep has been observed. Alkaloids are lipophilic and are easily absorbed; 10–20 mg may cause death in humans. 20 mg protoveratrine B, equivalent to 1–2 g of the dried rhizome, are lethal for humans; LD_{50} rabbit: 0.1 mg/kg p.o
SYMPTOMS Symptoms of *Veratrum* intoxication are similar to those of aconitine and include salivation, sneezing, burning and bitter taste, vomiting and bloody diarrhoea, bradycardia, hypotension, vertigo, headache, tremor, hallucinations, pain, itching and burning, followed by a complete anaesthesia and muscular paralysis. Death is caused by cardiac or respiratory arrest.
PHARMACOLOGICAL EFFECTS These alkaloids activate Na^+ channels and depolarise neuronal membranes; thus they disturb signal transduction in neurons and the heart. Cyclopamine causes malformation (cyclopean eye) when pregnant animals ingest the alkaloid (see aconitine).
FIRST AID First aid treatment is indicated if any parts of *Veratrum* have been ingested. It involves the induction of vomiting, the application of sodium sulphate or potassium permanganate and medicinal charcoal. **Clinical therapy:** Includes gastric lavage and artificial respiration, see aconitine.

Veratrum album L.

family: Melanthiaceae

vérâtre blanc (French); *Weißer Germer* (German); *veratro bianco* (Italian); *vedegambre, eléboro blanco* (Spanish)

Viscum album

mistletoe • European mistletoe

Viscum album

Viscum album fruits

PLANTS WITH SIMILAR PROPERTIES *Viscum* includes 65 species from tropical and temperate regions of the Old World, including *V. articulatum*; *Loranthus europaeus* is a related plant.

PLANT CHARACTERS Branching, woody perennials that grow on trees as semi-parasites. The leaves are leathery and yellowish green. Small flowers are followed by translucent white berries. There are three subspecies: one grows only on broadleaf trees, one mainly on pines and larches (*Pinus* and *Larix*) and one only on fir (*Abies*). Birds ingest the fruits but viscid tissue on the seeds prevents them from being swallowed. They then scrape the sticky seed off onto the bark of thin tree branches, where it germinates.

OCCURRENCE Europe and Asia.

CLASSIFICATION Cellular poison; moderately hazardous, II

ACTIVE INGREDIENTS All parts, especially fruits, are rich in lectins (0.1 %) (glycoproteins that can bind to specific sugar moieties; mistle lectin I–III; =Ml I–III). Of special interest are up to 0.1 % viscotoxins, i.e., polypeptides composed of 46 amino acids. Also present are mucilage, phenylpropanes and flavonoids. Secondary metabolites (SMs) of the host plant can influence the toxicological activity; mistletoes from maple, lime, *Nerium*, *Duboisia*, walnut and false acacia are quite toxic, since they sequester the SMs of their host plants.

UTILISATION Mistletoe was a sacred plant of the early Celts and Druids; and has survived up till now as a Christmas symbol. Mistletoe extracts and preparations are injected, mainly for the supportive treatment of cancer, degenerative inflammation, high blood pressure, poor circulation (as heart tonic), and as sedative. In Europe, berries have been used as birdlime to catch birds.

TOXICITY Children sometimes ingest mistletoe berries but the resulting gastrointestinal symptoms are usually transient and hardly fatal.

SYMPTOMS Berries are emetic and purgative; they cause strong thirst, bloody faeces, painful defecation and convulsions (when many fruits have been ingested). Extracts weaken respiration, cause gastrointestinal disorder, colic and sometimes urine retention.

PHARMACOLOGICAL EFFECTS Mistletoe preparations show immune-stimulating effects when injected at low concentrations but they are cytotoxic at higher concentrations. Viscotoxin has cardiotoxic properties in mammals (but not in birds) and resembles bee toxin. After injection, urticaria occurs that can develop into necrosis.

FIRST AID First aid treatment is only indicated if several fruits have been ingested. Treatment involves the induction of vomiting and the application of medicinal charcoal. In case of serious poisoning, see ricin.

Viscum album L.

family: Santalaceae (formerly Viscaceae)

gui blanc (French); *Mistel* (German); *vischio* (Italian); *muérdago* (Spanish)

Vitis vinifera

grape vine

Vitis vinifera (white grape)   *Vitis vinifera* (red grape)   *Vitis vinifera*

PLANTS WITH SIMILAR PROPERTIES A genus of 65 species in the N hemisphere. Besides the common grape vine, other vines have been cultivated, such as the North American *V. labrusca* and *V. riparia*, and the Asian *V. amurensis*. These and other species have been hybridised with *V. vinifera* or used as disease-resistant rootstocks.

PLANT CHARACTERS A deciduous woody climber (vine) with spirally coiled climbing tendrils, characteristic lobed leaves, tiny flowers and bunches of green, red or black berries.

OCCURRENCE Cultivated grapes probably originated in the Mediterranean area or the Middle East, from where they spread to all temperate regions of the world.

CLASSIFICATION Alcohol is mind-altering; slightly hazardous, III

ACTIVE INGREDIENTS The active ingredient in grapes is glucose, reaching concentrations of more than 20 % in ripe fruit. It is fermented with yeast (*Saccharomyces cerevisiae*) to form ethanol. Of pharmacological interest is the seed oil, which contains non-hydrolyzable or condensed tannins (procyanidins). Also important are the anthocyanins (mainly 3-glycosides of cyanidin and paeonidin), organic acids, tannins and other phenolic compounds present in red grapes and red wine.

UTILISATION Grape seed is an important commercial source of so-called botanical antioxidants (free-radical scavengers). Red grape leaves have been used in traditional medicine to treat diarrhoea, improve circulation and to control bleeding. Wine, especially red wine, is considered to be healthy if taken in moderation (see ethanol).

TOXICITY Lethal ethanol concentrations are 5–8 g/kg in adults and 3 g/kg in children.

SYMPTOMS Small doses of ethanol are stimulating, higher doses above a blood concentration (BAC) of 1.4 g/l (‰; per mille) are considered poisonous. See ethanol for a detailed description.

PHARMACOLOGICAL EFFECTS Oligomeric proanthocyanidins (OPC) are antioxidants and free-radical scavengers. The main molecular target for ethanol seems to be the GABA receptor. As an agonist, ethanol can enforce synaptic inactivation and chloride influx that are mediated by GABA. This activity antagonises agonistic reactions of acetylcholine, glutamate and serotonin that are mediated by sodium and potassium channels. Thus ethanol has an activity similar to that of barbiturates.

FIRST AID In case of serious intoxication, induce vomiting but avoid choking, keep patient warm. Control respiration and circulation. Gastric lavage with 2 % $NaHCO_3$ solution may be useful. In case of unconsciousness and danger of respiratory arrest, provide haemodialysis and artificial respiration. For further treatments, see ethanol.

Vitis vinifera L.   family: Vitaceae

vigne (French); *Weinrebe* (German); *vite* (Italian); *vid, parra* (Spanish)

Withania somnifera

winter cherry • ashwagandha

Withania somnifera fruits

PLANTS WITH SIMILAR PROPERTIES A genus of 10 species in Africa, Europe and Asia.

PLANT CHARACTERS A perennial, densely velvety shrublet of up to 1 m in height with a fleshy, often branched taproot and hairy leaves. The flowers are greenish yellow and inconspicuous. Characteristic are the bright red berries that are enclosed in a persistent (accrescent) bladdery or papery calyx.

OCCURRENCE *Withania somnifera* occurs over large parts of Africa, southeastern Europe, the Mediterranean area, Canary Islands, the Middle East and Asia. It is an important medicinal crop in India and other countries, where it is cultivated for the fleshy roots.

CLASSIFICATION Cell toxin; mind-altering, slightly hazardous, III

ACTIVE INGREDIENTS Withania leaves and roots are chemically very complex and a large number of compounds (more than 80 constituents) have been identified. Among them are withaferin A and related steroids with an ergostane skeleton (known as withanolides). They usually occur as free aglycones. Many alkaloids have been described, such as withasomnine. Different chemotypes of *W. somnifera* are known.

UTILISATION Ashwagandha is the Hindi name for the famous Ayurvedic drug derived from the fleshy roots of *W. somnifera* and known for its narcotic and diuretic properties. It was probably mentioned in the Indian Vedas, especially in the Atharvaveda, as esteemed medicine and as magic root, heal-all and aphrodisiac. It is amongst the most famous of all the plants used in Ayurvedic medicine and is still a popular sedative, hypnotic and adaptogenic tonic, used to counteract the symptoms of stress. The common name "Indian ginseng" is also widely used, as the roots not only resemble those of real ginseng (*Panax ginseng*), but are similarly used as general tonic to treat a very wide range of ailments. Leaf poultices are effective in the topical treatment of open wounds as well as septic, inflamed wounds and abscesses. It is also applied to treat haemorrhoids, rheumatism and syphilis. The fruits of *Withania coagulans* are known as panirband or vegetable rennet and are used to coagulate milk in cheesemaking.

TOXICITY No serious hazards are known. Overdoses may cause CNS disturbance.

SYMPTOMS Sedative and hypnotic effects have been ascribed to the alkaloids.

PHARMACOLOGICAL EFFECTS The withanolides and other *Withania* compounds have been the subject of numerous studies that demonstrated antibiotic, anti-inflammatory, cytotoxic, antitumour and cholesterol-lowering activities.

FIRST AID In case of overdosing, instil medicinal charcoal and sodium sulphate.

Withania somnifera (L.) Dunal

family: Solanaceae

withania (French); *Schlafbeere* (German); *witania, ginseng indiano* (Italian); *herba mora mayor, orval, bufera* (Spanish)

Xanthium strumarium

large cocklebur • rough cocklebur

Xanthium strumarium

Xanthium spinosum

PLANTS WITH SIMILAR PROPERTIES
A genus of 2 species (or up to 11 species, according to some authorities): *X. strumarium* (rough cocklebur, large cocklebur, Noogoora burr) and *X. spinosum* (spiny burweed, spiny cocklebur, clotbur, Bathurst burr). The toxic diterpenoid of *Xanthium* species are also present in other Asteraceae such as *Atractylis gummifera* (birdlime thistle) and *Callilepis laureola* (see the monograph).

PLANT CHARACTERS A robust annual of up to 0.8 m high. It has reddish stems and broadly heart-shaped to triangular leaves with coarsely toothed margins. Female flowers are arranged in pairs in a prickly involucre from which only the styles project. Fruits are enclosed in an accrescent involucre covered with hooks that aid dispersal by animals (these fruits are known as burs).

OCCURRENCE The plant is thought to have originated in America (perhaps the West Indies) but has become a cosmopolitan weed that is now found in practically all parts of the world.

CLASSIFICATION Cell toxin; moderately hazardous, II

ACTIVE INGREDIENTS Atractyloside (a glycoside of a kaurene diterpenoid) together with xanthinine and other sesquiterpene lactones.

UTILISATION Atractyloside has been employed in pharmacological research as inhibitor of ATP transporters (such inhibitors are rare). The burs contaminate wool and cause considerable losses to sheep farmers. *Atractylis* has a Mediterranean distribution (Europe and Africa) and has been used externally to treat boils and abscesses (the *shawk el- 'elk* of Arabic medicine).

TOXICITY Three cases of fatal poisoning occurred in China, after the buds of the plant were consumed (misidentified as food). Another fatality was that of a boy treated with *Atractylis gummifera* against intestinal worms. Livestock poisoning (cattle, sheep and pigs) is common. The lethal dose of *Xanthium* in pigs is 0.75–2 % of body weight (seedlings) or 0.3 % (seeds).

SYMPTOMS Human poisoning results in vomiting, coma and periodic spasms, followed by death within 2–4 days due to cardiac arrest. Autopsies show liver damage (centrilobular and midzonal necrosis). Applied on the skin, extracts cause contact dermatitis.

PHARMACOLOGICAL EFFECTS Atractyloside inhibits the transport of the phosphorylated nucleotides (ADP and ATP) across the inner mitochondrial membrane. Thus the export of ATP is inhibited so that cellular and muscular work stops because of lack of energy (see helenalin).

FIRST AID First aid treatment is indicated if larger quantities have been ingested: application of medicinal charcoal, sodium sulphate and detoxification as outlined under helenaline.

Xanthium strumarium L.

family: Asteraceae

glouteron (French); *Gewöhnliche Spritzklette* (German); *lappola comune* (Italian); *bardana menor* (Spanish)

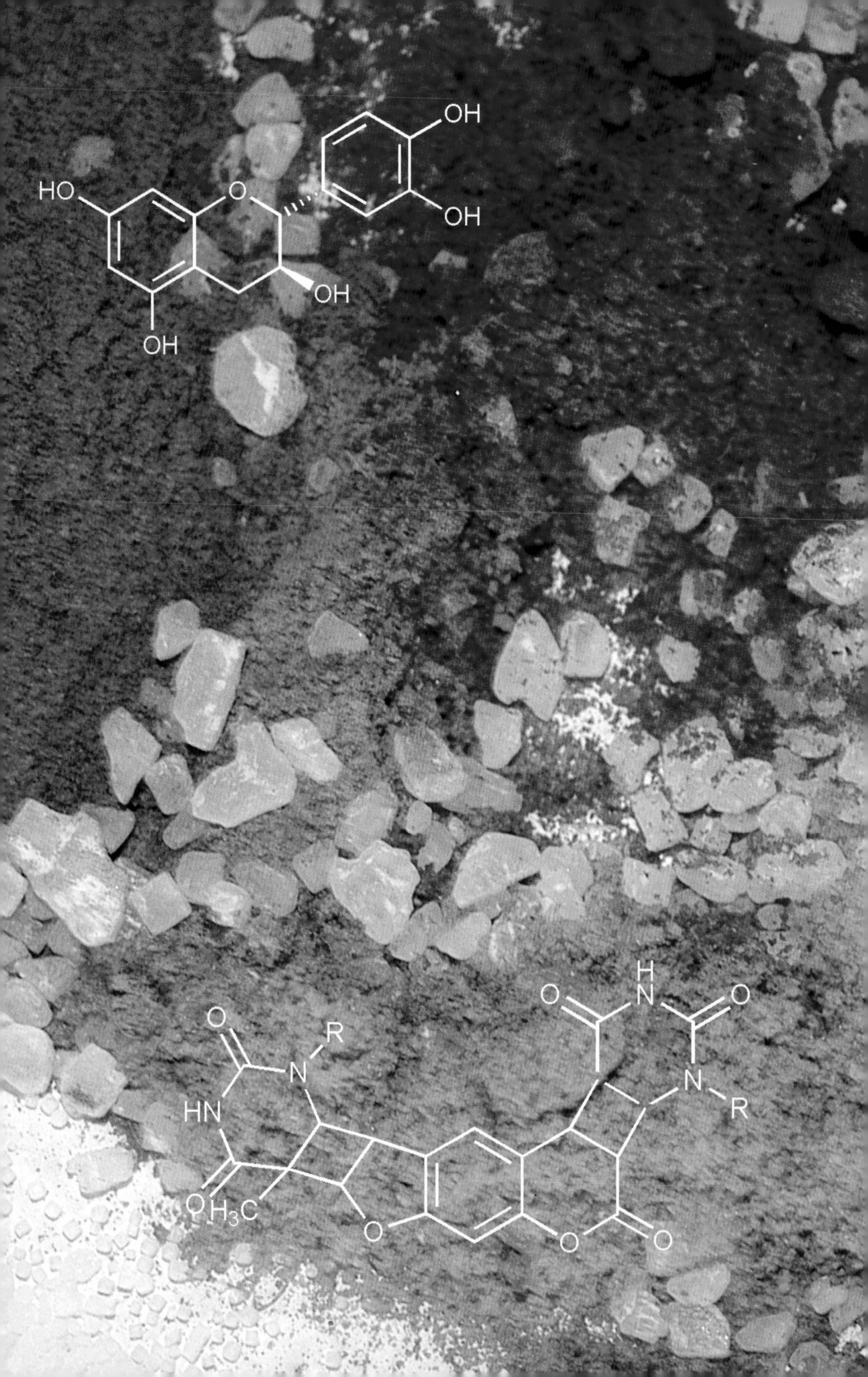
OH
HO
O
OH
OH
OH
O
R
N
HN
O
H3C
O
O
O
H
N
O
O
N
R
O

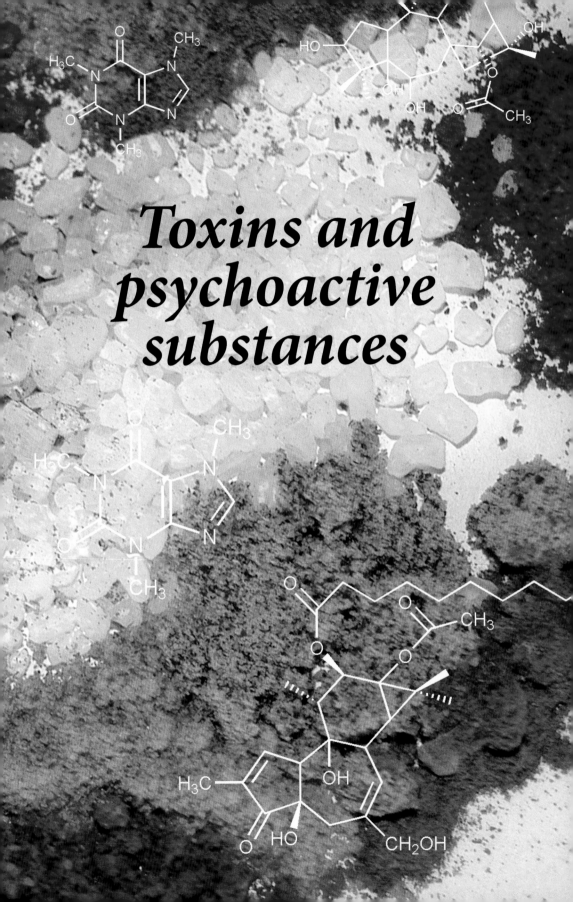

Toxins and psychoactive substances

How do poisons and mind-altering substances function?

Molecular modes of action

Toxins are substances that can negatively interfere with molecular targets in cells of humans or animals. Major cellular targets include

- the biomembrane
- proteins (including receptors, ion channels, enzymes, transporters, regulatory proteins, structure proteins, cytoskeleton, mitotic spindle (microtubules), transcription factors, hormones
- nucleic acids (DNA, RNA).

Toxins can act as agonists or antagonists at a given molecular target. If this happens to a crucial cellular target, severe negative consequences result. For example, if a compound is cytotoxic to individual kidney cells, the effect will usually damage the function of the kidney. Organ damage can be so serious that it leads to coma and death.

Acute toxins are those which disturb neuronal processes, inhibit or kill cells with high rates of protein synthesis, such as liver cells. Therefore, they are a prime target for many cytotoxic substances. In contrast, chronic exposure to mutagenic substances often does not lead to immediate death but can cause cancer or teratogenic effects.

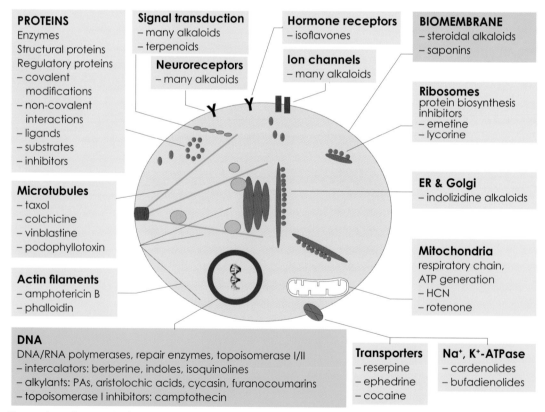

PROTEINS
Enzymes
Structural proteins
Regulatory proteins
– covalent
modifications
– non-covalent
interactions
– ligands
– substrates
– inhibitors

Signal transduction
– many alkaloids
– terpenoids

Neuroreceptors
– many alkaloids

Hormone receptors
– isoflavones

Ion channels
– many alkaloids

BIOMEMBRANE
– steroidal alkaloids
– saponins

Ribosomes
protein biosynthesis
inhibitors
– emetine
– lycorine

Microtubules
– taxol
– colchicine
– vinblastine
– podophyllotoxin

ER & Golgi
– indolizidine alkaloids

Actin filaments
– amphotericin B
– phalloidin

Mitochondria
respiratory chain,
ATP generation
– HCN
– rotenone

DNA
DNA/RNA polymerases, repair enzymes, topoisomerase I/II
– intercalators: berberine, indoles, isoquinolines
– alkylants: PAs, aristolochic acids, cycasin, furanocoumarins
– topoisomerase I inhibitors: camptothecin

Transporters
– reserpine
– ephedrine
– cocaine

Na⁺, K⁺-ATPase
– cardenolides
– bufadienolides

Examples of some toxins which modulate important cellular targets

Biomembrane

The biomembrane surrounds every living cell and functions as a permeation barrier. It prevents polar molecules from leaking out of the cell or unwanted molecules from entering a cell. Several SMs exist in nature which interfere with membrane permeability (Table p. 249). Most famous are saponins that occur widely in the plant kingdom and less commonly in animals; monodesmosidic saponins are amphiphilic and function basically as detergents that can solubilise biomembranes. With their lipophilic moiety they are anchored in the lipophilic membrane bilayer whereas the hydrophilic sugar part remains outside and interacts with other glycoproteins or glycolipids (Figure). As a result pores are generated in the membrane and make it leaky. This can be shown easily with red blood cells. If saponins are present, a haemolysis takes place, i.e. haemoglobin flows out of the cells. Also other lipophilic SMs, such as mono-, sesqui- and diterpenes can disturb membrane fluidity at higher concentration. Several animal and bacterial poisons have cytolytic properties (e.g. mellitin in bee venom). These compounds will interact with the lipophilic inner core of biomembranes represented by phospholipids and cholesterol (Figure). Such a membrane disturbance is unselective and therefore, SMs with such properties are toxic to all animal cells. Some of them also affect bacterial, fungal, and viral membranes.

Biomembranes harbour a wide set of membrane proteins, including ion channels, pumps and transporters for nutrients and intermediates, receptors, and proteins of signal transduction and the cytoskeleton. These proteins can only work properly if their spatial structure is in the right conformation. Membrane proteins with transmembrane domains are stabilised by the surrounding lipids. If lipophilic SMs dissolve in the biomembrane, they disturb the close interaction between membrane lipids and proteins thus changing the protein conformation. Usually a loss of function is the consequence. This property is known from anaesthetics that are small and lipophilic compounds. They inactivate ion channels and neuroreceptors and thus block neuronal signal transduction. Several of the small terpenoids can react in a comparable way; plants with essential oils are often employed in medicine as carminative drugs, i.e. a drug that relieves intestinal spasms. We suggest that such SMs inactivate ion channels and receptors that lead to relaxation of smooth muscles in the intestinal tissues. Or they can induce mind-altering effects.

It should be noted that animal cells express ABC transporters (p-gp, mdr protein) that actively pump out lipophilic compounds that have entered a cell. These pumps probably evolved in animals in response to poisonous compounds produced by plants. Together with detoxifying enzymes in the liver, they help the body to get rid of noxious chemicals.

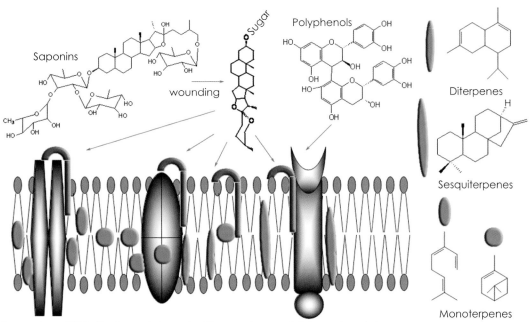

Examples of SMs that can interact with membrane lipids and proteins

243

Proteins and protein conformation

Proteins have multiple functions in a cell, ranging from catalytic enzymes, transporters, ion channels, receptors, microtubules, histones to regulatory proteins (signal molecules, transcription factors, etc.). Proteins can only work properly if they have the correct shape and conformation. Conformational changes also alter their properties and can prevent effective protein-protein cross talk. Protein activities are often regulated by phosphorylation or dephosphorylation. The addition or subtraction of such a bulky group induces a conformational change. Probably most SMs that have been found in nature interact with proteins in one or another way. Most SMs interfere with proteins in an **unselective** way, i.e. they affect any protein that they encounter (Figures). Such unselective interactions can be divided into those that involve **non-covalent** bonding and those with **covalent** bond formation.

In addition to the unselective interactions, many toxins are known to modulate protein activities in a **specific way** in that they can bind as a ligand to the active site of a receptor or enzyme; this binding has been termed "induced fit". In this case, the structure of a given toxin is often a mimic of an endogenous ligand. Well-studied examples include several alkaloids that are structural analogues of neurotransmitters, e.g. nicotine and hyoscyamine are mimics of acetylcholine; nicotine binds to nAChR or hyoscyamine to mAChR. The various steps in neuronal signalling and signal transduction provide central targets that are affected by several amines and alkaloids which are discussed on page 250.

An important class of toxins, the cardiac glycosides, inhibit Na^+,K^+-ATPase, one of the most important targets in animal cells, responsible for the maintenance of Na^+ and K^+ gradients. Cardiac glycosides bind to an extracellular loop of the protein and inhibits it. Cardiac glycosides can be divided in two classes, cardenolides and bufadienolides: **Cardenolides** have been found in Apocynaceae,

Examples of specific interactions of toxins with potential receptors

244

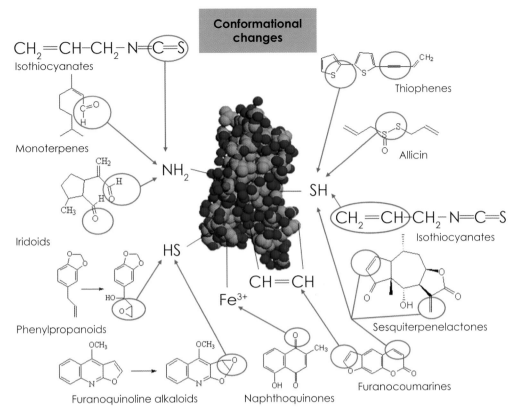

$CH_2=CH-CH_2-N=C=S$
Isothiocyanates

Monoterpenes

Iridoids

Phenylpropanoids

Furanoquinoline alkaloids

Conformational changes

NH_2

HS

Fe^{3+}

$CH=CH$

SH

Thiophenes

Allicin

$CH_2=CH-CH_2-N=C=S$
Isothiocyanates

Sesquiterpenelactones

Naphthoquinones

Furanocoumarines

Examples of toxins that can modify protein conformation by forming covalent bonds

Brassicaceae, Celastraceae, Ruscaceae/Convallariaceae, Ranunculaceae and Plantaginaceae/Scrophulariaceae. **Bufadienolides** occur in Crassulaceae, Hyacinthaceae and Ranunculaceae. Ouabain is shown as an example on page 244.

A number of diterpenes are infamous for their toxic properties (cytotoxicity, inflammation), such as phorbol esters of Euphorbiaceae and Thymelaeaceae (e.g. TPA p. 244). They specifically activate protein kinase C, which is an important key regulatory protein in animal cells. Diterpene forskolin acts as a potent activator of adenylyl cyclase. The sesquiterpene atractyloside inhibits ADP/ATP transport in mitochondria (Figure p. 244).

Microtubule formation is a specific target for the alkaloids vinblastine (from *Catharanthus roseus*), colchicine (*Colchicum autumnale*), maytansine (from *Maytenus ovatus*, *Putterlickia verrucosa*; Celastraceae) or the lignan podophyllotoxin (from *Podophyllum* and several *Linum* species). Taxol is a diterpene alkaloid (paclitaxel, taxol®) that can be isolated from several yew species (including the North American *Taxus brevifolia* and the European *Taxus baccata*). Taxol stabilises microtubules and thus blocks cell division in the late G2 phase; because of these properties, taxol has been used for almost 10 years with great success in the chemotherapy of various tumours. The cytotoxic secondary metabolites cause immunosuppression since they also block the multiplication of immune cells. They also inhibit microtubule formation in neuronal axons, which are necessary for the transport of synaptic vesicles to the synapses, causing neuronal disturbances. In general these metabolites can be regarded as highly poisonous.

Specific protein inhibitors can be found in the class of non-protein amino acids (NPAA) that often figure as anti-nutrients or anti-metabolites in many plants (e.g. in Fabaceae). Many NPAAs mimic protein amino acids and quite often can be considered to be their structural analogues that may interfere with the metabolism of a herbivore. For example, in ribosomal protein biosynthesis NPAAs can be accepted in place of the normal amino acid leading to defective proteins. NPAAs may competitively inhibit uptake systems (transporters) for amino acids. NPAAs can block amino acid biosynthesis by substrate competition or by mimicking end product mediated feedback inhibition of earlier key enzymes in the pathway.

Another example for a more specific inhibitor is HCN released from cyanogenic glucosides that are common SMs in plants and some invertebrates. HCN is highly toxic for animals or microorganisms due to the inhibition of enzymes of the respiratory chain (i.e. cytochrome oxidase) because it blocks the essential ATP production. HCN also binds to other enzymes containing heavy metal ions. In case of emergency, i.e. when plants are wounded by herbivores or other organisms, the cellular compartmentation breaks down and vacuolar cyanogenic glucosides come into contact with an active β-glucosidase of broad specificity, which hydrolyses them to yield 2-hydroxynitrile (cyanohydrine). 2-Hydroxynitrile is further cleaved into the corresponding aldehyde or ketone and HCN by a hydroxynitrile lyase.

A number of toxins are less specific for a particular target but can attack a certain type of macromolecules, such as proteins or nucleic acids. An important strategy is to form covalent bonds with a protein, often by binding to free amino-, SH or OH-groups (Figure p. 245). Among toxins with these properties we find aldehydes (found in several monoterpenes, iridoids) that bind to amino groups, SH reagents and compounds with activated double bonds such as sesquiterpene lactones, furanocoumarins, or phenylpropenes that couple to free SH groups or quinones that can attack metal proteins. The covalent modification leads to a conformational change in the protein and thus loss of its activity. SMs with reactive functional groups that are able to undergo electrophilic or nucleophilic substitutions are represented by isothiocyanates, allicin, protoanemonin, tulipalin, iridoid aldehydes, furanocoumarins, valepotriates, sesquiterpene lactones and SMs with active aldehyde groups, epoxide or terminal and/or exocyclic methylene groups.

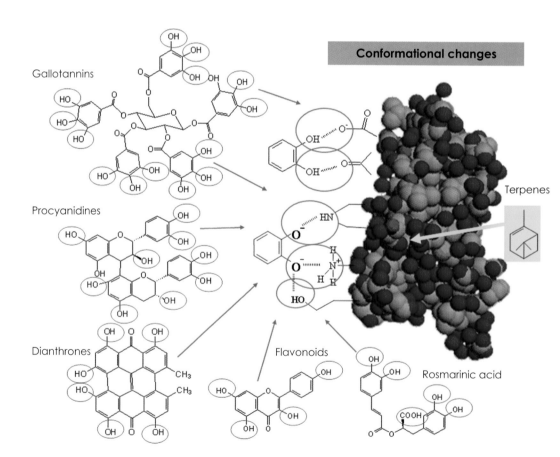

Examples of toxins that can modify protein conformation by forming non-covalent bonds

246

In several instances the reactive metabolites are not natively present in plants. Such "prodrugs" can be converted to active metabolites either by the wounding process (releasing metabolising enzymes) inside the producing organism or in the body of a herbivore/predator (after biotransformation in intestine or liver) (Table p. 14).

A major class of plant metabolites are phenolic substances (phenylpropanoids, flavonoids, catechins, tannins, lignans, anthraquinones). The phenolic OH-groups can partly dissociate under physiological conditions resulting in -O⁻ ions (p. 246). What polyphenols have in common is that they can interact with proteins by forming several hydrogen bonds and the much stronger ionic bonds with electronegative atoms or the positively charged side chains of basic amino acids (lysine, histidine, arginine). Individually, these non-covalent bonds are quite weak. But because several of them are formed concomitantly when a polyphenol interacts with a protein, a change in protein conformation is likely to occur which may lead to protein inactivation. Since most polyphenols are quite polar and therefore hardly absorbed after oral intake, they are usually not regarded as serious toxins.

If SMs bind to proteins of the human body they form new epitopes that can be attacked by the immune system. As a consequence, antibodies are formed against these modified proteins. If a person comes into contact again with such a toxin and if identical protein derivates are formed, an acute immune response can occur in the form of allergies or sometimes with life-threatening anaphylactic shock. Proteins in animal venoms that are injected into the body always function as novel antigens and induce the formation of antibodies. Well known are allergies against bee and wasp venoms in humans who have been stung before.

DNA, RNA

An important target in all organisms is **DNA, RNA** and the enzymes involved with DNA replication and transcription, but also DNA repair and DNA-topoisomerase. The translation of mRNA into protein in ribosomes is a basic target that is present in all cells. Protein synthesis can be affected by low-molecular-weight compounds but more prominently by several macromolecular peptides, including lectins (see ricin, abrin) and haemagglutinins. Inhibitors of these systems are often active against a wide range of organisms, such as bacteria, fungi and animal cells.

The DNA itself can be modified (**alkylation**) by compounds with reactive groups, such as epoxides (Figure p. 248). Infamous are pyrrolizidine alkaloids, aristolochic acid, cycasin, furanocoumarins and SMs with epoxy groups (often produced in the liver). Covalent modifications can lead to point mutations and deletion of single bases or several bases if the converted bases are not exchanged by repair enzymes.

Other SMs with aromatic rings and lipophilic properties **intercalate** DNA, which can lead to frameshift mutations (Figure). Intercalating alkaloids include emetine, sanguinarine, berberine, quinine, β-carboline and furanoquinoline alkaloids. Furanocoumarins combine DNA alkylation with intercalation. Furanocoumarins can intercalate DNA and upon illumination with UV light can form cross-links with DNA bases (p. 302), but also with proteins. They are therefore mutagenic and possibly carcinogenic. These compounds are abundant in Apiaceae (contents up to 4%), but also present in certain genera of the Fabaceae and Rutaceae.

Because frameshift mutations and non-synonymous base exchanges in protein coding genes alter the amino acid sequence in proteins, such mutations are usually deleterious for the corresponding cell. If they occur in germline cells such as oocytes and sperm cells, even the next generation is negatively affected either through malformations or protein malfunctions responsible for certain kinds of heritable health disorders or illnesses.

Interference with DNA, protein biosynthesis and related enzymes can induce complex chain reactions in cells. Among them figures apoptosis, which leads to programmed cell death, as an important process. Several alkaloids, flavonoids, allicin and cardiac glycosides have been shown to induce apoptosis in primary and tumour cell lines.

Some secondary metabolites bind to transcription factors and modulate their regulatory properties. Isoflavonoids are known to bind to steroid receptors in cells, which are known transcription factors and regulate oestrogen response genes. The steroid alkaloid cyclopamine from *Veratrum* species is an example of a poison that interferes with the activity of developmental genes (hedgehog gene). The formation of a cyclopian eye is a consequence (see p. 254).

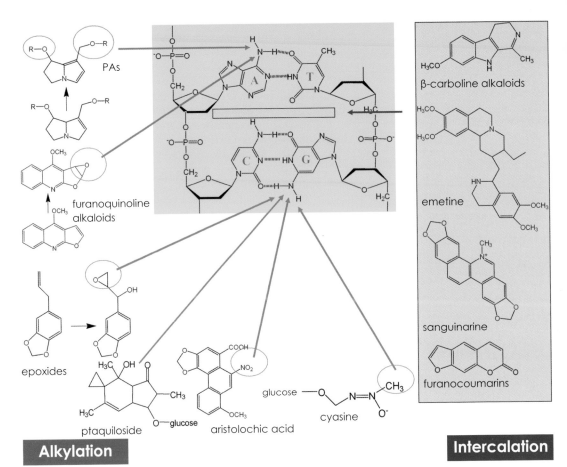

PAs

furanoquinoline alkaloids

epoxides

H₃C OH
ptaquiloside
O—glucose

aristolochic acid
COOH
NO₂
OCH₃

glucose —O—CH₂—N=N⁺—CH₃
O⁻
cyasine

β-carboline alkaloids

emetine

sanguinarine

furanocoumarins

Alkylation

Intercalation

Examples of toxins that can interact with DNA

Aristolochia clematitis with aristolochic acid

Senecio jacobaea with pyrrolizidine alkaloids

Examples of DNA active plants

Interaction of representative secondary metabolites with molecular targets

Target	Activity	SM (examples)
Biomembrane	membrane disruption disturbance of membrane fluidity inhibition of membrane proteins (change of protein conformation)	saponins small lipophilic SM small lipophilic SM
Proteins Non-selective interactions	non-covalent bonding (change of conformation) covalent bonding (change of conformation)	polyphenols such as phenylpropanoids, flavonoids, catechins, tannins, lignans, quinones, anthraquinones, some alkaloids isothiocyanates sesquiterpene lactones, allicin, protoanemonin, furanocoumarins iridoids, SM with aldehydes, SM with exocyclic CH_2 groups SM with epoxide groups
Specific interaction	inhibition of enzymes modulation of regulatory proteins inhibition of ion pumps inhibition of microtubule formation inhibition of protein biosynthesis inhibition of transporters modulation of hormone receptors modulation of neuroreceptors modulation of ion channels modulation of transcription factors	HCN from cyanogens, many structural mimics phorbol esters, caffeine cardiac glycosides vinblastine, colchicine, podophyllotoxin, taxol emetine, lectins non-protein amino acids genistein, many other isoflavonoids nicotine, many alkaloids, conotoxins, nereistoxin, argiotoxin, argiopin aconitine, many alkaloids; conotoxins, tetrodotoxin, saxitoxin, gonyautoxin, ciguatoxin, palytoxin cyclopamine, hormone mimics
DNA	covalent modifications (alkylation) (point mutations) intercalation (frameshift mutations) inhibition of DNA topoisomerase I inhibition of transcription	pyrrolizidine alkaloids, aristolochic acids, furanocoumarins, SM with epoxy groups planar, aromatic and lipophilic SM sanguinarine, berberine, emetine, qui- nine, furanocoumarins, anthraquinones camptothecin, berberine amanitine

How do neurotoxins and mind-altering substances work?

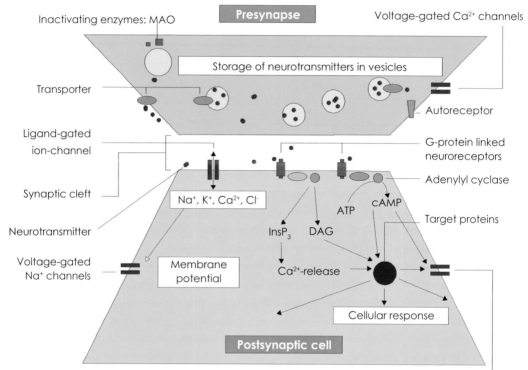

Main targets in neuronal signal transduction

In neuronal signal transduction an action potential is converted into a chemical signal in synapses (Figure) and at neuromuscular junctions. Two kinds of neuroreceptors are distinguished:
- Ligand-gated ion channels (ligands: acetylcholine, glutamic acid, GABA, serotonin, glycine)
- G-protein linked neuroreceptors (ligands: acetylcholine, serotonin, dopamine, noradrenaline)

When an action potential reaches the presynapse, voltage-gated calcium channels open. The influx of calcium ions induces the neurovesicles to fuse with the synaptic membrane and to release their neurotransmitters into the synaptic cleft. The neurotransmitters then bind to their corresponding neuroreceptor. When ligand-gated ion channels are activated, they open and ions, such as sodium and potassium ions, can enter the postsynaptic cell. This leads to an activation of voltage-gated sodium channels and a further depolarisation of the biomembrane. In a synapse the depolarisation can initiate a new action potential, whereas in muscle cells it stimulates voltage-gated calcium channels. The influx of calcium activates the actin-myosin system and thus muscle contraction. Activation of G-protein linked neuroreceptors leads to modulation of signal transduction pathways, which release either cAMP or $InsP_3$ and DAG as second messengers. The second messengers activate other ion channels or regulatory proteins, which generate a specific cellular response in the postsynapse.

During evolution some secondary compounds have been shaped and selected in such a way that they can effectively bind to neuroreceptors and either activate or inhibit them. A receptor and its natural ligand can be regarded as a lock and key, which have to be closely tuned in order to function properly. A natural product, which mimics a natural ligand would be a picklock, which could be used to manipulate, i.e. to either close or open the gate.

Neurotoxins, mind-altering or psychoactive compounds interfere with neuronal signalling and signal transduction (see Figure p. 250). Targets can be:

- The neuroreceptor itself:

    Agonists mimic the function of a neurotransmitter (acetylcholine, dopamine, noradrenaline, adrenaline, serotonin, GABA, glutamate, glycine, endorphines, peptides) by binding to its receptor and causing the normal response.

    Antagonists (often called "blocker") also bind to the receptor but act as an inhibitor of the natural ligand by competing for binding sites on the receptor, thereby blocking the physiological response.
- Voltage-gated Na^+, K^+ and Ca^{2+} channels (inhibition of action potential or repolarisation).
- The enzymes which deactivate neurotransmitters after they have bound to a receptor such as cholinesterase (inactivates acetylcholine), monoamine oxidase (inactivates neurotransmitters with amino groups) and catechol-O-methyltransferase (inactivates dopamine and noradrenaline).
- Transport processes, which are important for the uptake and release of the neurotransmitters into the presynapse or synaptic vesicles.
- Na^+, K^+, and Ca^{2+}-ATPases, which restore the ion gradients, must be considered in this category.
- Modulation of key enzymes of signal pathways:
    - adenylyl cyclase (making cAMP),
    - phosphodiesterase (inactivating cAMP or cGMP),
    - phospholipase C (releasing inositol phosphates such as IP_3 and diacylglycerol, DAG),
    - several protein kinases, such as protein kinase C or tyrosine kinase (activating other regulatory proteins or ion channels).

If neuronal signal transduction is completely blocked at any stage (especially of ion channels, Na^+, K^+-ATPase), the central nervous system can no longer control muscular activity, such as that of the heart, respiratory system or skeletal muscles. As a consequence neurotoxins often have paralytic properties and can cause death from respiratory or cardiac arrest. Also the CNS itself can be affected and neurotoxins cause numbness, unconsciousness and coma. Lower concentrations of neurotoxins often exhibit mind-altering properties, especially if neuroreceptors and corresponding enzymes are targeted. As described on page 10, SMs can act as stimulants, sedatives, hypnotics or hallucinogens depending on the part of the neuronal signal transduction which is affected.

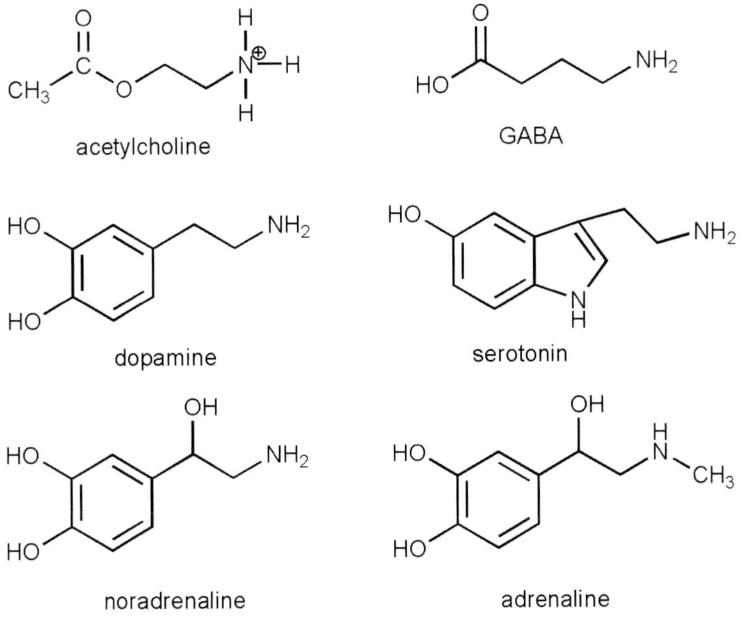

acetylcholine

GABA

dopamine

serotonin

noradrenaline

adrenaline

Structures of important neurotransmitters

Main classes of toxins and psychoactive compounds

In the context of this handbook we divide plant secondary metabolites into those which contain a nitrogen atom in their molecules and those without.

Secondary metabolites with nitrogen

In this large group we will discuss
- alkaloids (including amines)
- non-protein amino acids
- cyanogenic glucosides
- glucosinolates
- peptides (lectins, protease inhibitors)

Alkaloids

Alkaloids are secondary metabolites which contain one or several nitrogen atoms in their molecules. In most instances the N is heterocyclic and can react as a base. It is uncharged at alkaline pH values but becomes protonated under physiological conditions. The uncharged alkaloids are often lipophilic and can cross biomembranes by simple diffusion, whereas the protonated alkaloids cannot freely diffuse across membranes. About 20 000 alkaloid structures have been described; according to their ring structures they are classified in certain groups, such as indole, isoquinoline, quinoline, tropane, pyrrolizidine or quinolizidine alkaloids. Most alkaloids have pharmacological and toxicological relevance. Several are neurotoxins and/or mind-altering. A number of isolated alkaloids are used in medicine as therapeutic agents.

Aconitine, protoveratrine B and related terpene alkaloids

STRUCTURE The class of diterpenoid alkaloids comprises more than 450 structures. Aconitine ($C_{34}H_{47}NO_{11}$; MW 645.76); related compounds are picroaconitine, mesaconitine, lycaconitine ($C_{36}H_{48}N_2O_{10}$; MW 668.79), lycoctonine ($C_{25}H_{41}NO_7$; MW 467.61) and hypaconitine. Aconitine is rapidly absorbed after oral intake and distributed to all organs. Excretion via faeces and urine as the less toxic benzoylaconine and aconine. Some steroidal alkaloids have similar properties as aconitine: Protoveratrine A ($C_{41}H_{63}NO_{14}$; MW 793.96), protoveratrine B ($C_{41}H_{63}NO_{15}$; MW 809.96). These lipophilic alkaloids are easily and rapidly absorbed from skin or mucosal tissues.

OCCURENCE Diterpene alkaloids with similar properties to aconitine are common in *Aconitum*, *Delphinium*, and *Consolida* species (Ranunculaceae); they were also found in *Garrya* (Garryaceae), *Icacina* (Icacinaceae), *Inula royleana* (Asteraceae), *Spiraea japonica* (Rosaceae), *Anopterus* (Saxifragaceae) and *Erythrophleum* (Fabaceae). Aconitine has been detected in toxic honey; in this case bees had collected nectar and pollen from *Aconitum* plants. Protoveratrine and related steroidal alkaloids are common in *Veratrum* and *Zigadenus* species (Melanthiaceae). In the genus *Fritillaria* steroidal alkaloids such as imperialine are present, which constitute the active toxic principle.

CLASSIFICATION Neurotoxin, sodium channel activator; extremely hazardous, Ia

SYMPTOMS First effects of *Aconitum* and *Veratrum* poisoning set in after a few minutes; they include paraesthesia (burning, tingling, or numbness) of mouth and throat, followed by insensitivity of fingers and toes that spreads all over the body, accompanied by cold sweat, general coldness and severe diarrhoea. Further symptoms include general numbness, restlessness, nausea, strong vomiting, respiratory distress, muscular incoordination, convulsions, paralysis of tongue, facial and skeletal muscles, yellow-green vision, mydriasis and difficulties to accommodate, strong cardiac arrhythmia, bradycardia, hypotension, and severe pains. Death sets in after 30 minutes to three hours (depending on the dose) through respiratory or cardiac failure. High doses cause cardiac arrest. Patients remain conscious almost till death. Aconite is strongly psychedelic when smoked or absorbed through the skin.

TOXICITY Aconitine: LD_{50} mouse: 0.166 mg/kg i.v., 0.27 mg/kg s.c. 0.328 mg/kg i.p.; 1 mg/kg p.o.; LD_{50} cat: 0.07–0.13 mg/kg i.v.; LD_{50} for humans: 3–6 mg/kg or 2–15 g *Aconitum* tubers. LD_{50} for protoveratrine B: 20 mg, equivalent to 1–2 g of the dried rhizome are lethal for humans; LD_{50} rabbit: 0.1 mg/kg p.o.

Severe human intoxication from medicinal use of aconite has been reported; therefore the drug is no longer much used in modern medicine. But presently, TCM drugs with aconitine are available to patients in Western countries who are unaware of any toxicity hazard. Aconitine can penetrate the skin; care should be taken when handling this plant while gardening or flower arranging. The water in which aconite (cut flowers) has been imbibed, can become toxic. Livestock poisoning with *Aconitum, Delphinium* is a well-known problem in N America ("larkspur poisoning").

MODE OF ACTION These terpene alkaloids are potent activators of Na^+ channels (binding to receptor site 3) that are essential for neuronal signalling. If these ion channels are completely activated the action potential is no longer transmitted from nerves to muscles leading to a complete arrest of cardiac and skeletal muscles. Aconitine and protoveratrine B first activate and then paralyse the sensible nerve endings and neuromuscular junctions. In the heart aconitine causes extra systoles and fibrillation. In the CNS and spinal cord, centres for breathing, vomiting and temperature regulation are affected. Diterpenoid alkaloids from *Erythrophleum* (such as cassaidine, cassaine) exhibit digitalis-like effects on the heart and strong local anaesthetic activity. Ryanodine from *Ryania speciosa* (Salicaceae) blocks a calcium channel in the sarcoplasmic reticulum ("ryanodine receptor").

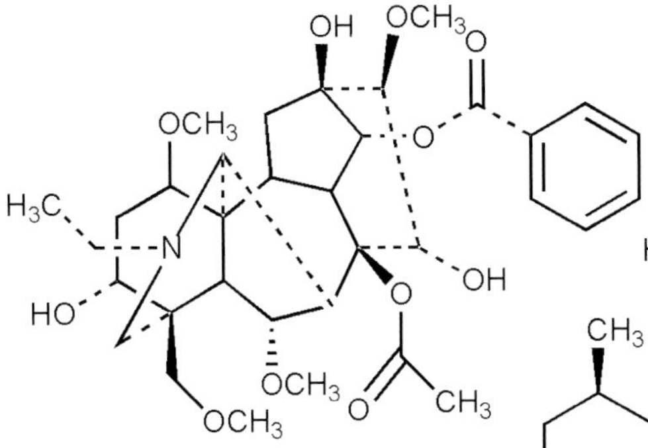

aconitine

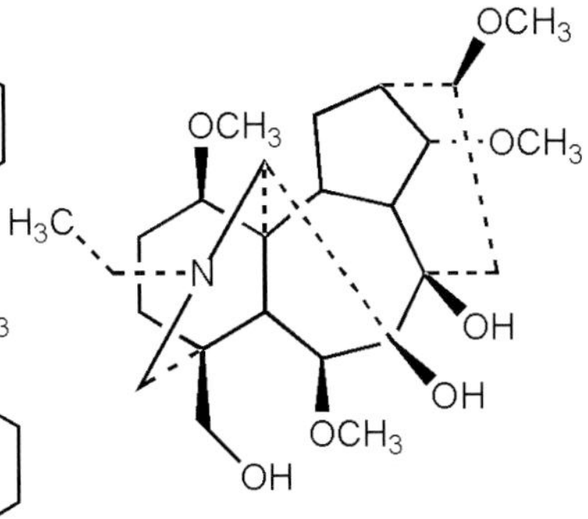

lycoctonine

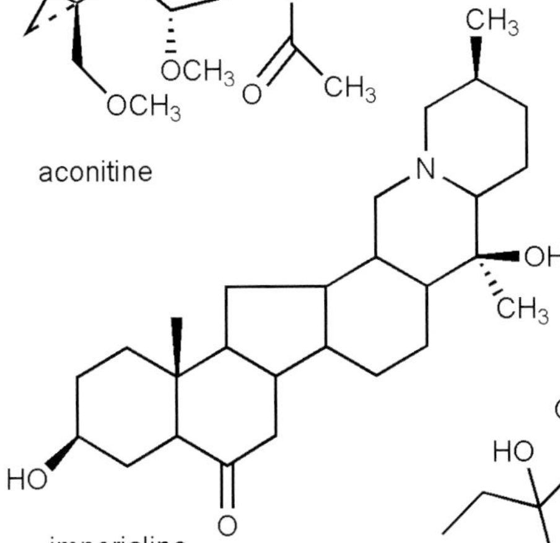

imperialine

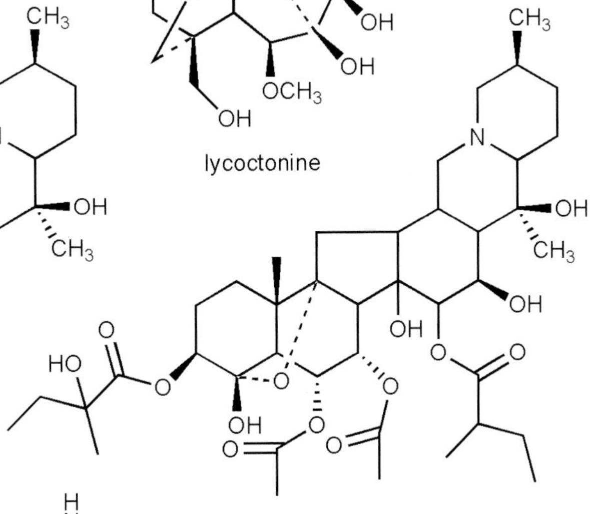

protoveratrine A

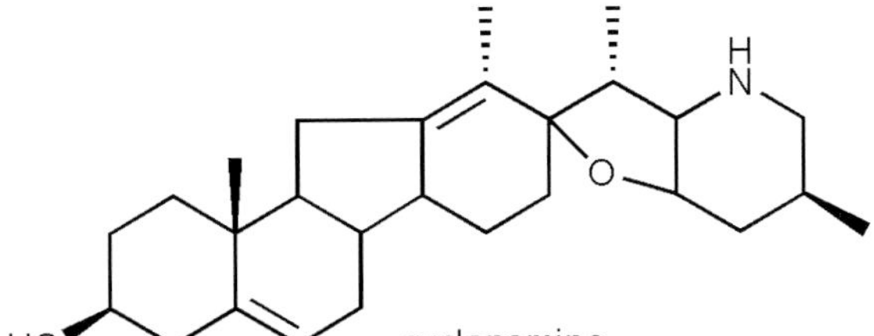

cyclopamine

USES (PAST AND PRESENT) Extracts from *Aconitum* have been widely used as arrow poison by early humans in Europe, Alaska and Asia. Since aconitine is also highly toxic when given orally it was used as a deadly poison to remove criminals, "disliked" contemporaries and for suicide.

Aconite has been mentioned as a deadly poison by Juvenal, Ovid, Plutarch and Theophrastos. The latter knew where the best aconite grew in Greece and that the plant was protected by the alkaloids against herbivores, such as sheep. In Rome, the cultivation of *Aconitum* was banned in the end (because it had been used too deliberately for murder) and the possession of the plant became a capital offence. On the Greek island of Chios aconite was allowed for euthanasia of old and infirm men.

A report from India says that wives used to get rid of their wealthy husbands by impregnating their shirts in an aconite extract (from *A. spicatum, A. ferox*). Since the alkaloids are lipophilic they would be taken up through the skin and would kill the husband straight away. It is said that the introduction of a custom to burn widows had suddenly increased life expectancy of men in the Himalaya region. Aconite was an ingredient of witch ointments together with tropane alkaloids and bufotenin.

Aconitum has been used medicinally (especially in China; TCM) as a painkiller; mostly externally as 1–2% solutions to treat neuralgic pains. Aconite roots have been ingested as a psychoactive drug, often leading to serious poisoning or even death.

The figure of a gigantic Cyclops (only having a single central eye) appears in Greek mythology. For example, in the Odyssey Ulysses had to outwit the cyclops Polyphem. Since mythical figures often have a rational base, it is tempting to correlate the ingestion of *Veratrum* alkaloids (for medicinal purposes or via milk from goats which had previously consumed *Veratrum* leaves) with the formation of a cyclopian eye. It can be shown experimentally that the alkaloids jervine and cyclopamine induce this malformation of the embryo if they have been ingested by female animals (e.g. sheep) during the early months of pregnancy. Another source for this myth could be fossil elephants and mammoths, which were certainly known to the Greeks. Feeding of *Veratrum viride* by pregnant sheep causes malformations of the head ("monkey face lamb") or extremities.

FIRST AID Induce vomiting by giving sodium sulphate or a 0.1% permanganate solution; then, instillation of sodium sulphate and medicinal coal. Clinical therapy should be provided as soon as possible. **Clinical therapy:** Gastric lavage (possibly with potassium permanganate) is advisable, instillation of medicinal charcoal and sodium sulphate, electrolyte substitution, control of acidosis with sodium bicarbonate. In case of spasms inject diazepam; for bradycardia atropine. Record electrocardiogram (ECG), if necessary, give antiarrhythmics, such as ajmaline or flecainide and magnesium salts. In case of serious hypotension, provide a drip injection with 5% glucose and inject vasoconstrictors. In case of respiratory paralysis provide artificial respiration. There are no specific antidotes.

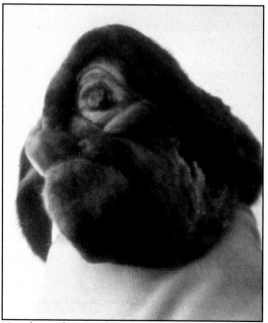

New-born sheep with cyclopian eye

Veratrum californicum as a source of cyclopamine

STRUCTURE Bufotenin (= *N,N*-dimethyl-5-hydroxytryptamine) ($C_{12}H_{16}N_2O$; MW 204.27), *N,N*-dimethyltryptamine (*N,N*-DMT) ($C_{12}H_{16}N_2$; MW 188.27), hordenine ($C_{10}H_{15}NO$; MW 165.24), psilocin ($C_{12}H_{16}N_2O$; MW 204.27).

tryptamine

N-methyltryptamine

N,N-dimethyltryptamine

serotonin

bufotenin

psilocybin (R = PO_3H_2)
psilocin (R = H)

N-methylserotonin

tyramine

hordenine

OCCURRENCE Bufotenin has been reported from *Anadenanthera, Mucuna pruriens* (Fabaceae), *Arundo donax* (Poaceae), *Banisteriopsis, Diplopterys* (Malpighiaceae) but also from some mushrooms of the genus *Amanita* and from skins of toads (*Bufo*).

Plants with *N,N*-DMT: *Mimosa tenuiflora* (= *M. hostilis*), *Anadenanthera peregrina*, several *Acacia* spp. (*A. polyacantha, A. cornigera, A. maidenii, A. nubica, A. phlebophylla, A. polyantha, A. senegal, A. simplicifolia*), *Calliandra* spp., *Desmodium* spp., *Mucuna pruriens* (Fabaceae), *Virola peruviana, V. elongata* (= *V. heiodora*) Epená, Yakée (Myristicaceae); *Banisteriopsis argentea, B. rusbyana* (Malpighiaceae); *Prestonia amazonica* (Apocynaceae), *Psychotria viridis, P. poeppigiana*, and *P. psychotriaefolia* (Rubiaceae); *Arundo donax, Phalaris arundinacea, Phragmites australis* (Poaceae) and *Zanthoxylum* spp. (Rutaceae).

Hordenine and other tyramines have been found in Cactaceae (*Ariocarpus, Coryphantha, Mammilaria, Obregonia, Pelecyphora, Solisia, Turbinicactus*) and in Poaceae (*Hordeum, Phalaris*).

Gramine ($C_{11}H_{14}N_2$; MW 175.25) has been found in several grasses and cereals but also in seeds of a few lupins (*Lupinus hispanicus*; Fabaceae).

Psilocin and its phosphate ester psilocybin are common ingredients of the sacred mushroom of Mexico "Teonanacatl" (*Psilocybe mexicana* and several other *Psilocybe* species; Strophariaceae); these alkaloids are also present in mushrooms of the genera *Conocybe, Copelandia, Galerina, Gymnopilus, Inocybe, Panaeolus* and *Pluteus*.

CLASSIFICATION Hallucinogen, mind-altering; moderately hazardous, II

SYMPTOMS Bufotenin is a hallucinogenic, psychotic substance that is especially active when given intravenously or intranasally. Signs of intoxication start immediately with hallucinations, anxiety, perceptual disturbances and reduction of awareness. The face becomes very red, the blood pressure increases and pupils become dilated. It is not clear how much of this compound can pass the blood-brain barrier; therefore vegetative effects are more pronounced than central ones.

N,N-**Dimethyltryptamine** (*N,N*-**DMT**). When injected intramuscularly (0.7 mg/kg) psychotropic effects set in after 2–3 min, reach a maximum after 10 min and last for almost 1 h. Psychotic reactions, which are accompanied by mydriasis, anxiety, perceptual distortions and hypertension, resemble those of LSD and mescaline. In schizophrenic patients the psychotic effects are enhanced. Oral applications of up to 350 mg do not cause hallucinogenic effects.

N,N-**Dimethyl-5-methoxytryptamine (5MeO-DMT)**. After smoking of 5–10 mg of the free base in cigarettes, hallucinogenic effects (visions, imaginations, reduction of awareness, feeling of being caged in and of isolation) set in after 1 min, reach a maximum after 2–3 min and last up to 20 min. 5–10 mg (i.m.) cause tremor and mydriasis.

Hordenine: The methyltyramines cause unpleasant hallucinations.

Gramine: Insect feeding inhibitor, which causes "Phalaris staggers" in sheep.

Psilocin and Psilocybin: The phosphate groups of psilocybin is removed in the body, which releases the active form, psilocin. Because of structural similarities with serotonin this amine is a strong hallucinogen, causing unusually pleasant sensation with physical and psychic relaxation.

TOXICITY LD_{50} (mouse i.p.): DMT, bufotenin, and 5MeO-DMT are 110, 290, and 115 mg/kg, respectively.

MODE OF ACTION The methylated tryptamines and derivatives are analogues of the neurotransmitter serotonin (5-hydroxytryptamine) and thus work as 5-HT agonists. They stimulate 5-HT receptors, and evoke psychedelic hallucinations and euphoric feelings.

USES (PAST AND PRESENT) Bufotenin is also present in toad skin, which has been used as a vital part of "witches' potion" in Europe. Indians of the Orinoco basin of Colombia and Venezuela, in the northern parts of the Brasilian Amazon and in northern parts of Argentina have used powders of Yopo as intoxicating drug. Indians apply free bases (produced by mixing the amines with alkaline ash or chalk from snails) nasally as a snuff (called yopo, niopo) or rectally as clyster. The free bases are readily absorbed and can pass the blood-brain barrier. The nasal and rectal routes are more effective than the oral route. *Arundo donax* has been dedicated to the god Pan; maybe the term "panic" describes the negative effects of *Arundo,* which might have been used together with *Peganum harmala* as intoxicant.

FIRST AID Administer medicinal charcoal and sodium sulphate. **Clinical therapy:** When larger doses have been ingested apply gastric lavage, medicinal charcoal, sodium sulphate, electrolyte substitution and sodium bicarbonate against acidosis. In severe cases, provide intubation and oxygen.

Bufotenin is present in skin glands of *Bufo bufo*

Caffeine and other purine alkaloids

STRUCTURE Main purine alkaloids are caffeine ($C_8H_{10}N_4O_2$; MW 194.20), theophylline ($C_7H_{10}N_4O_2$; MW 180.17), theobromine ($C_7H_{10}N_4O_2$; MW 180.17).

caffeine   theophylline   theobromine

OCCURRENCE *Coffea arabica* and other *Coffea* species (Rubiaceae); *Cola acuminata, Cola nitida, Theobroma cacao* (Malvaceae/Sterculiaceae), *Paullinia cupana* (Sapindaceae), *Ilex paraguariensis, I. cassine, I. guayusa* (Aquifoliaceae), and *Camellia sinensis* (Theaceae).

Other purines such as zeatin, which have been found in plants, mostly have plant growth regulating activities.

CLASSIFICATION Stimulant; slightly hazardous, III

SYMPTOMS Excessive amounts of coffee may lead to nervousness, headache, tremor, spasms, palpitations, high blood pressure, insomnia and indigestion. Caffeine is known to be addictive. The topical application of ointments with 30% caffeine can induce toxic effects.

TOXICITY Only high doses promote toxic effects. 150–200 mg/kg appears to be the LD_{50} for caffeine in humans; about 5 g caffeine has caused the death of a child.

MODE OF ACTION CNS stimulant inducing wakefulness and enhanced mental activity. Caffeine is an inhibitor of cAMP phosphodiesterase and an antagonist at adenosine receptors. As a consequence dopamine is released and many brain parts become activated. Caffeine is a cortical stimulant that increases the carbon dioxide sensitivity of the brain stem. It affects the cardiovascular system and has a positive inotropic action, resulting in tachycardia and an enhanced output. Theobromine and theophylline are cardiac stimulants, vasodilators and smooth muscle relaxants.

USES (PAST AND PRESENT) At present plants with caffeine are widely used by mankind, and our literature, music, arts and thus our culture (think of the Viennese coffee houses and their influence on artists and scientists) have certainly been positively influenced by caffeine and other purine alkaloids. The main active compound, however, is incorporated into numerous formulations used against fever, pain, and flu symptoms.

Tea was first described in 350 BC by Kuo P'o as a precious commodity of ancient China. Buddhist monks drank tea to sustain them through long hours of meditation. From China tea came to Japan where it became an integral part of the Japanese way of life. Tea was unknown to Europeans until 1550 and tea import started in the early part of the 17th century through the English East India Company. Commercial plantations were started in 1826 on Java by the Dutch and in 1836 in India by the British. Tea, which is rich in polyphenols, is often recommended in case of poisoning. This is a rational measure as the tannins can bind poisonous alkaloids and peptides.

Guarana has been long used by Amazonian Indians as a stimulating drug. It was esteemed as an elixir for permanent youth. This plant became better known to the world through Alexander von Humboldt who had enjoyed the drink or a paste made of maniok flour and guarana when travelling on the Rio Negro. The high caffeine content of guarana had been recognised early on and the plant is mainly used for "high energy" drinks and as a source of caffeine.

FIRST AID In case of overdose: Administration of sodium sulphate and medicinal charcoal. **Clinical therapy:** Gastric lavage (possibly with potassium permanganate), instillation of medicinal charcoal and sodium sulphate, diazepam in case of spasms.

Cytisine and related quinolizidine alkaloids

STRUCTURE More than 200 quinolizidine alkaloids (QA) are known. Most common are tetra-cyclic QA of the lupanine or anagyrine type, tricyclic QA of the cytisine type, bicyclic QA of the lupinine type, and more complex matrine alkaloids. Anagyrine ($C_{15}H_{20}N_2O$; MW 244.33), cytisine ($C_{11}H_{14}N_2O$; MW 190.24), lupanine ($C_{15}H_{24}N_2O$; MW 248.37); lupinine ($C_{10}H_{19}NO$; MW 169.27); N-methylcytisine ($C_{12}H_{16}N_2O$; MW 204.26); sparteine ($C_{15}H_{26}N_2$; MW 234.39). Ammodendrine ($C_{12}H_{20}N_2O$ MW 208.30) is included here, though it is a piperidine alkaloid, since it often co-occurs with QA.

sparteine

lupanine

angustifoline

anagyrine

cytisine

N-methylcytisine

lupinine

matrine

ammodendrine

OCCURRENCE Cytisine-type alkaloids are abundant in: *Ammodendron, Anagyris, Baptisia, Cytisus, Genista, Laburnum, Lupinus, Petteria, Sophora, Spartium, Thermopsis, Ulex* (Fabaceae), *Caulophyllum thalictroides, Leontice* (Berberidaceae). Lupanine-type alkaloids in: *Argyrolobium, Cadia, Calpurnia, Cytisus, Chamaecytisus, Lupinus, Lotononis, Ormosia, Virgilia* (Fabaceae). Lupinine-type in: *Lupinus, Lamprolobium* (Fabaceae), *Leontice* (Berberidaceae), and *Anabasis aphylla* (Amaranthaceae/Chenopodiaceae). Sparteine-type alkaloids in: *Cytisus, Hovea, Lupinus, Ormosia, Podopetalum, Piptanthus, Templetonia* (Fabaceae). Matrine-type in *Ammothamnus, Euchrestia, Goebelia, Sophora, Vexibia* (Fabaceae).

CLASSIFICATION Neurotoxin, mind-altering; highly hazardous, Ib

SYMPTOMS First effects of cytisine poisoning set in after 0.25–1 h and include burning of mouth and throat, intensive salivation, thirst, nausea followed by long hours of sometimes bloody vomiting, cold sweat, mydriasis, vertigo, excitation and confusion (with hallucinations, delirium), anxiety, tachycardia, trembling muscles, spasms and collapse. Death is caused by respiratory arrest. Symptoms are weaker in humans who are strong tobacco consumers.

Sparteine symptoms include headache, pressure in head, dizziness, sleepiness, eye flickering, double vision, difficulties of accommodation, palpitations, cardial pain, tingling in extremities, power loss in legs, moist and reddened skin. Strong intoxication leads to strychnine-type of paralysis, convulsions and death from suffocation while the heart is still beating. Symptoms start 20 min after intake of sparteine and reach a maximum after 4–5 h.

TOXICITY Cytisine: LD_{50} mouse: 101 mg/kg p.o., 1.7 mg/kg i.v., 9.4 mg/kg i.p.; LD_{50} cat: 3 mg/kg p.o.; dog 4 mg/kg p.o.; lupanine: LD_{50} rat: 177 mg/kg i.p. (death after 6–24 min), 1.6 g/kg (death after 2 min to 4 h); 13-hydroxylupanine: LD_{50} rat: 199 mg/kg i.p.; sparteine: LD_{50} mouse: 36 mg/kg i.p., 220 mg/kg p.o. Cytisine is transferred into milk of lactating cows and goats.

MODE OF ACTION Cytisine and lupanine are agonists at the nicotinic acetylcholine receptor, and similar to nicotine they react as a ganglion blocker in the vegetative nervous system. First effects lead to stimulation of sympathetic ganglia; later inhibiting effects dominate. QA affect the central nervous system, especially the medulla oblongata, centres for vomiting, vasomotor and respiration. Causes spasms and convulsions. Blood pressure is increased centrally and peripherally by vasoconstriction and by increase of cardiac activity. Respiration is first activated, later reduced; death is caused by respiratory arrest. Cytisine and *N*-methylcytisine have a higher affinity to nAChR and are stronger poisons than lupanine. Sparteine is an agonist at mAChR and blocks Na^+ channels; it causes bradycardia, a decrease of intracardial pressure, and arterial blood pressure. Anagyrine and ammodendrine show teratogenic properties in cattle and other livestock, leading to "crooked calf disease", accompanied by malformations such as arthrogrypose, torticollis, scoliose, and cleft palate.

USES (PAST AND PRESENT) Sparteine has been used medicinally to treat heart arrhythmia (Na^+ channel blocker) and during child birth (uterus contraction); since some patients are slow metabolisers of sparteine, intoxications have occurred. Some plants with cytisine and *N*-methylcytisine (such as *Cytisus canariensis, Spartium junceum,* and *Sophora secundiflora*) have been used as withdrawal drug for smokers and as analeptic. Sometimes they are also employed as mind-altering stimulant and hallucinogen.

FIRST AID Administration of sodium sulphate, medicinal charcoal, and plenty of warm tea or fruit juice. **Clinical therapy:** Gastric lavage (possibly with 0.1 % potassium permanganate), instillation of medicinal charcoal and sodium sulphate, electrolyte substitution, control of acidosis with sodium bicarbonate. In case of excitation and spasms inject diazepam and atropine; in case of severe intoxication, provide intubation and oxygen respiration.

Chamaecytisus supinus

Thermopsis caroliniana

Spartium junceum

Virgilia divaricata

Colchicine and related alkaloids

STRUCTURE Colchicine ($C_{22}H_{25}NO_6$; MW 399.45); colchicine is converted to lumicolchicine upon illumination.

colchicine

demecolcine

OCCURRENCE A typical alkaloid of plants in the genera *Colchicum, Gloriosa* and other Colchicaceae (or Liliaceae s. l.) (*Androcymbium, Anguillaria, Baeometra, Bulbocodium, Camptorrhiza, Dipidax, Iphigenia, Kreysigia, Littonia, Neodregea, Ornithoglossum, Sandersonia, Wurmbea*).

CLASSIFICATION Cell toxin; extremely hazardous, Ia

SYMPTOMS Colchicine has a bitter taste. After a latency period of 2–6 h symptoms can be seen. First effects are burning and prickling in mouth and throat, thirst, difficulties in swallowing, nausea, frequent and violent vomiting, and diuresis. Cell division becomes blocked after 10 h. After 12 to 24 h, colchicine poisoning mainly affects the gastrointestinal tract. Typical for colchicine poisoning are violent cholera-like bloody diarrhoea and abdominal pain. Central neuronal disturbance leads to shortness of breath, cyanosis, hallucinations, spasms, hypothermia, rapid weak pulse, tachycardia, hypotension, arrhythmia, renal failure (with haematuria, glucosuria), and epileptiform convulsions. Colchicine intoxication leads to paralysis and causes respiratory or cardiac arrest after 2 days; sometimes death is caused earlier due to circulatory collapse. If the acute poisoning has been survived, patients will suffer from hair loss, sometimes from permanent alopecia, and aplastic anaemia.

TOXICITY Poisoning can occur when plant parts are accidentally ingested (confusion with *Allium ursinum*, p. 29). Sometimes colchicine is added to intoxicating drugs as adulterant. Most cases of poisoning are caused through overdosing during gout treatment. Colchicine has been used for murder.
   LD_{50} rat: 1.7 mg/kg i.v., 4 mg/kg s.c.; mouse: 3.5 mg/kg; LD_{50} cat: 0.25 mg/kg i.v.; lethal dose for livestock: 1 mg/kg; for humans: 5 mg (children), 10–40 mg (adults); for a patient with kidney disorder 3 mg were lethal. Livestock (horses and pigs are more sensitive than sheep and cattle) are susceptible to colchicine poisoning; the alkaloids are transferred into the milk, which poses risks for milk consumers. Colchicine is mutagenic and can induce carcinomas. Androcymbine from *Androcymbium melanthoides* (Colchicaceae) has similar toxicity as colchicine.

MODE OF ACTION Colchicine has a high affinity for tubulin and blocks the polymerisation and depolymerisation of microtubules. Microtubules are necessary for cell division (they form the mitotic spindle apparatus) and intracellular transport of vesicles (important for cell mobility and the function of nerve cells). Colchicine inhibits the synthesis of collagen and activates collagenase. It stimulates the synthesis of prostaglandins, thereby enhancing disturbances of the GI tract. Colchicine can be classified as capillary poison.

USES (PAST AND PRESENT) Colchicine has been used against fast dividing cancer cells, but its toxicity prevents a general application, except for a few epidermal neoplasms and psoriasis. In spite of its toxicity colchicine and extracts from *Colchicum* are used in medicine today to treat acute gout and hereditary Mediterranean fever. Normally, macrophages would move to the site of gout inflammation, which is caused by a deposition of uric acid crystals. Since macrophages would release lactic acid there, the lowered hydrogen ion concentration would induce the precipitation of even more uric acid. This positive feedback mechanism is a vicious circle, since an increased inflammation will attract more macrophages, and more macrophages will increase uric acid crystal formation, etc. Since colchicine blocks the movement of macrophages, it can disrupt this mechanism.

Colchicine and colchicine-containing plants have been employed for suicide and murder since antiquity and still are. Colchicine is being used in agriculture to induce polyploidy in plants (which are usually bigger than the original diploid organism). Derivatives of colchicine, such as colcemide are used in cell biology to arrest cells in metaphase in order to visualise chromosomes.

Colchicum and *Gloriosa* are cultivated as garden and pot plants and can therefore be dangerous to children, pets and livestock.

FIRST AID Induce vomiting; administer sodium sulphate and 50–100 g medicinal charcoal. Keep patient in shock position and warm; provide warm tea or coffee. **Clinical therapy:** Gastric lavage (possibly with 0.1 % potassium permanganate), instillation of 30–50 g medicinal charcoal and sodium sulphate, plasma expander, electrolyte substitution, control of acidosis with sodium bicarbonate. In case of spasms inject diazepam; against vomiting give an anti-emetic such as triflupromazin; atropine against spasms (1 mg every 2 h, s.c.); in cases of severe intoxication, provide intubation and oxygen respiration. Do not give opium (which causes constipation) as colchicine is eliminated via the faeces; do not apply forced diuresis or haemoperfusion.

Colchicum autumnale (fruits)

Colchicum baytropium

Gloriosa superba

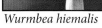
Wurmbea hiemalis

261

Ephedrine, cathinone and mescaline

STRUCTURE A number of phenylpropylamines occur in nature with pronounced pharmacological activity, including cathinone ($C_9H_{11}NO$; MW 149.19), cathine ($C_9H_{13}NO$; MW 151.21), ephedrine ($C_{10}H_{15}NO$; MW 165.24); mescaline ($C_{11}H_{17}NO_3$; MW 211.26).

L-ephedrine   D-pseudoephedrine

cathinone   norephedrine   mescaline

OCCURRENCE Ephedrine and related alkaloids occur in *Catha edulis*, *Maytenus krukovii* (Celastraceae) and commonly in *Ephedra* (Ephedraceae). Mescaline occurs in *Lophophora williamsii* and other cacti (*Pelecyphora*, *Trichocereus*).

CLASSIFICATION Mind-altering; moderately hazardous, II

SYMPTOMS First effects of the ephedrine-type alkaloids are dilated pupils, increase of blood pressure through vasoconstriction, enhanced and increased contraction of heart muscles, enhanced respiration, arrhythmia and hyperthermia. Higher doses can stimulate aggression and hyperactivity, sometimes with maniac, paranoid or psychotic behaviour. Further symptoms of overdosing are paleness, perspiration, anxiety, mydriasis, cold extremities, cyanosis, and tachycardia with weak pulse. Lethal poisoning is rare. Khat causes pain in testes and even impotence. Mescaline is a psychomimetic; it is a CNS depressant and hallucinogenic in high doses.

TOXICITY LD$_{50}$ mouse: racemic cathinone: 263 mg/kg (i.p.); ephedrine: 150 mg/kg i.p.; mescaline: LD$_{50}$ mouse: 212 mg/kg (i.p.); rat: 132 mg/kg (i.p.), 157 mg/kg (i.v.), 330–410 mg/kg (i.m.).

MODE OF ACTION Ephedrine is related to amphetamine and acts similarly to sympathomimetics. These alkaloids increase the release of noradrenaline and dopamine from catecholic synapses and inhibit their re-uptake. Thus they stimulate α- and β-adrenergic receptors in the peripheral and dopaminergic receptors in the central nervous system. The central stimulation is similar but weaker than that of amphetamine and includes an enhancement of the ability to concentrate and a decrease in the sensations of fatigue, hunger and pain. Ephedrine causes vasoconstriction, hypertension, bronchial dilatation, and heart stimulation. Although mescaline resembles dopamine and noradrenaline structurally, it apparently activates serotonin receptors in the CNS, similar to LSD.

USES (PAST AND PRESENT) Plants with ephedrine have been used as appetite suppressant. Khat is regularly consumed because it promotes mood brightening, mild excitation, euphoria and general arousal and helps against sleepiness. Ephedrine has been used medicinally to treat asthma, sinusitis and rhinitis, and as a mind-altering drug. *Ephedra* has been consumed as an aphrodisiac, which is more potent in women; in males it is stimulating but vasoconstriction can cause temporary impotence. Mescaline is famous for its mind-altering properties (see *Lophophora williamsii*).

FIRST AID Only after the ingestion of larger quantities; administer sodium sulphate and medicinal charcoal. **Clinical therapy:** Gastric lavage, instillation of medicinal charcoal and sodium sulphate, electrolyte substitution, control of acidosis with sodium bicarbonate. In case of spasms inject diazepam; in case of severe intoxication, provide intubation and oxygen respiration.

Ergot alkaloids

STRUCTURE Within the group of ergot alkaloids, two series can be distinguished: the clavine alkaloids and lysergic acid amides. Examples of clavine alkaloids are agroclavine ($C_{16}H_{18}N_2$; MW 238.33) and elymoclavine. Important lysergic acid amides include simple amides, such as ergine ($C_{16}H_{17}N_3O$; MW 267.33) and ergometrine ($C_{19}H_{23}N_3O_2$; MW 325.41), and more complex peptide alkaloids, such as ergotamine ($C_{33}H_{35}N_5O_5$; MW 581.68) and ergocristine ($C_{35}H_{39}N_5O_5$; MW 609.73). LSD (*N,N*-diallyllysergic acid amide; $C_{20}H_{25}N_3O$; MW 323.42) is a semi-synthetic compound derived from ergot alkaloids with strong hallucinogenic properties.

OCCURRENCE *Claviceps purpurea, C. microcephala, C. paspali* and more than 40 further members of this genus live as symbionts on grasses (tribes Festucaceae, Hordeae, Avenae, Agrosteae). Rye is especially affected among cereals. *Claviceps* is not a parasite but obviously a symbiotic organism. It takes nutrients from its host but provides chemical defence against herbivores as compensation. Field experiments have shown that such a fungal infection is an ecological advantage for grasses in the wild. A related fungus *Epichloe* also produces ergot alkaloids. Ergot alkaloids (such as agroclavine, chanoclavine, ergine, ergosine and ergometrine) are also common SMs of some genera of the Convolvulaceae (including *Argyreia, Ipomoea, Stictocardia tiliafolia, Turbina corymbosa*). A fungal symbiont is apparently the source of the alkaloids.

CLASSIFICATION Cytotoxin, neurotoxin, mind-altering; highly hazardous, Ib

SYMPTOMS Central effects of ergot alkaloid poisoning (caused by dopaminergic and serotoninergic activities) include vomiting, hyperthermia, mydriasis, and hallucinations. Further symptoms are diarrhoea, severe abdominal pain, headache, dizziness, anxiety, delirium, hypotension and cardiac arrest. Chronic use reduces blood circulation to the extremities, causing paraesthesia (skin prickling), numbness, strong pain, permanent muscle contraction, necrotisation and even gangrene.

clavines

lysergic acid amides

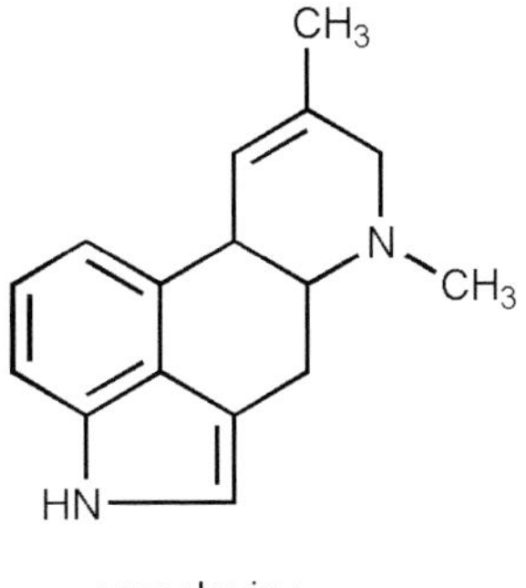

agroclavine

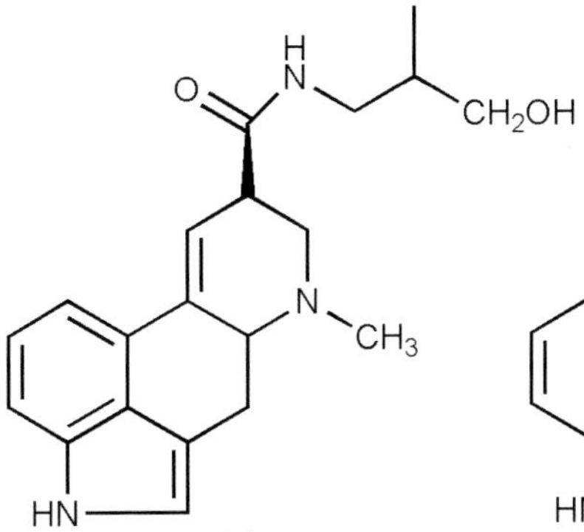

ergometrine

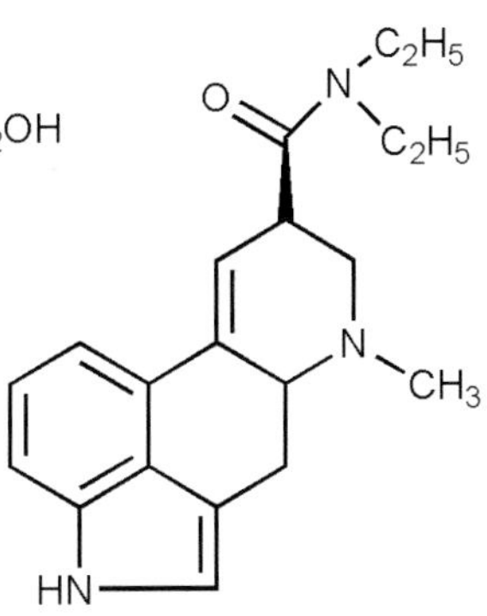

LSD

ergopeptides

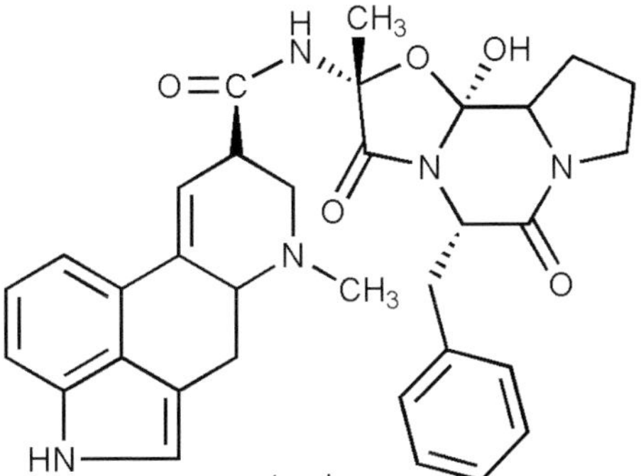

ergotamine

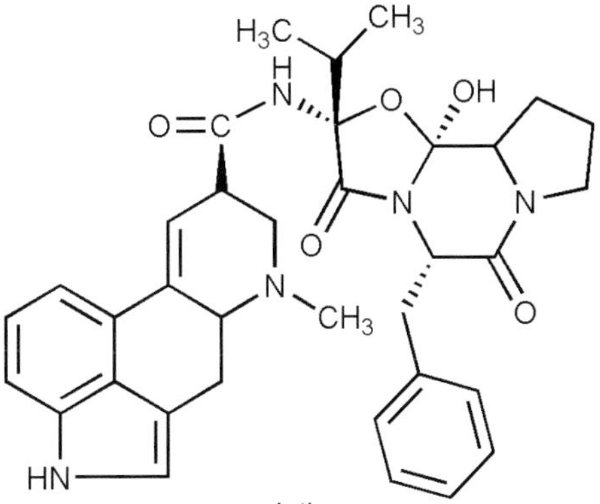

ergocristine

TOXICITY Lethal oral doses for humans are 5–10 g ergot and 10 mg ergotamine. About 0.1 % ergot in flour was considered as harmless, whereas 1 % was toxic and 8–10 % extremely hazardous. LD_{50} mouse (i.v.): ergotamine 6.0 mg/kg; ergometrine 0.15 mg/kg; rabbit: LD_{50} i.v.: ergocristin 1.9 mg/kg; ergocryptin 0.43 mg/kg, ergometrine 3.2 mg/kg. ergotamine 6–7.5 mg/kg; LSD LD_{50} mouse 46 mg/kg i.v.; humans 0.2 mg/kg.

MODE OF ACTION Ergot alkaloids interact as agonists, partial agonists but also antagonists with several of the important neuroreceptors, such as those for noradrenaline, serotonin and dopamine. Dramatic effects have been described, such as contraction of smooth muscles of peripheral blood vessels by blocking alpha-adrenergic receptors (causing gangrene), permanent contraction of uterine muscles, causing abortion. The alkaloids inhibit serotonin receptors but stimulate dopamine receptors. Ergometrine (an α-receptor agonist) and derivatives are used in obstetrics to stop bleeding after birth or abortion and ergotamine (antagonist at noradrenaline and 5-HT receptor; agonist at dopamine receptor) to treat migraine. Ergocornine reduces the secretion of prolactin and inhibits nidation as well as lactation. Elymoclavine is a prolactin release inhibitor. LSD, which is a synthetic derivate of ergot alkaloids, is one of the strongest hallucinogens.

USES (PAST AND PRESENT) Poisoning with ergot as a contaminant of flour is an old phenomenon. The toxic effects of ergot ("ergotism"), which has killed tens of thousands, have long been known: It has been speculated that the 1500 year old Eleusinian Mysteries of ancient Greece, were connected with ergot-induced hallucinations. The Parsees (about 350 BC) wrote of "noxious grasses that cause pregnant women to drop the womb and die in childbed". The Hebrews knew its hallucinogenic and toxic properties. Alarming mass intoxications (in AD 994 about 40 000 people were reported to have died because of ergotism or the "holy fire" as it was also called) became evident during medieval times when rye was grown in larger quantities and many people ingested contaminated flour. Mass poisoning was observed even in the 20th century, e.g., 11 319 cases of ergotism were observed during 1926/27 in a population of 506 000 in the vicinity of Sarapoul (near the Ural region). Whereas the incidence of ergotism has generally decreased in the 20th century, an increasing number of cases have been reported during the last decades when people have prepared their own "Müsli" from cereals of doubtful "organic" origin. The dramatic and cruel effects of ergotism can be seen in many paintings of the old masters. Since ergot was also used as an abortifacient, a number of poisoning cases can be attributed to deliberate overdosing. The hallucinogenic Mexican drug "ololiuqui" is composed of ergot alkaloids from *Turbina corymbosa, Ipomoea argyrophylla, I. tricolor* and other *Ipomoea* species.

FIRST AID Administration of medicinal charcoal and sodium sulphate. **Clinical therapy:** Gastric lavage, instillation of medicinal charcoal and sodium sulphate; forced diuresis, electrolyte substitution, control of acidosis with sodium bicarbonate (pH of urine 7.5). In case of spasms inject diazepam or hexobarbital (i.v.). Provide intubation and oxygen respiration in case of respiratory arrest or paralysis. In serious cases, provide haemodialysis, or peritoneal dialysis with forced diuresis. To overcome vasoconstriction, inject papaverine, other spasmolytics (Ronicol, Atriphos, Complamin) or beta blocker (Lopressor). Provide warmth, vessel massage, and a broad-spectrum antibiotic (in case of necrotic inflammation).

Claviceps purpurea on rye

Hyoscyamine, cocaine and other tropane alkaloids

STRUCTURE Hyoscyamine ($C_{17}H_{23}NO_3$; MW 289.38); related compounds with similar properties are scopolamine (hyoscine) ($C_{17}H_{21}NO_4$; MW 303.36) and littorine ($C_{17}H_{23}NO_3$; MW 289.38). Hyoscyamine and scopolamine can undergo racemisation; the racemic mixture is called atropine or atroscine, respectively. Further tropanes include hygrine ($C_8H_{15}NO$; MW 141.22), meteloidine ($C_{13}H_{21}NO_4$; MW 255.32), cocaine ($C_{17}H_{21}NO_4$; MW 303.36), cinnamoylcocaine ($C_{19}H_{23}NO_4$; MW 329.40).

hyoscyamine   scopolamine   meteloidine

cocaine   cinnamoylcocaine   hygrine

OCCURRENCE Common alkaloids in *Anthocercis, Atropa, Brugmansia, Datura, Duboisia, Hyoscyamus, Latua, Mandragora, Physalis, Physoclaina, Salpichroa, Scopolia, Schizanthus* (Solanaceae). Cocaine and related alkaloids occur in *Erythroxylum* (Erythroxylaceae). Alkaloids with tropane structures have also been detected in unrelated families, such as Convolvulaceae (*Convolvulus*), Brassicaceae, Dioscoreaceae, Elaeocarpaceae, Euphorbiaceae, Orchidaceae, Proteaceae (*Bellendena, Darlingia, Knightia*), Rhizophoraceae (*Bruguiera sexangula, Crossostylis ebertii*). Scopolamine can be transferred into the milk of lactating women.

CLASSIFICATION Neurotoxin, mind-altering; extremely hazardous, Ia

SYMPTOMS First poison effects set in 5 to 10 min after oral intake of material with tropane alkaloids. Adsorption is quick and almost complete. In case of weak poisoning, symptoms are dryness of mouth and throat accompanied by thirst, difficulties in swallowing, hoarseness, and dilated pupils that can no longer accommodate and are sensitive to light. Higher doses cause severe vegetative symptoms: psychomotoric unrest, excitation, redness of face (which later turns blue), hot flushes, hot dry skin, high fever, impaired vision with transient blindness, vomiting and headache, followed by an increase of systolic blood pressure and tachycardia with irregular heartbeat. After 15 min muscles start to fail, patients are out of balance and can no longer stand upright. Generalised spasms and epileptiform seizures have also been recorded. Furthermore, patients who have ingested more than 3 mg atropine show unrest, excitation, euphoria, an urge to move, chattiness, visions and are even psychotic having visual, long-lasting (several days) hallucinations, tantrum, and delirium that can lead to coma. Severe intoxication can lead to central respiratory arrest and death. Scopolamine has sedating properties at lower doses but causes excitation at higher levels.

When atropine is taken regularly at a dose of more than 3.6 mg/day, dependence can result. This leads to internal secretory disturbances, reduction of cognitive capacity and even imbecility.

Cocaine stimulates the pleasure centres of the brain, and leads to addiction and the long-term consequences of drug abuse. Symptoms of cocaine ingestion include euphoria, urge to speak, enhanced libido, loss of inhibitions, hallucinations, dilated pupils; higher doses cause vasoconstriction, hypertension, tachycardia and even death from respiratory arrest. Schizophrenic episodes also can occur.

TOXICITY Atropine: LD_{50} mouse: 400 mg/kg p.o.; 90 mg/kg i.v.; 2 g/kg s.c.; 250 mg/kg i.p.; lethal oral dose for adult humans: 10 mg; for children a few mg. Scopolamine: LD_{50} mouse: 163 mg/kg i.v.; 328 mg/kg i.p.; LD_{100} for humans: >100 mg.

Poisoning can occur from overdosing in medicinal applications, accidental ingestion of plant material (e.g. *Atropa* berries), or deliberately from exploiting possible hallucinogenic effects or for suicide and murder.

Since tropane alkaloids can be transferred via the placenta to the embryo and to the milk, they should not be used during pregnancy and the lactating period.

Most animals are also susceptible to tropane alkaloids, others such as thrushes, rabbits and guinea pigs are highly tolerant because they have esterases that detoxify atropine.

The administration of 1–2 g (p.o.) and 0.2–0.3 g s.c. cocaine can be lethal for humans. Sensitive persons have died after snorting 30 mg cocaine.

MODE OF ACTION Hyoscyamine and the other tropane alkaloids are antagonists at the muscarinic acetylcholine receptor (they block the binding of acetylcholine to mAChR) and have therefore parasympatholytic properties. These alkaloids block smooth muscles, which leads to spasmolysis and loss of motility in several organs (GI tract, bladder, bronchia), inhibition of glandular secretions (salivary, bronchial, sweat glands), tachycardia, at the eye mydriasis and accommodation disturbance. Hyoscyamine produces central excitation (with hallucinations), at higher doses a central paralysis is more dominant. Scopolamine shows stronger CNS sedating properties even at lower concentrations than hyoscyamine. Within 24 h about 85–88% of the absorbed alkaloids are eliminated via the urine (50% unchanged).

Cocaine is a local anaesthetic and euphoric. It blocks Na^+ channels in neurons, enhances release of dopamine and noradrenaline and inhibits their re-uptake. When the re-uptake is blocked by cocaine, dopamine concentration in the synaptic gap is increased leading to a prolonged activation of dopamine receptors. Since dopamine is involved with the control of pleasure response, cocaine stimulates the so-called pleasure centres.

USES (PAST AND PRESENT) Plants, extracts and pure tropane alkaloids have a long history of magic and murder. They have been taken since antiquity to generate hallucinations and intoxication. They were also ingested for suicide and murder. Atropine has been used medicinally for the treatment of spasms of smooth muscles in the gastrointestinal and urinary tract, gall ducts and bronchia. Atropine and scopolamine are locally employed at the eye as mydriatic and cycloplegic to facilitate inspections and diagnosis. Atropine has also been taken to treat bradycardia, arrhythmia and hyperhidrosis. Hyoscyamine and especially scopolamine are used as premedication for narcosis because of their sedating properties; atropine is given as an antidote in case of poisoning with parasympathomimetics. Scopolamine is applied as transdermal plaster to treat seasickness.

Medicinally, leaves of *Datura* were used in the 19th and 20th century as herbal cigarettes to treat patients with asthma and other respiratory conditions.

Coca leaves are an old traditional masticator of the Andes region (known as mate de coca), to relieve hunger and fatigue. Cocaine was the world's first commercial anaesthetic which no longer has much medicinal relevance but that has become a popular illicit drug used by millions of people.

FIRST AID Induce vomiting; administration of sodium sulphate and medicinal charcoal; for cocaine see *Erythroxylum coca*.

Clinical therapy: Parasympathomimetics (i.e. inhibitors of cholinesterase) such as physostigmine (adults 1–2 mg i.v. or i.m.; children 0.5 mg) can be injected as antidotes. Provide gastric lavage (possibly with water or 0.1% potassium permanganate) as soon as possible, instillation of medicinal charcoal and sodium sulphate and even forced diuresis. In case of excitation inject diazepam; patients need to be observed all the time if they show delirium and hallucination. Provide intubation and oxygen respiration in case of respiratory arrest and plasma expander when shock occurs. If the bladder is paralysed, a urine catheter is advisable. Reduce temperature by applying cold compresses but not antipyretics.

Indole alkaloids

STRUCTURE Indole alkaloids derive from tryptamine as a precursor. In case of monoterpene indole alkaloids tryptamine is combined with secologanin. Indole alkaloids comprise a large group of compounds with more than 1 200 known structures. Several of them are relevant in a medicinal or toxicological context.

Examples: Ajmaline ($C_{20}H_{26}N_2O_2$; MW 326.44); ajmalicine ($C_{21}H_{24}N_2O_3$; MW 352.44); reserpine ($C_{33}H_{40}N_2O_9$; MW 608.70), vinblastine ($C_{46}H_{58}N_4O_9$; MW 811.00), strychnine ($C_{21}H_{22}N_2O_3$; MW 334.42), gelsemine ($C_{20}H_{26}N_2O_4$; MW 358.44), camptothecin ($C_{20}H_{16}N_2O_4$; MW 348.36).

A closely related alkaloid group are the harman or β-carboline alkaloids. Examples are: harmaline ($C_{13}H_{14}N_2O$; MW 214.27), harmine ($C_{13}H_{12}N_2O$; MW 212.25), harman ($C_{12}H_{10}N_2$; MW 182.23).

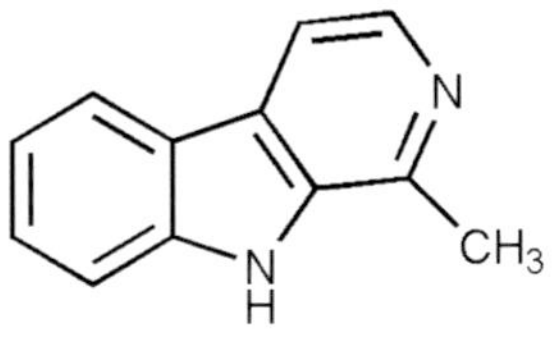

harman 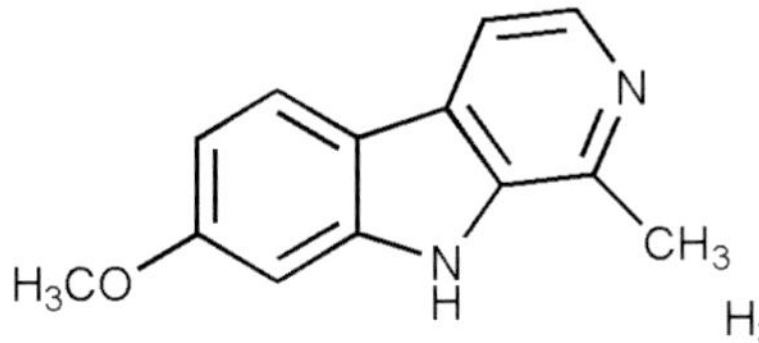   harmine 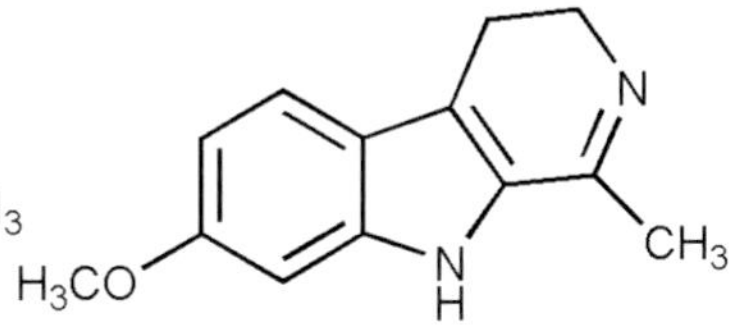   harmaline

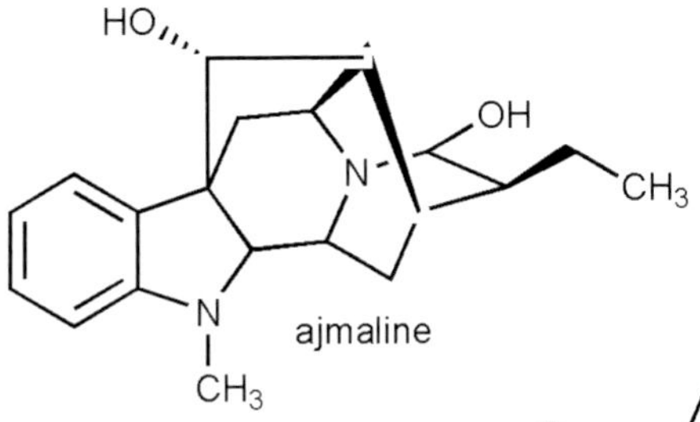

ajmaline 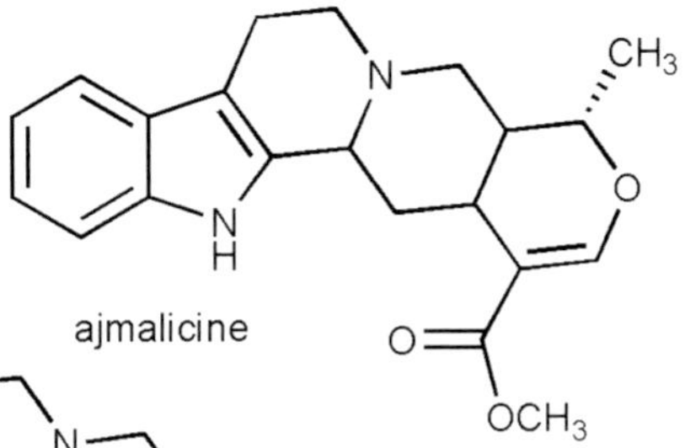   ajmalicine

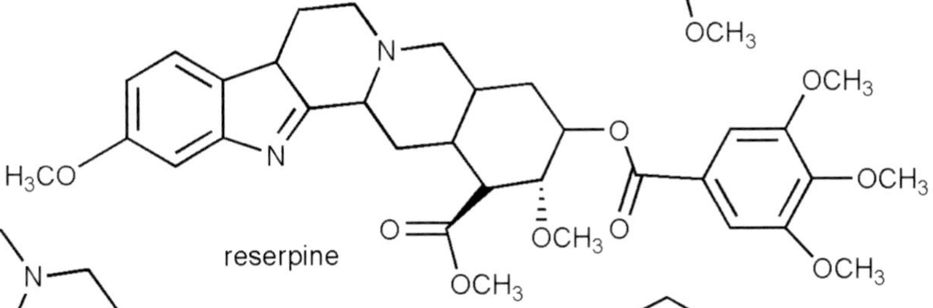

reserpine

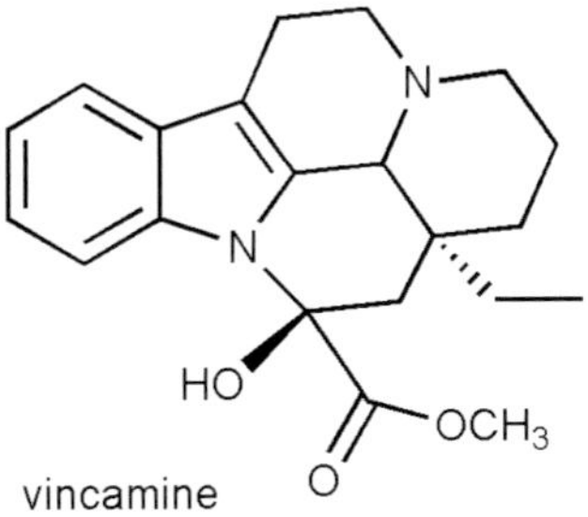

vincamine 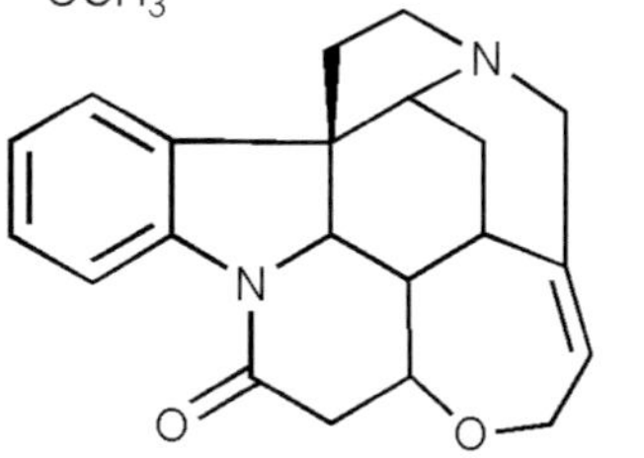   strychnine

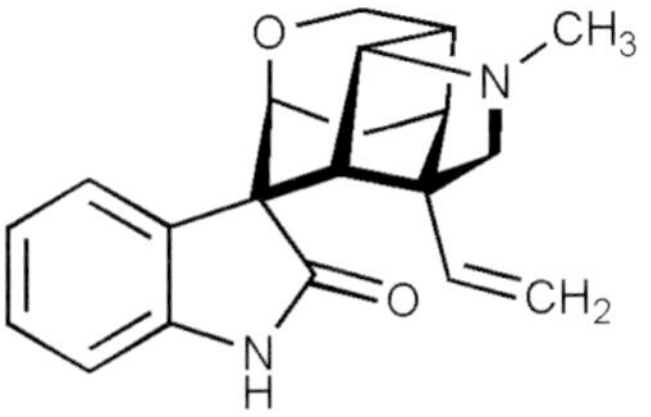

gelsemine 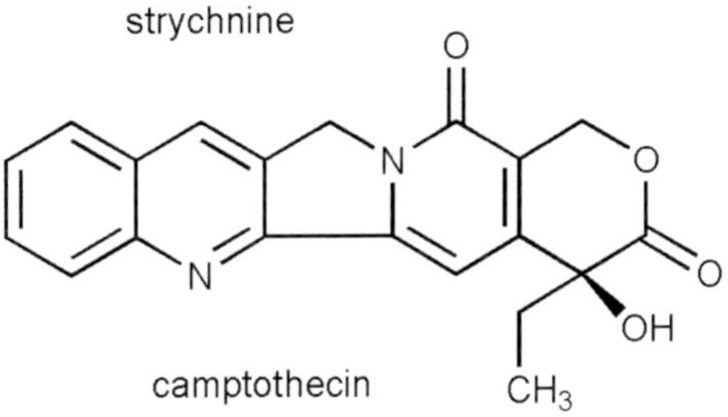   camptothecin

267

OCCURRENCE Indole alkaloids occur mainly in four plant families – the Apocynaceae (*Alstonia, Amsonia, Aspidosperma, Bleekeria, Bonafousia, Cabucala, Callichia, Catharanthus, Conopharyngia, Gabunia, Geissospermum, Hedranthera, Hunteria, Melodinus, Ochrosia, Peschiera, Picralima, Rauvolfia, Rhazya, Tabernaemontana, Tonduzia, Vallesia, Vinca, Voacanga*), Loganiaceae (*Strychnos*), Gelsemiaceae (*Gelsemium, Mostuea*), and Rubiaceae (*Adina, Antirhea, Borreria, Cephaelis, Cinchona, Corynanthe, Evodia, Hedyotis, Hortia, Mitragyna, Pausinystalia, Pogonopus, Psychotria, Uncaria*). Other families include Rutaceae (*Flindersia, Murraya*), Simaroubaceae (*Amaroria, Quassia*), and Verbenaceae (*Clerodendron*). *Catharanthus roseus* is the only source for dimeric Vinca alkaloids used in cancer therapy. Simple indole alkaloids of the physostigmine type are found in *Physostigma* (Fabaceae).

Harman or β-carboline alkaloids occur in Polygonaceae (*Calligonum*), Eleagnaceae (*Eleagnus*), Fabaceae (*Petalostylis*), Malpighiaceae (*Banisteriopsis*), Rubiaceae (*Singickia*), Symplocaceae (*Symplocos*), Zygophyllaceae (*Peganum, Zygophyllum*) and Rutaceae (*Clausena, Murraya*).

Camptothecin (formally a quinoline alkaloid, but derived from the tryptamine/secologanine pathway) occurs in *Camptotheca acuminata* (Cornaceae), *Nothapodytes foetida, Pyrenacantha klaineana, Merrilliodendron megacarpum* (Icacinaceae), *Ophiorrhiza pumila, O. mungos* (Rubiaceae), *Ervatamia heyneana* (Apocynaceae) and *Mostuea brunonis* (Gelsemiaceae).

CLASSIFICATION Cell and nerve poisons, mind-altering; highly to moderately hazardous, Ib–II

SYMPTOMS Symptoms of toxic indole alkaloids include supraventricular tachycardia, cardiac fibrillation, hypotension, cerebral spasms, coma, cardiac and respiratory arrest.

TOXICITY Several indole alkaloids are used medicinally. Overdosing can cause serious poisoning. Toxiferine I and II from *Strychnos* are neuromuscular blocking agents, thus highly toxic and used as an arrow poison. Physostigmine is highly toxic and calabar beans were used as an ordeal poison in West Africa. Gelsemine and gelsemicine are CNS active and highly toxic.

MODE OF ACTION Ajmaline blocks sodium channels and has therefore anti-arrhythmic properties because it lowers cardiac excitability. It has negative inotropic properties. Ajmalicine has a pronounced dilatatoric activity in blood vessels, which causes hypotension. Akuammine has local anaesthetic properties (similar to those of cocaine).

Toxiferine I and II and related alkaloids are strong inhibitors of nicotinic AChR at the neuromuscular plate and cause paralysis of muscle cells. Ibogaine from *Tabernanthe iboga* is a CNS stimulant with anticonvulsant and hallucinogenic properties.

Physostigmine, eseridine and related compounds are strong inhibitors of cholinesterase with wide-ranging parasympathetic activities.

β-Carboline alkaloids are inhibitors of MAO and agonists at serotonin receptors. Since they enhance serotonin activity, they exhibit substantial hallucinogenic activities.

Dimeric Vinca alkaloids (vincristine, vinblastine, leurosine) inhibit tubulin polymerisation and intercalate DNA. As a consequence they effectively block cell division. Reserpine and related alkaloids inhibit transporters for neurotransmitters at vesicle membranes and thus act as an antihypertensive and tranquilliser.

Strychnine is a CNS stimulant and extremely toxic. Strychnine inhibits a glycine receptor (an inhibiting neuroreceptor) and thus increases excitability of all parts of the CNS. Strychnine is especially active in the spinal cord which has a high density of glycine receptors. Even small sensory signals can cause symmetric muscular spasms.

Gelsemine activates acetylcholine receptors (especially mAChR and nAChR) and inhibits cholinesterase and the synaptic re-uptake of dopamine, noradrenaline and serotonin.

USES (PAST AND PRESENT) Ajmaline is used medicinally to treat tachycardial arrhythmia, extra systoles, fibrillation and angina pectoris. Ajmalicine is used as a tranquilliser, antihypertensive to improve cerebral blood circulation. Reserpine and related alkaloids are used as antihypertensive and tranquillising drugs. Physostigmine is used as a miotic in eye treatments. Mesembrine, a simple indole alkaloid from *Mesembryanthemum* (= *Sceletium*) (Aizoaceae), a narcotic with cocaine-like activities, has been used as an antidepressant. *Strychnos* seeds were mainly used to kill rodents and other pests, but also as arrow and ordeal poison. Nux vomica is sometimes used for criminal purposes and responsible for accidental deaths. Several S American *Strychnos* species (*S. castelnaei, S. toxifera*) are a source for curare; quaternary bisindole alkaloids, such as C-toxiferine I and others; they are used in surgery to paralyse muscular movements.

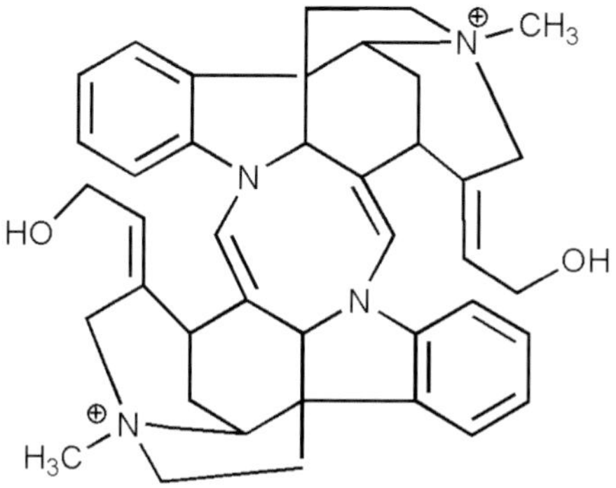

toxiferine I

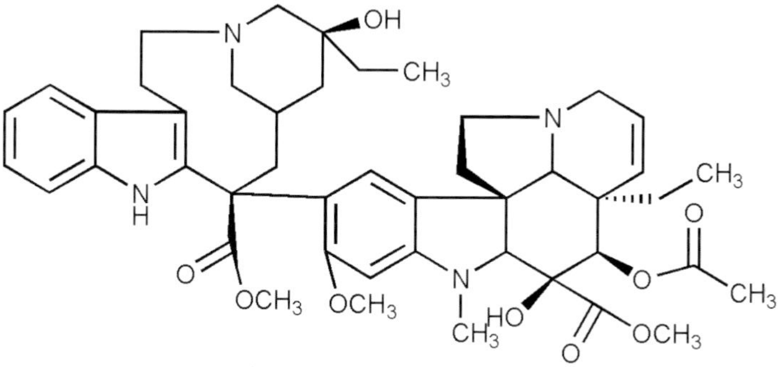

vinblastine

The harman alkaloids (harmin and others) are known hallucinogenic substances. Plants containing them are famous drugs in S America (including ayahuasca). In the Old World, these alkaloids occur in *Peganum harmala*; the motif of a flying carpet is probably associated with intoxication from Peganum alkaloids. The alkaloids are said have been used by the Nazis for brainwashing.

FIRST AID Induce vomiting; in case of children use "sirup of ipecac"; then administration of sodium sulphate and medicinal charcoal; (see also individual plant species).

Clinical therapy: Gastric lavage (possibly with 0.1% potassium permanganate), instillation of medicinal charcoal and sodium sulphate, possibly electrolyte substitution, control of acidosis with sodium bicarbonate. In case of spasms give diazepam; in severe intoxication, provide artificial respiration (intubation and oxygen supply) and forced diuresis. Infuse a glucose solution. Record electrocardiogram (ECG).

Rauvolfia vomitoria

Psychotria emetica

Isoquinoline alkaloids

STRUCTURE Isoquinoline alkaloids that derive from the amino acid tyrosine represent one of the largest groups of secondary metabolites. It can be grouped into benzylisoquinoline, bisbenzylisoquinoline, phthalide isoquinoline, aporphine, morphinane, protoberberine, protopine, benzophenanthridine and Erythrina alkaloids. Several bioactive isoquinoline alkaloids are known from plants. Important examples are: papaverine ($C_{20}H_{21}NO_4$; MW 339.40), tubocurarine ($C_{37}H_{42}N_2O_6$; MW 610.76), berbamine ($C_{37}H_{40}N_2O_6$; MW 608.74), noscapine ($C_{22}H_{23}NO_7$; MW 413.43), biculluline ($C_{20}H_{17}NO_6$; MW 367.36), bulbocapnine ($C_{19}H_{19}NO_4$; MW 325.37), magnoflorine ($C_{20}H_{24}NO_4$; MW 342.42), berberine ($C_{20}H_{18}NO_4$; MW 336.37), protopine ($C_{20}H_{19}NO_5$; MW 353.38), emetine ($C_{29}H_{40}N_2O_4$; MW 480.65), chelerythrine ($C_{21}H_{18}NO_4$; MW 348.37), chelidonine ($C_{20}H_{19}NO_5$; MW 353.36), sanguinarine ($C_{20}H_{14}NO_4$; MW 332.34), erythroidine ($C_{16}H_{19}NO_3$; MW 273.34).

benzylisoquinoline

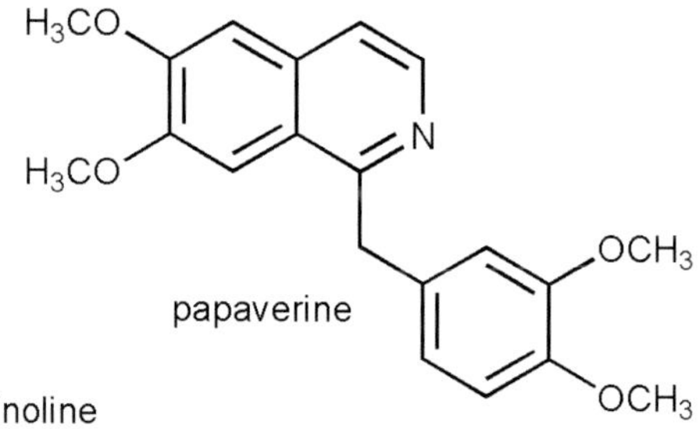

papaverine

bisbenzyl isoquinoline

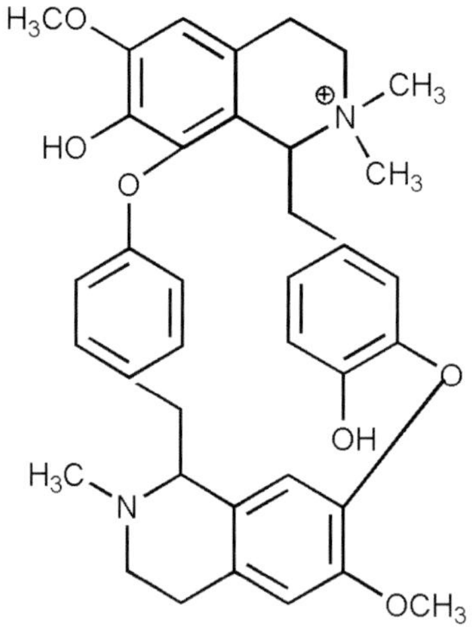

tubocurarine

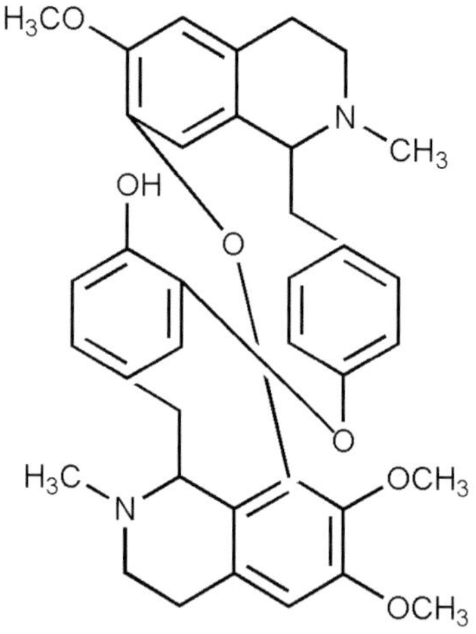

berbamine

OCCURRENCE Isoquinoline alkaloids are common in genera of the Papaveraceae (*Adlumia, Argemone, Bocconia, Chelidonium, Corydalis, Dicentra, Dicranostigma, Eschscholzia, Fumaria, Glaucium, Macleaya, Meconopsis, Papaver, Romneya, Sanguinaria, Stylophorum*), Ranunculaceae (*Aconitum, Coptis, Hydrastis, Thalictrum, Zanthorrhiza*), Berberidaceae (*Berberis, Leontice, Mahonia, Nandina*), Annonaceae (*Artabotrys, Asimina, Coelocline, Desmos, Guatteria, Isolona, Phaeanthus, Xylopia*), Monimiaceae (*Atherosperma, Boldea, Daphnandra, Doryphora, Monimia, Nemuaron, Peumus*), Menispermaceae (*Abuta, Anisocycla, Arcangelisia, Chondrodendron, Cocculus, Cissampelos, Duguettia, Jateorrhiza, Pycnarrhena, Sciadotenia, Sinomenium, Stephania, Tiliacora*), Lauraceae (*Albertisia, Actinodaphne, Beilschmidia, Cassytha, Cryptocarpa, Laurelia, Laurus, Litsea, Machylus, Nectandra, Ocotea, Phoebe, Sassafras*), Ancistrocladaceae (*Ancistrocladus*), Aristolochiaceae (*Aristolochia*), Euphorbiaceae (*Croton*), Rutaceae (*Evodia, Fagara, Toddalia, Zanthoxylum*), Nymphaeaceae (*Nelumbo*), Celastraceae (*Euonymus*), Magnoliaceae (*Liriodendron, Michelia*), Rhamnaceae (*Retanilla, Rhamnus*), Alangiaceae (*Alangium*), Hernandiaceae (*Gyrocarpus, Hernandia, Illigera*), Sapindaceae (*Pteridophyllum*), Buxaceae (*Buxus*), Rubiaceae (*Cephaelis, Psy-*

chotria), Fabaceae (*Erythrina*), Apiaceae (*Heracleum*), Cactaceae (*Carnegiea, Cephalocereus, Lophocereus, Lophophora*), and Amaranthaceae/Chenopodiaceae (*Salsola*). When present in latex, the alkaloids are complexed by chelidonic or meconic acid. Sanguinarine has been discovered in milk of lactating cows that were feeding on plants with this alkaloid.

phthalid isoquinoline

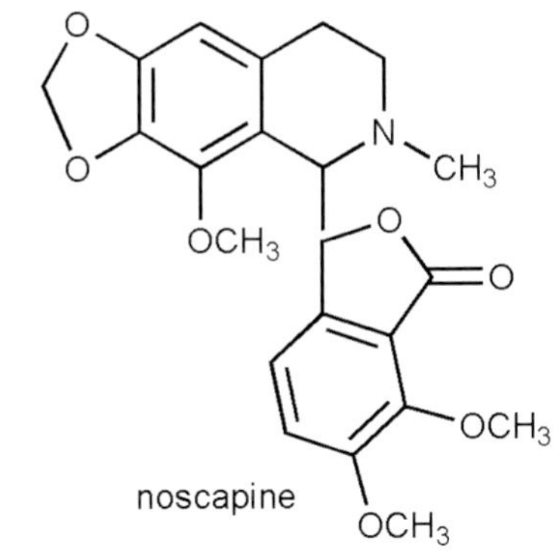

noscapine

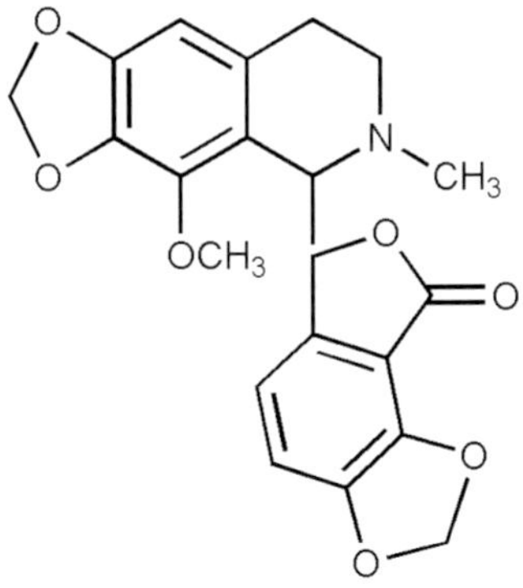

bicuculline

aporphine

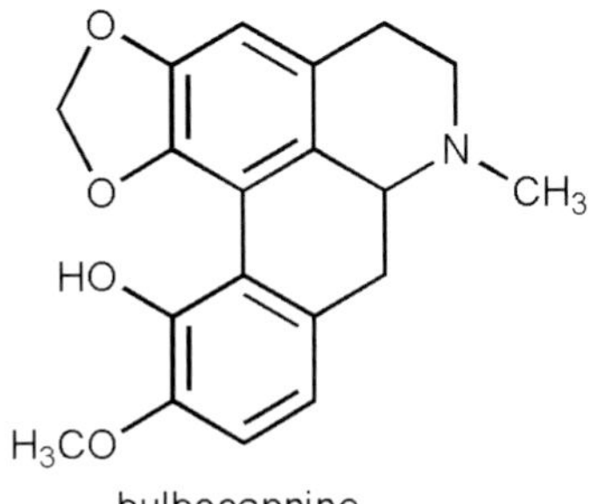

bulbocapnine

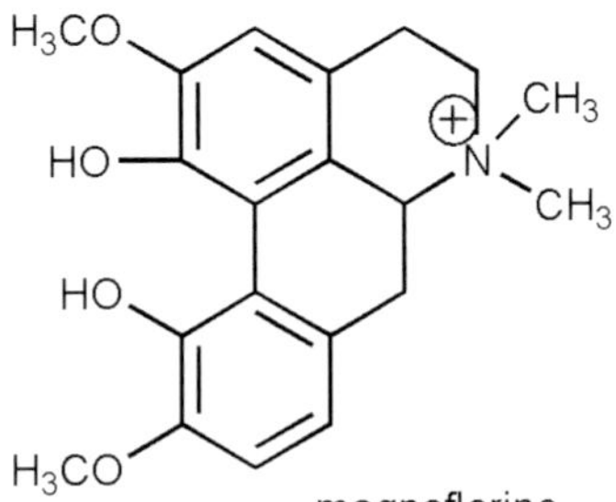

magnoflorine

CLASSIFICATION Cell toxins, neurotoxins; highly to moderately hazardous, I-II

SYMPTOMS First symptoms of *Chelidonium* poisoning (as a representative for many plants with isoquinoline alkaloids) include vomiting and gastroenteritis with bloody diarrhoea. Higher doses have central paralytic activities, which concern the vasomotor and respiratory centre. Other effects include atony of smooth muscles of lung, gut and coronary vessels, hypertension, hypoglycaemia, inflammation and burning in mouth and throat. Paralysis of extremities, diuresis, dizziness, stupor, arrhythmia, and bradycardia also occur. Death is caused by respiratory arrest in collapse. On the skin, the *Chelidonium* alkaloids can cause inflammation, blisters and ulcers. The alkaloids also exhibit antibacterial, antifungal, and antiviral properties. Phthalide isoquinoline alkaloids, such as adlumine are convulsant, cardiac depressant and intestinal stimulant. Berberine and other protoberberines cause cardiac damage, dyspnoea, and hypotension. Bulbocapnine from *Corydalis* causes catalepsy, sedation and potentiates hypnotics.

TOXICITY Many isoquinoline alkaloids, especially bisbenzylisoquinolines, protoberberine and benzophenanthridines exhibit cytotoxic properties. Bisbenzylisoquinolines are often toxic (dauricine: LD_{50} i.p. mice 6 mg/kg), cytotoxic, CNS depressant, anaesthetic, and respiratory paralytic. Bisbenzylisoquinoline alkaloids from *Chondrodendron* and other Menispermaceae, such as tubocurarine, are muscle relaxant (inhibition of nAChR) and extremely poisonous when injected. Berbamine is highly toxic with curare-like effects.

LD_{50} values: Chelerythrine: mouse; 95 mg/kg s.c.; chelidonine: mouse: 34.6 mg/kg i.v.; guinea pig: 2 g/kg s.c.; sanguinarine: mouse: 19.4 mg/kg i.v., 102 mg/kg s.c.; erythroidine: mouse i.p. 29.5 mg/kg; glaucine, an aporphine alkaloid: mouse i.v. 4.8 mg/kg; isocorydine: rat i.p. 10.9 mg/kg; mecambrine: mouse 4.1 mg/kg; berberine: humans: 27.5 mg/kg; isothebaine: humans 26 mg/kg. Emetine: lethal oral dose for humans: 1 g.

271

protoberberine

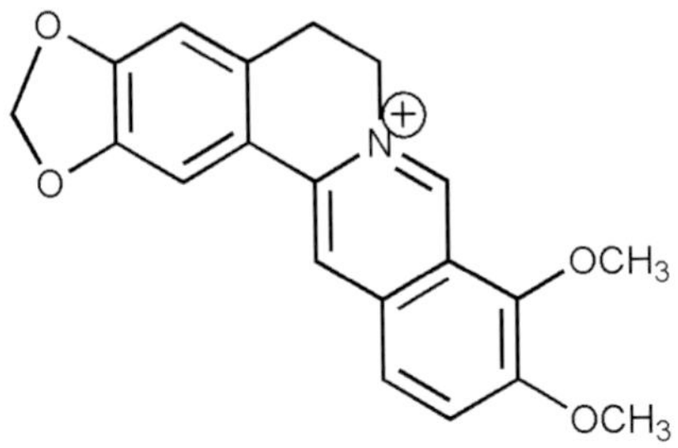

berberine

protopine

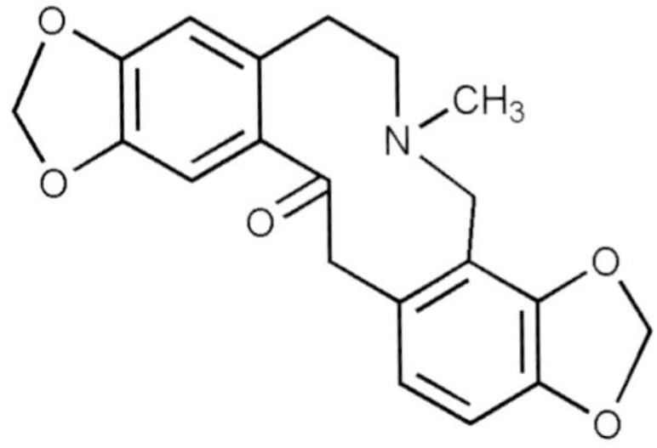

protopine

emetine

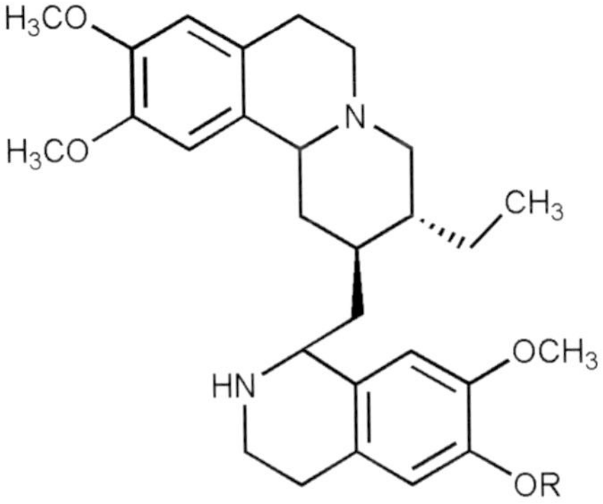

emetine (R = CH₃)
cephaeline (R = H)

Erythrina alkaloids

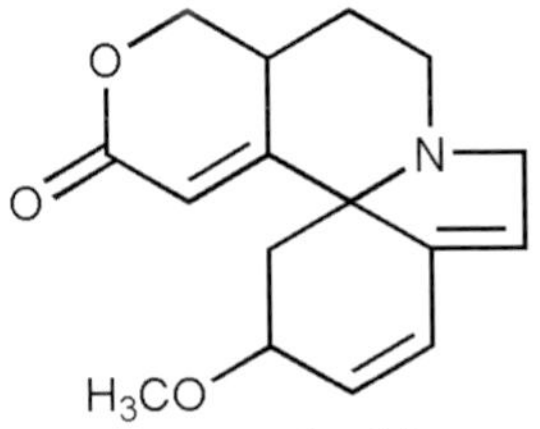

erythroidine

benzophenanthridine

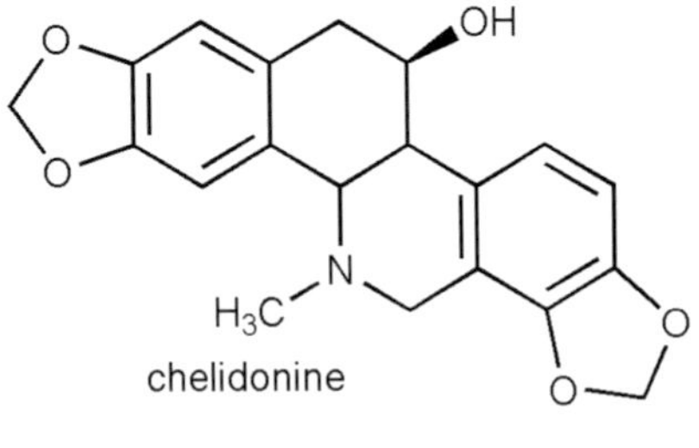

chelidonine

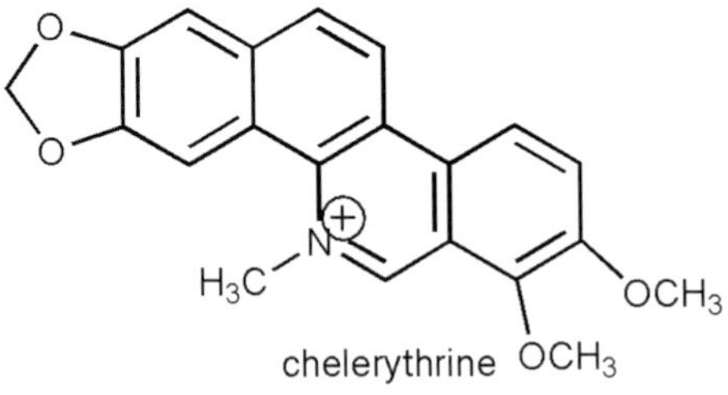

chelerythrine

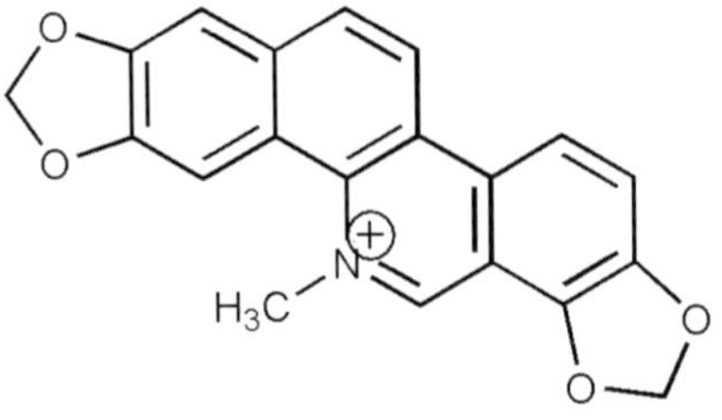

sanguinarine

272

MODE OF ACTION Chelidonine and noscapine inhibit microtubule polymerisation. Several of the protoberberine and benzophenanthridine alkaloids (such as sanguinarine) intercalate DNA and inhibit DNA topoisomerase I and several other enzymes. Cytotoxic, apoptotic and antimicrobial activities have been described. Metabolising enzymes, such as cytochrome oxidase p450 are inhibited by alkaloids with a methylene dioxy bridge that is common in this group. Argemonine shows analgesic and anti-arrhythmic activities. Aromoline (a bisbenzylisoquinoline) has membrane stabilising properties. Bicuculline acts as a GABA antagonist.

Emetine is a potent inhibitor of protein biosynthesis, intercalator and inducer of apoptosis. Erythroidine and other Erythrina alkaloids, several aporphine (such as magnoflorine) and bisbenzylisoquinoline alkaloids (such as rodiasine, macoline, tubocurarine) block signal transduction at the neuromuscular plate. Laudanidine, a simple benzylisoquinoline, shows strychnine-like activities with convulsions and paralysis. Oxyacanthine, a bisbenzylisoquinoline alkaloid, acts as an adrenaline antagonist and causes vasodilation. Papaverine inhibits phosphodiesterase and thus acts as smooth muscle relaxant, vasodilator, and spasmolytic. Protopine is another smooth muscle relaxant and sedative.

USES (PAST AND PRESENT) *Chelidonium majus* has been used in traditional medicine and phytomedicine as cholagogue, spasmolytic, diuretic and analgesic drug or to treat warts. Chelidonine has been employed as a painkiller to treat abdominal pain, and to treat spasms and asthma. Extracts of *Sanguinaria canadensis* have been included in mouthwashes and toothpaste (see *Sanguinaria canadensis*). Extracts of *Eschscholzia californica*, which are rich in aporphine, protoberberine and benzophenanthridine alkaloids, have been employed as a mild psychoactive drug to induce euphoria. *Alangium lamarckii* with alamarine and alangicine (that are structurally related to emetine) has been used in India against leprosy, other skin diseases, tonic, laxative and anthelmintic. The aporphine boldine is used to treat hepatic dysfunction and cholelithiasis. Emetine and cephaeline have been used as emetic, expectorant and anti-amoebic. Cepheranthine, a bisbenzylisoquinoline from *Stephania* has been used to treat tuberculosis and leprosy. Erythrina alkaloids have been used as curare substitute. Tubocurarine and other bisbenzylisoquinolines from *Chondrodendron* and *Ocotea* have been used traditionally as arrow poison but also in surgery to paralyse muscles.

FIRST AID Induce vomiting; administration of medicinal charcoal and sodium sulphate. Contact of latex with skin: wash with Roticlean (polyethylene glycol 400) or water; in case of skin blisters, cover with sterile cotton and apply locally locacorten foam; (also consult individual species monographs). **Clinical therapy:** Gastric lavage (possibly with 0.1% potassium permanganate), instillation of medicinal charcoal and sodium sulphate; electrolyte substitution, control of acidosis with sodium bicarbonate (pH of urine 7.5). In case of spasms inject diazepam (i.v.). Provide intubation and oxygen respiration in case of respiratory arrest or paralysis. Check diuresis and kidney function.

Dicentra spectabilis   *Thalictrum aquilegifolium*

Lycopodium alkaloids

STRUCTURE More than 150 structures have been described from Lycopodium alkaloids. Examples are: cernuine ($C_{16}H_{26}N_2O$; MW 262.40), lycopodine ($C_{16}H_{25}NO$; MW 247.38), and selagine ($C_{15}H_{18}N_2O$; MW 242.32).

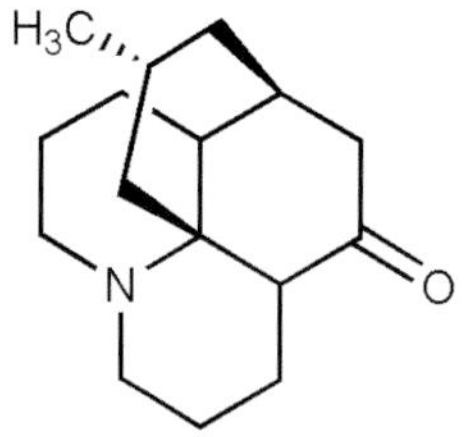

lycopodine

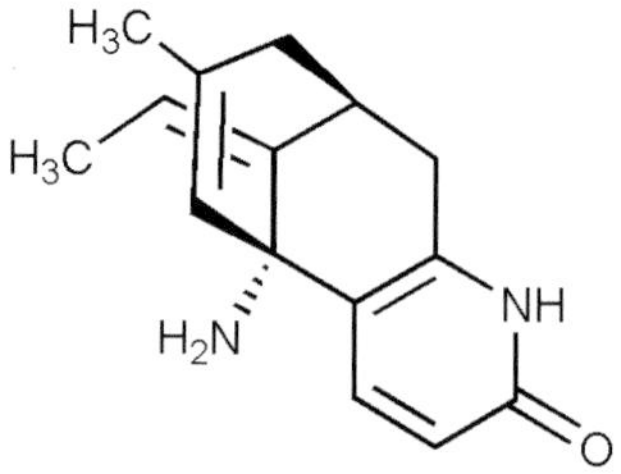

selagine

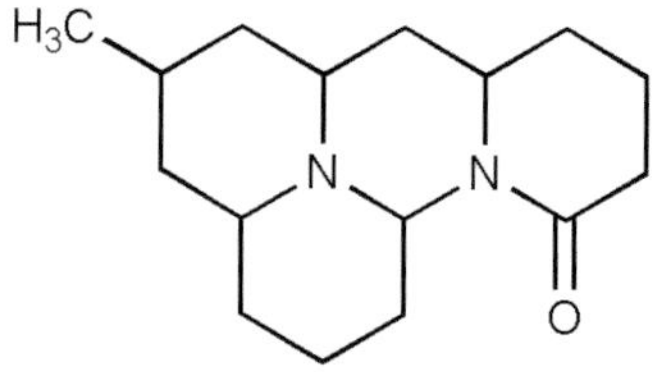

cernuine

OCCURRENCE The Lycopodium alkaloids are abundant in clubmosses of the genera *Lycopodium, Lycopodiella, Huperzia* and *Phyloglossum* (Lycopodiaceae).

CLASSIFICATION Neurotoxin; moderately hazardous, II

SYMPTOMS Lycopodium alkaloids cause vomiting, nausea, dizziness, staggering and coma.

TOXICITY About 200 mg of *L. clavatum* is lethal for mice and frogs. Spores can cause allergic asthma.

MODE OF ACTION Lycopodium alkaloids can bind to cholinergic neuroreceptors and cause paralysis of CNS and skeletal muscles, but enhance respiratory activities, uterine contractions and intestinal motility.

USES (PAST AND PRESENT) *Lycopodium* has been used in traditional medicine to treat various skin disorders and as a tonic (in TCM).

FIRST AID Administration of sodium sulphate, medicinal charcoal. **Clinical therapy:** Gastric lavage (possibly with 0.1% potassium permanganate), instillation of medicinal charcoal and sodium sulphate. In case of excitation and spasms inject diazepam; in case of severe intoxication, provide intubation and oxygen respiration.

Lycopodium annotinum

Lycopodium gnidoides

Lycorine and other Amaryllidaceae alkaloids

STRUCTURE Biogenetically these alkaloids derive from tyrosine. They share a common 15-carbon nucleus. More than 300 structures have been described from this group. Examples of some of these alkaloids are:

Ambelline ($C_{18}H_{21}NO_5$; MW 331.37), lycorine ($C_{16}H_{17}NO_4$; MW 287.32), galanthamine ($C_{17}H_{21}NO_3$; MW 287.36), haemanthamine ($C_{17}H_{19}NO_4$; MW 301.35), narciclasine ($C_{14}H_{13}NO_7$; MW 307.27), pancratistatine ($C_{14}H_{15}NO_8$; MW 325.28).

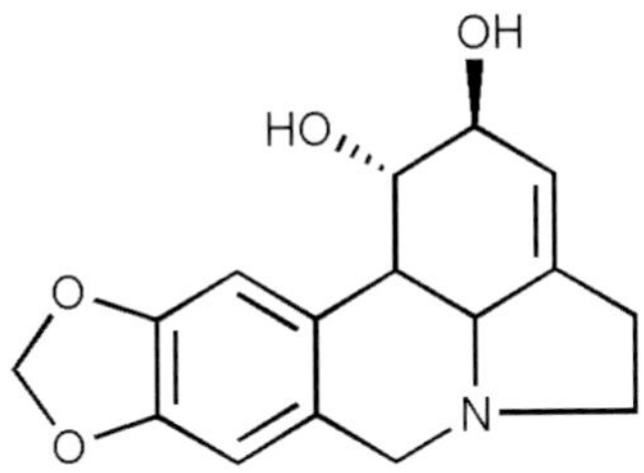

lycorine

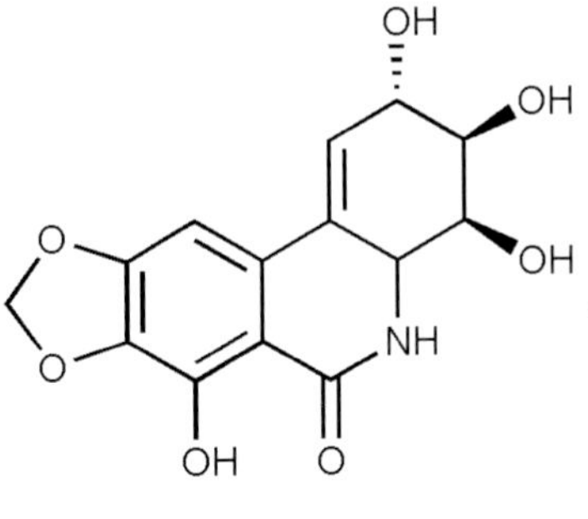

narciclasine

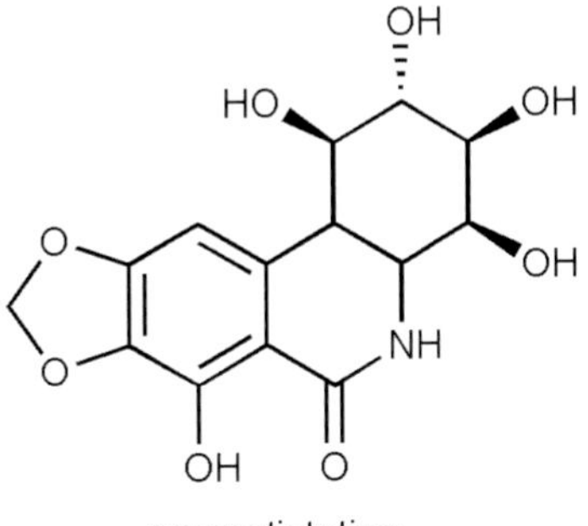

pancratistatine

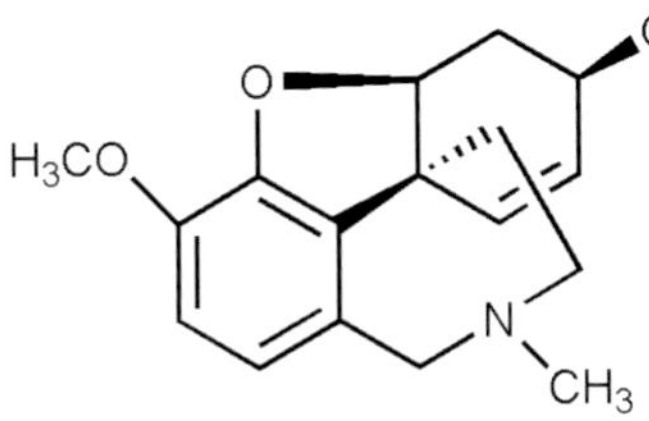

galanthamine

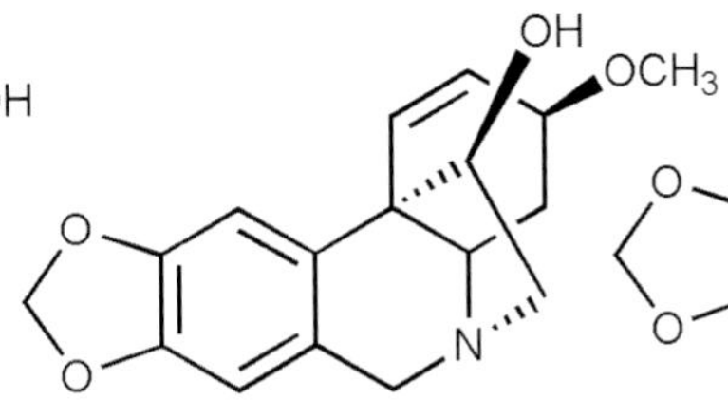

haemanthamine

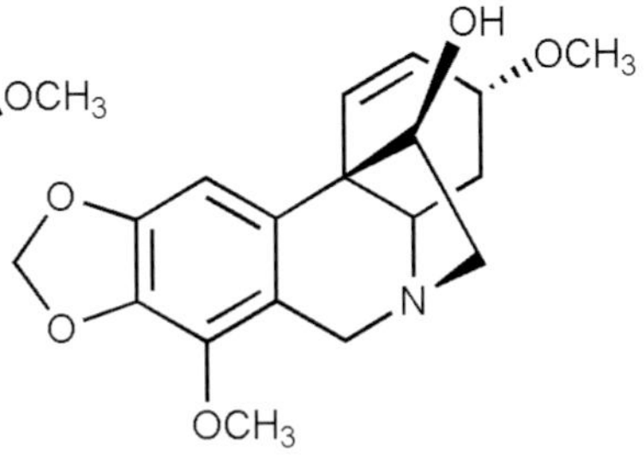

ambelline

OCCURRENCE Widely distributed among Amaryllidaceae genera: *Amaryllis, Boophone, Clivia, Crinum, Galanthus, Leucojum, Hippeastrum, Narcissus, Pancratium* and *Zephyranthes*. Bulbs have high alkaloid contents, especially in the epidermis of the outer scales and in mucilage-rich cells with raphides. These raphides easily come into contact with skin when handling bulbs and cause dermatitis (daffodil itch; lily rash). The raphides also enhance the uptake of alkaloids. Galanthamine occurs in the genera *Crinum, Galanthus, Hippeastrum, Hymenocallis, Leucojum, Lycoris, Narcissus, Pancratium* and *Ungernia*.

CLASSIFICATION Cell toxin; highly hazardous, Ib–II

SYMPTOMS Lycorine and other Amaryllidaceae alkaloids increase heart rate and systolic blood pressure. A peripheral vasoconstriction and increase of blood glucose levels were also observed in animal experiments. Symptoms of poisoning include salivation, perspiration, nausea, severe vomiting and strong diarrhoea. Central paralysis, convulsions and spasms, liver damage, depression of respiration and hypotension can lead to collapse and death from respiratory arrest. Oxalate crystals and raphides can cause local inflammation and damage of mucosal tissues in mouth and throat. Ambelline, caranine, crinamine are very toxic causing respiratory arrest.

TOXICITY A few grams of daffodil bulbs can be lethal. A woman who had eaten a bouquet of daffodils, died after 2 days. LD_{50} in dogs: lycorine: 41 mg/kg; crinamine: 10 mg/kg p.o.

MODE OF ACTION Lycorine and narciclasine inhibit ribosomal protein biosynthesis. By binding to the 60S subunit it inhibits peptide elongation. Since protein biosynthesis is a basic biochemical process, lycorine can inhibit healthy and tumour cells, and replication of animal viruses. Lycorine also inhibits the biosynthesis of vitamin C. The oxalate crystals can prick cells and may enhance the absorption of the alkaloids. Galanthamine inhibits cholinesterase and can be regarded as a potent parasympathomimetic. In addition, it shows analgesic properties. Ambelline also has analgesic properties, but it is too toxic for medicinal use.

USES (PAST AND PRESENT) Since several Amaryllidaceae have beautiful flowers, many of them are cultivated as pot or garden plants. Therefore, an accidental poisoning of children and pets is possible, if they ingest plant material. Some bulbs resemble onions. If they are stored in the wrong place and are not adequately labelled, they can be mistaken for onions and eaten instead. Galanthamine, which is isolated from *Galanthus woronowii, Leucojum aestivum, Narcissus pseudonarcissus* and *N. nivalis*, has been introduced as a therapeutic to treat Alzheimer's disease. *Galanthus* might have been the famous "moly" of Ulysses (see Introduction).

FIRST AID Give medicinal charcoal and sodium sulphate. Allow patient to drink plenty of tea and give shock prophylaxis. **Clinical therapy:** After ingestion, gastric lavage (possibly with potassium permanganate), electrolyte substitution, control acidosis with sodium bicarbonate (urine pH 7.5) and kidney function; in case of spasms give diazepam; in severe cases supply oxygen and intubation.

Leucojum aestivum

Narcissus poeticus

Galanthus nivalis

Hippeastrum x *hortorum*

Morphine and other morphinane alkaloids

STRUCTURE This group of SMs contains several alkaloids with a morphinane skeleton. Important members are:

Morphine ($C_{17}H_{19}NO_3$; MW 285.35), heroin ($C_{21}H_{23}NO_5$; MW 369.41), codeine ($C_{18}H_{21}NO_3$; MW 299.37), thebaine ($C_{19}H_{21}NO_3$; MW 311.38).

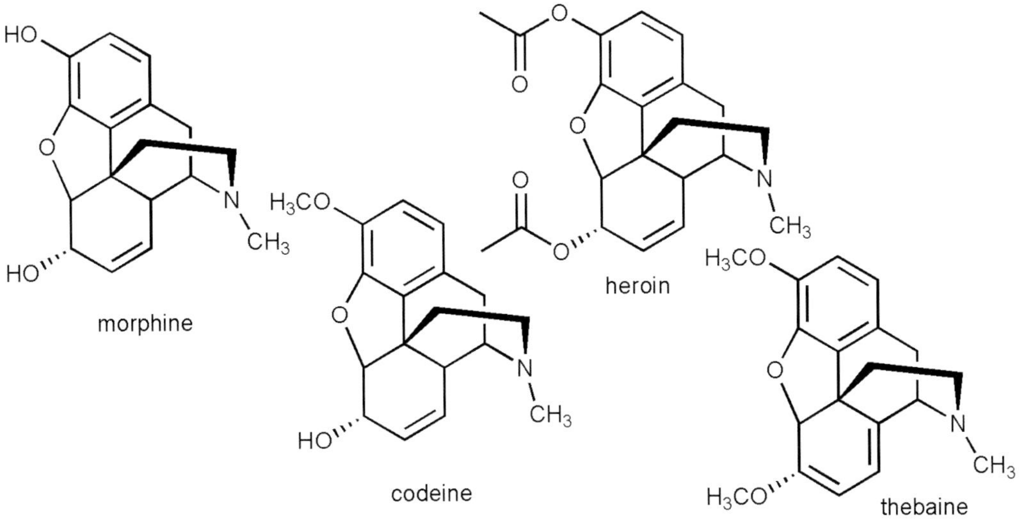

morphine

heroin

codeine

thebaine

OCCURRENCE Morphinane alkaloids are typical of members of the genus *Papaver*; especially rich are *P. somniferum* and *P. bracteatum*. Salutaridine has additionally been isolated from *Croton salutaris* (Euphorbiaceae) and related species.

CLASSIFICATION Mind-altering; extremely hazardous, Ia

SYMPTOMS Morphine causes central analgesia (dose 5–10 mg), euphoria (after several applications) and sedation. Further symptoms are strong miosis, serious respiratory depression, vomiting, bradycardia, perspiration, itching, reddening of face, vertigo, dizziness, motoric unrest, hypothermia, cyanosis, weakness, constipation, inhibition of nociceptive reflexes, sleepiness; overdoses cause coma and death after 7–12 h from respiratory arrest. Chronic use causes bradycardia, permanent sleepiness, miosis, weight loss, constipation, loss of libido, impotence, trembling, ataxia, slurry speech, pale and dry skin and hair loss.

TOXICITY Morphine and other morphinane alkaloids exhibit addictive properties. 200–400 mg morphine, given orally, are lethal for humans, 100–200 mg after parenteral application. Heroin: 50–75 mg are lethal. Drug addicts can tolerate much higher doses. Thebaine causes convulsions at higher doses.

MODE OF ACTION Morphine is usually administered parenterally for medical and narcotic purposes. The oral route is less important. Morphine is readily absorbed from mucosal cells and distributed within the body. It is mainly eliminated via the urine, less via gall and faeces. Morphine is a powerful agonist of endorphine receptors in the brain and other organs with sleep-inducing, analgesic and hallucinogenic effects. Morphine is highly addictive. Codeine is an effective painkiller (though less active than morphine, but also less addictive); it sedates the cough centre and has thus antitussive activity. Opiates reduce the release of the neurotransmitters acetylcholine, noradrenaline and dopamine in parts of the CNS. They inhibit acetylcholine mediated membrane depolarisation and adenylate cyclase. Morphine induces the release of histamine, which causes skin reddening and itching. Heroin has a good bioavailability and can pass the blood-brain barrier. It can be metabolically converted into morphine.

USES (PAST AND PRESENT) Poppy was known in antiquity (Egypt, Persia, Arabia and Greece) and the Sumerians used it as early as 4000 BC. Poppy was dedicated to the gods of death

"Thanatos", of sleep "Hypnos" and of dreams "Morpheus". Dioscorides gave an excellent description of the medicinal properties of opium: a pea-sized piece of opium kills pain effectively, induces sleep, activates digestion, silences cough and stomach troubles. Dioscorides also described the use of poppy seeds (which are rich in oil but almost devoid of alkaloids) for baking bread (as is done today) and the adulteration of opium with latex of *Papaver rhoeas, Glaucium flavum* or other lactiferous plants. Opium was mixed with hemlock (see *Conium maculatum*) in ancient Greece for murder or to execute the condemned. The world demand for opiates is presently nearly 200 tons of alkaloid per year. In 2007 about 9 000–10 000 tons of opium have been produced, 92 % came from Afghanistan. Heroin has replaced morphine today as an illegal addictive drug since it has a stronger mind-altering effect (as it crosses the blood-brain barrier more effectively). Pure alkaloids are nowadays used in standardised modern medicines intended for oral and parenteral use – mainly to treat intense pain (e.g. in cancer patients) and as antitussive medicines.

FIRST AID Place patient in recovery position, empty mouth and throat from vomit, supply oxygen, intubation and artificial respiration. Inject morphine antidote naloxon (0.4 mg) every 5 min until patient wakes up. Withdrawal therapy for drug addicts with methadone possible. **Clinical therapy:** After ingestion, gastric lavage (possibly with potassium permanganate), instil medicinal charcoal and sodium sulphate. Apply artificial respiration; provide electrolyte substitution, control acidosis with sodium bicarbonate (urine pH 7.5) and kidney function. In case of spasms give diazepam, against arrhythmia lidocaine, and naloxon as antidote (initially 0.4 mg i.v.; might be repeated 1 or 2 times). Apply a bladder catheter in case of muscle contraction.

Papaver somniferum

Papaver argemone

Papaver rhoeas

Glaucium flavum

Pyrrolidine and piperidine alkaloids

STRUCTURE Pyrrolidines and piperidines comprise a heterogenous group of alkaloids with some infamous and toxic members. Examples are: anabasine ($C_{10}H_{14}N_2$; MW 162.24), arecoline ($C_8H_{13}NO_2$; MW 155.20), castanospermine ($C_8H_{15}NO_4$; MW 189.22), coniine ($C_8H_{17}N$; MW 127.23), deoxymannojirimycin ($C_6H_{13}NO_4$; MW 163.18), lobeline ($C_{22}H_{27}NO_2$; MW 337.47), nicotine ($C_{10}H_{14}N_2$; MW 162.24), pelletierine ($C_8H_{15}NO$; MW 141.22), piperine ($C_{17}H_{19}NO_3$; MW 285.35), sedamine ($C_{14}H_{21}NO$; MW 219.33), and swainsonine ($C_8H_{15}NO_3$; MW 173.22).

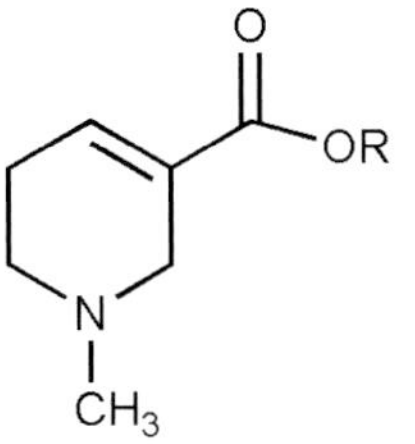

arecaidine (R = H)
arecoline (R = CH₃)

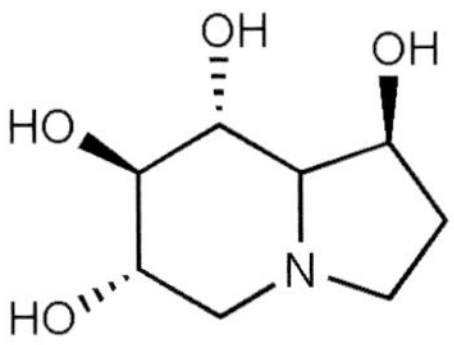

castanospermine

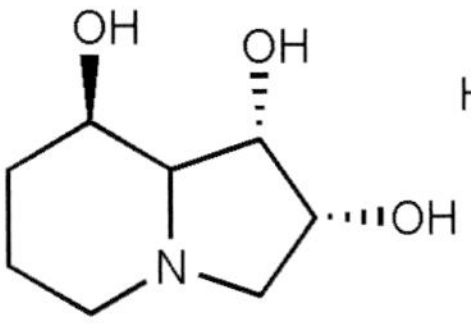

swainsonine

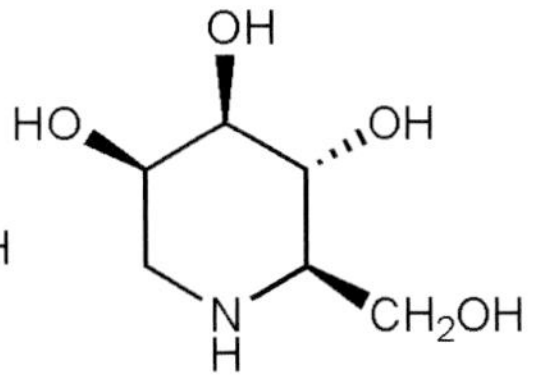

deoxymannojirimycin (DMJ)

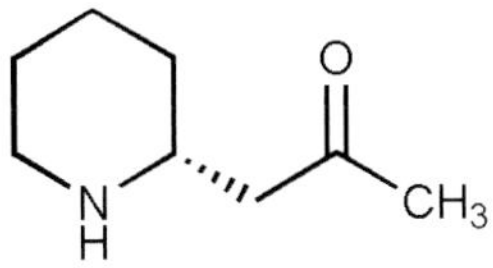

coniine

pelletierine

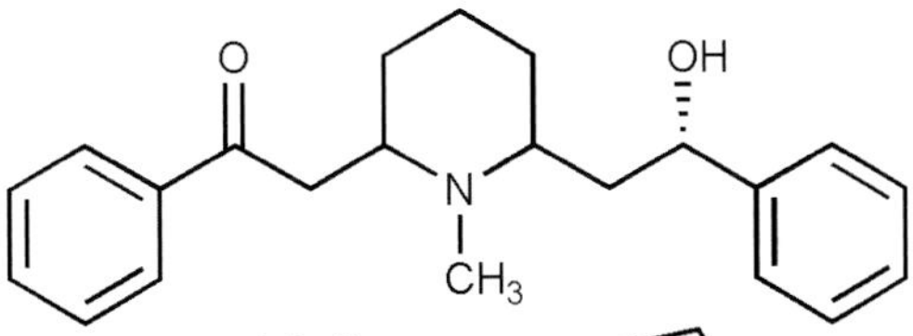

lobeline

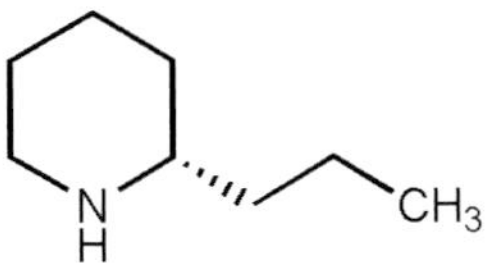

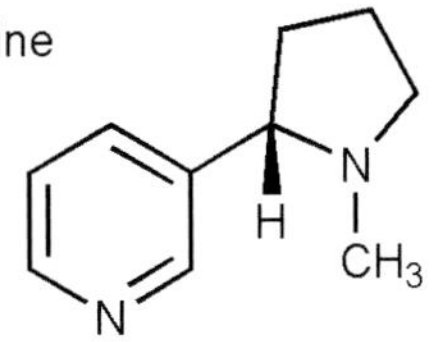

nicotine

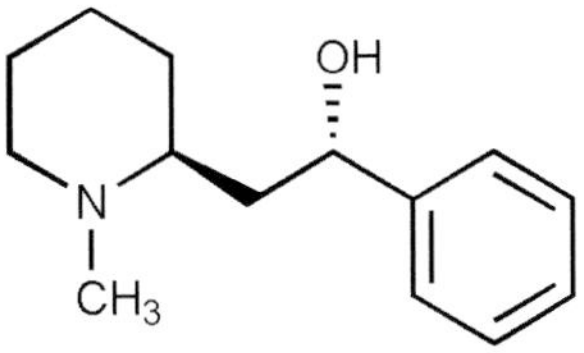

sedamine

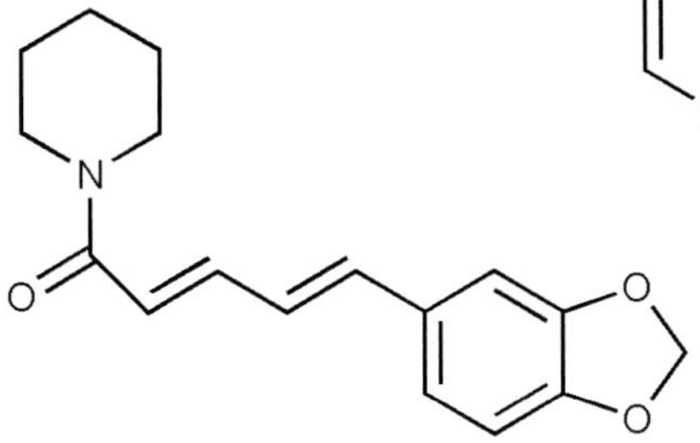

piperine

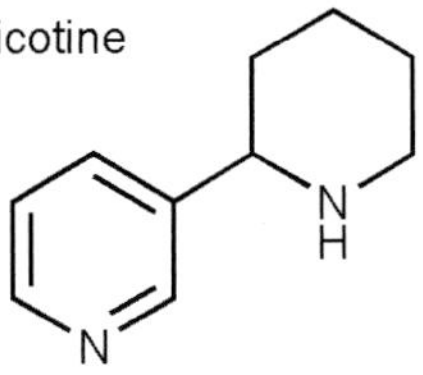

anabasine

OCCURRENCE Anabasine has been detected in *Anabasis* (Amaranthaceae/Chenopodiaceae), *Nicotiana*, *Duboisia* (Solanaceae), *Zinnia*, *Zollikoferia* (Asteraceae), *Alangium* (Alangaceae). Arecoline and arecaidine are the active principles in *Areca catechu* (Arecaceae). Castanospermine, DMJ and swainsonine have been detected in Fabaceae *(Castanospermum, Lonchocarpus, Swainsonia)* and some microorganisms. Coniine, γ-coniceine and *N*-methylconiine have been detected in *Conium maculatum* (Apiaceae), *Aloe gililandii*, *A. ballyi*, *A. ruspoliana* (Asphodelaceae), *Sarracenia flava* (Sarraceniaceae). Lobeline is a typical metabolite of *Lobelia* and *Campanula medium* (Campanulaceae). Nicotine and related alkaloids have been found in *Nicotiana* (Solanaceae), *Asclepias syriaca* (Apocynaceae), *Equisetum arvense* (Equisetaceae) and *Sedum acre* (Crassulaceae). Pelletierine has been described from *Punica* (Lythraceae/Punicaceae), *Duboisia* (Solanaceae) and *Sedum acre* (Crassulaceae). Piperine is the pungent principle of *Piper nigrum* and other *Piper* species (Piperaceae). Sedamine is accumulated by *Sedum acre* and related pungent *Sedum* species (Crassulaceae).

CLASSIFICATION Neurotoxins, mind-altering; highly to moderately hazardous, Ia–II

SYMPTOMS The symptoms of anabasine and nicotine poisoning are similar and appear rapidly. They include nausea, weakness, headache, salivation, convulsions, confusion, difficulty in breathing and hypertension. In extreme cases, paralysis of the respiratory muscles leads to rapid death. Seeds of *Castanospermum* cause severe gastroenteritis and have caused death in livestock. *Conium* alkaloids cause paralysis of motor nerve endings which leads to burning sensation in mouth and throat, salivation, drowsiness, trembling, nausea, vomiting and diarrhoea, tachycardia, mydriasis, breathing difficulty and finally asphyxia followed by mental disturbance and ascending paralysis, which starts at the tip of arms and legs and ends with respiratory failure and death. Sedamine has a pungent and acrid taste.

TOXICITY Anabasine exhibits acute and subacute toxicity. Conium alkaloids are extremely toxic and teratogenic in livestock. Nicotine is highly toxic causing respiratory arrest; lethal dose for humans (adults): 40–60 mg p.o. Coniine: LD_{50} mouse: 19 mg/kg i.v., 100 mg/kg p.o.. Lethal dose (p.o.) in humans: 0.5–1 g.

MODE OF ACTION Many of the pyrrolidine and piperidine alkaloids can bind to cholinergic neuroreceptors (nAChR and/or mAChR) and cause paralysis of CNS and skeletal muscles, but enhance respiratory activities and intestinal motility. They affect both sympathetic and parasympathetic nerves. Arecoline and arecaidine exhibit parasympathetic activities and act as a central stimulant. Castanospermine inhibits α- and β-glucosidases, DMJ and swainsonine mannosidases that are important to trim oligosaccharide residues of glycoproteins in the endoplasmic reticulum and Golgi apparatus.

USES (PAST AND PRESENT) *Areca catechu* nuts have been used in SE Asia as a stimulating narcotic (together with *Piper betle* leaves; see *A. catechu*) but also against worms and as laxative in livestock. *Castanospermum australe* has very large seeds. Australian aborigines eat the seeds as food after soaking and leaching them in water. Castanospermine has been used as a potential anti-HIV agent. *Conium maculatum* has been used in antiquity to execute criminals; a famous victim was the philosopher Socrates (see page 18). Lobeline has been used in asthma therapy, as anti-smoking agent and analeptic. Nicotine is a CNS stimulant with addictive and tranquillising properties; before the availability of synthetic insecticides, nicotine was widely used to control insects in agriculture (see *N. tabacum*). Pelletierine has been used against intestinal tapeworms. *Piper* fruits are widely used as hot spice and sometimes as insecticide.

FIRST AID Give medicinal charcoal and sodium sulphate. Allow the patient to drink plenty of tea or water and keep warm. After contact of alkaloids with skin, wash with water and soap.
Clinical therapy: After ingestion of larger doses, gastric lavage (possibly with 0.1 % potassium permanganate), electrolyte substitution, control acidosis with sodium bicarbonate (urine pH 7.5); in case of spasms give diazepam; in severe cases supply oxygen respiration and intubation.

Areca catechu   *Sedum acre*

Quinoline alkaloids

STRUCTURE Quinoline alkaloids derive from anthranilic acid. Several of them have some importance in medicine, such as quinine, quinidine and cinchonidine. The alkaloids of interest are acronycine ($C_{20}H_{19}NO_3$; MW 321.38), arborine ($C_{16}H_{14}N_2O$; MW 250.30), cinchonidine ($C_{19}H_{22}N_2O$; MW 294.40), quinine ($C_{20}H_{24}N_2O_2$; MW 324.43), fargarine ($C_{13}H_{11}NO_3$; MW 229.24), dictamnine ($C_{12}H_9NO_2$; MW 199.21), skimmianine ($C_{14}H_{13}NO_4$; MW 259.27), peganine ($C_{11}H_{12}N_2O$; MW 188.23), vasicinone ($C_{11}H_{10}N_2O_2$; MW 202.22).

OCCURRENCE Quinoline alkaloids occur in Rutaceae (*Acronychia, Adiscanthus, Aegle, Afraele, Atalantia, Balfourodendron, Bauerella, Casimiroa, Choisya, Cusparia, Dictamnus, Esenbeckia, Euodia, Fagara, Flindersia, Galipea, Geijera, Glycosmis, Haplophyllum, Heliettia, Lunasia, Melicope, Monniera, Murraya, Orixa, Ptelea, Ruta, Sarcomelicarpa, Skimmia, Teclea, Xylocarpus, Zanthoxylum*), Acanthaceae (*Adhatoda, Anisotes*), Rubiaceae (*Cinchona, Remija*), Oleaceae (*Olea, Ligustrum*), Zygophyllaceae (*Nitraria, Peganum*), Plantaginaceae (*Linaria*), Araliaceae (*Mackinlaya*), Asteraceae (*Echinops*), Saxifragaceae/Hydrangeaceae (*Dichroa, Hydrangea*), Calycanthaceae (*Calycanthus*), Apocynaceae (*Aspidosperma*), Malvaceae (*Sida*), Brassicaceae (*Lunaria*).

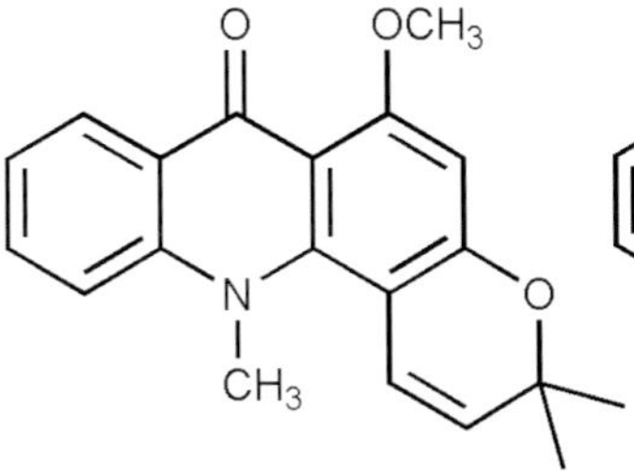

acronycine

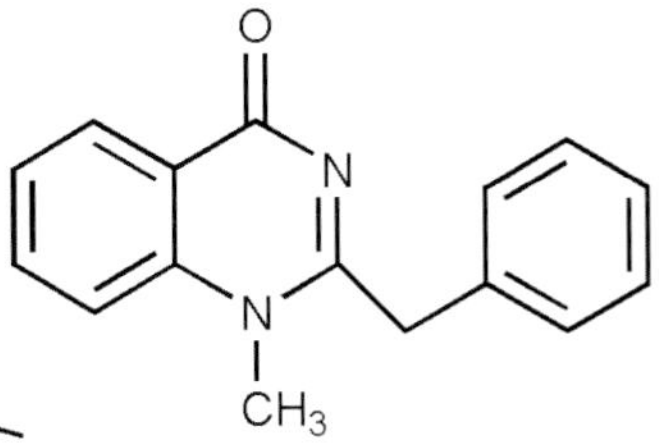

arborine

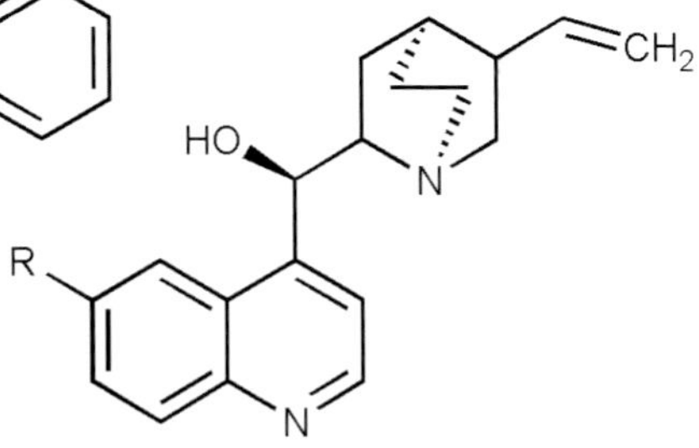

cinchonidine (R = H)
quinine (R = OCH$_3$)

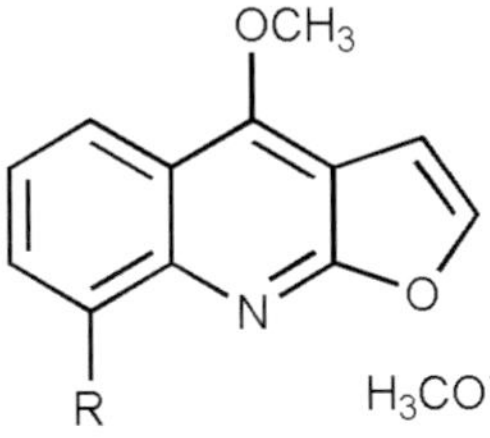

fagarine (R = OCH$_3$)
dictamnine (R = H)

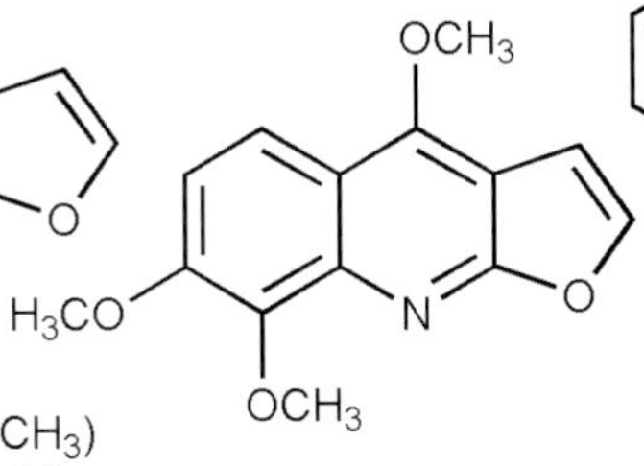

skimmianine

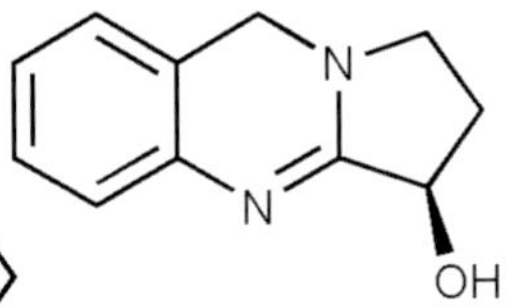

peganine

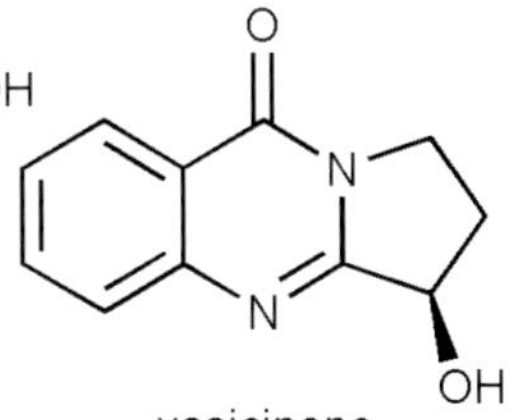

vasicinone

CLASSIFICATION Cell toxin, mutagen; moderately hazardous, Ib–II

SYMPTOMS Acronycine and related compounds are CNS depressant, strongly cytotoxic and phototoxic. When skin that has been in contact with furoquinolines, such as fagarine, dictamnine or skimmianine, is exposed to sunlight severe burns can occur with blister formation, inflammation and necrosis. Several alkaloids are CNS depressant.

TOXICITY Because of alkylating and intercalating properties, several quinoline alkaloids are mutagenic and carcinogenic (e.g. acronycine). Peganine and other quinolines exhibit abortifacient properties. LD_{50}: lunacrine: mouse i.v. 80 mg/kg. Calycanthine from *Calycanthus* and *Palicourea* is highly toxic and causes violent convulsions, paralysis and cardiac arrest.

MODE OF ACTION Most quinoline alkaloids intercalate DNA and thus cause frameshift mutations. Furanoquinolines can be activated by light and can form covalent bonds with DNA bases, such as thymine (see furanocoumarins). This explains their cytotoxicity, antibacterial and antifungal properties. Arborine inhibits peripheral cholinergic activities and is a central hypotensive agent. Spasmolytic activity is also a common effect. Quinidine additionally inhibits Na^+ channels. Peganine and vasicine (and related compounds) show cholinergic activity. Dictamnine is a strong muscle contractant. Kokusaginine enhances the concentrations of noradrenaline and dopamine in the brain.

USES (PAST AND PRESENT) Cinchonidine and quinine have been used as antimalarial. Quinidine has anti-arrhythmic properties. Febrifugine from *Dichroa febrifuga* is 100 times more potent as antimalarial than quinine but too toxic. Quinine is very bitter and is used as a bittering agent in the food industry.

FIRST AID Administration of sodium sulphate, medicinal charcoal. When skin or eyes have been in contact with the furanoquinolines, wash with water or polyethylene glycol. Treat skin with cortisone ointments, when inflamed; eyes with anti-allergic eyedrops (opticrom). **Clinical therapy:** Gastric lavage (possibly with 0.1 % potassium permanganate), instillation of medicinal charcoal and sodium sulphate. In case of excitation and spasms inject diazepam; in case of severe intoxication, provide intubation and oxygen respiration.

Skimmia x *foremanii*

Calycanthus floridus

Zanthoxylum americanum

Ptelea trifoliata

Senecionine and related pyrrolizidine alkaloids

STRUCTURE PAs (of which more than 200 are known) consist of a necine base (heliotridine, retronecine) that carries 1, 2 or cyclic esters. In plants, they often occur as *N*-oxides that can be reduced in the animal body to free alkaloids. Examples of simple monoesters are: heliotrine and supinidine; for open diesters: symphytine ($C_{20}H_{31}NO_6$; MW 381.48) and lasiocarpine ($C_{21}H_{33}NO_7$; MW 411.50); for cyclic esters: senecionine ($C_{18}H_{25}NO_5$; MW 335.41) and senkirkine ($C_{19}H_{27}NO_6$; MW 365.43).

PA monoesters

heliotrine

supinidine

PA diesters (open)

symphytine

lasiocarpine

PA diesters (cyclic)

senecionine

senkirkine

OCCURRENCE Widely present in plants of the family Boraginaceae (*Alkanna, Amsinckia, Anchusa, Arnebia, Asperugo, Borago, Cordia, Cynoglossum, Echium, Heliotropium, Lappula, Lithospermum, Lindelofia, Messerschmidia, Myosotis, Rindera, Solenanthus, Symphytum, Tournefortia, Trichodesma*), Fabaceae (tribe Crotalarieae: *Crotalaria, Lotononis*), Asteraceae (subfamily Senecioneae: *Adenostyles, Brachyglottis, Cacalia, Conoclinum, Doronicum, Echinacea, Emilia, Eupatorium, Farfugium, Ligularia, Petasites, Senecio, Tussilago*). Less common in Apocynaceae, Celastraceae, Elaeocarpaceae, Euphorbiaceae, Orchidaceae, Poaceae, Ranunculaceae and Sapotaceae. PAs can be transferred into honey by bees collecting nectar and pollen from PA plants and into milk of cows, sheep and goats feeding on PA plants. PAs are not degraded substantially during storage, silage or during rumen digestion in ruminants (cows). Several insects are known that have specialised on PA plants, which accumulate PAs as acquired defence compounds.

CLASSIFICATION Cell toxin, neurotoxin, mutagen; moderately hazardous, II

SYMPTOMS PAs are metabolically activated in the liver. Therefore, this organ is primarily affected by PA toxicity. Symptoms include loss of appetite, weariness and abdominal pain. Long-term damage includes veno-occlusion (the so-called Budd-Chiari syndrome), with thrombosis of the hepatic vein, leading to cirrhosis of the liver, necrosis, megalocytoses and hyperplasia. The last forms can lead to adenoma and carcinoma. Metabolites that reach the lung can cause pneumonia, emphysema, oedema, pulmonal arterial hypertension and acute heart failure (cor pulmonale). PAs and PA metabolites are eliminated via the urinary tract. Megalocytosis of kidneys and bladder tumours have been observed as a consequence. The pancreas and gastrointestinal tract can also be affected.

The toxic and carcinogenic properties of PAs have been detected in livestock feeding on PA plants but also in humans drinking PA containing "health teas" for several months. In one incident a baby died from veno-occlusive disease (VOD) after the mother drank a tea with PA-rich *Petasites* leaves (mistaken for *Tussilago*) during pregnancy. In another similar case, a baby exposed to *Petasites* tea developed serious VOD. PAs also affect several neuroreceptors, such as acetylcholine, serotonin and adrenergic receptors. Ingestions of high doses lead to neuronal disturbance. Heliotrine causes neuromuscular block and respiratory failure. Senecionine acts as anticholinergic.

TOXICITY PA poisoning can be caused by drinking herbal teas ("bush teas"), ingesting TCM drugs that have become fashionable in Western countries or by consuming comfrey (*Symphytum*) as a salad. In Afghanistan and India, contamination of cereals with seeds of *Heliotropium* and *Crotalaria* has caused serious poisoning and even death. PA doses, which cause liver disease upon ingestion after more than 2 weeks, are 0.7–1.5 mg/kg/d for adults and 0.01–0.06 mg/kg/d for children. Macrocyclic diesters are more toxic than open-ringed monoesters. Monocrotaline: LD_{50} mouse: 166–170 mg/kg p.o., rat 71 mg/kg p.o.; senecionine: LD_{50} mouse: 64 mg/kg i.v., rat 85 mg/kg i.p.; lasiocarpine: LD_{50} mouse: 85 mg/kg i.v., rat: 88 mg/kg i.v., 72–79 mg/kg i.p.: symphytin LD_{50} mouse: 300 mg/kg i.p.; seneciphylline: rat 77 mg/kg i.p.; retrorsine: rat 35 mg/kg i.p.

Serious poisoning of livestock has been recorded, especially from North America, Brazil and Australia; normally animals avoid eating plants with PAs, but in times of food shortage, e.g. during droughts when no other plant is available, PA plants may be consumed. About 50% of feed-related deaths in livestock are caused by PA ingestion.

MODE OF ACTION N-Oxides are reduced to free base by intestinal microorganisms. The free base can be absorbed by free diffusion and is transported to the liver. Metabolising enzymes, such as CYP3A4, either hydroxylate them or form PA-*N* oxides. PAs that carry a double bond in 1–2 position become activated by this detoxification step. After loss of a molecule water, reactive pyrroles (dehydropyrrolizidines) are formed that can alkylate DNA-bases. In the presence of nucleophilic groups, such as HS, OH or NH_2 (which are common in proteins) and OH or NH_2 (common in nucleic acids), a nucleophilic substitution (SN2-reaction) takes place. In this process the necine esters are split and the ester part is substituted with DNA-bases. In case of diesters, crosslinks can be formed between adjacent or between bases of both DNA strands. These alkylations can lead to mutation and cell death (especially in the liver). Furthermore, mutations can lead to malformations in pregnant animals and humans, and to cancer of liver, kidneys and lungs. Monoesters are monofunctional alkylants and usually cause acute intoxication whereas bifunctional diesters or cyclic esters are more toxic and have mutagenic, teratogenic and carcinogenic properties.

USES (PAST AND PRESENT) Several PA-containing plants are used in traditional phytomedicine to treat bleeding or diabetes or generally as herbal tea (*Senecio, Petasites, Heliotropium, Crotalaria*); *Symphytum officinale* and other Boraginaceae to treat wounds, broken or injured bones. The drug registration agencies have limited the daily intake of PA to less than 1 µg. Others, such as comfrey *Symphytum* x *uplandicum*, are regularly supplied on local markets as "healthy" salad ingredients; usually neither seller nor customer knows about the inherent hazard. Honey from bees that collected nectar and pollen near fields with PA plants can contain substantial amounts of PA; a check for PAs should be part of the quality control.

FIRST AID Give medicinal charcoal to absorb the alkaloids; administration of plenty of warm black tea, and of sodium sulphate. Provide shock prophylaxis (keep patient at quiet place and warm). In case of blisters of skin or mucosa, apply sterile cover. **Clinical therapy:** In case of ingestion of larger doses: gastric lavage (possibly with potassium permanganate), instillation of medicinal charcoal and sodium sulphate, electrolyte substitution, control of acidosis with sodium bicarbonate. In case of spasms inject diazepam; against nausea antiemetics such as triflupromazin. For severe intoxication, provide intubation and oxygen therapy. Observe kidney/liver function and blood coagulation capacity. In case of VOD, apply therapies as for patients with liver cirrhosis. There are no specific antidotes.

Solanine and other steroid alkaloids

STRUCTURE Steroidal alkaloids of *Solanum* species consist of a lipophilic steroid moiety and a hydrophilic oligosaccharide chain. The alkaloids are heat stable. Two groups are distinguished: a spirosolane type, with soldulcidine and tomatidine; a solanidane type with solanine and chaconine. Typical substances are: α-solanine ($C_{48}H_{73}NO_{15}$; MW 904.12), α-chaconine ($C_{45}H_{73}NO_{14}$; MW 852.09) and solasonine ($C_{45}H_{73}NO_{16}$; MW 884.09).

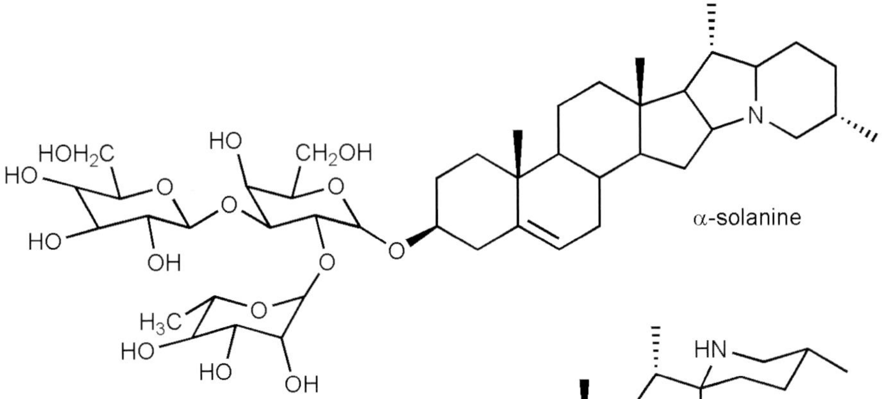

α-solanine

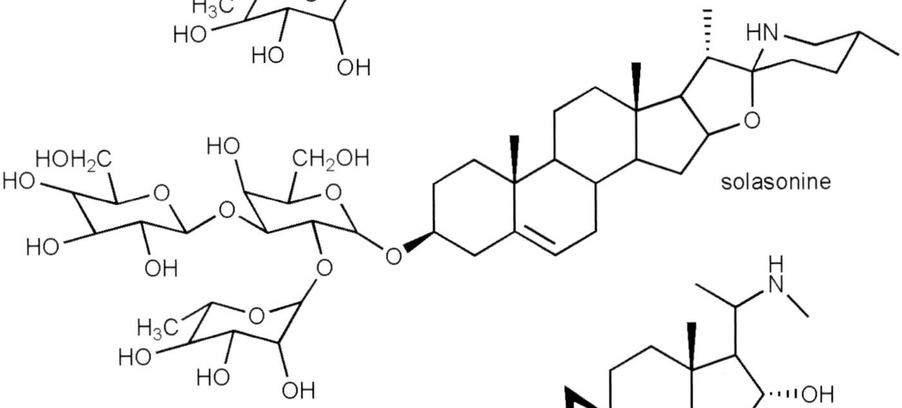

solasonine

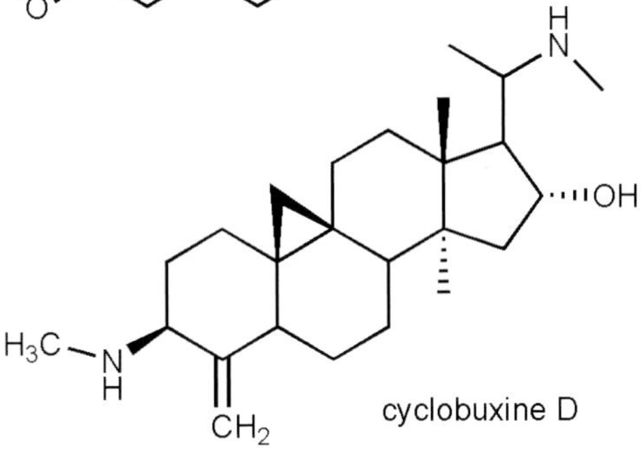

cyclobuxine D

OCCURRENCE Steroidal alkaloids are produced by four unrelated plant families: Apocynaceae, Buxaceae, Liliaceae and Solanaceae. They are especially widely distributed within the very large genus *Solanum* that includes potato and other food plants; these alkaloids are also present in *Lycopersicon* (tomato), *Withania* (Solanaceae) and *Notholirion hyacinthum, Veratrum stenophyllum* (Melanthiaceae). All green parts of *Solanum* plants accumulate high levels of steroidal alkaloids. Fruits of several *Solanum* species are green when immature but turn red when ripe (tomato; *Solanum dulcamara*). Whereas green fruits are toxic (this is also true for tomatoes!), alkaloid levels are very low in red fruits of some (e.g. *Solanum dulcamara*) but not all species (African eggplant, S. *macrocarpon*). Alkaloids are also found in tubers; when tubers are stored in light, at high temperatures, start sprouting, or become wounded, alkaloid levels can increase substantially. Therefore, potato tubers should be stored correctly and, since the alkaloids are located in the skin, potatoes should be peeled and boiled in water. Plants of the genus *Buxus* contain a series of steroidal alkaloids, such as cyclobuxine D, buxamine E, which are quite toxic and strongly purgative. Steroidal alkaloids also occur in the Apocynaceae (*Funtumia, Holorhaena, Kibatalia, Paravallaris*). Steroidal alkaloids of Liliaceae s.l. (*Fritillaria, Veratrum, Zigadenus*) have been treated under "aconitine" because of similar properties.

CLASSIFICATION Cell toxin, mind-altering; moderately hazardous, Ib–II

SYMPTOMS Symptoms of glycoalkaloid intoxication include a serious irritation of mucosa in mouth, throat (scratching and burning) and GI tract, and positive inotropic heart activity. *Solanum* alkaloids enhance tonus and contraction of the uterus, can damage kidneys, and CNS causing paralysis and death from central respiratory arrest. Further symptoms are headache, vomiting, strong diarrhoea, vertigo, dizziness, apathy, tremor, fever, respiratory depression, tachycardia, haematuria. Neurological symptoms include hallucinations, anxiety, restlessness and visual disturbance. On mucosa, the alkaloids cause reddening, blister formation and necrosis. Symptoms start 4–19 h after ingestion and can last for 3–6 days. *Buxus* alkaloids are strong purgatives and antiarrhythmic agents.

TOXICITY Solanine: LD_{50} mouse: 42 mg/kg i.p.; rabbit 20–30 mg/kg i.p. The minimum toxic dose of solanine in humans is 2–5 mg/kg; lethal oral dose for adults is about 300–500 mg. Some *Solanum* alkaloids have shown teratogenic (causing malformations) and embryotoxic properties in animal experiments. Therefore, alkaloid levels in potato and tomato should be adequately monitored. In case of severe diarrhoea with abdominal pain an MD should not only consider *Salmonella* infection, Botulinus poisoning, or appendicitis, but should take poisoning from potato alkaloids into consideration.

MODE OF ACTION Steroidal alkaloids behave like saponins. They can insert their lipophilic steroid moiety into the lipid bilayer of biomembranes complexing cholesterol, whereas they bind to glycolipids and glycoproteins with their sugar side-chain. This behaviour causes tension and transient holes can form in the membrane, which thus becomes leaky. If the dose is high enough, a general cytotoxic effect can be seen. This property also explains the strong skin irritation seen on mucosa and the antibacterial and antifungal properties known from saponins.

The alkaloids also contain a tertiary nitrogen atom, which is protonated under physiological conditions. This part of the molecule can interact with neureceptors. In addition, the alkaloids inhibit choline esterase that breaks down ACh in the synapse. Therefore, the *Solanum* alkaloids cause some neuronal effects (including hallucination). If injected, the steroidal alkaloids exhibit similar properties as k-strophantoside and cause death from central paralysis of the CNS.

USES (PAST AND PRESENT) *Solanum* alkaloids have been used in agriculture as an insecticide. Several *Solanum* species are part of traditional medicine. The Solanaceae and the genus *Solanum* have several species that are used as food plants. *Solanum dulcamara* had been used by Germanic tribes as narcotic; it was placed into the bed of children to protect them against evil spirits and magic. Extracts were included in witches' potions.

FIRST AID Give medicinal charcoal to absorb the alkaloids; administration of plenty of warm black tea, and of Na^+ sulphate. **Clinical therapy:** In case of ingestion of larger doses: gastric lavage, instillation of medicinal charcoal and sodium sulphate, electrolyte substitution, control of acidosis with sodium bicarbonate. In case of spasms inject diazepam. In case of severe poisoning, provide intubation and oxygen respiration. Observe kidney/blood functions.

Solanum dulcamara

Solanum (Lycopersicon) esculentum

Amygdalin, other cyanogenic glucosides and HCN

STRUCTURE Cyanogenic glucosides (CG) comprise a widely distributed group of SM with about 60 structures with similar pharmacological properties. Amygdalin ($C_{20}H_{27}NO_{11}$, MW 457.44); prunasin ($C_{14}H_{17}NO_6$, MW 295.30); linamarin ($C_{10}H_{17}NO_6$, MW 247.25).

Cyanogenic glucosides are stored as glucosides in the vacuoles of seeds, leaves and roots. When a plant is wounded (e.g. by herbivores or during food processing) the cellular compartmentation breaks down. This brings the cyanogenic glucosides into contact with an enzyme (β-glucosidase, emulsin) which immediately splits the glucosides. After spontaneous hydrolysis or enzymatic cleavage by nitrilase, three compounds are set free: a sugar, an aldehyde and most importantly, the very toxic hydrocyanic acid (HCN). Cyanogenic glucosides can thus be classified as prefabricated defence compounds that release the toxin only after enzymatic activation. Cyanogenic glucosides can also be cleaved and converted to HCN to some degree by gastric acid in the stomach. Prunasin occurs in all plant parts, whereas amygdalin is restricted to seeds.

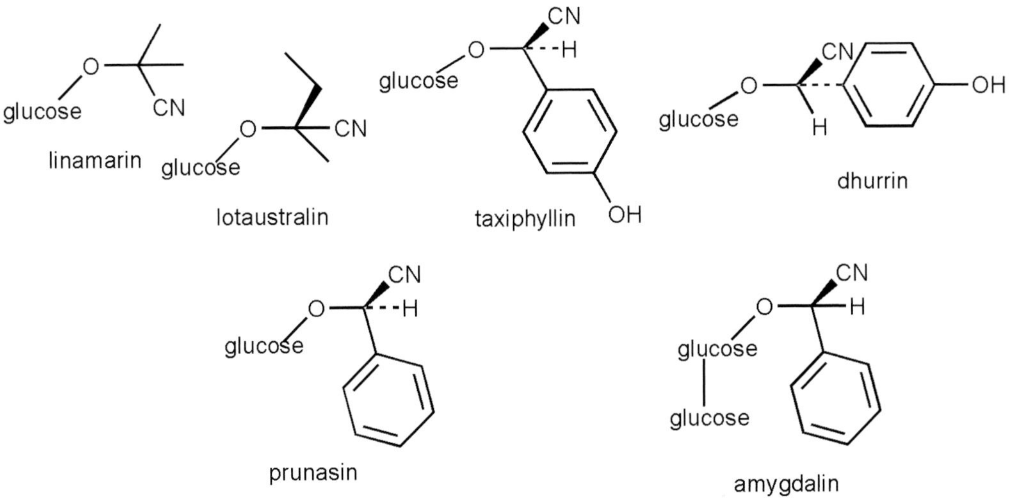

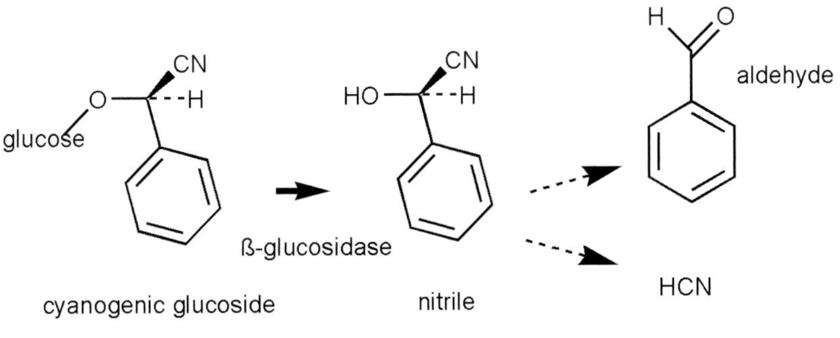

OCCURRENCE Cyanogenic glucosides are widely distributed defence compounds in plants (more than 2 000 cyanogenic species in 100 families), especially abundant in Rosaceae (*Amelanchier, Cotoneaster, Cydonia, Malus, Prunus, Pyrus, Sorbaria, Sorbus*), Fabaceae (*Acacia, Dorycnium, Holocalyx, Lotus corniculatus, Ornithopus, Phaseolus lunatus, Piptadenia, Tetragonolobus, Trifolium repens*), Euphorbiaceae (*Acalypha indica, Manihot esculenta, Phyllanthus*), Caprifoliaceae (*Lonicera*), Poaceae (*Dendrocalamus, Hordeum, Sorghum*), Campanulaceae (*Campanula*), Dioscoreaceae (*Dioscorea bulbifera*), Linaceae (*Linum usitatissimum*), Asteraceae (*Anthemis altissima, Gerbera jamesonii*); Lamiaceae (*Perilla frutescens*), Passifloraceae (*Adenia, Deidamia, Passiflora*), Ranunculaceae (*Aquilegia*), Sapindaceae (*Cardiospermum, Heterodendron*), Juncaginaceae (*Triglochin maritima*). Cyanogenic glucosides also occur in ferns, fungi, algae and bacteria and even some animals (defence secretions of arthropods).

CLASSIFICATION Cellular poison; highly hazardous, Ib

SYMPTOMS First signs of poisoning include burning and scratching of the throat, upper respiratory tract and the eyes, followed by severe headache, anxiety, dyspnoea, extensive saliva and tear formation, vomiting, gastroenteritis, vertigo, increased feeling of weakness, spasms, coma and death by respiratory arrest. Death comes quickly after the ingestion of large doses of HCN. The colour of skin and lips is pink at the beginning but develops blue coloration (cyanosis) in later stages. If patients have ingested bitter almonds or other Rosaceae, their breath smells typically of benzaldehyde. HCN can be detected in breath and in blood by special assays. If a patient has survived HCN intoxication, symptoms usually disappear without further consequences. Long-term diseases related to dietary cyanide intake include (i) konzo, an upper motorneuron disease characterised by irreversible but non-progressive symmetric spastic paraparesis with an abrupt onset; (ii) tropical ataxic neuropathy (TAN), a term used to describe several neurological syndromes whose clinical features include optical atrophy, angular stomatitis, sensory gait ataxia and neurosensory deafness; (iii) goitre and cretinism, which are caused by iodine deficiency, can be considerably aggravated by a continuous dietary cyanide exposure. These diseases occur in countries where cassava is consumed as a staple food and dietary intake of protein and/or iodine is inadequate.

TOXICITY HCN: LD_{50} mouse: 4.2 mg/kg p.o.; rabbit: 2.7 mg/kg; rat: 10–15 mg/kg p.o.; guinea pig: 3 mg/kg, p.o. For humans 1 mg/kg HCN or 2 mg/kg KCN or NaCN are lethal; thus 50 mg HCN or 150–250 mg KCN are a lethal dose. Potentially toxic is plant material containing more than 20 mg HCN/100 g FW.

Although leaves of several plants also accumulate CG, they are hardly eaten by humans and therefore not important from a toxicological perspective. However, seeds of several *Prunus* and *Malus* species contain relatively high amounts of CG. Ingestion of these seeds can be dangerous. For example, deadly for humans are 1 seed of bitter almonds/kg body weight as well as 1 g/kg body weight crushed seeds of apples/pears.

Unintended poisoning is not uncommon in cases when food plants with cyanogenic glucosides are not properly prepared (such as bamboo shoots, *Manihot*, *Phaseolus lunatus*, *Prunus dulcis*).

MODE OF ACTION HCN is quickly absorbed from the GI tract and the skin by free diffusion and can enter all cells of the body. HCN forms a complex with enzymes containing Fe^{3+}. One of the most important enzymes of the mitochondrial respiratory chain is cytochrome oxidase, which requires iron ions (Fe^{3+}) for its enzymatic activity. Blocking this enzyme leads to an immediate stop of cellular respiration; thus ATP can no longer be made. As the demand for ATP is high in the body, HCN blocks all cellular processes that require energy. Cells accumulate lactate under these conditions, which causes a pronounced lactacidosis. A lack of energy quickly leads to paralysis of the brain, especially of the breathing centre. Death occurs within half an hour to 1 hour after ingestion of plant material with cyanogenic glucosides. HCN can be converted to rhodanide with help of the enzyme rhodanase. Giving thiosulphate in case of poisoning can enhance this process. The injection of Fe^{3+} can help to overcome HCN poisoning because the formation of the HCN-Fe^{3+} complex is reversible.

USES (PAST AND PRESENT) HCN, plants with cyanogenic glucosides or laurel-cherry water have often been used for murder and suicide. Laetrile, an apparently ineffective anticancer drug based on amygdalin, which was and still is widely used in the USA, has led to several cases of severe HCN poisoning. It had been termed vitamin B_{17}. TCM uses amygdalin as an antitussive agent. HCN has been used in agriculture to kill insects in closed environments; it is also a common chemical in organic synthesis.

FIRST AID Quick action is required. Induce vomiting and perform gastric lavage, possibly with 300 ml of 0.2% potassium permanganate, then 5% thiosulfate or 1% H_2O_2 solution; or inject sodium thiosulfate (10%; 10–20 ml i.v.), which helps the body to detoxify HCN to rhodanide. In severe cases (patient unconscious): immediately inject 3.25 mg/kg 4-dimethylaminophenol (DMAP) i.v. (otherwise i.m.); then inject sodium thiosulphate (10%; 100 ml i.v.). Control acidosis with sodium bicarbonate. In severe cases, provide intubation, oxygen respiration, Cobalt-EDTA (Kelo-cyanor). Treat lactacidosis with sodium bicarbonate or Tris-buffer (THAM). Check for lung oedema.

Mustard oils / glucosinolates

STRUCTURE More than 80 different glucosinolates have been found in higher dicotyledonous plants. Glucosinolates are stored as glucosides in the vacuoles of seeds leaves, and roots. When a plant is wounded (e.g. by herbivores or during food processing), the cellular compartmentation breaks down. This brings the glucosinolates into contact with an enzyme (thioglucosidase, myrosinase) which immediately splits the glucosides. After further chemical rearrangements, 3 compounds are set free: a sugar, sulphate and most importantly, the reactive isothiocyanate (mustard oils). Sometimes other metabolites are generated, such as thiocyanates, nitriles or goitrines. Glucosinolates can be classified as prefabricated defence compounds that release the toxin only after enzymatic activation.

Glucosinolate	Isothiocyanate
glucocapparin	methyl isothiocyanate
sinigrin	allyl isothiocyanate
glucotropaeolin	benzyl isothiocyanate

OCCURRENCE Common secondary metabolites of the Brassicales, which comprise among others the families Brassicaceae (*Alyssum, Armoracia, Brassica, Conringia, Crambe, Iberis, Lepidium, Nasturtium, Raphanus, Sinapis, Wasabia*), Capparaceae (*Capparis, Cleome*), Tropaeolaceae (*Tropaeolum*), Resedaceae (*Reseda*), Moringaceae, Tovariaceae and Stegnospermataceae. All plant parts may accumulate glucosinolates, but seeds and roots are often especially rich in these allelochemicals. Concentrations are in the range of 0.1–0.2% FW in leaves, up to 0.8% in roots and up to 8% in seeds. Glucosinolates serve as feeding deterrents against insects and vertebrates. In addition, the isothiocyanates exhibit antibacterial and antifungal properties.

CLASSIFICATION Cellular poison, irritant; moderately to highly hazardous, Ib–II

SYMPTOMS Mustard oils are volatile and pungent and strong irritants to skin, mucosal tissues of mouth, throat, GI tract and eye; because of their lipophilicity, they are easily absorbed through the skin. They cause burns of first and second degree. Ingestion of larger doses causes burning, abdominal pain, vomiting, convulsions, unconsciousness, decrease of blood pressure and even death. When applied (25 mg/kg) to male rats for 2 years, animals developed papillomas of the bladder; thus a tumour promoting activity cannot be ruled out. Nitriles cause damage to kidneys and liver. Thiocyanates and goitrins are thyreostatic and can promote the development of a goitre.

Intoxication of livestock (especially horses and ruminants) sometimes occur when they are fed on rape seed cake or other food items with mustard oil. Ingestion of fresh plants leads to gastro-enteritis, inflammation of kidneys, colic and diarrhoea.

TOXICITY Serious skin damage has become less common since medicinal applications of mustard oil on skin are hardly used in practice. Poisoning of livestock is more common, especially when rape seed cake (which is left over after oil pressing) is fed to animals. LD_{50}: Allylisothiocyanate: Mouse: 108.5 mg/kg p.o., 69.5 mg/kg after inhalation; rats (p.o.) 25 to 200 mg/kg.

MODE OF ACTION Mustard oils, like capsaicin, activate TRP channels (transient receptor potential family of ion channels). In addition, isothiocyanates are reactive compounds that can form covalent bonds with SH-, NH_2-groups of proteins. The modified proteins change their conformation and intrinsic activity. If many proteins are treated in such a way, cells die and inflammation starts, usually resulting in blister formation. Goitrin (5-vinyl-2-oxazolidinethione), which derives from progoitrin in most brassicas, inhibits the incorporation of iodine into thyroxine precursors and interferes with its secretion.

USES (PAST AND PRESENT) Plants with glucosinolates are often used as spices or food plants; some of them have been employed in traditional medicine to treat rheumatism and bacterial infections. Methylisothiocyanate is used in agriculture to kill pathogens in the soil.

FIRST AID Contact with eyes or skin: wash with polyethylene glycol or water and soap. In case of skin blisters, cover with sterile cotton and locally apply corticoid ointment. Eye: apply 1–2 drops of Chibro-Kerakain as a painkiller; then wash for 10 min with water and sodium bicarbonate. Consult an ophthalmologist. Ingestion of larger doses: Drink plenty of water and then give medicinal charcoal. Observe respiration; provide shock prophylaxis. **Clinical therapy:** After overdosing: Gastric lavage with 0.1 % permanganate, instil medicinal charcoal and sodium sulphate. Provide plasma expander when patient suffers from shock; possibly intubation and artificial respiration. Check circulation, kidney and liver function. If mustard oils have been inhaled, treat with dexa- methason spray.

Iberis amara

Nasturtium officinale

Cleome spinosa

Tropaeolum tuberosum

Non-protein amino acids (NPAAs)

STRUCTURE Whereas proteins of all organisms are built up by the 20 common *L*-amino acids, more than 600 other amino acids have been discovered in plants that are not building blocks of proteins (therefore called "non-protein amino acids").

Protein amino acid	NPAA	Occurrence
Arginine	canavanine indospicine	many legumes *Indigofera*
Proline	azetidine-2-carboxylic acid	*Polygonatum, Convallaria, Drimia, Delonix*
Ornithine	canaline	*Canavalia ensiformis*
Asparagine	3-cyanoalanine	*Vicia*
Lysine	*S*-aminoethylcysteine	*Acacia*
Glutamine	albizziine	*Albizia*
Methionine	*Se*-methylselenocysteine selenocystathionine	*Astragalus, Oonopsis* *Stanleya, Lecythis, Astragalus, Neptunia*
Cysteine	djenkolic acid S-methylcysteine sulphoxide	*Pithecolobium, Albizia, Mimosa, Acacia,* *Brassica*
Alanine	β-alanine	*Iris, Lunaria*
Tyrosine	mimosine	*Leucaena*

A special case can be found in *Allium* species (Alliaceae); the NPAA alliin is converted into a reactive metabolite allicin ($C_6H_{10}OS_2$, MW 162.28) that can bind to SH-groups of proteins. Several other SH reagents are present in garlic or onion oil: a related derivative without an *S*-oxide group is diallyl disulphide ($C_6H_{10}S_2$, MW 146.28), which exhibits insecticidal and antifungal properties. Another product of garlic oil is diallyl sulphide ($C_6H_{10}S$, MW 114.21), a strong irritant to eyes and skin. Propanethial *S*-oxide (C_3H_6SO, MW 90.15) derived from *S*-propenylcysteine *S*-oxide occurs in onion (*Allium cepa*) and is responsible for the main lachrymatory activity when onions are cut or bruised.

OCCURRENCE NPAAs are especially abundant in the family Fabaceae but are also present in the seeds of several monocots (Alliaceae, Iridaceae, Hyacinthaceae), Cucurbitaceae, Euphorbiaceae, Resedaceae, Sapindaceae and Cycadaceae. NPAAs are also toxic components of some fungi (e.g. coprine in *Coprinus* species). Organs rich in these metabolites are seeds (Fabaceae) or rhizomes (monocots). Concentrations in seeds can exceed 8% of dry weight and up to 50% of the nitrogen present can be attributed to them. Since NPAAs are often (at least partly) remobilised, during germination they certainly function as nitrogen storage compounds in addition to their role as defence chemicals against herbivores and microbial pathogens.

CLASSIFICATION Cell toxin, neurotoxin; moderately hazardous, II–III

SYMPTOMS *S*-Methylcysteine sulphoxide, which occurs in *Brassica* species, is converted in the rumen of ruminants to dimethyldisulphide. This metabolite has haemolytic activities and can cause anaemia ("kale poisoning").

Symptoms of neurolathyrism are muscular weakness, paralysis of the legs and, occasionally, the arms, bladder and bowel, tremors, convulsions, psychic disturbances and even death. *Lathyrus sativus* has been imported to India as a lentil substitute and neurolathyrism was common in times of famine in India. *L. odoratus* causes a bone disorder (osteolathyrism). Jackbean, *Canavalia ensiformis*, contains L-canavanine and causes haemorraghic enteropathies.

L-Indospicine from *Indigofera spicata* causes liver cirrhosis and malformations in animals.

Mimosine from *Leucaena leucocephala* is a tyrosine mimic and causes of hair loss, weight loss, anorexia, general malaise, eye inflammation and foetal malformations in animals.

Se-Methylselenocysteine and selenocystathionine cause selenosis with infertility, abdominal pain, nausea, vomiting and diarrhoea. About 1 to 2 weeks after consumption, a reversible loss of scalp and body hair occurs. Acute selenium poisoning causes "blind staggers" in livestock, characterised by staggering, foaming at the mouth, strong pains and respiratory arrest within 24 h.

Hypoglycine and its derivates occur in akee fruits from the West Indies (*Blighia sapida;* Sapindaceae) and *Billia hippocastanum* (Sapindaceae). They are very toxic and cause a syndrome known as "vomiting sickness". Symptoms include hypoglycaemia, violent retching, vomiting, convulsions, coma and even death.

Allicin that is released from alliin is a reactive sulphhydryl reagent that can bind to SH-groups of proteins; allergenic reactions are a plausible consequence in members of the genus *Allium*.

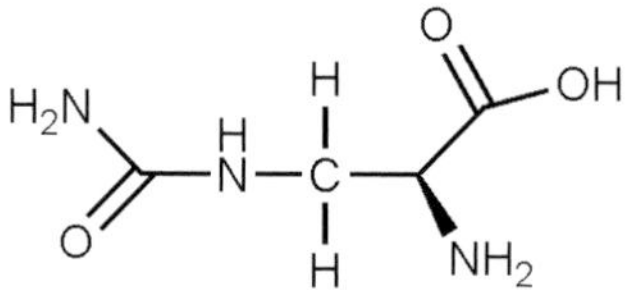

L-albizziine 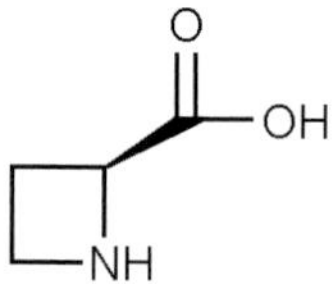   L-azetidine 2-carboxylic acid 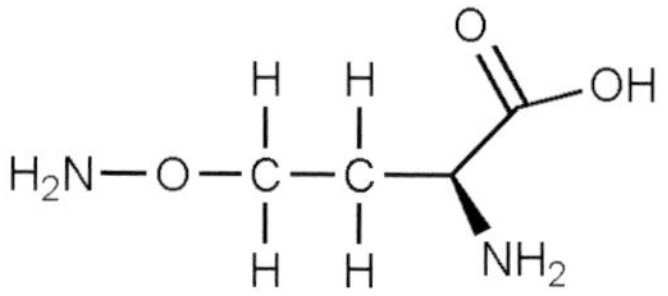   L-canaline

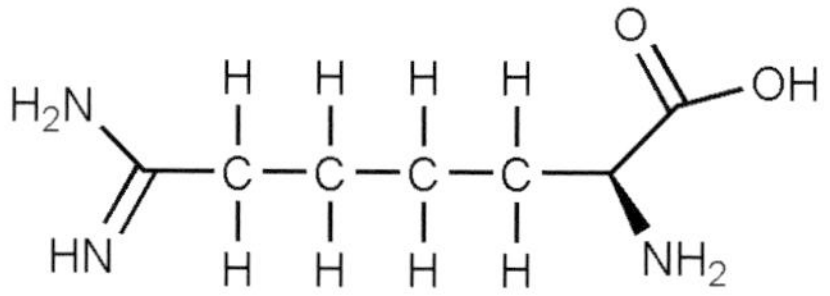

L-indospicine

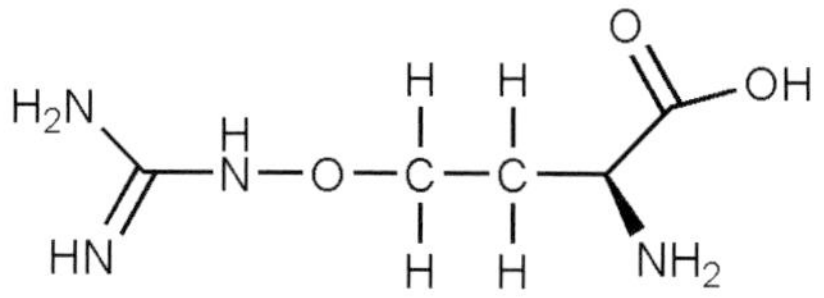

L-canavanine 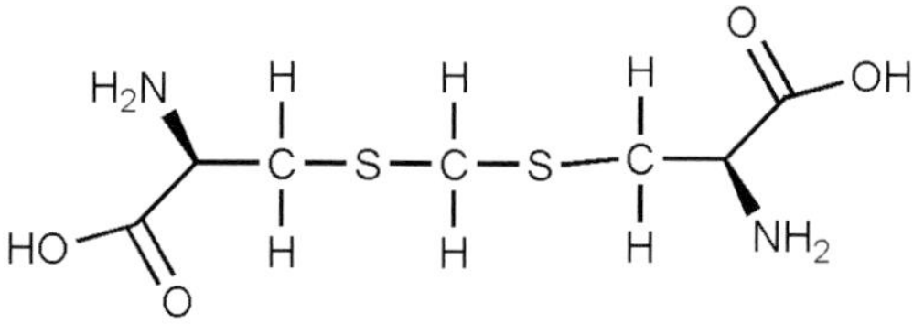   djenkolic acid

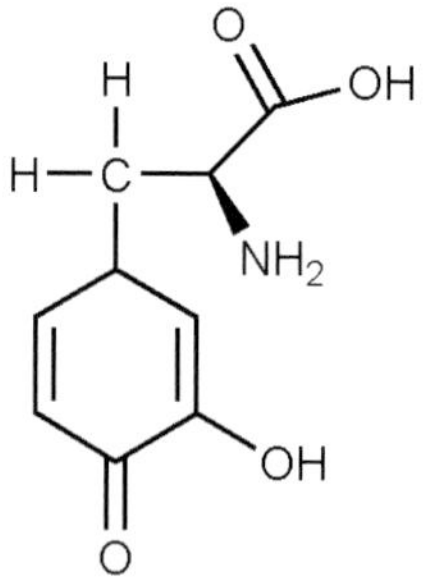

L-mimosine

alliin   alliinase   allicin

TOXICITY Some NPAAs from *Lathyrus latifolius, L. sativus*, and *L. sylvestris* and *Vicia* species such as L-β-cyanoalanine, L-α-γ-diamino butyric acid, α-amino-β-oxalyl aminopropionic acid β-*N*-(γ-L-glutamyl)-aminopropionic acid, cause a neuronal disorder (neurolathyrism) in humans and livestock when eaten in large quantities. If seeds of *L. sativus*, which contain up to 2.5% α-amino-β-oxalyl aminopropionic acid, are not detoxified by extensive leaching in water, neurolathyrism can occur.

Indospicine shows teratogenic, abortifacient and hepatotoxic properties in animals. A cleft palate and dwarfism have been observed in rats.

3-Methylamino-L-alanine from *Cycas* species (Cycadales) is very toxic to animals, causing convulsions and even death.

If cattle or horses feed on large quantities of onions they can develop haemolytic anaemia with "Heinz bodies" in the erythrocytes. Sheep even died when grazing on *Allium ursinum*.

MODE OF ACTION NPAAs often figure as antinutrients or antimetabolites. Many non-protein amino acids resemble protein amino acids and quite often can be considered to be their structural analogues; they may interfere with the metabolism of humans, animals, even microbes and plants:
- In ribosomal protein biosynthesis NPAAs can be accepted in place of the normal amino acid and incorporated into proteins, which often become disturbed in their function.
- NPAAs may competitively inhibit uptake systems for amino acids in the gut.
- NPAAs can inhibit amino acid biosynthesis by substrate competition or by mimicking end product mediated feedback inhibition of earlier key enzymes in the pathway.
- NPAAs may affect other targets, such as DNA-, RNA-related processes (canavanine, mimosine), receptors of neurotransmitters, inhibit collagen biosynthesis (mimosine), or β-oxidation of lipids (L-hypoglycine).

In garlic and onions alliin is converted enzymatically upon tissue damage into allicin, a diallylsulphide. Allylsulphide is quite reactive and can form disulphide bridges with SH-groups of proteins. This would explain the wide range of pharmacological activities (antidiabetic, antihypertensive, antithrombotic and antibiotic properties) that were attributed to garlic. Mimosine is teratogenic and inhibits DNA synthesis.

USES (PAST AND PRESENT) Plants with NPAAs are usually not eaten by humans or used as food for livestock. Some plants with NPAAs are employed in phytomedicine, such as *Sutherlandia frutescens* (contains canavanine). Garlic and onion are widely consumed as spice and vegetables; the allicin is thought to have health promoting properties.

FIRST AID In case of overdose: Induce vomiting, administration of sodium sulphate and medicinal charcoal. **Clinical therapy:** Gastric lavage (possibly with potassium permanganate), instillation of medicinal charcoal and sodium sulphate, diazepam in case of spasms.

Allium sativum

Polygonatum multiflorum

Abrin, ricin other lectins and peptides

STRUCTURE Ricin, abrin, phasin and other lectins are composed of two peptide chains, a haptomer (B-chain) and an effectomer (A-chain). Abrin subunits have a molecular weight of 63 and 67 kDa. Ricin is a mixture of four lectins (RCL1-4). RCL 4 has a structure similar to that of abrin and a MW of 62 kDa. These lectins are very stable molecules and can resist degradation by proteases.

OCCURRENCE Lectins are small glycosylated proteins which are common in seeds of several plants, such as abrin in *Abrus precatorius*, phasin in *Phaseolus vulgaris*, robin in *Robinia pseudoacacia* (Fabaceae), ricin in *Ricinus communis* (Euphorbiaceae), modeccin in *Adenia digitata* (Passifloraceae). Less toxic lectins occur in seeds of several plants, especially of legumes and mistletoe. Some of them contribute to allergic properties of a plant, such as peanut lectin (PNA) in peanut seeds (*Arachis hypogaea*), ragweed pollen allergen (Ra5) from *Ambrosia artemisiifolia*. In plants, seed lectins serve as defence compounds and nitrogen storage compounds that are remobilised during germination.

Seeds of several plants accumulate other small peptides such as **protease inhibitors**. They inhibit the activity of intestinal proteases, such as trypsin and chymotrypsin.

Some plants are rich in **hydrolytic proteases**, such as bromelain in *Ananas comosus*, ficin in *Ficus glabrata*, papain in *Carica papaya*. In case of wounds these proteases can produce adverse effects. Since lectins and the other peptides are proteins, they can be destroyed by heat, i.e. cooking in water at more than 65 °C.

CLASSIFICATION Cell poison; highly hazardous, Ia

SYMPTOMS First toxic effects of lectin poisoning set in after a latency period of some hours, even days. Symptoms include nausea, vomiting, serious gastroenteritis with bloody diarrhoea (with lipid granules), general weakness, tachycardia, abdominal pains, and severe fluidity loss that can cause circulatory collapse. In case of severe intoxication further symptoms are mydriasis, spasms of hands and legs, fever, signs of liver, kidney and pancreas failure. Death is caused by paralysis of central nervous breath centre or uraemia after 3–4 days, sometimes even later. If humans are exposed to lectins via the skin or lungs, allergic reactions, including anaphylactic shock have been reported. Abrin has a strong ophthalmologic effect, causing opacity of lens, even blindness, glaucoma, inflammation and hypertrophy of upper eyelids, and conjunctivitis.

TOXICITY Ricin: LD_{50} mouse: 0.1 µg/kg i.p.; lethal oral dose for adult humans: 1 mg/kg; 5–6 seeds of *Ricinus* have killed children, 20 seeds adults. Abrin: LD_{50} mouse: 40 µg/kg i.p., 0.7 µg/kg i.v.; rabbit: 0.06 µg/kg i.v. The oral LD_{50} in mice is reported to be 2 mg/kg body weight. About 0.5 g of *Abrus* seeds (ca. 2 seeds) can be fatal in humans; a single seed has caused death in children. 60 g will kill a horse.

MODE OF ACTION The lectins bind to cells via the haptomer (haemagglutinating activity) and become internalised by endocytosis. Once in the cell they have an affinity for ribosomes and the A-chain (which has *N*-glycosidase activity) blocks ribosomal protein translation by inactivating elongation factors EF1 and EF2. A cell that no longer is able to make proteins will die. Lectins are toxic when taken orally, but more toxic when applied intramuscularly or intravenously. Lectins are among the most toxic peptides produced in nature. Other toxic peptides are found in the venom of snakes, spiders, other animals and in some bacteria (causing whooping cough, cholera or botulism).

USES (PAST AND PRESENT) Accidental poisoning from seeds that have been ingested is not uncommon. Seeds of *Ricinus* and *Abrus* look beautiful and have been incorporated into necklaces, toys, masks and ornaments. If seeds have been chewed by children, intoxication might occur. Pure ricin was used to murder a Bulgarian secret agent in London, who was pricked with a needle containing ricin.

FIRST AID Induce immediate vomiting; administration of sodium sulphate and medicinal charcoal. Provide shock prophylaxis (calm, warmth, hot drinks); maybe intubation and cardiac pressure therapy. **Clinical therapy:** Patients should be treated in a hospital even if symptoms are not apparent. Gentle gastric lavage. Electrolyte substitution, control of acidosis with sodium bicarbonate. Control fluidity loss and kidney function, possibly plasma expander. In case of spasms give diazepam (i.v.). In severe cases, artificial respiration and oxygen supply. Patients need to be monitored for several days. Specific antidotes are not available.

Secondary metabolites without nitrogen

This class comprises a heterogenic group, which will be structured by clustering together:
- phenolics
- terpenoids
- alcohol
- quinones
- polyacetylenes
- acids

Phenolics

Most phenolic compounds derive from an aromatic amino acid, such as phenylalanine, tyrosine or tryptophan. The aromatic ring often carries one or several hydroxyl groups. Since these are phenolic hydroxyl groups they can dissociate under physiological conditions and form negatively charged phenolate ions. As mentioned before (p. 247) these hydroxyl groups and the phenolate ions can form hydrogen bonds and ionic bonds with proteins and DNA and thus can change their activity. Other functional groups, such as terminal methylene groups, epoxide or aldehyde groups can add further pharmacological properties.

Flavonoids, isoflavones, alkyl phenols, tannins, lignans and other phenolics

STRUCTURE AND OCCURENCE Compounds with phenolic hydroxyl groups are widely distributed in the plant kingdom.

Phenylpropanoids derive from phenylalanine and tyrosine via desamination and represent building blocks for a wide variety of phenolic compounds. Phenylpropanoids can occur as simple compounds (as phenylacrylic acids, phenylacrylic aldehydes or phenylallyl alcohols) that differ by their degree of hydroxylation and methoxylation.

Examples of simple phenols and phenylpropanoids

Simple phenols. The simple phenol (C_6H_6O) occurs freely in *Elsholtzia*, *Perovskia* (Lamiaceae), *Paedera* (Rubiaceae), *Gossypium* (Malvaceae), *Paeonia* (Paeoniaceae), *Populus* (Salicaceae) and *Morus* (Moraceae). Other simple phenols include *p*-cresol, catechol, 2,6-dimethoxyphenol, gallic acid, guaiacol, orcinol, 4-methylcatechol, phloroglucinol, pyrogallol and resorcinol. They have 2 or 3 phenolic hydroxyl groups, which are either free or methylated.

Alkyl and alkenyl phenols. Some phenols carry long alkyl and alkenyl side chains. They cause severe allergic skin reactions. Alkyl and alkenyl phenols such as urushiol are abundant in Anacardiaceae (*Anacardium, Holigarna, Lithrae, Mangifera, Mauria, Metopium, Rhus, Schinus, Semecarpus*), Hydrophyllaceae (*Phacelia, Turricula*), Proteaceae (Grevillea), *Ginkgo*, and *Philodendron*.

Simple phenylpropanoids with terminal methylene groups are treated under "myristicin", phenols with quinone function under "hydroquinone".

alkylphenols

alkenylphenols

anacardic acid

bilobol

grevillol

urushiol III

Examples of alkyl and alkenyl phenols

Phenylpropanoids can condense with a **polyketide moiety** to form stilbenes, flavonoids, isoflavones, chalcones, catechins, xanthones and anthocyanins. These compounds are characterised by two aromatic rings that carry several phenolic hydroxyl or methoxyl groups. In addition, they often occur as glycosides and are stored in the vacuole. The aglycones figure as the active metabolite in most instances.

Stilbenes are widely distributed among gymnosperms and angiosperms.

Isoflavones, of which more than 600 structures are known, are typical secondary metabolites of Fabaceae (especially in the subfamily Papilionoideae) and many of them exhibit phytoestrogenic activities, thus mimicking the activity of the female sex hormone estradiol. Isoflavones have limited distribution in Asteraceae, Iridaceae (*Iris*), Myristicaceae (*Osteophleum, Virola*), Amaranthaceae/Chenopodiaceae (*Spinacia*), Moraceae (*Maclura*) and Rosaceae (*Cotoneaster*). A few isoflavonoids are toxic, such as rotenone, which has been found in *Derris, Tephrosia, Lonchocarpus* (Fabaceae) and *Verbascum thapsus* (Scrophulariaceae). Other isoflavones are known phytoalexins with antifungal properties; they are induced after infection with pathogens.

Biflavonyls are typical for gymnosperms; they rarely occur in angiosperms. Some of them are inhibitors of phosphodiesterase.

Anthocyanins. The colour of anthocyanins depends on the degree of glycosylation, hydrogen ion concentration and the presence of certain metals (e.g. aluminium ions) in the vacuole. Parallel to a change in pH of the vacuole in developing flowers, a colour change from pink to dark blue can be observed in several species of the Boraginaceae (e.g. *Symphytum, Echium*). Anthocyanins, which are present in almost all angiosperm genera, are hardly of toxicological relevance.

Non-hydrolysable tannins – catechins form a special class of flavonoids, which often dimerise or even polymerise to form procyanidins and oligomeric procyanidins. The conjugates (which cannot be hydrolysed; "non-hydrolysable tannins") are characterised by a large number of phenolic hydroxyl groups.

Hydrolysable tannins. Another important group of tannins is hydrolysable (so-called hydrolysable tannins). They represent esters of gallic acid and sugars; in addition several moieties of gallic acid can be present that are also linked by ester bonds. Gallic acid can additionally be condensed with catechins.

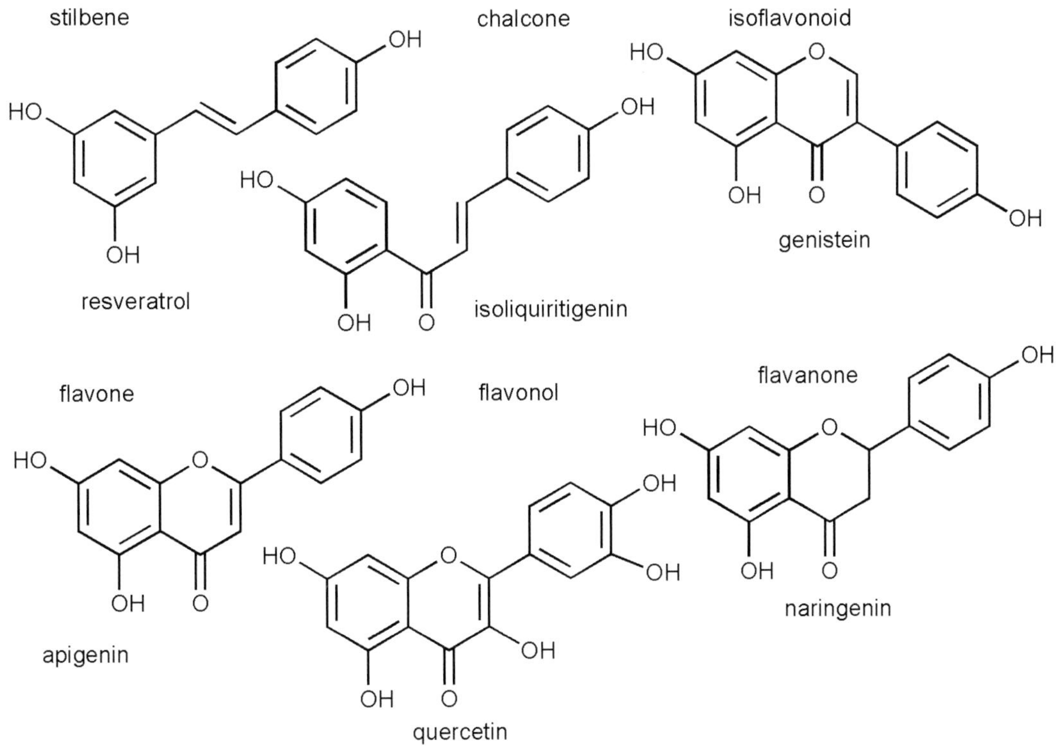

Examples of stilbenes, chalcones, isoflavonoids and flavonoids

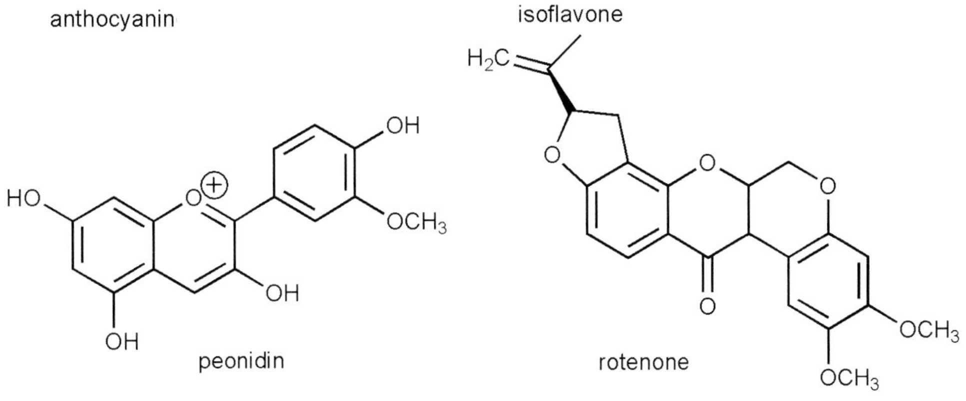

Examples of anthocyanins and rotenoids

The gallotannins are widely distributed in plants, often in bark, leaves and fruits. Especially rich in gallotannins are galls induced by various gall-forming insects. Gallotannins contain a large number of phenolic hydroxyl groups so that they can form stable protein-tannin complexes and thus interact with a wide variety of protein targets in microbes and animals (see also catechins; last paragraph).

Lignans, of which more than 200 structures are known, are dimers of phenylpropanoids. Many of them have pronounced pharmacological and toxicological properties. Two types are widely distributed: podophyllotoxin-like and pinoresinol-like structures.
Compounds with a podophyllotoxin skeleton are found in *Linum* (Linaceae), *Austrobaileya* (Austrobaileyaceae), *Amyris, Haplophyllum* (Rutaceae), *Cleistanthus, Phyllanthus* (Euphorbiaceae), *Bursera* (Burseraceae), *Polygala* (Polygalaceae), *Hyptis* (Lamiaceae), *Diphylleia, Podophyllum* (Podophyllaceae), *Anthriscus* (Apiaceae), *Hernandia* (Hernandiaceae), *Callitris, Juniperus, Libocedrus* (Cupressaceae), *Justicia* (Acanthaceae) and *Sesbania* (Fabaceae).

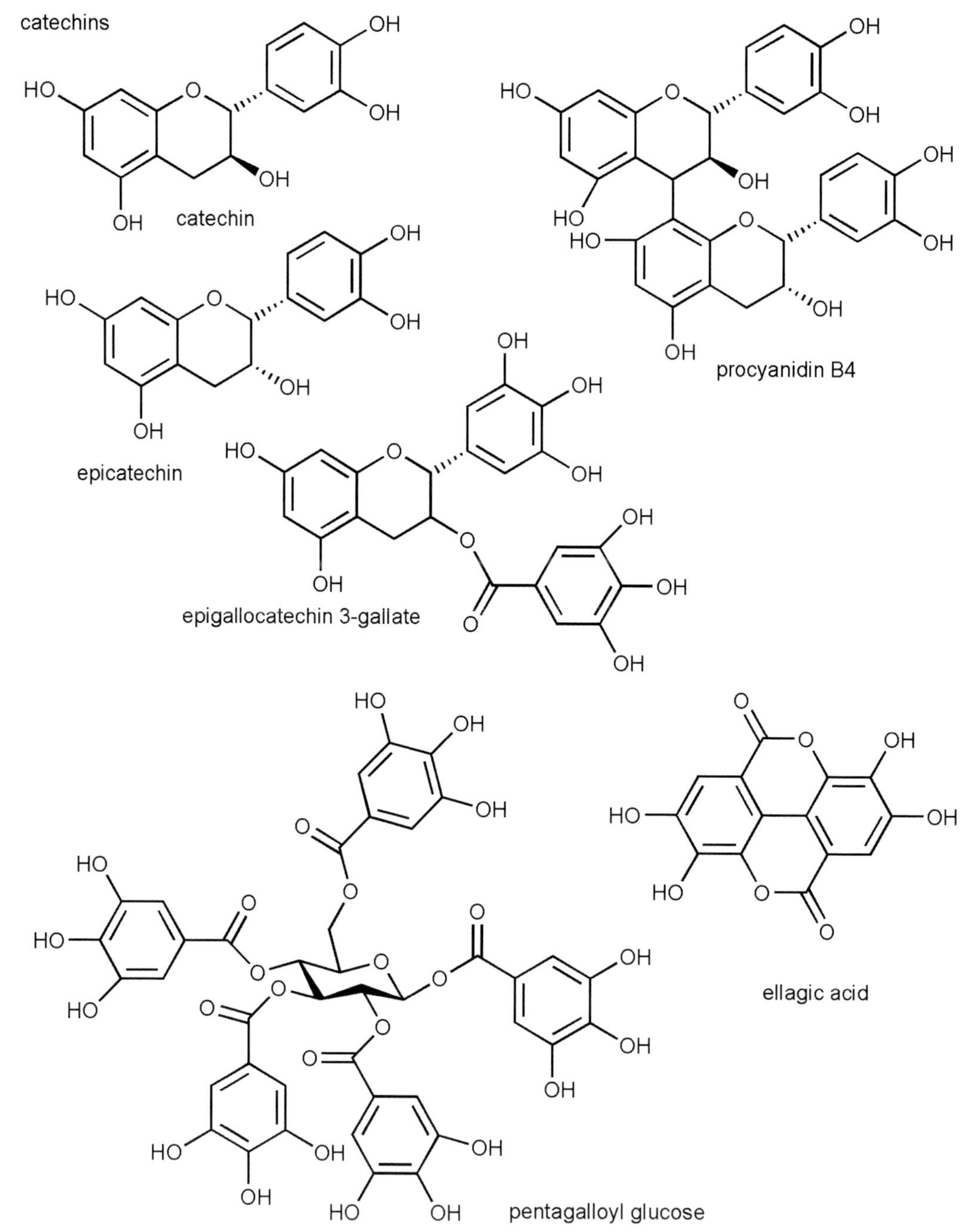

catechins

catechin

epicatechin

procyanidin B4

epigallocatechin 3-gallate

ellagic acid

pentagalloyl glucose

Examples of tannins

Compounds with a piniresinol skeleton occur in *Acanthopanax, Eleutherococcus* (Araliaceae), *Liriodendron* (Magnoliaceae), *Paulownia, Penstemon* (Scrophulariaceae), *Fraxinus, Olea* (Oleaceae), *Eucommia* (Eucommiaceae), *Allamanda* (Apocynaceae), *Eucalyptus, Virola* (Myrtaceae), *Araucaria* (Araucariaceae), *Evodia, Fagara, Haplophyllum, Ruta, Zanthoxylum* (Rutaceae), *Litsea* (Lauraceae), *Magnolia* (Magnoliaceae), *Phryma* (Phrymaceae), *Picea, Pinus, Tsuga,* (Pinaceae), *Sesamum* (Pedaliaceae), *Styrax* (Styracaceae), *Populus* (Salicaceae) and *Fagus* (Fagaceae).

CLASSIFICATION Skin irritant, cytotoxin; moderately to slightly hazardous, II–III

Examples of lignans

podophyllotoxin

α-peltatin

pinoresinol

SYMPTOMS Alkyl phenols: extremely allergenic compounds that are responsible for over 2 million poisoning instances (*Rhus* dermatitis) in USA. Inflamed skin areas rapidly develop dermatitis with swelling, reddening, fever, pain and unpleasant itching. *Rhus* dermatitis is long-lasting and can persist for weeks and months. Even years later, parts of the affected skin can develop a relapse of dermatitis. Contact with the eye is extremely hazardous and can lead to blindness. After oral ingestion of *Rhus* parts, a strong irritation can be seen on mucosa of mouth, throat and GI tract.

Symptoms of oral **phodophyllotoxin** poisoning include a strong irritation of mucosa and eyes, serious gastroenteritis with abdominal colic, vomiting, long-lasting sometimes bloody diarrhoea, cough, tachycardia arrhythmia, paraestesia in legs, polyneuritis, damage of central and peripheral nervous system, ileus, vaginal bleeding, nephritis, spasms, delirium, coma and death from respiratory arrest.

TOXICITY Quercetin: LD_{50} mouse: 160 mg/kg p.o. Several polyphenols can induce mutations in Ames test, but also show antimutagenic properties because of their high antioxidative activities.

Simple phenols are strongly irritating; 2–4 g are toxic, 10–15 g lethal in humans. Rotenone: LD_{50} mouse i.p. 2.8 mg/kg. About 300–600 mg podophyllin is a lethal dose in humans. Podophyllotoxin: LD_{50} rats: 8.7 mg/kg i.v., 15 mg/kg i.p.

MODE OF ACTION Phenolic hydroxyl groups in polyphenols can dissociate under physiological conditions and generate O^- ions. The polyphenols can interact with proteins by forming several hydrogen bonds and ionic bonds. Thereby they modulate the conformation of proteins (such as enzymes, receptors, ion channels), so that proteins cannot perform their function as free proteins. Polyphenols exhibit pleiotropic effects as they will interact with any protein they encounter. Cellular effects are among others inhibition of cell growth by inducing apoptosis and blocking of telomerase. Many phenolics are bitter or astringent, antioxidant; they exhibit antibacterial, antifungal and antiviral activities.

Simple phenolics can pass lipid membranes and can denature proteins, ion channels, enzymes and lipids whereas larger tannin molecules are not absorbed from the GI tract. The higher the number of phenolic hydroxyl groups, the stronger the astringent and denaturing effect. Tannins form stable protein–tannin complexes and inhibit enzymatic activities very effectively.

Stilbenes, such as resveratrol, have antioxidant, antibacterial and antifungal activities, but are hardly toxic for humans. Isoliquiritigenin and glyceollin II inhibit mitochondrial monoamine oxidase and uncouple mitochondrial oxidative phosphorylation.

The alkyl and alkenyl phenols such as urushiol can dive into biomembranes or proteins with their lipophilic side chain exposing the phenolic moiety with the reactive hydroxyl groups to the external side so that they can form non-covalent and even covalent bonds with membrane proteins.

Rotenone inhibits the mitochondrial respiratory chain and is therefore highly toxic. The chalcone *O*-glycoside phloridzin from *Malus* (Rosaceae) and *Kalmus, Pieris, Rhododendron* (Ericaceae) inhibits glucose transport at biomembranes. It can cause glycouria by interfering with tubular glucose reabsorption.

Podophyllotoxin and related lignans are antimitotic and strongly purgative. They also exhibit antiviral, antitumour, and piscicidal activities. Pinoresinol and related compounds are inhibitors of cAMP phosphodiesterase and are cytotoxic, insecticidal and immune modulating.

Tannins are strong antioxidants, with anti-inflammatory, antidiarrhoeal, cytotoxic, antiparasitic, antibacterial, antifungal and antiviral activities. Ingestion of large amounts can cause disturbance of GI tract and of liver metabolism in humans.

USES (PAST AND PRESENT) Most medicinal plants contain some sort of polyphenol as active principle. Several polyphenols are used as antioxidant and radical scavengers in drugs and nutraceuticals. Some have anti-inflammatory, antidepressant, hypotensive, antihepatotoxic (silybin, silandrin, silychristin), antimutagenic, antispasmodic and expectorant activities. Whereas many phenolics are bitter (e.g. naringin, neoeriocitrin, neohesperidin), some have a sweet taste, such as the dihydroflavonols taxifolin 3-O-acetate, 6-methoxytaxifolin and 6-methoxyaromadendrin 3-O-acetate.

Rotenone and related compounds are used as insecticides and piscicides. It can be used as a fish poison because rotenone has a low oral toxicity in humans.

Fruits of the S American *Schinus molle* and *Schinus terebinthifolius* are used as spices because of a peppery taste (pink pepper). They can irritate mucosa.

FIRST AID After overdose, instil medicinal charcoal. **Clinical therapy:** In severe cases: gastric lavage (possibly with 0.1% potassium permanganate), instillation of medicinal charcoal and sodium sulphate, electrolyte substitution; possibly artificial respiration and oxygen therapy.

Alkyl phenols: Contact with eyes or skin: wash with water, 0.1% potassium permanganate or polyethylene glycol; in case of skin blisters, cover with sterile cotton and locally with corticoid oinment. Itching can be treated with antihistaminic medicine (see *Rhus*).

Glycine max

Podophyllum hexandrum

Linum flavum

Trifolium vulgare

Coumarins and furanocoumarins

STRUCTURE Phenylpropanoids serve as building blocks for coumarins and furanocoumarins (FC) of which over 700 structures have been determined. Simple coumarins are stored as coumaroylglucosides (e.g. melitoside) in the vacuole. Upon tissue damage these glucosides are hydrolysed by a β-glucosidase. After isomerisation and formation of a lactone ring, simple coumarins are generated.

melitoside (in vacuole) → β-glucosidase → spontaneous → coumarin: umbelliferone

psoralen, bergapten, xanthotoxin, imperiatorin, angelicin, pimpinellin, visnadin

Furanocoumarins are divided into linear and angular structures.

OCCURRENCE Whereas simple coumarins occur in more than 100 plant families, furanocoumarins are more restricted but widely distributed in Apiaceae (genera *Agasyllis, Apium, Ammi, Anethum, Angelica, Anthriscus, Athamanta, Cachrys, Cnidium, Coelopleurum, Cymopterus, Daucus, Ferula, Foeniculum, Heracleum, Levisticum, Libanotis, Ligusticum, Meum, Myrrhis, Pastinaca, Peucedanum, Petroselinum, Pimpinella, Prangos, Selinum, Seseli, Smyrniopsis, Sphenosciadium, Zizia*), Rutaceae (*Aegle, Chloroxylon, Citrus, Fagara, Flindersia, Haplophyllum, Heliettia, Luvunga, Myrtopsis, Phebalium, Poncirus, Psilopeganum, Ptelea, Ruta, Skimmia, Thamnosma, Xanthoxylum*), Fabaceae (*Bituminaria, Coronilla, Psoralea*), Fagaceae (*Castanopsis*), Moraceae (*Brosimum, Ficus*), Pittosporaceae, Solanaceae (*Lycopersicon*), Rosaceae (*Fragaria*), Asteraceae (*Artemisia, Trichocline*), and Thymelaeaceae. FC occur in all plant organs and are localised in excretory ducts or on leaves surfaces (as in Rutaceae).

FC help plants as defence chemicals against insects and fungal pathogens and are synthesised under stress conditions (phytoalexins). If *Pastinaca sativa* or *Apium graveolens* are stored wounded, infected or under stress conditions, they can generate high FC contents that might be toxic upon consumption. FC accumulate in peel of citrus fruits, not in fruit flesh. Essential oils do not contain FC, whereas pressed oils do (FC content in citrus oil 35 mg/kg). Therefore, fruit juices and cosmetics that include pressed citrus oils can contain phototoxic FC and produce photodermatitis. Coumarin levels can reach concentrations of up to 2% in plants and are common in the Apiaceae (most genera), and in certain genera of Fabaceae (e.g. *Dipteryx odorata, Melilotus officinalis*), Poaceae (e.g. *Anthoxanthum odoratum*), and Rubiaceae (e.g. *Galium odoratum*).

CLASSIFICATION Cell toxin, phototoxic; moderately to slightly hazardous, II–III

SYMPTOMS FC mainly cause photodermatitis: accidental skin contact with toxic furanocoumarins (e.g. when gardeners cut hogweed, *Heracleum mantegazzianum,* on a sunny day) results in itching and blistering, about 40–50 h after exposure. Sunlight is required to develop blisters; therefore, affected skin parts should be kept away from the sun. The blisters (similar to third degree burns) heal slowly and leave dark patches on the skin for several weeks (therefore FC are sometimes cosmetically used in case of vitiligo). High dermal and oral doses can cause damage to liver and kidneys and can be lethal.

   Higher doses of coumarins cause headache, vertigo, sleepiness and nausea. Chronical ingestion can also lead to liver damage. Simple coumarins can be dimerised to dicoumarol by fungi in mouldy hay. The dicoumarol is a vitamin K antagonist, that has caused "sweet clover disease" in animals, leading to death from internal bleeding. Chronic intake of coumarin tea can reduce blood coagulation.

TOXICITY Furanocoumarins and coumarins are lipophilic compounds that can easily be absorbed through the skin and GI tract. They are known phototoxic substances and allergens. Chronic oral application of FC in rats and dogs (up to 100 mg/kg) induced disturbance of liver; higher doses cause disturbances of adrenal and thyroid glands and reproduction.

MODE OF ACTION When furanocoumarins have entered the skin and cells, they can intercalate DNA. Furanocoumarins have highly reactive double bonds in the furan and lactone ring. When activated by sunlight (UV 310–360 nm), they form bivalent (linear FC) or monovalent (angular FC) crosslinks with pyrimidine bases of both DNA strands but also with many proteins (enzymes, ion channels). This effect leads to apoptotic cell death and also to mutations (causing skin cancer, also kidney tumours in animals). Some FC inhibit metabolising enzymes, such as CYP1A, CYP2B and CYP3A. In the liver, FC are metabolised to epoxides, which are highly reactive intermediates and can bind to proteins and DNA. Furan rings and SMs with furan rings can be opened upon oxidation by cytochrome oxidase (CYP2E1) in the liver, exposing 1 or 2 functional aldehyde groups that can couple to amino groups of proteins and nucleotides. Therefore, higher doses of FC are cytotoxic for inner organs. When FC bind to proteins, allergies can result.

monoadduct with thymine

monoadduct with thymine

bi-adduct with thymine

USES (PAST AND PRESENT) Furanocoumarins (e.g. 8-MOP) are employed medicinally to treat psoriasis and vitiligo; since FC can kill proliferating keratocytes upon UV exposure, this treatment brings some relief for psoriasis patients. Since the FC therapy can induce precanceroses and melanoma it has been more and more replaced by alternative therapies. FC are also used in cosmetics as tanning lotion. Several Apiaceae have been employed in traditional medicine against infections or as spasmolytic. Extracts from *Heracleum sphondylium* have been used medicinally, as well as in liqueurs and other alcoholic drinks in E Europe.

Plants with coumarins are aromatic plants that have been used to flavour beverages (e.g. *Galium odoratum*; May punch; 3 g leaves/ 1 l) or food (e.g. tonka beans, *Dipteryx odorata*).

FIRST AID Since furanocoumarins cause serious skin inflammation, the therapy is similar to treating burns. Avoid exposure to sunlight. Apply sterile tissue on inflamed blisters; external treatment with corticoids and PVP iodine. After ingestion of higher doses, instil medicinal charcoal and sodium sulphate.

Angelica archangelica

Apium graveolens

Pastinaca sativa

Poncirus trifoliata

Myristicin and other phenylpropanoids

This group of compounds comprises simple derivates of tyrosine, often with a methylene-dioxy bridge and terminal methylene group in the propyl side chain. The terminal methylene group makes them highly reactive, so that they are treated here in a special monograph.

STRUCTURES Examples: myristicin ($C_{11}H_{12}O_3$, MW 192.22), safrole ($C_{10}H_{10}O_2$, MW 162.19), eugenol ($C_{10}H_{12}O_2$, MW 164.21), apiole ($C_{12}H_{14}O_4$, MW 222.24), β-asarone ($C_{12}H_{16}O_3$, MW 208.26), elemicin ($C_{12}H_{16}O_3$, MW 208.26), estragole ($C_{10}H_{12}O$, MW 148.21).

myristicin

safrole

eugenol

estragole

apiole

β-asarone

elemicin

OCCURRENCE Myristicin occurs in *Myristica fragrans* (Myristicaceae), *Apium graveolens, Daucus carota, Levisticum scoticum, Oenanthe, Pastinaca sativa, Petroselinum crispum* (Apiaceae) and *Cinnamomum glanduliferum* (Lauraceae).

β-Asarone is a major ingredient of *Asarum* species (Aristolochiaceae), *Piper angustifolium* (Piperaceae), and in *Acorus calamus* (Araceae).

Safrole in *Piper auritum* (Piperaceae), *Cinnamomum camphora, Ocotea preciosa, Sassafras albidum* (Lauraceae), *Illicium religiosum* (Illiciaceae), *Eremophila longifolia* (Myoporaceae), *Ocimum basilicum* (Lamiaceae), *Juniperus scopulorum* (Cupressaceae) and *Myristica fragrans* (Myristicaceae).

Eugenol in *Syzygium aromaticum*) *Pimenta dioica* (Myrtaceae), *Cinnamomum, Sassafras* (Lauraceae), *Origanum majorana* (Lamiaceae), *Achillea fragrantissima, Artemisia* (Asteraceae), *Piper betle* (Piperaceae), *Rosa rugosa, Geum* spp. (Rosaceae) and *Myristica fragrans* (Myristicaceae).

Apiole: *Crithmum maritimum, Petroselinum crispum* (Apiaceae), *Cinnamomum camphora, Ocotea* spp. (Lauraceae), *Piper angustifolium* (Piperaceae).

Elemicin: *Canarium commune* (Burseraceae), *Cinnamomum glanduliferum* (Lauraceae), *Cymbopogon procerus* (Poaceae), *Boronia pinnata, Zieria smithii* (Rutaceae), *Melaleuca bracteata* (Myrtaceae), *Dalbergia spruceata, Monopteryx uaucu* (Fabaceae).

Estragole: *Persea gratissima* (Lauraceae), *Artemisia dracunculus, Solidago odora, Tagetes florida* (Asteraceae), *Agastache, Feronia elephantum, Orthodon* (Lamiaceae), *Dictamnus* (Rutaceae), *Piper betle* (Piperaceae).

CLASSIFICATION Cell toxin, mind-altering; slightly hazardous, III

SYMPTOMS First symptoms after overdosing include nausea, vomiting, abdominal pain, pressure in chest, restlessness, tremor, heavy perspiration, diuresis, dizziness, delirium and even death.

TOXICITY Phenylpropanoids with a terminal methylene group can form covalent bonds with DNA and are therefore mutagenic and possibly carcinogenic. β-Asarone has been shown to be carcinogenic in animals. The abortifacient properties of parsley (*Petroselinum crispum*) are partly due to apiole, which is cytotoxic in higher doses.

MODE OF ACTION Myristicin inhibits MAO, which induces an increase of biogenic amine neurotransmitters, such as dopamine, serotonin and noradrenaline. Psychotropic effects resemble those of amphetamine.

Phenylpropanoids with a terminal methylene group (such as myristicin, safrole, eugenol, estragole, apiole and elimicin) can react with SH-groups of proteins. In the liver, these compounds are converted to epoxides, which can alkylate proteins and DNA. Therefore, they are potentially mutagenic and tumours (liver) have been observed in animal experiments.

USES (PAST AND PRESENT) Myristicin has been used in agriculture as an insecticide. *Asarum* and *Petroselinum* have been used as abortifacient. Eugenol has pronounced antimicrobial properties and is used medicinally against infections (e.g. in dentistry). Several medicinal plants with phenylpropanoids with a terminal methylene group have been employed in traditional medicine to treat microbial infections.

FIRST AID After ingestion of an overdose, instil medicinal charcoal and plenty of water. **Clinical therapy:** In severe cases: gastric lavage (possibly with 0.1 % potassium permanganate), instillation of medicinal charcoal and sodium sulphate, electrolyte substitution; possibly artificial respiration and oxygen therapy; in case of excitation and spasms inject diazepam and/or atropine.

Asarum europaeum

Acorus calamus

Crithmum maritimum

Quinones

Quinones can be divided into:
- hydroquinones and naphthoquinones
- anthraquinones and naphthodianthrones.

Since quinones can interact with various proteins they sometimes show substantial pharmacological and toxicological activities.

Hydroquinones and naphthoquinones

STRUCTURE Simple hydroquinones occur as glycosides, such as arbutin, in many plants. When the sugar moiety is cleaved by glucosidase, hydroquinone is released that can easily be oxidised to p-benzoquinone ($C_6H_4O_2$, MW 108.10). Examples of other benzoquinones are primin ($C_{12}H_6O_3$, MW 198.18) and geranylbenzoquinone ($C_{16}H_{20}O_2$, MW 244.34). Examples of naphthoquinones are: plumbagin ($C_{11}H_8O_3$, MW 188.19), lapachol ($C_{15}H_{14}O_3$, MW 242.28), juglone ($C_{10}H_6O_3$, MW 174.16).

arbutin

primin

geranylbenzoquinone

plumbagin

juglone

lapachol

OCCURRENCE Benzoquinones are widely distributed in genera of the Ericaceae, such as *Arbutus, Arctostaphylos, Calluna, Erica* and *Vaccinium*. Also in *Bergenia* (Saxifragaceae), *Pyrus* (Rosaceae), *Origanum* (Labiaceae), *Protea* (Proteaceae), *Xanthium* (Asteraceae), *Pinus* (Pinaceae), *Pimpinella, Petroselinum* (Apiaceae), *Ardisia, Embelia, Myrsine, Rapanea* (Myrsinaceae), *Rauvolfia* (Apocynaceae), *Adonis* (Ranunculaceae), *Cordia* (Boraginaceae) and *Oxalis* (Oxalidaceae). Benzoquinones are also present in the defence secretion of many beetles, such as the bombardier beetle.

In Hydrophyllaceae (*Phacelia tanacetifolia, Turricula, Wigandia*) prenylated quinones/hydroquinones are present that have lipophilic geranyl-, farnesyl-, or geranyl-geranyl side chains, such as geranylbenzoquinone. They have similar properties as urushiol in *Rhus toxicodendron*. Primulaceae/Myrsinaceae (genera *Anagallis, Dionysia, Glaux, Primula*) and Orchidaceae (genus *Cymbidium*) have similar compounds; primin is responsible for a serious contact dermatitis caused by *Primula obconica*. Similar quinones have been found in several members of the genus *Iris* (Iridaceae). *Grevillea robusta* (Proteaceae) accumulates the allergenic alkylphenol grevillol, a resorcinol derivative.

The naphthoquinone plumbagin and related compounds are present in *Drosera, Dionaea* (Droseraceae), *Aristea, Sisyrynchium, Sparaxis* (Iridaceae) and *Pera* (Euphorbiaceae) and appear to be responsible for the antitussive effects. In several members of the Bignoniaceae (*Catalpa, Haplophragma, Kigelia, Paratecoma, Radermachera, Tabebuia, Tectona*), Malvaceae (*Hibiscus*), Fabaceae (*Diphysa*) and Proteaceae (*Conospermum*) naphthoquinones of the lapachol type are characteristic secondary metabolites. *Juglans, Carya* (Juglandaceae), *Lawsonia* (Lythraceae), *Impatiens* (Balsaminaceae) and *Lomatia* (Proteaceae) contain naphthoquinones of the juglone type. The red-coloured alkannin and shikonin are used as natural dyes. Alkannin has been found in *Alkanna, Arnebia, Macrotomia, Plagiobothrys* (Boraginaceae), shikonin in *Echium, Lithospermum,* and *Onosma* (Boraginaceae). Further families include Aristolochiaceae (*Aristolochia*), Pyrolaceae (*Chimaphila, Pyrola*) and Ebenaceae (*Diospyros, Euclea, Moba*).

CLASSIFICATION Cell toxin, skin irritant; slightly hazardous, III

SYMPTOMS Quinones mostly affect skin, mucosal tissues, especially the eye. On skin they can cause discoloration, erythema and blisters. When these compounds come into contact with the eye, they can cause conjunctivitis and ulcerations of the cornea. Inflammation and blisters can be generated; in serious instances, a patient can become blind.

Geranylbenzoquinone can anchor in biomembranes with the lipophilic side chain and can interfere with proteins and other molecules on the membrane surface. It is likely that allergenic proteins are generated that can lead to contact dermatitis, which is a known phenomenon in the USA where *Phacelia* species are common, or from gardeners handling *Primula obconica*.

Plant material with juglone induces paralysis, colic, anorexia, lethargy in horses (black walnut toxicosis). Juglone is a potent skin irritant.

TOXICITY The immediate toxicity is moderate; LD_{50} for *p*-benzoquinone in rats: 130 mg/kg p.o. Naphthoquinones can cause serious mucosal irritation when ingested (e.g. as health tea) or when inhaled (dust from sawing trees from Bignoniaceae or Juglandaceae).

MODE OF ACTION Hydroquinones can be oxidised to quinones under physiological conditions. Quinones and naphthoquinones are redox reagents that can bind to enzymes or interact with proteins containing Fe^{2+}/Fe^{3+}, such as cytochromes and haemoglobin. Alkylated quinones can be regarded as haptens that form novel antigens when bound to proteins. They are the cause of dermatitis observed in gardeners who often handle corresponding plants.

USES (PAST AND PRESENT) Drugs with arbutin are used in traditional medicine to treat bacterial infections of the urinary tract, because the hydroquinone shows antibiotic properties. The defence secretion of a beetle (*Palembus ocularis*; Tenebrionidae) with hydroquinones has been used in S & C America to treat asthma and several other inflammatory disorders. A tea from *Tabebuia impetiginosa* ("Lapacho or Inka tea"), used by S American Indians, has been introduced in Europe as a general health tea and even for the treatment of cancer. Consumption of this tea can cause serious irritation of mucosal tissues, e.g. in the GI tract.

FIRST AID In case of skin contamination: wash affected parts with water and soap or with polyethylene glycol 400; then apply sterile dressing. Eye contact: first, place 1–2 drops of a local painkiller into the eye, then flush eyes under running water for 10 min. Consult an ophthalmologist. After oral ingestion: give plenty of water to drink. **Clinical therapy:** In serious instances: provide gastric lavage with at least 10 g medicinal charcoal, then medicinal charcoal and sodium sulphate. In order to avoid the formation of methaemoglobin thionine, methylene blue, toluidine blue or vitamin C can be given.

Arbutus unedo

Phacelia tanacetifolia

307

Anthraquinones and naphthodianthrones

STRUCTURES In plants, anthracene compounds are usually present as mono- or diglycosides. Microorganisms, fungi and marine animals usually produce free anthraquinones. Anthracenes can exhibit the oxidation stages of anthrone, anthranol and anthraquinone. Oxidative coupling reactions can lead to dimeric compounds. Whereas anthrone glycosides are abundant in living cells, anthraquinones dominate in dead or metabolically less active plant parts (bark, seeds) or excretions (*Aloe*). Depending on the degree of substitution in the aromatic rings, anthraquinones can be divided into chrysophanol and rubiadin type (in Rubiaceae). Anthraquinones and dianthrones have a yellow or red colour; in alkaline solutions they form reddish phenolates that are easily soluble in water. Dimeric anthraquinones are also present in *Hypericum perforatum* (Clusiaceae/Hypericaceae) and contribute to its pharmacology.

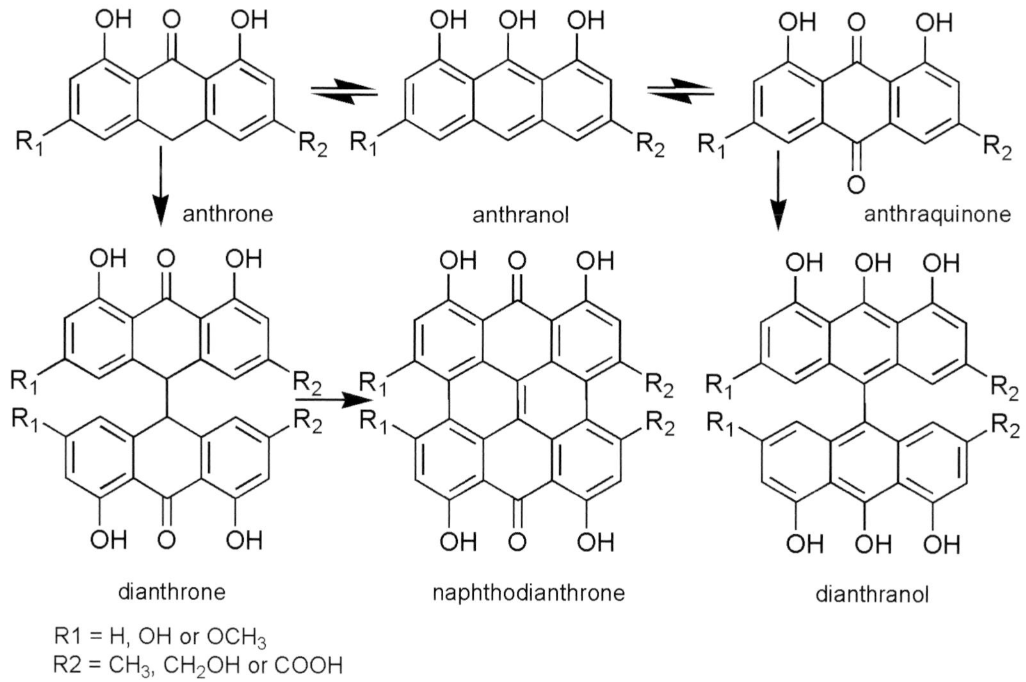

anthrone   anthranol   anthraquinone

dianthrone   naphthodianthrone   dianthranol

R1 = H, OH or OCH$_3$
R2 = CH$_3$, CH$_2$OH or COOH

OCCURRENCE Anthracene derivatives are common in certain genera of the Asphodelaceae (*Aloe, Asphodelus, Eremurus, Haworthia, Kniphofia*), Clusiaceae/Hypericaceae (*Harungana, Hypericum, Psorospermum*), Fabaceae (*Andira, Cassia, Ferreira, Gleditsia, Vatairea*), Gesneriaceae (*Streptocarpus*), Xanthorrhoeaceae (*Xanthorrhoea*), Myrsinaceae (*Myrsine*), Polygonaceae (*Fagopyrum, Oxygonum, Rheum, Rumex, Polygonum*), Rhamnaceae (*Karwinskia, Maesopsis, Rhamnus, Ventilago*), Rubiaceae (*Asperula, Cinchona, Coprosma, Coelospermum, Galium, Hymenodictyon, Morinda, Plocama, Relbunium, Rubia, Sherardia*), Plantaginaceae/Scrophulariaceae (*Digitalis, Scrophularia*) and Verbenaceae (*Tectona*). Anthraquinones are common defence compounds in lichens (*Caloplaca, Lecidea, Nephroma*). Anthraquinones are transferred into the milk of lactating women.

CLASSIFICATION Cell toxin, mutagen; slightly hazardous, II

SYMPTOMS Anthrones and dianthrones are irritating on skin and mucosal membranes; overdoses cause serious disturbance of GI tract, sometimes with bloody diarrhoea, and irritation of kidneys. Also disturbances of heart, circulation and kidneys have been reported. Death can be caused by severe loss of water and electrolytes.

TOXICITY These compounds can also interact with DNA and can probably cause mutations. Acute toxicity is low; oral LD$_{50}$ of sennosides in rats and mice was higher than 5 g/kg. LD$_{50}$ i.v: sennidin 46 mg/kg, rhein 25 mg/kg. Chronic application change colon morphology; gut mucosa can become black in colour (pseudomelanosis coli).

MODE OF ACTION Anthranoids of the emodin type (1,8-dihydroxyanthracene derivatives) are present as β-glycosides. These are merely the prodrugs that are neither absorbed nor hydrolysed in the upper gastrointestinal tract, but are only broken down into anthrones when they reach the colon. The resulting anthrones are the active laxatives that act on the colon motility, resulting in faster bowel movements. Molecular targets include a chloride channel, Na$^+$, K$^+$-ATPase. The synthesis of a prostaglandin PGE$_2$, histamine, serotonin is stimulated and gastrointestinal hormones are released. Anthrones enhance peristalsis and the secretion of water and inhibit its absorption in the colon. The anthraquinones from *Andira* inhibit mitochondrial metabolism. Anthraquinones from *Rubia* can complex calcium ions. Anthraquinones also have phenolic OH-groups and can therefore interact with many proteins in the human body. Some anthraquinones can intercalate DNA and are thus potentially mutagenic. Some anthraquinones have therefore antibacterial, antiviral and cytotoxic properties.

Dimeric naphthodianthrones, such as hypericin, can exhibit phototoxic properties.

USES (PAST AND PRESENT) Anthraquinones and drugs with these SMs (especially *Aloe, Cassia, Rhamnus, Rheum*) have been used in medicine as laxatives for several thousand years. Their use is not encouraged because of severe side effects. *Hypericum perforatum* is an old medicinal plant which was used against various infections. In modern medicine extracts with hyperforin and hypericin are successfully employed to treat depression.

FIRST AID Give medicinal charcoal and sodium sulphate. Allow the patient to drink plenty of tea or water. **Clinical therapy:** In case of serious overdose: gastric lavage (possibly with 0.1 % potassium permanganate), electrolyte substitution, control acidosis with sodium bicarbonate (urine pH 7.5); in case of spasms give diazepam; in severe cases supply intubation and oxygen respiration.

Aloe ferox

Rhamnus frangula

Galium verum

Rubia peregrina

Polyacetylenes

STRUCTURE Polyacetylenes (acetylenes or polyines) are reactive unsaturated hydrocarbons (13–17 carbons) with several double and triple bonds (ethylenic groups). Aethusin ($C_{13}H_{14}$; MW 170.26); cicutoxin ($C_{17}H_{22}O_2$; MW 258.36); falcarinol ($C_{17}H_{24}O$; MW 244.38), falcarindiol ($C_{17}H_{24}O_2$; MW 260.38).

In some plants, polyines exist in which oxygen or sulphur (thiophenes) have been added to the triple bonds and secondary ring formations have occurred. Examples for thiophenes: 5-(3-buten-1-ynyl)-2,2'-bithienyl; BBT) ($C_{12}H_8S_2$; MW 216.33), α-terthienyl ($C_{12}H_8S_3$; MW 248.39).

falcarinol

aethusin

cicutoxin

thiophenes

5-(3-buten-1-ynyl)-2,2-bithienyl (BBT)

α-terthienyl

OCCURRENCE Mixtures of polyacetylenes are found in high concentrations in *Aethusa cynapium, Cicuta virosa, Cicuta douglasii, Chaerophyllum temulentum, Oenanthe crocata, Sium latifolium* (Apiaceae), in Araliaceae (*Dendropanax, Fatsia, Hedera, Schefflera*) and Asteraceae. Several Apiaceae, Asteraceae, Campanulaceae, Oleaceae, Pittosporaceae and Santalaceae accumulate low amounts of polyacetyles, which are not relevant in a toxicological context. Some plants produce polyacetylenes after fungal infection (so-called phytoalexins). Falcarinol occurs in *Chaerophyllum, Daucus, Falcaria, Levisticum, Sium* (Apiaceae) and *Hedera, Schefflera* (Araliaceae). Polyacetylenes are common toxins of many mushrooms (Basidiomycetes) and some red algae (*Chondria, Laurentia*).

Thiophenes are known from *Dahlia, Eclipta, Flaveria, Rudbeckia, Tagetes, Tessaria, Porophyllum* (Asteraceae). They exhibit a wide range of phototoxic activities against bacteria, fungi, viruses, insects and human cells. They can cause erythema and hyperpigmentation when applied on the skin. Several *Tagetes* species were used in pre-Columbian times as mind-altering stimulants.

CLASSIFICATION Cell toxin, neurotoxin; highly hazardous, Ib

SYMPTOMS First symptoms of polyacetylene poisoning include salivation, burning sensation in mouth and throat. The skin gets white and cold perspiration occurs; body has a dark colour and appears to be inflated. Furthermore, nausea and vomiting, headache, abdominal pain, kidney damage, dilated pupils, coma and dyspnoea are observed; paralysis starts from legs and arms; death from respiratory arrest. Symptoms resemble hemlock poisoning (see coniine). Cicutoxin affects the breath and vasomotoric centre of the brain. After an initial excitation, a paralysis of all important centres follows, leading to a central respiratory arrest. Cicutoxin causes epileptiform seizures, which are accompanied by trismus (tetanus of jaw muscles) and opisthotonus (spasms of back muscles). During seizures patients are unconscious and show no corneal reflexes; their pupils are maximally dilated, breathing is irregular, foam is formed at the mouth, mucosal tissues are cyanotic (blue colour); pulse is weak and rapid. Respiration and consciousness can return after a seizure. Seizures resume after 15 min until the patient is completely exhausted. Death occurs within a few hours during a seizure due to suffocation or after a seizure due to respiratory arrest.

TOXICITY Polyines are volatile and unstable compounds that disappear in dried plant material. Therefore, fresh material is mostly relevant for poisoning. Ingestion of rhizomes (e.g. accidentally as salad or in search of hallucinogenic effects) can cause serious poisoning and several cases of death have been reported. About 15 kg fresh plant material of *Aethusa cynapium* is lethal for cattle. 2–3 pieces of rhizomes can kill cattle and horses. LD_{50} mouse: Cicutoxin: $LD_{50} = 9.2$ mg/kg, i. p.; aethusin: LD_{50} 93.3 mg/kg, i. p.; aethusanol A: LD_{50} 100.8 mg/kg, i. p. 2–3 g of *Cicuta virosa* rhizome appears to be lethal. Oenanthetoxin: mouse 1.2 mg/kg i.p.; mouse, cat, guinea pig: 0.05–0.15 mg/kg p.o. Falcarinol: Mouse: 100 mg/kg i.p. Thiophenes exhibit pronounced nematocidal, antimicrobial and herbicidal activities.

MODE OF ACTION Polyacetylenes are highly lipophilic compounds that are easily absorbed. They can pass the blood-brain barrier and can affect neuronal cells and receptors. Polyacetylenes can directly influence membrane fluidity. They have triple bonds and are therefore very reactive compounds that can form covalent bonds with important macromolecules of the cell, such as proteins (including important receptors, ion channels, structural proteins and enzymes). These covalent modifications can change protein confirmation and thus their activity. Interactions with neuronal elements can cause CNS symptoms and cell death. They affect mainly the medulla oblongata and thus the centres for breathing and vasomotoric activity. Polyacetylenes in general show antibiotic, antiviral, insecticidal and cytotoxic properties (potential anticancer agents against leukaemia). Falcarindiol exhibits antifungal and analgesic activities. Falcarinol is a powerful allergen, since it binds to proteins and thus forms novel antigens. It causes contact dermatitis.

USES (PAST AND PRESENT) *Cicuta* rhizome and fruits have been used in traditional medicine to treat skin disorders, rheumatism and gout. Some of the plants with polyacetylenes have been used in traditional medicine because of their antibiotic properties.

FIRST AID Administration of sodium sulphate and medicinal charcoal. **Clinical therapy:** Gastric lavage under anaesthesia because of severe seizures (with 0.1% potassium permanganate), instillation of medicinal charcoal and sodium sulphate, electrolyte substitution, control of acidosis with sodium bicarbonate (pH of urine should be 7.5). In case of spasms inject diazepam or thiobarbiturates (penthotal, trapanal); in case of serious poisoning, provide artificial respiration (intubation and oxygen supply), haemodialysis and haemoperfusion. In case of shock symptoms give plasma expander. Provide glucose infusions, but no phenothiazines or analeptics.

Falcaria vulgaris

Tagetes patula

311

Terpenoids

Terpenoids are built from C_5 units such as dimethylallyl-pyrophosphate and isopentenyl-pyrophosphate. Depending on the number of C_5 units we distinguish between:

- monoterpenes (C_{10})
- sesquiterpenes (C_{15})
- diterpenes (C_{20})
- triterpenes (C_{30}) (including steroids, cardiac glycosides, saponins)
- tetraterpenes (C_{40}).

Terpenoids are usually lipophilic compounds that can interfere with lipophilic parts of biomembranes and proteins. Depending on other functional groups they can interact with proteins and/or DNA. Several groups, such as mono-, sesqui-, di-, and triterpenes are relevant in a toxicological context.

Monoterpenes

STRUCTURE Monoterpenes, of which more than 600 structures are known, have 10 C-atoms. They can occur as open, cyclic or bicyclic compounds. Most of them are highly lipophilic, volatile and have/possess 1 or 2 functional groups. Monoterpenes occur as linear molecules or with 1 or 2 ring structures.

Examples are limonene ($C_{10}H_{16}$; MW 136.24); α- and β-thujone ($C_{10}H_{16}O$; MW 152.23), ascaridole ($C_{10}H_{16}O_2$; MW 168.24), borneol ($C_{10}H_{18}O$; MW 154.25), camphene ($C_{10}H_{16}$; MW 136.24), camphor ($C_{10}H_{16}O$; MW 152.24), 1,8-cineole ($C_{10}H_{18}O$; MW 154.25), citral ($C_{10}H_{16}O$; MW 152.24), citronellal ($C_{10}H_{18}O$; MW 154.25), carvacrol ($C_{10}H_{14}O$; MW 150.22), p-cymene ($C_{10}H_{14}$; MW 134.22), linalool ($C_{10}H_{18}O$; MW 154.25), menthol ($C_{10}H_{20}O$; MW 156,27), myrcene ($C_{10}H_{16}$; MW 136.24), nerol ($C_{10}H_{18}O$; MW 154.25), ocimene ($C_{10}H_{16}O$; MW 136.24), pinene ($C_{10}H_{18}O$; MW 136.24), pinocarvone ($C_{10}H_{14}O$; MW 150.22), sabinol ($C_{10}H_{16}O$; MW 152.24), terpineol ($C_{10}H_{18}O$; MW 154.24) and thujone ($C_{10}H_{16}O$; MW 152.24).

linear monoterpenes

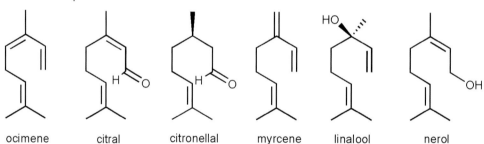

| ocimene | citral | citronellal | myrcene | linalool | nerol |

monocyclic monoterpenes

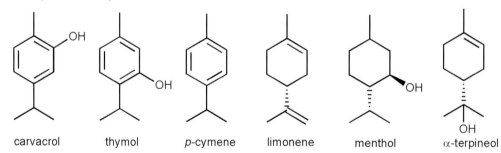

| carvacrol | thymol | p-cymene | limonene | menthol | α-terpineol |

OCCURRENCE Widely present in Asteraceae, Apiaceae, Burseraceae, Dipterocarpaceae, Lamiaceae, Lauraceae, Myricaceae, Myristicaceae, Poaceae, Rutaceae, Verbenaceae, and conifers (Pinaceae, Cupressaceae). Monoterpenes also occur in some bryophytes and fungi. Thujone is one of the more toxic monoterpenes; it occurs in *Artemisia, Salvia officinalis, Chrysanthemum* (= *Tanacetum*) and *Thuja*. Ascaridole is a toxic constituent of *Chenopodium ambrosioides* var. *anthelminticum*.

bicyclic monoterpenes

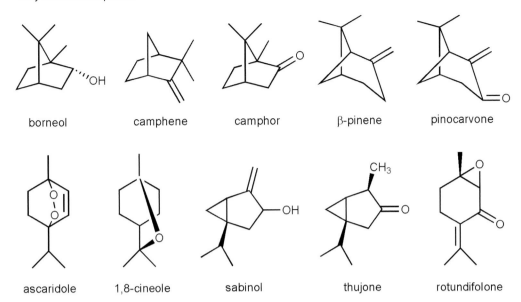

borneol　　camphene　　camphor　　β-pinene　　pinocarvone

ascaridole　　1,8-cineole　　sabinol　　thujone　　rotundifolone

CLASSIFICATION Cell toxin, neurotoxin, mind-altering; slightly to moderately hazardous, II–III

SYMPTOMS Applied to the skin, monoterpenes and aliphatic hydrocarbons can irritate and sensitise it, which causes hyperaemia and in some cases contact dermatitis. Ingestion of high doses of essential oil or isolated compounds cause hyperthermia, salivation, colonic spasms, vertigo, headache, confusion, GI disturbance, vomiting; many monoterpenes have nephro- and liver-toxic properties. High doses of monoterpenes cause narcotic effects, which can lead to degeneration of peripheral and central neurons, polyneuropathy and muscular atrophy (especially upon chronic exposure). Euphoria and hallucinations can also occur.

Symptoms of **thujone** poisoning include mydriasis, visual disturbance, headache, gastroenteritis with severe vomiting, convulsions and damage of kidney, heart and liver. Death is caused by circulatory and respiratory arrest. Chronic ingestion of thujone initially leads to CNS stimulation, optical and acoustical hallucinations, later to deep depression, epileptic seizures and loss of personality. Vincent van Gogh was an absinthe addict; it can be speculated that his later life and art were influenced by chronic thujone poisoning. He was also overdosed with digitalis and therefore saw and painted the sun with a large halo (side effect of cardiac glycosides).

Thujone, camphor, pinene, pulegone, pinocampton and perillaketone are more hazardous than other monoterpenes and have negative effects at higher doses, such as skin irritant, neuro-, hepato- and bronchotoxic properties.

Sabinene and sabinol (from *Juniperus sabina*) are thought to represent the active principle of abortifacient oils.

TOXICITY Whereas plants that produce essential oils usually are not poisonous, isolated oil however has toxicological relevance. Limonene and other monoterpenes; LD_{50} rat: limonene 5.3 g/kg p.o., 5 g/kg topical; p-cymene: 4.75 g/kg p.o.; about 50 g of monoterpenes can be lethal for humans. Camphor: 1 g is lethal for children, 20 g for adults. Eucalyptus oil (mainly 1,8 cineole): 30 and more ml oil can cause serious poisoning; terpentine oil from Pinaceae is rich in α-pinene and β-pinene and can be lethal when 60–120 ml are ingested. Thujone is a pronounced neurotoxin; it induces epileptic seizures and serious psychic disturbances. It irritates mucosa and shows cytotoxic effects on liver and kidney cells. Thujone has abortifacient properties. LD_{50} α-thujone mouse: 87.5 mg/kg s.c.; β-thujone mouse 442.4 mg/kg s.c.

MODE OF ACTION The lipophilic monoterpenes tend to dissolve in cellular membranes. There, they can interact with phospholipids and membrane proteins. Depending on their inner membrane concentration the monoterpenes can enhance the fluidity and permeability of membranes. If they interact with lipophilic domains of proteins, they can induce conformational changes that can alter the biochemical properties of ion channels, transporters or receptors. High doses of small lipophilic SMs will therefore produce some sort of psychological disturbance, as do simple narcotics or ethanol: e.g., higher doses of camphor (6–10 g) (present in *Cinnamomum camphora*) cause psychomotoric unrest and hallucinations, as well as dizziness, nausea, vomiting, anxiety, inhibition of neuronal activity; inflammation of stomach mucosa and nephritis.

Thujone contains a cyclopropane ring, which makes the molecule highly reactive. It is likely that thujone can alkylate important proteins of the neuronal signal transduction, therefore causing neuronal disorder. It has been speculated that thujone binds to tetrahydrocannabinol receptors in the brain.

Sabinene and sabinol are reactive monoterpenes with a highly reactive cyclopropane ring and an exocyclic methylene group, which can form covalent bonds with SH-groups of proteins. Other monoterpenes with reactive cylopropane rings include 3-carene, umbellulone, chrysanthemic acid, cinerin, rotundifolone and pyrethrins.

Monoterpenoids with a peroxide bridge, such as ascaridole, are reactive compounds, which can alkylate proteins.

Monoterpenes with exocyclic or terminal methylene groups, as in camphene, pinocarvone or in linalool, can bind to SH groups of proteins and thus change their conformation.

Monoterpenes with phenolic hydroxyl groups (such as thymol and carvacrol) or with aldehyde function (such as citral, citronellal), which can bind to proteins, have pronounced antiseptic properties and are active against bacteria and fungi.

USES (PAST AND PRESENT) Essential oils are used in aromatherapy and in phytomedicine to treat rheumatism, infections (bacterial, fungal), cold, unrest, flatulence, intestinal spasms, as stomachic and to improve taste. Essential oils are part of many perfumes and of some natural insect repellents.

FIRST AID After ingestion of larges doses: instillation of sodium sulphate and medicinal charcoal. **Clinical therapy:** Provide gastric lavage (possibly with 0.1% potassium permanganate), instillation of medicinal charcoal, sodium sulphate and polyethylene glycol. In case of excitation inject diazepam. Provide intubation and oxygen respiration in case of respiratory arrest and plasma expander when shock occurs. In case of thujone overdose: treat with atropine against colic and diazepam against spasms; control liver, blood and kidney functions. In case of severe poisoning apply haemoperfusion with medicinal charcoal and amberlite, and haemodialysis.

Chrysanthemum (Tanacetum) vulgare

Thuja occidentalis

314

Terpenes and phenylpropanoids with aldehyde groups (*including iridoid glucosides*)

Some monoterpenes and phenylpropanoids contain highly reactive aldehyde groups, which can form covalent bonds with proteins and DNA. They are of special relevance for toxicology and pharmacology and are therefore treated separately.

STRUCTURE Several secondary metabolites contain aldehyde groups as a relevant pharmacophore.

Examples are: safranal ($C_{10}H_{14}O$, MW 150.22), cinnamaldehyde (C_9H_8O, MW 132.16), *p*-anisaldehyde ($C_8H_8O_2$, MW 136.15), polygodial ($C_{15}H_{22}O_2$, MW 234.34), and iridoid glycosides: aucubin ($C_{15}H_{22}O_9$, MW 346.34), catalpol ($C_{15}H_{22}O_{10}$, MW 362.34) and iridodial ($C_{10}H_{16}O_2$, MW 168.24).

Iridoid glucosides, such as aucubin and harpagoside are cleaved by β-glucosidase into an unstable aglycone. The lactol ring can open and produces a functional dialdehyde. Such dialdehydes can also occur as a free molecule, e.g. iridodial in *Myoporum* (Myoporaceae) and in the defence secretion of ants, or dolichodial in *Teucrium marum* (Lamiaceae). Catalpol has an epoxide ring in addition. The tetraterpene crocin from *Crocus sativus* is stored as a glycoside. After hydrolysis, reactive aldehydes are generated, such as picrocrocin and safranal. Safranal can bind to proteins and DNA and therefore has mutagenic properties. SMs with furan rings can be opened upon oxidation by cytochrome oxidase (CYP2E1) in the liver, exposing 1 or 2 functional aldehyde groups.

cinnamaldehyde

safranal

citronellal

H_3CO *p*-anisaldehyde

polygodial (R = H)
warburganal (R = OH)

euphroside

aucubin

catalpol

iridodial

OCCURRENCE Aldehydes of simple monoterpenes are widely distributed as components of essential oil in Lamiaceae. *p*-Anisaldehyde has been found in *Ptelea* (Rutaceae), *Agastache* (Lamiaceae), *Magnolia* (Magnoliaceae), *Vanilla* (Orchidaceae), *Pinus* (Pinaceae), *Cassia* (Fabaceae), *Pimpinella* (Apiaceae) and *Illicium* (Illiciaceae). Cinnamaldehyde has been found in *Cinnamomum* (Lauraceae), *Hyacinthus* (Hyacinthaceae), *Narcissus* (Amaryllidaceae), *Lavandula, Pogostemon* (Lamiaceae) and *Commiphora* (Burseraceae).

Iridoid glycosides (secoiridoids, secologanin derivates) with more than 200 structures are widely distributed in the families Apocynaceae, Bignoniaceae, Gentianaceae, Globulariaceae, Lamiaceae, Loganiaceae, Menyanthaceae, Monotropaceae, Plantaginaceae, Rubiaceae, Scrophulariaceae, Valerianaceae and Verbenaceae. Aucubin has been found in *Plantago* (Plantaginaceae), *Aucuba japonica* (Cornaceae), *Euphrasia, Rhinanthus, Veronica* (Scrophulariaceae) and *Ajuga* (Lamiaceae), catalpol in *Catalpa* (Bignoniaceae), *Veronica* (Scrophulariaceae), *Plantago* (Plantaginaceae) and *Buddleja* (Buddlejaceae). Harpagoside and harpagide have been found in *Harpagophytum procumbens* (Pedaliaceae), *Scrophularia* (Scrophulariaceae) and *Lamium* (Lamiaceae).

The dialdehydes polygodial and warburganal have a peppery taste and were found as the active principle in *Polygonum hydropiper* (Polygonaceae), *Drimys lanceolata* (Winteraceae) and *Warburgia* (Canellaceae).

CLASSIFICATION Cell toxin; moderately hazardous, II–III

SYMPTOMS Aldehydes irritate skin and mucosal tissue of the mouth, throat and eyes. Inflammation with blister formation can result. They can cause itching, nausea, vomiting and dyspnoea. In a higher dose they exhibit narcotic activities. Aldehydes can also induce allergic reactions because proteins carrying a new alkyl group form new epitopes that can generate an immune response.

In false morel mushroom (*Gyromitra esculenta*) a toxic aldehyde is present (gyromitrin) that causes gastroenteritis with vomiting, abdominal cramps, headache, cyanosis, jaundice, convulsions and coma.

TOXICITY Toxicological data are hardly available for the natural products, but exist for organic chemicals such as formaldehyde, acetaldehyde and glutaraldehyde. They can be considered as model compounds. LD_{50}: Acetaldehyde: rat 1.9 g/kg (p.o.); glutaraldehyde: rat 134 mg/kg p.o.; 60 ml formaldehyde (p.o.) are lethal for humans.

MODE OF ACTION Aldehydes can easily form Schiff's bases with free amino groups of proteins under physiological conditions. The modified proteins can change their conformation and usually lose their intrinsic property. Aldehydes have pleiotropic activity, because they interact with any protein they encounter. Aldehydes can be used to inactivate proteins in animal cells, microorganisms and viruses. If aldehydes reach the cell nucleus, they can also alkylate DNA bases and are therefore potentially mutagenic and carcinogenic. The iridoid catalpol has a reactive epoxide ring in addition (which can bind to proteins and DNA); mutagenic properties have been reported. Some iridoids have antimicrobial, cytotoxic and antitumour activities. Monoterpenes with aldehyde function (such as citral, citronellal) have pronounced antiseptic properties and are active against bacteria and fungi.

USES (PAST AND PRESENT) Organic aldehydes are used in organic synthesis, as preserving and disinfecting agents, for fixation of microscopic specimens, leather and gelatine. *Valeriana* is used as a sedative and tranquilliser. Several plants with iridoid glucosides are used in traditional medicine to treat rheumatism, constipation and microbial infections.

FIRST AID If skin has been contaminated with aldehydes, wash it with water and soap or a diluted ammonia solution (1 %). In case of eye contact, place 1–2 drops of Chibro-Kerakin into the eye, then wash for 10 min with water and apply Isogutt; consult an ophthalmologist. In case of oral uptake, drink plenty of water and instil medicinal charcoal and sodium sulphate as a laxative.

Plantago media

Warburgia salutaris

Euphrasia officinalis

Harpagophytum procumbens

Sesquiterpenes and sesquiterpene lactones

This structurally highly diverse group of secondary metabolites contains many substances with pronounced biological activities. From a toxicological and pharmacological point of view sesquiterpene lactones are especially relevant as they are chemically very reactive.

STRUCTURE Sesquiterpene lactones often have a furan moiety, 1 to 3 exocyclic methylene and epoxide groups, which are important for biological activity. More than 3 000 structures, most of them from Asteraceae, have been described. Examples are: helenalin ($C_{15}H_{18}O_4$, MW 262.31), autumnolide ($C_{15}H_{20}O_5$, MW 280.33), lactucin ($C_{15}H_{16}O_5$, MW 276.29), lactupicrin ($C_{23}H_{22}O_7$, MW 410.43), parthenolide ($C_{15}H_{20}O_3$, MW 248.33), cynaropicrin ($C_{19}H_{22}O_6$, MW 346.39), picrotoxinin ($C_{15}H_{16}O_6$, MW 292.29), artemisinin ($C_{15}H_{22}O_5$, MW 282.34), santonin and ($C_{15}H_{18}O_3$, MW 246.31).

helenalin autumnolide lactucin

lactucopicrin parthenolide cynaropicrin

OCCURRENCE Sesquiterpenes and sesquiterpene lactones are widely distributed in Asteraceae (e.g. in the genera *Achillea, Ambrosia, Anaphalis, Anthemis, Arnica, Artemisia, Arctium, Arctotis, Baileya, Balduina, Baltimora, Cacalia, Calea, Calocephalus, Carpesia, Centaurea, Chaenactis, Chromolaena, Chrysanthemum, Cichorium, Cnicus, Cynara, Dicoma, Dugaldia, Elephantopus, Encelia, Enhydra, Eremanthus, Eriophyllum, Eupatorium, Gaillardia, Geigera, Grossheimia, Helenium, Helianthus, Homogyne, Hymenoxys, Inula, Isocarpha, Iva, Jurinea, Lactuca, Liatris, Ligularia, Lychnophora, Matricaria, Melampodium, Mikania, Moquinia, Onopordum, Oxylobus, Parthenium, Petasites, Podanthus, Psilostrophe, Saussurea, Senecio, Smallanthus, Stokesia, Tanacetum, Telekia, Tithonia, Ursinia, Vanillosmopsis, Vernonia, Viguiera, Wedelia, Xanthium, Xeranthemum, Zaluzania, Zexmenia, Zinnia*), but are also common in some Apiaceae (*Laser, Laserpitium, Thapsia*), Lamiaceae (*Glechoma*), Illiciaceae (*Illicium*), Coriariaceae (*Coriaria*), Magnoliaceae (*Liriodendron, Magnolia, Michelia*), Menispermaceae (*Anamirta*), Euphorbiaceae (*Hyaenanche*), Lauraceae (*Laurus nobilis, Lindera*), gymnosperms (*Cupressaceae*), and a few mosses (*Frullania*). Ptaquiloside has been detected in some ferns such as *Pteridium aquilinum, Cheilanthes, Dennstaedtia, Histiopteris, Hypolepis,* and *Pteris*.

CLASSIFICATION Cell toxins, neurotoxins, allergenic; moderately hazardous, II

SYMPTOMS Sesquiterpene lactones can bind to various proteins and are therefore potentially allergenic. Gardeners who have to handle plants from these genera often suffer from serious contact dermatitis. Allergic symptoms include erythema, swellings and inflammation on hands, lower arms and neck. Pollen with sesquiterpene lactones can cause allergic reactions; a known example is the introduced weed *Parthenium hysterophorus* in India. Contact allergies have been observed being caused by *Achillea, Ambrosia, Anthemis, Arnica, Artemisia, Chrysanthemum, Cichorium, Cynara, Helenium, Helianthus, Hymenoclea, Inula, Iva, Lactuca, Parthenium, Tanacetum* and *Taraxacum*. Absinthin from *Artemisia absinthium* can cause nervousness, convulsions or even death. Overdoses of picrotin and picrotoxinin cause extreme excitation of CNS with nausea, vomiting, miosis, bradycardia, spasms, cyanosis, dyspnoea and respiratory arrest. Coriamyrtin (a sesquiterpene with a picrotoxin skeleton carrying 2 epoxide rings) from *Coriaria* (Coriariaceae) and mellitoxin (a related structure from *Hyaenanche globosa*, Euphorbiaceae) are highly toxic to animals; they can also cause strong excitation of the CNS. Helenalin causes paralysis of voluntary and cardiac muscles, also fatal gastroenteritis.

Santonin from various *Artemisia* species is the active principle for the treatment of intestinal worms. Because of high toxicity this application has been discontinued in most countries of the world.

TOXICITY Ptaquiloside can alkylate DNA and is therefore mutagenic and carcinogenic. Picrotoxinin from *Anamirta cocculus* inhibits GABA receptors and can cause severe spasms. Lethal dose: 2–3 g seeds or 20 mg picrotoxinin. Lethal dose of santonin in children: 60–300 mg. Several plants with sesquiterpene lactones, such as *Geigera, Hymenopsis* and the mushroom *Coriaria* are toxic to livestock and have caused mass poisoning.

picrotoxinin

artemisinin

santonin

ptaquiloside

MODE OF ACTION Sesquiterpene lactones can alkylate proteins by binding to free SH-groups via 1 or 2 exocyclic methylene groups and by the en-on configuration in the furan ring. Some sesquiterpene lactones (e.g. acroptilin, arctolide, artecanin, arteglasin A, autumnolide, baileyin, canin, elephantin, elephantopin, eupachloroxin, euparotin, eupatoroxin, glaucolide, graminiliatrin, ludovicin A, melampodin, michelenolide, viscidulin B) have 1 or 2 epoxide functions which are highly reactive. The activated compounds can bind covalently to DNA and proteins. Some sesquiterpenes have furan rings which can be opened upon oxidation by cytochrome oxidase (CYP2E1) in the liver, exposing 1 or 2 functional aldehyde groups that can couple to amino groups of proteins and nucleotides. Because these alkylations are unspecific, many proteins can be affected. Alkylated proteins can change their conformation and are no longer able to properly interact with substrates, ligands or other protein. If DNA becomes alkylated in such a process, mutations can be a result. Sesquiterpene lactones also bind glutathione (via SH-groups) and can deplete its content in the liver. As a consequence, these terpenoids exhibit broad biological activities, including cytotoxic, antibiotic, anthelminthic, antiphlogistic, phytotoxic, insecticidal and antifungal properties. When sesquiterpene lactones couple to proteins, these can become antigens and induce the formation of antibodies. These antibodies can cause allergic reactions upon further exposition to sesquiterpenes.

318

Therefore, many plants causing contact dermatitis contain sesquiterpene lactones.

Picrotoxinin from *Anamirta cocculus* inhibits GABA receptors and can cause severe spasms. Artemisinin from *A. annua* has a reactive peroxide bridge. It can react with proteins and DNA; it is toxic to protozoan parasites. Many sesquiterpene lactones have a bitter taste.

USES (PAST AND PRESENT) *Anamirta cocculus*, which contains picrotoxinin, has been used to treat vertigo. Young shoots of *Pteridium aquilinum* with mutagenic ptaquiloside have been eaten as vegetable in Japan and caused a substantial incidence of bladder cancer. Several plants with sesquiterpene lactones have been used in traditional medicine or phytotherapy (*Achillea, Arnica, Artemisia, Matricaria, Parthenium*) because they exhibit anti-inflammatory, expectorant, antibacterial, antifungal and antiparasitic properties. Artemisinin is presently used and developed to treat malaria.

FIRST AID Administration of medicinal charcoal, sodium sulphate, and plenty of warm tea or fruit juice. Contact allergies can be treated with glucocorticoids and antihistaminics (see species monographs). **Clinical therapy:** Gastric lavage, instillation of medicinal charcoal and sodium sulphate, electrolyte substitution, control of acidosis with sodium bicarbonate. In case of excitation and spasms inject diazepam; in case of severe intoxication, provide intubation and oxygen respiration.

Helianthus salicifolius

Ambrosia artemisiifolia

Laurus nobilis

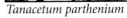
Tanacetum parthenium

319

Diterpenes

Among diterpenoids, phorbol esters and grayanotoxin are of special toxicological relevance.

Phorbol esters

STRUCTURE Phorbol esters can be divided into those with a tigliane moiety and others with a ingenane (occurring in Euphorbiaceae) or daphnane skeleton; daphnanes are common in Thymelaeaceae but also occur in Euphorbiaceae. The phorbol moiety ($C_{20}H_{28}O_6$; MW 364.44) is linked to 1 or 2 long-chained esters so that the phorbol esters resemble diacyl glycerol. Tigliane type: 12-tetradecanoylphorbol 13-acetate ($C_{36}H_{56}O_8$; MW 616.84); daphnane type: mezerein ($C_{38}H_{38}O_{10}$; MW 654.72).

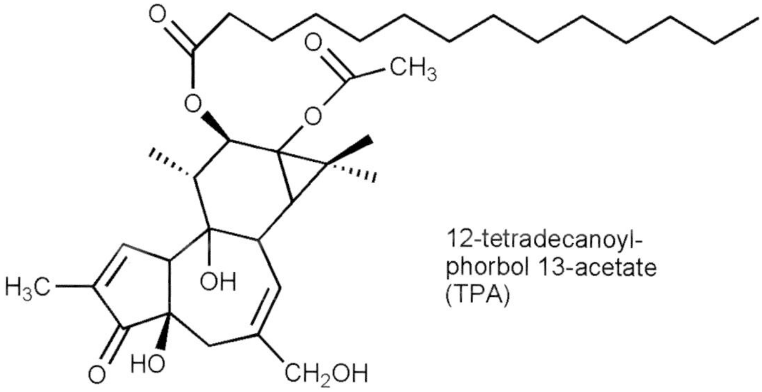

12-tetradecanoyl-phorbol 13-acetate (TPA)

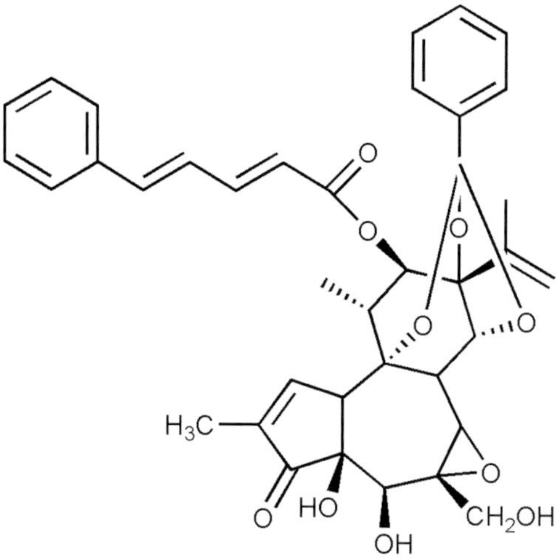

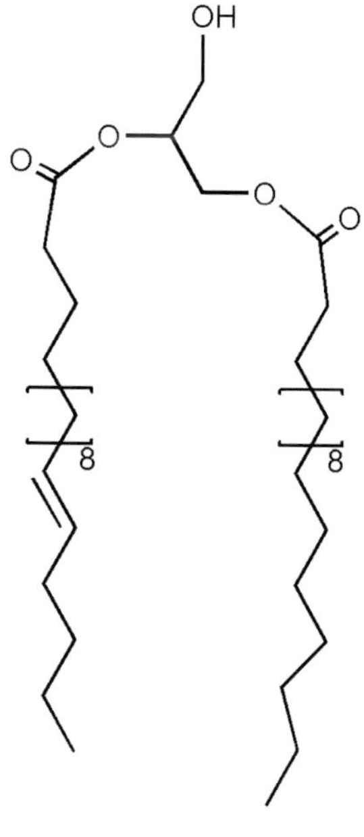

diacylglycerol (DAG)

mezerein

OCCURRENCE Common toxin in latex of Euphorbiaceae (especially in *Aleurites, Baliospermum, Codiaeum, Croton, Euphorbia, Excoecaria, Jatropha, Hippomane, Hura, Sapium, Synadenium*), in fruits and aerial parts of Thymelaeaceae (*Daphne, Daphnopsis, Gnidia, Pimelea, Synaptolepis*).

CLASSIFICATION Cell toxin, co-carcinogen, skin irritant; extremely hazardous, Ia

SYMPTOMS Plants with phorbol esters are strong purgatives; they induce drastic diarrhoea 5–10 min after ingestion. They are also potent skin irritants and lead to painful inflammation especially of mucosal tissue and of the eye (can cause intermittent blindness). Strong urticaria and blisters with oedematic swelling can develop. Skin can be destroyed and necrotic tissue will be shed off. Symptoms after ingestion of phorbol esters include: burning and scratching in mouth and throat (with urticaria and inflammation), enhanced saliva production, difficulties in swallowing, vomiting, agitation and impaired consciousness, dilated pupils, abdominal pain, watery and painful sometimes bloody diarrhoea, vertigo, arrhythmia, serious kidney inflammation, delirium and in severe cases death after 1–3 days (collapse). Extremities are cold and covered by cold sweat.

TOXICITY Plants with phorbol esters promote an adverse response even in low doses, such as strong skin irritation or diarrhoea. 4 seeds or 20 drops of the oil of *Croton tiglium* are lethal for a human, 15 can kill a horse. 10–12 berries of *Daphne mezereum* can be lethal.

MODE OF ACTION Phorbol esters resemble the endogenous signal compound diacylglycerol (DAG), which activates protein kinase C (PKC). PKC is an important regulatory enzyme in cellular signal transduction. It phosphorylates many cellular proteins (thus changing their activity) from ion channels, over enzymes to transcription factors. Phorbol esters are agonists of PKC and can thus trigger many cellular responses, including activation of cell division (co-carcinogen). Some derivatives of mezerein are cytotoxic and antileukaemic.

USES (PAST AND PRESENT) Several Euphorbiaceae and Thymelaeaceae are grown as ornamental plants and therefore accidental poisoning (internal and external) of children and gardeners can occur. Pets and livestock can also be affected.

In Curacao, where a beverage is produced from *Croton flavens* (Welensali tea), a high incidence of oesophagus cancer has been recorded. Cancer formation is linked to phorbol esters such as welensali factor F1. Some plants (such as *Daphne mezereum*) have been used in traditional medicine as laxative and blister forming drug.

Phorbol esters of *Hura crepitans* have been used as fish poison.

FIRST AID Immediately instil 200 ml paraffinum liquidum, a generous quantity of medicinal charcoal and sodium sulphate. Provide plenty of tea. Contact with eyes or skin: wash with polyethylene glycol or water; in case of skin blisters, cover with sterile cotton and apply locally a corticoid medicine.**Clinical therapy:** Gastric lavage (possibly with 0.1 % potassium permanganate), instillation of medicinal charcoal, sodium sulphate and 20 ml calcium gluconate (i.v.); electrolyte substitution, plasma expander, control of acidosis with sodium bicarbonate. In case of spasms inject diazepam. Provide intubation and oxygen respiration in case of respiratory arrest. Check kidney function. Salivation and vomiting can be treated by injecting atropine.

Jatropha podagrica   *Codiaeum variegatum*

Andromedotoxin or grayanotoxin I

STRUCTURE Andromedotoxin (synonyms: acetylandromedol, grayanotoxin-I, rhodotoxin) ($C_{22}H_{36}O_7$; MW 412.53) and asebotoxin II ($C_{23}H_{36}O_6$; MW 408.54) are examples of toxic diterpenes. Other toxic diterpenoids include atractyloside ($C_{30}H_{46}O_{16}S_2$; MW 803.01).

grayanotoxin I   asebotoxin II

atractyloside

OCCURRENCE Andromedotoxin and related compounds are common in Ericaceae, especially in the genera *Chamaedaphne, Gaultheria, Kalmia, Ledum, Leucothoe, Lyonia, Pernettya, Pieris* and *Rhododendron*. Since the toxins occur in nectar and pollen, bees can transfer the substances into honey. Atractyloside has been found in *Atractylis gummifera* and *Callilepis laureola* (Asteraceae).

CLASSIFICATION Neurotoxin; extremely hazardous, Ia

SYMPTOMS First symptoms are painful irritation of mucosal tissues in mouth and stomach, salivation, cold sweat and cold skin, as well as nausea, vomiting and diarrhoea. Central symptoms include dizziness, headache, fever, intoxication; visual disturbance and even blindness have been observed. Paralysis of sensitive nerve terminals can lead to paraesthesia of legs, arms and mouth. Some patients suffer from progressive muscular weakness, difficulties of coordination, spasms and paralysis of extremities. Hypotension, bradycardia and arrhythmia can progress to cardiac and respiratory arrest and even death. Most poisonings with andromedotoxin are not very serious and not lethal if patients have been treated adequately.

TOXICITY Andromedotoxin is a lipophilic compound that is easily absorbed from the mouth and throat or the GI tract. The compounds are rapidly inactivated and eliminated via faeces and urine. A few leaves or flowers of plants with grayanotoxin are enough for a lethal dose in humans. LD_{50} in mice: 1.28 mg/kg i.p.; guinea pig: 1.3 mg/kg i.p.; rat: 2–3 mg/kg p.o. Bees and livestock have been poisoned by andromedotoxin. Atractyloside: LD_{50} rat 431 mg/kg i.m.

MODE OF ACTION Andromedotoxins are inhibitors of sodium channels (similar to aconitine), which bind to receptor site II of the channel. Sensitive membranes remain depolarised and can no longer transmit action potentials. After initial excitation nerve cells soon become paralysed. The depolarisation leads to an enhanced level of intracellular calcium ions, which cause a positive inotropic reaction in heart muscles. Bradycardia and hypotension result. Atractyloside is a specific inhibitor of ATP/ADP transport across mitochondrial membranes and thus blocks energy supply of cells and the body.

USES (PAST AND PRESENT) Poisoning can occur when honey with grayanotoxins is consumed. An early record of this phenomenon was reported by Xenophon in antiquity. In Turkey a bitter and potentially toxic "Pontic honey" is produced, which contains nectar and pollen from *Rhododendron ponticum* and *R. luteum*. *Rhododendron* and other Ericaceae are common ornamental plants that are regularly found in gardens and parks.

FIRST AID Administration of sodium sulphate and medicinal charcoal. **Clinical therapy:** In serious cases: gastric lavage (possibly with 0.1 % potassium permanganate), instillation of medicinal charcoal and sodium sulphate, possibly electrolyte substitution, control of acidosis with sodium bicarbonate. In case of spasms give diazepam, in case of bradycardia atropine. If necessary, provide shock treatment by giving plasma expander and artificial respiration.

Kalmia latifolia

Rhododendron hirsutum

Rhododendron calophytum

Pernettya mucronata

Steroids and triterpenes

In this group, cardiac glycosides, cucurbitacins and saponins are of special interest from a toxicological point of view.

Cardiac glycosides

A special case of steroidal saponins are cardiac glycosides that inhibit Na^+, K^+-ATPase and are therefore strong toxins but also useful in medicine to treat heart problems.

STRUCTURE Two groups of cardiac glycosides are distinguished: Cardenolides and bufodienolides. They have a sugar side chain at C_3 (R_1 in structures).

Typical **cardenolides** are convallatoxin ($C_{29}H_{42}O_{10}$; MW 550.66), lanatoside A ($C_{49}H_{76}O_{19}$; MW 69.14), lanatoside C ($C_{49}H_{76}O_{20}$; MW 985.14), cymarin ($C_{30}H_{44}O_9$; MW 548.68), oleandrin ($C_{32}H_{48}O_9$; MW 576.73), digitoxin ($C_{41}H_{64}O_{13}$; MW 764.96), digoxin ($C_{41}H_{64}O_{14}$; MW 780,96), gitoxin ($C_{41}H_{64}O_{14}$; MW 780.96), k-strophantine ($C_{41}H_{64}O_{19}$; MW 860.96), adonitoxin ($C_{29}H_{42}O_{10}$; MW 550.66), thevetin ($C_{42}H_{64}O_{19}$; MW 872.97), purpurea glycoside A ($C_{47}H_{74}O_{18}$; MW 927.11), evomonoside ($C_{29}H_{44}O_8$; MW 20.70), helveticoside ($C_{29}H_{42}O_9$; MW 534.66) and ouabain = g-strophanthin ($C_{29}H_{44}O_{12}$; MW 584.67).

Typical **bufadienolides** are scillaren ($C_{36}H_{52}O_{13}$; MW 692.81), proscillaridin A ($C_{30}H_{42}O_8$; MW 530.64) and hellebrin ($C_{36}H_{52}O_{15}$; MW 724.81).

cardenolides

digitoxigenin: R_1 = H; R_5 = CH_3
gitoxigenin: R_1 = H; R_5 = CH_3; R_6 = OH
oleandrigenin: R_1 = H; R_5 = CH_3; R_6 = O-acetyl
digoxigenin: R_1 = H; R_5 = CH_3; R_7 = OH
evonogenin: R_1 = H; R_2 = OH; R_3 = OH; R_5 = CH_3
strophantidol: R_1 = H; R_2 = OH; R_5 = CH_2OH
adonitoxigenin: R_1 = H; R_5 = CHO; R_6 = OH
strophantidin: R_1 = H; R_2 = OH; R_5 = CHO
ouabagenin: R_1 = H; R_2 = OH; R_3 = OH; R_4 = OH; R_5 = CH_2OH

OCCURRENCE Cardenolides: *Digitalis* (Plantaginaceae/Scrophulariaceae), *Convallaria* (Ruscaceae/Convallariaceae), *Acokanthera, Adenium, Alafia, Apocynum, Cerbera, Hunteria, Nerium, Strophanthus, Thevetia,* (Apocynaceae), *Asclepias, Calotropis, Cryptostegia, Cynanchum, Gomphocarpus, Periploca, Sarcostemma, Xysmalobium* (Apocynaceae/Asclepiadoideae), *Adonis* (Ranunculaceae), *Euonymus, Lophopetalum* (Celastraceae), *Cheiranthus, Erysimum* (Brassicaceae), *Ornithogalum, Rhodea,* (Hyacinthaceae), *Coronilla/Securigera* (Fabaceae), *Antiaris, Castilloa, Maquira, Naucleopsis* (Moraceae), *Corchorus olitorius* (Malvaceae/Tiliacae).

Bufadienolides: *Bowiea, Drimia, Scilla, Urginea* (Hyacinthaceae), *Cotyledon, Kalanchoe, Tylecodon* (Crassulaceae), *Eranthis, Helleborus* (Ranunculaceae), *Homeria, Moraea* (Iridaceae), *Melianthus* (Melianthaceae), *Thesium* (Santalaceae).

CLASSIFICATION Heart poison; extremely to highly hazardous, Ia–Ib

SYMPTOMS First effects of poisoning with cardiac glycosides are nausea, vomiting, weakness, a pronounced bradycardia with heartbeat below 40 per minute; both ventricular and supraventricu-

bufadienolides

scillarenin: R_1 = H; R_3 = CH_3
bufalin: R_1 = H; R_3 = CH_3; no double bond (db)
bufotalin: R_1 = H; R_3 = CH_3; R_4 = O-acetyl
hellibrigenin: R_1 = H; R_2 = OH; R_3 = CHO, no db
scillirosidin: R_1 = H; R_3 = CH_3; R_5 = O-acetyl; R_6 = OH

lar extrasystoles (arrhythmia) which can lead to ventricular fibrillation. AV conductor becomes blocked and the blood pressure rises with 140 heartbeats/minute. Patients appear confused and are sometimes psychotic. They have headaches and hallucinations, are tired and suffer from insomnia and changed colour vision (it is likely that Vincent van Gogh, who had painted the sun with a coloured halo, was overdosed with digitalis). Death is caused by cardiac arrest through ventricular fibrillation after a few minutes upon i.v. injection, or hours/days after oral ingestion. Cardiac glycosides are very bitter and normally cause spontaneous vomiting.

TOXICITY Poisoning with cardiac glycosides is quite common since cardiac glycosides are used in heart medicine and because of their narrow therapeutic window (1 500 intoxications per year in Germany). In patients with kidney insufficiency, cardiac glycosides can accumulate.

Digitoxin: LD_{50} guinea pig: 60 mg/kg p.o.; cat: 0.18 mg/kg p.o.; gitoxin: LD_{50} mouse: 6.4 mg/kg i.p., cat: 0.5–0.8 mg/kg i.v.; hellebrin: LD_{50} cat: 0.1 mg/kg p.o.; adonitoxin: LD_{50} cat: 0.191 mg/kg i.v.; lanatoside A: LD_{50} cat: 0.36–0.38 mg/kg i.v.; proscillaridin A: LD_{50} cat: 0.14 mg/kg i.v.; 0.73 mg/kg p.o.; ouabain: LD_{50} cat: 150 µg/kg i.v.; rat 14 mg/kg i.v.; oleandrin LD_{50} cat: 200 µg/kg i.v.; purpurea glycoside A: LD_{50} cat: 0.3–0.5 mg/kg i.v.; calotropin LD_{50} cat i.v. 110 µg/kg; helveticoside LD_{100} in cats 104 µg/kg i.v.

Lethal dose for humans (i.v.): digitoxin 587 µg/kg; digoxin 314 µg/kg; ouabain 117 µg/kg; k-strophantin 110 µg/kg; proscillaridin 120 µg/kg; p.o: ouabain 75 mg.

MODE OF ACTION Cardiac glycosides inhibit a most important ion pump in the human body, the Na^+, K^+-ATPase. This ion pump builds up the Na^+ and K^+ gradients of all cell membranes that are essential for active transport and for neuronal signalling. Blocking Na^+, K^+-ATPase will inhibit the activity of muscles and nerve cells.

The varying structures hardly influence the binding of cardiac glycosides at Na^+, K^+-ATPase; however their absorption differs. Polar molecules are absorbed very slowly, but once inside the body their action is rapid. Lipophilic glycosides easily diffuse into the body but because of binding to plasma proteins they remain in the system for prolonged periods of time.

At therapeutic doses, the inhibition of Na^+, K^+-ATPase indirectly increases the intracellular Ca^{++} concentrations in heart muscles that trigger muscle contraction (positive inotropic activity).

In summary, cardiac glycosides slow down heartbeat and exhibit positive inotropic, positive bathmotropic, weakly negative chronotropic and dromotropic heart activity.

USES (PAST AND PRESENT) The deadly action of cardiac glycosides has been known since antiquity. Oleander was frequently used for murder in the Roman empire. Even today, cardiac glycosides are frequently employed for murder and suicide. In Africa they form the major source of arrow poisons and ordeal poisons. Cardiac glycosides are an important heart medicine to treat heart insufficiency (up to NYHA III and IV) and hypertonia. These beneficial activities were discovered in 1785 in *Digitalis purpurea* by Withering, who found that digitalis helped to treat the symptoms of a weak heart, "dropsy" (oedema that causes swollen legs). Cardiac glycoside containing extracts are used in traditional medicine as diuretic and expectorant (saponin effect).

Extracts from many heart poison plants were successfully used as arrow poisons in several civilisations (Africa, S. America, Asia) (see Neuwinger, 1998, for a detailed and comprehensive coverage).

FIRST AID Induce vomiting; administration of sodium sulphate and 1 g/kg medicinal charcoal. **Clinical therapy:** Gastric lavage (possibly with 0.1% potassium permanganate), instillation of medicinal charcoal; provide heart pacemaker (in severe cases) and use a defibrillator in case of ventricular fibrillations. Inject Digitalis antidote BM (Fab antibody fragments in case of pure glycosides), give 3 x 4 g colestyramine to bind the glycoside during enterohepatic cycling; haemoperfusion and forced diuresis with Lasix or Osmofundin; electrolyte (K⁺) substitution; phenytoin, lidocaine (xylocain) in case of ventricular extra systoles; against sinus bradycardia and AV block: atropine or orciprenaline (Alupent). Avoid substances that affect AV conductor, such as ajmaline, procainamide, β-blocker or quinidine. The degree of poisoning can be estimated by determining the potassium level in serum: above 5 mval/l serious poisoning has taken place.

Asclepias syriaca

Homeria flaccida

Eranthis hyemalis

Strophanthus boivinii

326

Cucurbitacin E and related triterpenes

STRUCTURE Cucurbitacins are a class of toxic triterpenoids that occur as glycosides or free aglycones (which represent the active principle).

Examples are: Cucurbitacin E ($C_{32}H_{44}O_8$; MW 556.70), cucurbitacin I ($C_{30}H_{42}O_7$; MW 514.67); cucurbitacin J ($C_{30}H_{42}O_7$; MW 516.68); bryonin ($C_{48}H_{66}O_{18}$; MW 930.99).

OCCURRENCE Widely spread secondary metabolite in members of the Cucurbitaceae, for example in the genera *Acanthosicyos, Brandegea, Bryonia, Citrullus, Coccinia, Cucumis, Cucurbita, Ecballium, Echinocystis, Fevillea, Hemsleya, Lagenaria, Luffa, Momordica, Solena*, and in *Gratiola officinalis* (Plantaginaceae/ Scrophulariaceae), *Begonia* (Begoniaceae), *Anagallis arvensis* (Myrsinaceae), *Phormium tenax* (Phormiaceae), *Desfontainia* (Desfontainiaceae), *Crinodendron* (Elaeocarpaceae) *Tropaeolum majus* (Tropaeolaceae), *Purshia tridentata* (Rosaceae), and *Iberis amara, Lepidium sativum* (Brassicaceae). Cucurbitacins are stored as glycosides in the vacuole; upon wounding a β-glucosidase (elaterase) releases the active aglycone. Some cucurbitacins occur as bidesmosides, such as bryosides in *Bryonia*.

cucurbitacin B

cucurbitacin D

β-glucosidase

free cucurbitacin

monodesmosidic cucurbitacin in vacuole

glucose

Some plants such as *Ajuga* (Lamiaceae), *Vitex* (Verbenaceae), *Brassica* (Brassicaceae), *Lychnis* (Caryophyllaceae), *Rhaponticum, Serratula* (Asteraceae), *Diploclisia* (Menispermaceae), *Achyranthes, Pfaffia* (Amaranthaceae), *Podocarpus* (Podocrapaceae), and many ferns (Pteridophyta) accumulate steroids that are identical or similar to insect moulting hormones (**ecdysone**, 10-hydroxyecdysone); these compounds inhibit insect development but are not toxic for mammals. Mammalian steroidal sex hormones are less commonly produced by plants, examples of **oestrogens** are *Phaseolus vulgaris, Glycine max* (Fabaceae), *Salix* (Salicaceae), *Phoenix dactylifera* (Arecaceae), *Punica granatum* (Lythraceae/Punicaceae), of **androgens**: pollen of *Pinus sylvestris* (Pinaceae). See isoflavonoids for nonsteroidal phytoestrogens.

CLASSIFICATION Cytotoxin; highly hazardous, Ib (cucurbitacins)

SYMPTOMS Contact with sap of fresh plant material with cucurbitacins leads to skin reddening, painful inflammation and blister formation. First effects of poisoning are disorders of the gastrointestinal tract, including vomiting and bloody diarrhoea. Mucosal skin shows ulceration, sometimes even perforation. Further symptoms are headache, tachycardia, spasms, liver and kidney damage with oligouria, bloody urine and icterus. Uterine bleeding can cause abortion. Death occurs in collapse by respiratory arrest.

TOXICITY LD$_{50}$ mouse: 340 mg/kg p.o.; lethal dose for humans (p.o.): 0.6 g fruit sap of *Ecballium elaterium* and 3 g of *Citrullus colocynthis*. 40 berries of *Bryonia alba* are lethal for adults, 15 for children. Lethal dose for cucurbitacins: cat 1 mg/kg i.p., rabbits 6 mg/kg. Cucurbitacin B has an LD$_{50}$ of 0.5 mg/kg (when injected into a rabbit) or when given orally to mice, the LD$_{50}$ is 5 mg/kg.

MODE OF ACTION Free cucurbitacins are highly cytotoxic as they block mitosis in metaphase by inhibiting microtubule formation. Glycosides of cucurbitacin E and J are eliminated via the enterohepatic pathway.

USES (PAST AND PRESENT) Drugs with cucurbitacins have been used to treat malaria, as emetic or anaesthetic (now obsolete), and in traditional medicine as diuretic, abortifacient and drastic laxative. Cucurbitacins irritate intestinal mucosa and cause release of water into the gut lumen. This in turn activates gut peristalsis and promotes diarrhoea. Due to cytotoxicity, cucurbitacins inhibit tumour growth *in vitro* and *in vivo*. Cucurbitacins have been used to treat nasopharyngeal carcinoma. For topical use, *Bryonia* cucurbitacins are included in creams and ointments to treat rheumatism and muscle pain.

A number of vegetables come from Cucurbitaceae; the bitter cucurbitacins have been removed through selection of non-bitter varieties in most cases.

FIRST AID Administration of medicinal charcoal and sodium sulphate; provide plenty of liquid (e.g. tea). Contact with skin: wash with polyethylene glycol 400 or water; in case of skin blisters, cover with sterile cotton and apply locally corticoid ointment. **Clinical therapy:** Gastric lavage, instillation of medicinal charcoal and sodium sulphate; electrolyte substitution, control of acidosis with sodium bicarbonate (pH of urine 7.5). In case of spasms inject diazepam (i.v.). Provide intubation and oxygen respiration in case of respiratory arrest or paralysis. Check diuresis and kidney function.

Anagallis arvensis

Lagenaria siceraria

Luffa cylindrica

Momordica charantica (flower and fruit)

Saponins

STRUCTURE Triterpenes and steroids (biogenetically derived from C-30 precursors) can occur as free compounds but more often as saponins (with one to several sugar molecules attached to them). Whereas free triterpenes and steroids are lipophilic compounds, the glycosidic saponins are more water soluble and are stored in the vacuole. Saponins are mainly stored as inactive furostanol glycosides or triterpene bidesmosides in the vacuole (the active compounds would destroy the tonoplast of the producing plants!). When attacked by microbes or herbivores these compounds are converted into spirostanol glycosides or triterpene monodesmosides that can attack biomembranes.

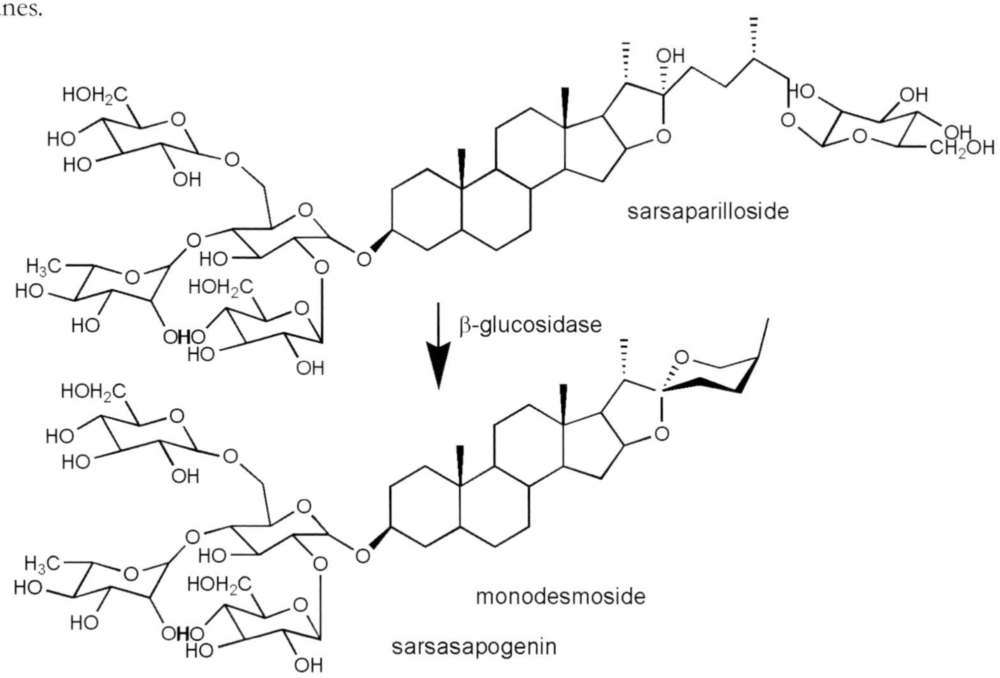

Example of a steroidal saponin that is converted to a monodesmosidic spirostanol

OCCURRENCE Saponins are widely distributed in the plant kingdom and approximately 70% of all plants produce them. **Steroidal saponins** are abundant in monocots of the families Agavaceae, Alliaceae, Asparagaceae, Ruscaceae/Convallariaceae, Dioscoreaceae, Liliaceae, Poaceae, Smilacaceae, Trilliaceae and Zingiberaceae but also occur in some dicots (Fabaceae, Scrophulariaceae, Solanaceae).

Triterpene saponins mainly occur in angiosperm families such as the Aquifoliaceae, Apiaceae, Araliaceae, Asteraceae, Caryophyllaceae, Chenopodiaceae, Cucurbitaceae, Fabaceae, Melanthiaceae, Phytolaccaceae, Polygalaceae, Primulaceae, Ranunculaceae, Rhamnaceae, Rosaceae, Sapindaceae, Sapotaceae, Styracaceae and Theaceae. Gymnosperms are apparently without saponins. Some saponins are present in ferns (such as the sweet-tasting osladin in *Polypodium vulgare*).

CLASSIFICATION Cell poison; moderately hazardous, II

SYMPTOMS Symptoms include gastrointestinal disturbance, nausea, gastritis, abdominal pain, sleepiness, vomiting, diarrhoea, arrhythmia, paralysis and kidney damage.

Horse chestnut: Symptoms of strong GI disturbance, which is caused by the saponins, can be seen after ingestion of one seed. They include flushed, reddened skin, thirst, mydriasis, oedema, vomiting, diarrhoea, hypotension, unconsciousness and collapse.

TOXICITY Polar saponins are absorbed to a limited degree, but higher doses can lead to serious toxic effects, which can be seen in Liliaceae or Agavaceae with steroidal saponins from the genera *Agave, Asparagus, Paris* and *Polygonatum*. However, if the saponins co-occur with oxalate raphides, they can easily enter cells and the body and cause internal disturbances.

Example of a bidesmosidic triterpene saponin that is converted to a monodesmoside

MODE OF ACTION Monodesmosidic saponins are amphiphilic compounds (i.e., they have a hydrophilic sugar and a lipophilic terpene moiety). They can complex cholesterol in biomembranes with their lipophilic terpenoid moiety and bind to surface glycoproteins and glycolipids with their sugar side chain. Saponins can increase the fluidity of the membranes, which can lead to uncontrolled efflux of ions and metabolites and receptors, or even to cell leakage resulting in cell death. This activity can easily be demonstrated with erythrocytes that lose their haemoglobin through haemolysis when in contact with saponins. A conformational change of membrane proteins (e.g. ion channels, transporters) can occur when the lipophilic compounds dive into the biomembrane and insert close to membrane proteins. If Na^+, K^+ or Ca^{++} channels are affected, a disturbance of signal transduction may result.

This membrane activity is rather unspecific and saponins show antimicrobial and cytotoxic activities against a wide set of organisms, ranging from bacteria, fungi, to insects and vertebrates. Cardiac glycosides and steroidal alkaloids from *Solanum* species also react as saponins at high doses (see cardiac glycosides and solanine).

Because of general toxicity, an internal use of saponins in medicine refers to low doses: e.g. saponins affect the *Nervus vagus* in the stomach and reflectorily induce a secretion of water in the lungs thus leading to secretolytic effects. Therefore, plants containing saponins are widely used in traditional medicine as secretolytic drugs. Higher doses cause an emetic effect or toxicity.

Steroids, triterpenes and saponins of plants sometimes structurally resemble endogenous hormones, e.g. glucocorticoids. The anti-inflammatory effects reported from many drugs could be due to such a corticomimetic effect. The triterpenes or steroids could directly interfere with phospholipase or they could act at the level of nuclear receptors (a target of corticoids) that modulate gene activity. A pronounced anti-inflammatory activity has been reported for glycyrrhizic acid from *Glycyrrhiza glabra* (Fabaceae), a triterpene saponin with a sweet taste. This saponin and others have structural similarities to glucocorticoids and bind to hormone receptors, and inhibit Δ-5β-reductase, which catalyses the inactivation of aldosterone. This leads to an elevated level of aldosterone in the body, causing water retention and oedema, serious hypertension and heart insufficiency. Such symptoms have been observed after consumption of large amounts of liquorice.

Cimicifugoside, a triterpene saponin from *Cimicifuga* (Ranunculaceae) inhibits nuceloside transport in mammalian cells.

USES (PAST AND PRESENT) Saponins have been used as a detergent for washing clothes. Since saponins are highly toxic for fish (inhibition of respiration), they have been traditionally employed for fishing – saponin extracts are poured into water and fishermen wait to collect the dead fish. Saponins also kill water snails and have been employed to eliminate snails in tropical waters that transmit human parasites, such as *Schistosoma* (causing schistosomiasis, also known bilharzia).

Steroidal saponins are used in the preparation of steroid hormones.

FIRST AID Medicinal charcoal. Wash eyes; skin inflammation can be treated with corticoid ointment. **Clinical therapy:** Gastric lavage (possibly with potassium permanganate), instillation of medicinal charcoal and sodium sulphate, electrolyte substitution, control of acidosis with sodium bicarbonate. In case of spasms and colic give diazepam and atropine; in severe cases, provide intubation and oxygen respiration.

Agave americana

Asparagus officinalis

Paris quadrifolia

Saponaria officinalis

331

Small reactive metabolites
Ethanol

STRUCTURE Ethanol is a product of anaerobic fermentation. It has a simple structure C_2H_5OH (MW 46.07).

OCCURRENCE Produced by yeasts, such as *Saccharomyces cerevisiae*, when growing on a sugar-rich food source, such as juices from grapes (*Vitis vinifera*), apples (*Malus domesticus*), sugar cane (*Saccharum officinarum*), sugar beet (*Beta vulgaris* var. *vulgaris*), agave (*Agave mexicana*), palm bleeding sap (*Cocos nucifera*) or on a starch source that has been digested with amylase (rice, barley, corn, potato).

CLASSIFICATION Mind-altering; slightly hazardous, III

SYMPTOMS Small doses are stimulating, while doses above a blood alcohol concentration (BAC) of 1.4 g/l (‰; per mille) are considered toxic. Such a level can be reached by consuming 100 g ethanol in a short time. At high BAC values a depression of central functions sets in.
Ethanol effects:

- 0.1–0.5 ‰: no visible or apparent consequences
- 0.3–1.2 ‰ (30–50 ml ethanol): mild euphoria, sociability, chattiness, enhanced self-confidence, loss of inhibitions, decrease of concentration and power of judgement
- 0.9–2.5 ‰ (40–100 ml ethanol): emotional lability, decrease or loss of inhibitions, judgement, memory and articulation, prolonged reaction times and disturbance of motoric and sensoric coordination
- 1.8–3 ‰ (60–150 ml): disorientation and confusion, dizziness, strong emotional states (anxiety, fear, sorrow), disturbance of sensoric and motoric ability, slurred speech
- 2.7–4.0 ‰: apathy, increased paralysis, reduced response to any stimulation, inability to stand or walk, vomiting, involuntary urination and defecation, reduced consciousness, deep sleep and stupor
- 3.5–5 ‰ (150–300 ml ethanol): complete unconsciousness, coma, hypothermia, circulatory and respiratory irregularities
- above 4.5 ‰: death from respiratory arrest.

Further symptoms are reduction of muscular tonus, hyperventilation, tachycardia, hypertension, enhanced diuresis, hypoglycaemia, libido, also nausea and vomiting. Long-term overdosing with ethanol leads to physical and psychological dependence, disturbance of erythropoiesis, nephritis, pancreatitis, gastritis, polyneuropathies, myocardiopathy, dementia, fatty liver and liver cirrhosis. An enhanced prevalence of liver and rectum carcinomas has been observed in alcoholics. Alcoholism (ethanol consumption above 75 g/day) during pregnancy can lead to severe malformations including cerebral damage.

TOXICITY Ethanol is easily absorbed after oral ingestion, but also via the pulmonal and dermal route. It is distributed into all organs and can cross the blood-brain barrier. Ethanol is degraded by alcohol dehydrogenase (ADH) into the reactive acetaldehyde and further by aldehyde dehydrogenase (ALDH) to acetic acid. Ethanol degradation is about 0.1 g/kg/h in men and 0.085 g/kg/h in women. Lethal ethanol concentrations are 5–8 g/kg in adults and 3 g/kg in children.

MODE OF ACTION Ethanol has depressant activity in the CNS. Since ethanol is soluble in lipids, it is likely that it dissolves in the cell membrane. If it accumulates in the neighbourhood of ion channels or other membrane proteins it can change their conformation and thus their activities (similar to the activity of some lipophilic anaesthetics). This activity interferes with inhibiting systems first, later with excitable ones. A more important target seems to be the GABA receptor. As an agonist ethanol can enforce synaptic inactivation and chloride influx which are mediated by GABA. This activity antagonises agonistic reactions of acetylcholine, glutamate and serotonin which are mediated by sodium and potassium channels. Thus ethanol has a similar activity as barbiturates.

USES (PAST AND PRESENT) Humans discovered alcoholic fermentation more than 5 000 years ago. First alcoholic drinks were beer and wine. Most ethanol produced today is used for alcoholic drinks (more than 10 litre per person per year in Western countries); other applications are in chemical, cosmetic and pharmaceutical industries as solvent or raw material, preservation and disinfection or for bioethanol in cars. Alcoholic fermentation can produce maximally 18 % alcohol. By distillation, drinks with higher ethanol content can be produced (such as vodka, whiskey, rum, tequila, schnapps, grappa).

FIRST AID In case of serious intoxication, induce vomiting but avoid choking, keep patient warm. Control respiration and circulation. Gastric lavage with 2% $NaHCO_3$ solution. In case of unconsciousness and danger of respiratory arrest, provide haemodialysis and artificial respiration. To induce ethanol degradation, one may infuse 40 ml 40% laevulose solution with vitamin C. In case of strong excitation inject diazepam (10 to 50 mg). Chronic poisoning (delirium tremens) can be treated with clomethiazol or chlorodiazepoxide for sedation, or with a single physostigmine injection (2 mg; i.m.). See handbooks of toxicology and pharmacology for further treatment of alcoholic poisoning and of alcoholics.

Vitis vinifera – grapes ready for pressing

Agave for tequila production

Beta vulgaris var. *vulgaris*

Saccharum officinarum

Oxalic acid, oxalates and other organic acids

STRUCTURE Oxalic acid ($C_2H_2O_4$, MW 90.04) is a simple dicarboxylic acid, which can be present as a free acid or as a salt (e.g. water soluble potassium oxalate). In the Araceae and Liliaceae oxalic acid is often deposited as hardly soluble calcium oxalate crystals that can form sharp needles (raphides), which makes such plants particulary hazardous. Oxalic acid is a strong acid and powerful reducing agent.

Other organic acids with hazardous properties include formic acid (CH_2O_2, MW 46.03), glycolic acid ($C_2H_4O_3$, MW 76.05), glyoxylic acid ($C_2H_2O_3$, MW 74.04), malonic acid ($C_3H_4O_4$, MW 104.07), monofluoroacetic acid ($C_2H_3O_2F$, MW 78.04) and shikimic acid ($C_7H_{10}O_5$, MW 174.16).

formic acid   **glycolic acid**   monofluoroacetic acid

glyoxylic acid   oxalic acid   shikimic acid

OCCURRENCE Oxalic acid and oxalates are widely distributed in plants at low concentrations (<0.1 %); in some plants higher concentrations (up to 10–20 %) are reached and deposited as raphides. Examples are: *Fagus sylvatica* seeds (Fagaceae), *Arum, Aglaonema, Alocasia, Anthurium, Arisarum, Caladium, Calla, Colocasia, Dieffenbachia, Dracunculus, Epipremnum, Monstera, Philodendron, Spathiphyllum, Zantedeschia* and other Araceae, *Rumex acetosa, Rheum* x *hybridum* and other Polygonaceae, *Oxalis, Averrhoa carambola*, and other Oxalidaceae, *Mesembryanthemum, Psilocaulon* (Aizoaceae/Mesembryanthemaceae), *Amaranthus* (Amaranthaceae), *Spinacia oleracea, Halogeton glomeratus, Kochia scoparia, Chenopodium* species (Amaranthaceae/Chenopodiaceae), *Opuntia* (Cactaceae), *Portulaca oleracea, Portulacaria afra* (Portulacaceae), *Aeschmea, Cryptanthus, Tillandsia cyanea, Vriesea splendens* (Bromeliaceae), *Begonia* (Begoniaceae), *Parthenocissus quinquefolium, Vitis vinifera* (Vitaceae), *Tamus communis* (Dioscoreaceae), *Equisetum palustre* (Equisetaceae), *Hyacinthus* (Hyacinthaceae), *Galanthus* (Amaryllidaceae).

Formic acid is found in stinging hairs of nettles such as *Urtica, Girardinia, Laportea*, Urticaceae and *Loasa,* Loasaceae but is also present in low concentrations in many leaves and roots. Glycolic acid and glyoxylic acid are found in unripe fruits (grapes) and young green leaves. Malonic acid is sometimes accumulated in higher concentrations (e.g. in *Apium graveolens*, Apiaceae; *Phaseolus coccineus*, Fabaceae; *Beta vulgaris*, Amaranthaceae/Chenopodiaceae; and *Hordeum vulgare*, Poaceae).

Monofluoroacetic acid has been found as a toxic acid in *Dichapetalum cymosum* (Dichapetalaceae), and several members of the Australian genera *Oxylobium* and *Gastrolobium* (Fabaceae). Shikimic acid is widely present in low concentrations; higher amounts are found in fruits of *Illicium verum* (Illiciaceae) and in some fruits, such as gooseberry.

CLASSIFICATION Cytotoxin; moderately to highly hazardous, Ib–II

SYMPTOMS The sharp oxalate raphides of Araceae are potent irritants of skin and mucosal tissues; they can penetrate cells and cause necrosis. The release of histamine causes itching, burning, salivation and severe inflammation. Especially painful are inflammation of eyes, when plant sap is spilled into them. Blepharospasm, conjunctivitis and keratitis can result. Gardeners can develop contact dermatitis when regularly handling Araceae.

First symptoms after ingestion are burning of the mouth, throat and oesophagus, followed (within 1 hour) by nausea, vomiting, gastrointestinal disorder, strong, sometimes bloody diarrhoea, arrhythmic pulse, bradycardia, hypotension, strong clonic spasms, collapse of circulation, dyspnoea, cyanosis, coma and death by cardiac arrest or fatal kidney damage. Symptoms of kidney damage (when water-soluble oxalates were ingested) include albinuria, haemoglobinuria and haematuria because kidney tubules become blocked by Ca-oxalate crystals. This causes necrotic nephritis and loss of kidney function, which can be seen by the occurrence of oedema (especially of eyelids) and hypertension. Also liver damage can occur. Chronic exposure to plants with oxalic acid can lead to damage of blood vessels plus gangrene, gastrointestinal problems and albinuria. Patients with reduced kidney function and kidney stones are especially vulnerable and should avoid plants with oxalic acid or else they can develop more kidney stones and nephritis.

TOXICITY Oxalic acid and oxalates are not metabolised in the body; they are partly eliminated via the kidneys and urinary tract. Oxalic acid and its salts are caustic to skin and mucosal tissue because of high acidity and reduction power. About 5–15 g of oxalic acid or oxalates are lethal for humans. Ingestion of plants rich in oxalic acid is however seldom fatal, although a few cases have been reported after ingestion of large amounts of *Rumex, Oxalis* and *Rheum*. Animal poisoning has been reported; ruminants can apparently detoxify oxalates through their rumen microorganisms. Formic acid is a strong acid with acute toxicity. Glycolic, glyoxylic and malonic acid are irritant to skin and mucous membranes. Monofluoroacetic acid is highly poisonous with lethal doses of 2–5 mg/kg in humans.

MODE OF ACTION Acidity, reduction power of oxalic acid and raphides can harm mucosal tissues of the mouth and throat but also the gastrointestinal tract. These tiny pricks can activate mastcells that release histamine. Histamine induces local swelling and pain. Raphide perforations can enhance the entry of other noxious chemicals into cells, such as proteases, lectins or saponins; in *Begonia* oxalates increase the toxicity of cucurbitacin B, present in tubers. In Bromeliaceae raphides mediate the absorption of proteolytic enzymes.

Oxalic acid forms insoluble salts with calcium. If calcium oxalate is deposited in kidney tubules, kidney tissue becomes damaged. By depletion of calcium in the heart, the heart muscles can be damaged and its contractibility is reduced. In the blood, coagulation is also hampered by Ca^{++} depletion. In the CNS oxalic acid causes excitation and at higher doses a central paralysis.

Monofluoroacetic acid is an inhibitor of Krebs cycle enzymes and in consequence blocks cellular respiration. Shikimic acid is a powerful mutagen.

USES (PAST AND PRESENT) Plants with oxalic acid have a sour taste and some are consumed as vegetables, such as rhubarb or sorrel.

FIRST AID Give calcium to precipitate oxalic acid as insoluble calcium oxalate, which can no longer be absorbed from the gastrointestinal tract and which can be eliminated via the faeces; milk or chalk are good sources of calcium (see species monographs). **Clinical therapy:** Gastric lavage (with 1% potassium permanganate and calcium salts, such as 1% calcium gluconate, calcium lactate, $CaCl_2$ solution), instillation of calcium gluconate and medicinal charcoal. To avoid calcium depletion give calcium parenterally. Allow plenty of liquids. Control kidney function and induce diuresis with furosemid. In case of kidney failure apply haemodialysis or peritoneal dialysis. Control acidosis with sodium bicarbonate. In case of spasms inject diazepam. In severe cases, provide intubation and oxygen respiration. If oxalates came into contact with eyes or skin: wash with polyethylene glycol 400 or water; in case of skin blisters, cover with sterile cotton and locally with corticoid ointment. In case of necrotic mucosal inflammation provide glucocorticoids, antihistamines and anti-infectives (such as chamomile); local analgesics against pain.

Rumex acetosa

Zantedeschia aethiopica

Protoanemonin, tulipalin and parasorbic acid

STRUCTURE The glucoside ranunculin ($C_{11}H_{15}O_8$; MW 275.24) is accumulated in the vacuole. Upon wounding and tissue decompartmentation, the glucoside is hydrolysed by a β-glucosidase. The highly reactive protoanemonin is then set free. This compound can be inactivated by forming a dimeric anemonin. Dried plants mostly contain the less toxic anemonin. Pure protoanemonin is a yellow fluid with a pungent smell. In tulip, the glycosides tuliposides A,B (tuliposide A: $C_{11}H_{17}O_8$; MW 277.26; tuliposide B: $C_{11}H_{15}O_9$; MW 293.26) are present (1–2.4 % FW in stems), which upon wounding release the reactive lactone tulipalin with an exocyclic methylene group. Another example of a reactive lactone is parasorbic acid

OCCURRENCE Ranunculin is a characteristic metabolite of Ranunculaceae, especially rich are members of the genera *Anemone, Clematis, Helleborus, Knowltonia, Myosurus, Pulsatilla* and *Ranunculus*. Tuliposide (up to 0.1 %) has been found in the genera *Alstroemeria, Bornarea, Erythronium, Fritillaria, Gagea, Notholirion, Lilium* and *Tulipa*. *Narthecium ossifragum* (Nartheciaceae) contains nartheside A and B, which are converted into the reactive narthogenins; nartheside A, B have similar properties as protoanemonin and tulipalin. Parasorbic acid is present in fruits (up to 0.3 %) of *Sorbus aucuparia, S. aria, S. americana, S. sambucifolia* and *S. tianshanica* (Rosaceae).

ranunculin   protoanemonin   anemonin

tuliposide A   tulipalin A

parasorbic acid

CLASSIFICATION Cell toxin, allergen; moderately hazardous, II

SYMPTOMS Protoanemonin and tulipalin are strong irritants of skin and mucosal tissue (mouth, throat, bronchia) and cause gastroenteritis with vomiting, bloody diarrhoea and nephritis. Upon contact with the skin, skin reddening, itching and blister formation can occur followed by inflammation with necrotic wounds. Also the CNS can be affected. Upon ingestion of plants rich in protoanemonin central excitations can be observed, that are followed by central paralysis. Death is caused by respiratory arrest.

Tulipalin is responsible for a serious dermatitis in gardeners who regularly handle tulip bulbs or come into contact with tulip saps ("tulip fingers", "tulip nails"). It forms an itching and burning excema of hands and finger tips. Nails become weak. When plants with tuliposide are ingested, gastrointestinal symptoms are usually observed.

Parasorbic acid is a lactone with a pungent smell and skin irritating properties. It causes gastroenteritis and nephritis upon ingestion of larger amounts. In animals mydriasis and ataxia have been observed as symptoms.

TOXICITY Poisoning with protoanemonin can occur when plant extracts are ingested in traditional medicine (*Anemone, Pulsatilla*) or when leaves or flower buds of *Caltha palustris* or *Ficaria*

verna are eaten as salad or substitute for capers. Animal poisoning has been reported when large amounts of *Ranunculus* were included in the feed. Protoanemonin: LD_{50}: mouse 190 mg/kg i.p.; dog 20 mg/kg, p.o. About 30 plants of *Anemone* are said to be lethal for humans. Parasorbic acid: LD_{50} mouse: 750 mg/kg i.p.; about 90 kg fresh fruits are lethal for humans.

MODE OF ACTION Tulipalin and protoanemonin have a highly reactive extracyclic methylene group that can form covalent bonds with free SH-groups of proteins or glutathione. Therefore, cytotoxic and allergenic effects can occur. Protoanemonin can also alkylate DNA and is therefore mutagenic. It exhibits antibacterial and antifungal properties. Tuliposide and tulipalin have cytotoxic and fungitoxic properties.

USES (PAST AND PRESENT) Some plants with protoanemonin are used in traditional medicine to treat infections and cold. Many of them have beautiful flowers and are often grown as ornamental plants.

FIRST AID If large amounts (more than 2 tulip bulbs) were consumed, instil medicinal charcoal, sodium sulphate and induce vomiting. Allergic symptoms and skin wounds can be treated with local application of corticoid ointment.

Anemone canadensis

Clematis integrifolia

Tulipa sylvestris

Narthecium ossifragum

Quick guide to poisonous plants, mind-altering plants and fungi

Species with a monograph are printed in **bold**.

Occurrence:
"orn. plant" – plants used indoors or in gardens as ornamental plants.

Toxicity Class:
Class Ia: extremely hazardous; Ib: highly hazardous; II: moderately hazardous; III: slightly hazardous.
IN =inflammatory; CT = cell toxin; NT = neurotoxin; HP = heart poison, MA = mind-altering; MP = medicinal plant; MU = mutagenic;
GI = disturbance of GI tract; AP = animal poison.

Therapy: species or toxin monograph containing more information on symptoms, toxicity, modes of action and treatment.

Comment: A number of scientific names and affiliations have changed recently because of findings from new DNA studies (according to APG-II).

Scientific name (common name); Family	Occur-rence	Hazardous plant parts	Active principle	Toxicity Class	Symptoms & Therapy
Abrus precatorius (rosary bean); Fabaceae	tropics	leaves, seeds	abrin a-d (lectin); triterpene saponins (abrusosides)	Ia CT, NT, GI	Onset of symptoms delayed by several hours; vomiting, diarrhoea, acute gastroenteritis, chills, convulsions, death from heart failure. **Therapy:** Abrin
Acacia sieberiana var. *woodii* (paper-bark thorn), *A. caffra*, *A. berlandieri*, *A. georginae*; Fabaceae	Africa, America, Australia	aerial parts	cyanogenic glucosides: proacacipetalin, 3-hydroxyheteroden-drin	Ib CT, GI, AP	When seeds are crushed they release HCN which is a strong poison; inhibits cellular respiration; young shoots may be grazed and cause poisoning of animals. **Therapy:** Amygdalin
Acacia simplicifolia, A. confusa, A. phlebophylla, A. polyantha, A. maidenii, A. campylacantha and related species; Fabaceae	Australia, Africa	leaves, bark	*N,N*-DMT, MMT, tryptamine	III MA	*N,N*-DMT stimulates serotonin receptors; hallucinogenic, psychedelic. **Therapy:** Bufotenine
Acalypha hispida (copper leaf), *A. indica*; Euphorbiaceae	Australia, Asia; common orn. plant	aerial parts	acalyphin (cyanogenic glucoside)	II CT, NT, GI	Irritation and inflammation of stomach and intestinal mucosa, skin irritation; after substantial ingestion: CNS disturbance, HCN poisoning. **Therapy:** Amygdalin

Scientific name (common name); Family	Occurrence	Hazardous plant parts	Active principle	Toxicity Class	Symptoms & Therapy
Acokanthera oppositifolia (= *A. ouabaio*) (bushman's poison), *A. oblongifolia, A. schimperi, A. longiflora;* Apocynaceae	Africa south of Sahara, as orn. plant in tropics	stems, leaves, seeds	ouabain, acovenoside A, spectabilin, acospectoside A, acobioside A, acokantherin and other cardiac glycosides	Ia CT, HP, GI, AP, MP	Cardiac glycoside poisoning; nausea, salivation, purging and exhaustion, with the usual respiratory and cardiac abnormalities; high doses lead to immediate death through heart failure; famous as arrow poisons; livestock poisoning. **Therapy:** Cardiac glycosides
Aconitum napellus (monkshood), *A. lycoctonum, A. carmichaelii, A. chinense, A. ferox, A. reclinatum, A. uncinatum, A. columbianum* and related species; Ranunculaceae	Europe, Asia, N & S America; orn. plant	all parts, especially tubers	aconitine, mesaconitine, lycoctonine and other terpene alkaloids (up to 2% in tubers)	Ia NT, MA, AP, MP	Aconitine activates Na$^+$ channels and is thus a strong nerve and muscle poison; causes numbness, paralysis; strongly psychedelic when smoked or absorbed through the skin. **Therapy:** Aconitine
Acorus calamus (sweet flag); Acoraceae	Europe, Asia, N America	rhizome	β-asarone; decadienal, caryophyllene	II–III MU, MA	β-asarone as active principle: a stimulant claimed to be hallucinogenic at high doses; has been used as aphrodisiac; mutagenic and carcinogenic. **Therapy:** Myristicin
Actaea spicata (bane berry), *A. pachypoda* (= *A. alba), A. rubra, A. arguta* and related species; Ranunculaceae	Europe, Asia, N America; orn. plants	all parts, especially berries and roots	magnoflorine and other isoquinoline alkaloids; ranunculin is probably absent	III GI, AP, IN	Ingestion of fruits is rarely fatal but leads to disturbance of the GI tract; livestock poisoning has been recorded; juice is irritant on skin and mucous membranes causing blistering and ulceration. **Therapy:** Oxalic acid
Adenia digitata (wild passion flower), *A. glauca, A. gummifera, A. volkensii;* Passifloraceae	S Africa	roots, aerial parts	modeccin (toxic lectin); tetraphyllin B (cyanogenic glucoside)	Ia CT, GI	Vomiting, diarrhoea and convulsions; kidney and liver damage; a modeccin injection causes multi-system organ failure. **Therapy:** Abrin
Adenium multiflorum (impala lily), *A. obesum* (desert rose), *A. boehmianum;* Apocynaceae	Asia, Africa; orn. plant worldwide	aerial parts, lacticifers	obebioside; echubioside, honghelin, somalin and other cardiac glycosides and steroid glycosides	Ia CT, HP, GI, NT	Cardiac glycoside poisoning, affecting mainly livestock; cytotoxic properties; used as arrow and fish poison. **Therapy:** Cardiac glycosides
Adonis vernalis (pheasant's eye), *A. aestivalis* and related species; Ranunculaceae	S & C Europe, Asia; orn. plant	all parts	adonitoxin and other cardiac glycosides (cardenolides)	Ib CT, HP, GI, MP	Symptoms of cardiac glycoside poisoning; diuresis, irritation of GI tract with strong diarrhoea, cardiac arrest, livestock poisoning more common (especially horses, sheep). **Therapy:** Cardiac glycosides

Scientific name (common name); Family	Occurrence	Hazardous plant parts	Active principle	Toxicity Class	Symptoms & Therapy
Aesculus hippocastanum (horse chestnut), *Aesculus pavia, A. californica, A. flava, A. glabra, A. sylvatica*; Hippocastanaceae / Sapindaceae	C Europe; N America	seeds, fruit coat	triterpene saponins (aescin); esculin (coumarin)	III CT, IN, GI, MP	Young children sometimes eat seeds which are used as toys; ingestion can cause disturbance of GI tract; contact allergy; livestock poisoning. **Therapy:** Saponins.
Aethusa cynapium (fool's parsley); Apiaceae	Europe; naturalised N America	all parts	polyacetylenes (aethusin, aethusanol A & B)	Ib CT, NT, GI, IN	Burning sensation in mouth; vomiting, diarrhoea, headache, tachycardia, mydriasis; generalised seizures and convulsions; coma, respiratory arrest (*Cicuta*). **Therapy:** Polyacetylenes
Agapanthus africanus (African lily); Agapanthaceae	S Africa; orn. plant	rhizome	yuccagenin and agapanthagenin (steroidal saponins)	II–III CT, GI	Saponins have cytolytic properties; substantial ingestion causes GI tract and kidney disturbance. **Therapy:** Saponins.
Agaura salicifolia; Ericaceae	E Africa	leaves	diterpene (similar to andromedotoxin)	Ib CT, NP, AP	Probably similar symptoms as with andromedotoxin intoxication; livestock poisoning (sheep, goats). **Therapy:** Andromedotoxin
Agave tequilana, A. americana, A. lechuguilla and related species; Agavaceae	C America; widely cultivated and introduced	all parts	steroid saponins such as hecogenin, oxalic acid, 8 % sugar	II, CT, IN, AP, MA	Strong skin and mucosal irritant, kidney disturbance; source of tequila, mescal; pulque is sometimes mixed with intoxicating herbs (*Turbina, Sophora, Datura, Lophophora, Psilocybe*); livestock poisoning under drought conditions. **Therapy:** Ethanol and saponins.
Ageratum conyzoides (billy-goat weed), *A. houstonianum;* Asteraceae	tropical America; introduced to Africa	aerial parts	cyanogenic glucosides, coumarins; precocoene I (juvenile hormone activity)	II CT, GI	Substantial ingestion can cause disturbance of GI tract; livestock poisoning (cows, rabbits). **Therapy:** Amygdalin
Aglaonema commutatum (Chinese evergreen); Araceae	Indonesia; orn. plant	all parts	Ca^{2+}-oxalate, saponins, toxic peptides	Ib CT, IN, GI	Irritation of skin and mucous membranes, GI tract, kidney and CNS disturbance. **Therapy:** Oxalic acid and saponins
Agrostemma githago (corn cockle); Caryophyllaceae	S Europe; orn. plant	all parts, seeds	githagin, agrostemmic acid (triterpene saponins), toxic lectins	Ib CT, GI IN	Mucosal irritation; dizziness, vomiting, diarrhoea, respiratory distress, headache, pains in spine, tachycardia; paralysis, coma and death. **Therapy:** Saponins
Ailanthus altissima (tree-of-heaven); Simaroubaceae	China; orn. tree	seeds, bark	quassin, ailanthin; indole alkaloids	II CT, GI, IN	Skin irritant; vomiting, gastroenteritis with severe diarrhoea, vertigo, headache, paralysis. **Therapy:** Indole alkaloids

Scientific name (common name); Family	Occurrence	Hazardous plant parts	Active principle	Toxicity Class	Symptoms & Therapy
Albizia saman (= *Samanea saman*); Fabaceae	SE Asia	leaves, seeds	saponins, lectins	II CT, GI	Substantial ingestion can cause GI tract disturbance and multi-organ failure. **Therapy: Abrin**
Albizia versicolor, *A. tanganyicensis*, *A. anthelmintica*; Fabaceae	S Africa; orn. plants	seeds, bark	4'-O-methylpyridoxine, 5'-O-acetyl-4'-O-methylpyridoxine (alkaloids)	II CT, GI, NT, AP	Animal poisoning (cattle, sheep): antivitamin B_6 activity: hypersensitivity, intermittent convulsions, high temperature; livestock may ultimately die from heart failure; human deaths from overdoses of *A. anthelmintica*. **Therapy: Pyrrolidine alkaloids**
Aleurites fordii (tung nut), *A. cordata*, *A. moluccana*, *A. trisperma* and related species; Euphorbiaceae	SE Asia, West Indies & C America; Hawaii	fruits, seeds, aerial parts	12-O-palmityl-13-O-acetyl-16-hydroxyphorbol (diterpene ester); saponins	II CT, IN, GI	Phorbol esters activate protein kinase C; therefore co-carcinogenic; probably associated with nasopharyngeal tumours in S China; vomiting; severe gastroenteritis, intestinal colic, electrolyte imbalance; skin irritant. **Therapy: Phorbol esters**
Alkanna tinctoria; Boraginaceae	Mediterranean; orn. plant	roots	alkannin (red dye); 7-angeloylretronecine, triangularine and other PAs	II CT, NT, MU	PAs are mutagenic and carcinogenic; substantial ingestion can cause GI tract and CNS disturbances. **Therapy: Senecionine**
Allamanda cathartica (allamanda), *A. neriifolia*; Apocynaceae	S & C America; orn. plant	fruits, sap of stems and leaves	plumericin (indole alkaloid)	II CT, GI	Causes strong disturbance of GI tract hence the name; especially cramps and diarrhoea; dehydration; electrolyte abnormalities. **Therapy: Anthraquinones**
Allium sativum (garlic), *A. ursinum*, *A. canadense*, *A. cepa*, *A. vineale* and related species; Alliaceae	Asia; cultivated worldwide	bulbs	alliin (NPAA) is cleaved by allinase to allicin and various allylsulphides	III GI, MP	Allicin and allylsulphides can bind to proteins; at high doses toxic effects can be seen (especially in children); irritation of GI tract, vomiting, sweating; nausea, colics, diarrhoea; can cause allergies. **Therapy: Non-protein amino acids**
Alocasia macrorrhiza (elephant's ear), *A. indica*, *A. odora*; Araceae	Asia; orn. plant in tropics	aerial parts	Ca^{2+}-oxalate; toxic peptides	II CT, IN	Severe inflammation of the mouth and throat; numbness of the oral cavity, swollen lips, difficulty in swallowing, salivation, and abdominal pain, possible asphyxiation. **Therapy: *Dieffenbachia***
Aloe ferox (Cape aloe, bitter aloe), *A. vera* and other *Aloe* species; Asphodelaceae	S Africa; orn. plant in subtropics	aerial parts, leaf exudate	aloin and other anthraquinones	II GI, MP	Strong laxative causing gastroenteritis, at high doses intestinal bleeding, enhances menstrual and uterine bleeding (abortifacient), kidney disturbance, hypertrophy of intestinal tissues. **Therapy: Anthraquinones**

Scientific name (common name); Family	Occurrence	Hazardous plant parts	Active principle	Toxicity Class	Symptoms & Therapy
Alstonia scholaris (milky pine), *A. angustifolia* and related species; Apocynaceae	India, SE Asia	bark, especially seeds	dictamnine, echitamine, alstovenine, reserpine and other monoterpene indole alkaloids; *A. angustifolia* with yohimbine	II NT, MA, MP	Alstovenine is a MAO inhibitor, explaining hallucinogenic, stimulant and psychedelic properties; yohimbine has aphrodisiac properties; bark extracts used in Ayurvedic medicine. **Therapy:** Indole alkaloids
Alstroemeria spp. (Peruvian lily); Alstroemeriaceae	Chile, Peru; common orn. plant	all parts, especially bulbs	tulipalin, tuliposide A,	III IN, GI, AP	Gastrointestinal disturbance; serious skin allergen; animal poisoning. **Therapy:** Protoanemonin
Amanita muscaria (fly agaric); *A. pantherina, A. cothurnata, A. gemmata, A. smithiana, A. strobiliformis;* Amanitaceae	northern hemisphere	fruiting body	ibotenic acid, muscimol, muscarine, muscazone	Ib–II NT, MA	Muscimol is a strong parasympatholytic, psychoactive and hallucinogenic agent; bradycardia, rapid breathing, inebriation, manic behaviour, delirium, deep sleep. **Therapy:** *A. muscaria*
Amanita phalloides (death cup), *A. verna, A. virosa, A. bisporigera, A. ocreata;* Amanitaceae	northern hemisphere	fruiting body	cyclopeptides, amatoxins, amanitine, phalloidin, phallotoxins	Ia CT, NT, GI	Deadly poison; 6–24 h latent period before first symptoms occur. Diarrhoea, severe abdominal cramps, nausea, vomiting, liver and kidney failure; even death. **Therapy:** *A. phalloides*
Amaryllis belladonna (belladonna lily) and cultivars; Amaryllidaceae	S Africa; widely cultivated orn. plant	all parts, especially bulbs, seeds	mainly ambelline, also lycorine, caranine, acetylcaranine and undulatine and other isoquinoline alkaloids	Ib GI, NT, AP	Analgesic similar to morphine; nausea, vomiting, strong perspiration, diarrhoea; kidney trouble; also known animal poison (sheep); used as arrow poison. **Therapy:** Lycorine and *Amaryllis*
Amelanchier ovalis (June berry), *A. alnifolia* and related species; Rosaceae	Europe; orn. plant	seeds	prunasin and other cyanogenic glucosides (CG) (50 ppm)	III CT, GI, AP	Substantial ingestion can cause gastrointestinal disturbances; animal poisoning when feeding on *A. alnifolia* with high CG content in leaves. **Therapy:** Amygdalin
Ammocharis coranica; Amaryllidaceae	S Africa	bulbs, seeds	lycorine, but also acetylcaranine, caranine, crinamine and other alkaloids	Ib–II CT, NT, GI	Substantial ingestion can cause GI tract disturbance with vomiting and diarrhoea, heavy sweating, and even death. **Therapy:** Lycorine
Amsinckia intermedia (tarweed) and related species; Boraginaceae	N America; orn. plant	aerial parts	pyrrolizidine alkaloids (amsinckine)	II CT, NT, MU, AP	PAs are mutagenic and carcinogenic, affecting liver and other organs; livestock poisoning (horses, pigs, cattle). **Therapy:** Senecionine

Scientific name (common name); Family	Occurrence	Hazardous plant parts	Active principle	Toxicity Class	Symptoms & Therapy
Anacardium rhinocarpus, A. occidentale (cashew nut); Anacardiaceae	tropical America, SE Asia; cultivated in tropics	stem bark, fruits, seed shell, sap	cardol, anacardic acid; acuja oil	II CT, IN, GI	Skin and mucosal irritant, contact dermatitis, blister formation; gastroenteritis and paralysis of jaws; used as fish poison. **Therapy:** *Rhus*
Anadenanthera peregrina (yopo), *A. colubrina*; Fabaceae	tropics of S America	seeds	*N,N*-DMT, 5-MeO-DMT, bufotenine, traces of β-carboline alkaloids	II NT, MA	Bufotenine, *N,N*-DMT, 5-MeO-DMT stimulate serotonin receptors; strong hallucinogen. **Therapy:** Bufotenine
Anagallis arvensis (scarlet pimpernel); Myrsinaceae	Europe, Asia, America, Australia	all parts, especially roots	cucurbitacin B, arvenin I and arvenin II (cucurbitacins), triterpene saponins, oxalates	II CT, GI, AP,	Diuretic, substantial ingestion causes gastrointestinal disturbance, trembling, kidney damage, slightly narcotic, mainly livestock poisoning (sheep, calves). **Therapy:** Saponins and cucurbitacins
Anamirta cocculus (fish berry); Menispermaceae	E Asia; Sri Lanka	fruits	picrotoxin (picrotoxigenin, picrotin); menispermine (isoquinoline alkaloids)	Ib NT, MA, MP	Stimulates motoric elements of CNS, occurrence of spasms, headache; vomiting, sleepiness, coma. **Therapy:** Sesquiterpenes
Anchusa officinalis (common bugloss), *A. caerulea* and related species; Boraginaceae	Europe; orn. plants	all parts	lycopsamine and other PAs	III CT, NT, MU, GI	Mainly livestock poisoning; PAs are mutagenic and carcinogenic; neurotoxic, liver (veno-occlusive disease) and lung damage. **Therapy:** Senecionine
Andira araroba, A. inermis; Fabaceae	S America	bark, seeds	chrysarobin (an anthraquinone), andirine (alkaloid)	II CT, GI, IN	Skin and mucosal irritant with blister formation; vomiting, diarrhoea, nephritis, cytotoxic; used as fish poison. **Therapy:** Anthraquinones
Andromeda polifolia (wild rosemary); Ericaceae	C & E Europe	leaves, flowers	possibly andromedotoxin, iridoid glycosides	II CT, NT	Burning sensation in mouth; salivation. **Therapy:** Andromedotoxin
Anemone nemorosa, A. ranunculoides, A. occidentalis, A. patens, A. canadensis, A. coronaria and related species; Ranunculaceae	Europe, Asia, N America; orn. species	all parts	ranunculin is converted enzymatically to protoanemonin	II CT, NT, IN, GI, AP	Nausea, CNS, GI tract and kidney disturbance; mutagenic; blistering, ulceration and inflammation of skin; livestock poisoning. **Therapy:** Protoanemonin and *Pulsatilla*
Anthemis cotula (stinking mayweed); Asteraceae	Europe, Africa, America	aerial parts	anthecotulide and other sesquiterpenes	III IN, CT, GI	Skin irritant; contact allergen; can cause anaphylactic shock. **Therapy:** Sesquiterpenes

Scientific name (common name); Family	Occurrence	Hazardous plant parts	Active principle	Toxicity Class	Symptoms & Therapy
Anthurium scherzianum x *hortulanum* (flamingo flower), *A. wildenovii, A. x roseum;* Araceae	tropical America; orn. plant	young leaves	Ca^{2+}-oxalate raphides	II CT, IN, GI	Mucosal irritant; severe swelling of throat and mouth; kidney disturbance, nausea, vomiting, diarrhoea. **Therapy:** Oxalic acid
Antiaris toxicaria (upas tree); Moraceae	Malaysia, Africa, Thailand	aerial parts, latex	cardioactive glycosides (toxicariosides, antiarin derivates)	Ia CT, HP	Na^+, K^+-ATPase inhibition; symptoms of cardiac glycoside poisoning; used as arrow poison. **Therapy:** Cardiac glycosides
Apocynum cannabinum (Indian hemp), *A. androsaemifolium;* Apocynaceae	N America; orn. plant	all parts, especially roots	cymarin, cynamarin, strophanthidine, and other cardiac glycosides (cardenolides)	Ib CT, NT, HP, GI	Substantial ingestion causes cardiac glycoside intoxication; gastrointestinal disturbance, cardiac arrest; toxic to livestock. **Therapy:** Cardiac glycosides
Aquilegia vulgaris (columbine) and other species; Ranunculaceae	Europe, Asia, N America; common orn. plant	all parts, seeds	cyanogenic glucoside	II CT, MP	High doses can induce dizziness, mydriasis, unconsciousness, respiratory problems; ingestion of seeds has been fatal to children. **Therapy:** Amygdalin
Aralia spinosa (prickly ash) and related species; Araliaceae	N America	aerial parts; bark, fruits	triterpene saponins	II CT, GI, IN	Skin and mucous membrane irritation; disturbance of GI tract; possible livestock poisoning. **Therapy:** Saponins
Archidendron jiringa; Fabaceae	SE Asia	seeds	djenkolic acid (NPAA)	II CT	NPAAs interfere with amino acid metabolism. **Therapy:** Non-protein amino acids
Arctostaphylos uva-ursi (common bearberry) and related species; Ericaceae	Europe, Asia, N America	leaves	arbutin, methylarbutin; hydroquinone; tannins	III CT, GI, IN	Hydroquinone is a strong irritant, especially to mucous membranes of GI tract; can induce vomiting, uterus contractions. **Therapy:** Hydroquinones
Ardisia crenata (spiceberry); Myrsinaceae	Asia	roots, fruits	saponins (cyclamiretin A)	II-III CT, GI	Skin and mucous membrane irritation; disturbance of GI tract. **Therapy:** Saponins
Areca catechu (areca nut palm); Arecaceae	SE Asia; cultivated in tropical rain-forest	seeds	arecoline, arecaidine, guvacine, guvacoline	II NT, MA, MP	Arecoline activates mAChR; therefore psychoactive; induces salivation, bradycardia, respiratory and cardiac arrest; betel chewing can lead to oral cancer. **Therapy:** Pyrrolidine alkaloids and *Areca*

Scientific name (common name); Family	Occurrence	Hazardous plant parts	Active principle	Toxicity Class	Symptoms & Therapy
Argemone mexicana (prickly poppy). *A. albiflora*, *A. ochroleuca*, *A. glauca*, *A. intermedia* and related species; Papaveraceae	C America; widely grown orn. plant	aerial parts, seeds	sanguinarine, berberine, protopine and several other isoquinoline alkaloids	II CT, NT, MA, MP, GI	Traditionally smoked instead of *Cannabis* in Mexico; aphrodisiac and euphoric effects; narcotic, blurred vision, emetic, sedative; GI tract disturbance; coma. **Therapy:** Isoquinoline alkaloids and *Argemone*
Argyreia nervosa (Hawaiian wood rose) and related species; Convolvulaceae	India, Australia; orn. plant	seeds	ergot alkaloids (0.3 %) such as agroclavine, ergine, chanoclavine I, lysergic acid, ergometrine and others	Ib–II NT, MA	Strongly psychedelic (4–8 seeds), similar to LSD, colourful visions. **Therapy:** Ergot alkaloids
Ariocarpus fissuratus (living rock) and related species; Cactaceae	Texas, Mexico	aerial parts	hordenine, *N*-methyltyramine, and other tyramine derivatives	II NT, MA	Mimics noradrenaline and dopamine; unpleasant psychoactive properties. **Therapy:** Bufotenine
Arisaema triphyllum (green dragon), *A. dracontium*; Araceae	Europe, America; orn. plant	whole plant	Ca²⁺-oxalate raphides; toxic peptides	II CT, IN, GI	Irritation of mucous membranes with oedema and blistering, burning sensation of lips and mouth; GI tract disturbance. **Therapy:** Oxalic acid
Aristolochia clematitis (birthwort). *A. grandiflora*, *A. petersiana* and related species; Aristolochiaceae	Mediterranean, N America, Asia, Africa; orn. plant	aerial parts	aristolochic acid and related alkaloids; magnoflorine	II CT, GI, MU, MP	Aristolochic acid has mutagenic and carcinogenic properties; ingestion can induce vomiting, disturbance of GI tract; tachycardia; hypotension, respiratory arrest. **Therapy:** *Aristolochia*
Armoracia rusticana (horseradish); Brassicaceae	Asia; widely cultivated	roots	glucosinolates release the active isothiocyanates upon enzymatic hydrolysis	III CT, IN, GI	Substantial ingestion can cause nausea, bloody vomiting, and diarrhoea. **Therapy:** Mustard oils
Arnica montana, *A. chamissonis*, *A. fulgens*, *A. cordifolia* and related species; Asteraceae	Europe, N America	aerial parts, flowers	helenaline and other sesquiterpene lactones	II CT, IN, MP, GI	Skin irritation; severe gastroenteritis with diarrhoea and stomach pains, headache, vertigo, bradycardia, anxiety, strong heartbeat; abortifacient. **Therapy:** Sesquiterpenes
Artemisia absinthium (wormwood) and related species; Asteraceae	Europe, Asia	aerial parts	thujone, absinthin and other terpenoids	II–III CT, NT, MA, MP	Essential oil (especially thujone) mixed with alcohol has hallucinogenic effects similar to THC; headache, nausea; high doses cause spasms and paralysis leading to coma and death; had been a popular abortifacient. **Therapy:** Monoterpenes

345

Scientific name (common name); Family	Occurrence	Hazardous plant parts	Active principle	Toxicity Class	Symptoms & Therapy
Artemisia cina (wormseed); Asteraceae	E Europe	flowers, seeds	santonin, artemisin and other sesquiterpenes	Ib CT, NT, GI, MP	Disturbance of GI tract, active on CNS and spinal nerves; irritation of motoric nerves induces unconsciousness and epileptic convulsions; respiratory arrest. **Therapy:** *Artemisia cina*
Arum maculatum (cuckoo pint, lords and ladies), *A. italicum*, *A. palaestinum* and related species; Araceae	Europe, Asia; common orn. plant	aerial parts, fruits	aroin, cyanogenic glucosides; saponins, Ca^{2+}-oxalate	Ia CT, IN, GI	Skin irritant; blister formation, burning sensation in mouth and throat; cardiac arrhythma, CNS disturbance, spasms; low body temperature, internal bleeding, disturbance of GI tract. **Therapy:** *Arum* and oxalic acid
Arundo donax (giant reed); Poaceae	Europe, Asia, Africa, naturalised in New World	roots, rhizomes	*N,N*-DMT, 5-MeO-DMT, bufotenine	II NT, MA	Stimulates neuronal serotonin receptors; potentially hallucinogenic; has been dedicated to the god Pan; it is possible that the term "panic" describes the negative effects of a trip; might have been used together with *Peganum harmala* as intoxicant. **Therapy:** Bufotenine
Asarum europaeum; Aristolochiaceae	Europe	all parts	β-asarone, camphor	II CT, MU, GI	Substantial ingestion can induce disturbance of GI tract and CNS. **Therapy:** Myristicin
Asclepias tuberosa (milkweed), *A. syriaca*, *A. fruticosa* (= *Gomphocarpus fruticosus*), *A. physocarpa* and related species; Apocynaceae (Asclepiadoideae)	N America; Africa; orn. plant	aerial parts, latex	asclepiadin, gomphoside, afroside and other cardiac glycosides and saponins	II CT, HP, GI, NT, AP	Symptoms of cardiac glycoside intoxication; disturbance of GI tract, weakness, spasms, seizures; known animal poisoning (sheep, cattle, goats, horses, fowl). **Therapy:** Cardiac glycosides and *Asclepias*
Asparagus officinalis (asparagus); Asparagaceae	Europe; widely cultivated	red fruits	fruits contain several steroid saponins (spirostanol and furostanol glycosides)	II–III CT, GI	Cytotoxic; gastrointestinal disturbances; no severe intoxication. **Therapy:** Saponins
Aspergillus flavus, *A. parasiticus*; Trichocomaceae	worldwide (grows on food items)	mycelia	aflatoxin	Ia CT, GI, MU	Highly carcinogenic, can cause liver disturbance and liver cancer. **Therapy:** Coumarins
Aspidosperma quebracho-blanco (white quebracho); Apocynaceae	S America	bark	yohimbine, quebrachamine and other monoterpene indole alkaloids	Ib–II NT, IN	Alpha-sympatholytic properties; local anaesthetic; emetic; vasoconstriction; skin irritant. **Therapy:** Indole alkaloids

Scientific name (common name); Family	Occurrence	Hazardous plant parts	Active principle	Toxicity Class	Symptoms & Therapy
Astragalus molissimus (locoweed), *A. didymocarpus, A. lentiginosus, A. wootonii, A. calycosus* and related species; Fabaceae	America, Europe, Asia	aerial parts	hydroxyalkaloids (swainsonine)	II CT, NT	The hydroxyalkaloids inhibit glucosidases and other hydrolases in lysosomes and Golgi apparatus; induce structural changes in nerve cells; locoism: livestock act in an irrational and aggressive manner, clumsy gait, seizures, loss of coordination, death. **Therapy:** Non-protein amino acids
Astragalus pectinatus (poison milkvetch), *A. racemosus* and related species; Fabaceae	N America	aerial parts	NPAAs with selenium	II CT, NT	Selenium-containing NPAAs are mimics of cystein; substantial ingestion can induce breathing difficulty, depression, coma, death; blind staggers in livestock. **Therapy:** Non-protein amino acids
Atropa belladonna (deadly nightshade) and related species; Solanaceae	Europe, Asia; orn. plant	all parts	hyoscyamine, scopolamine and other tropane alkaloids	Ia NT, MA, MP	mAChR antagonist, parasympatholytic properties; hallucinogenic in various forms; aphrodisiac; mydriasis; hyperthermia, inhibition of salivation; death from respiratory arrest. **Therapy:** Hyoscyamine and *Atropa*
Aucuba japonica (Japanese aucuba) and related species; Cornaceae	SE Asia; orn. plant	aerial parts, fruits	aucubin and other iridoid glucosides	II–III CT, GI	Disturbance of GI tract; diarrhoea; colic; large doses induce cerebral bleeding. **Therapy:** Terpenes and phenylpropanoids with aldehyde groups
Azadirachta indica (neem tree); Meliaceae	Asia; mainly tropical countries	fruits, leaves	tetranortriterpenoids (limonoids), azadirachtin	III, NP, MP	Extracts are used as natural insecticides; overdosing causes Reye syndrome, azadirachtin is an antagonist of the insect hormone ecdysone. **Therapy:** Saponins
Baccharis pteronioides; Asteraceae	S America	aerial parts	sesquiterpene lactones	II CT, AP	Toxic to sheep, cattle. **Therapy:** Sesquiterpenes
Bahia oppositifolia; Asteraceae	C America	aerial parts	cyanogenic glucosides	Ib–II CT, NT, AP	Symptoms of HCN poisoning; livestock poisoning (sheep, cattle). **Therapy:** Amygdalin
Balanites aegyptiaca; Zygophyllaceae	N Africa	bark	saponins	II–III CT, GI	Fish poison; skin and mucous membrane irritation; disturbance of GI tract. **Therapy:** Saponins
Balanophora abbreviata; Balanophoraceae	SE Asia	stem, tuber	Ca²⁺-oxalate	II CT, IN	Irritation of mucous membranes and skin, burning sensation. **Therapy:** Oxalic acid
Balansia cyperi and related fungi; Clavicipitaceae	Europe	aerial parts	symbiotic fungus of Cyperaceae; ergot alkaloids, such as ergobalansine	Ib NT, MA	Ergot alkaloids affect dopamine, noradrenaline and serotonin receptors, therefore psychoactive; psychedelic and hallucinogenic effects. **Therapy:** Ergot alkaloids

Scientific name (common name); Family	Occurrence	Hazardous plant parts	Active principle	Toxicity Class	Symptoms & Therapy
Bambusa arundinacea (bamboo); Poaceae	Asia	shoots	cyanogenic glucosides	Ib–II CT	Cyanogenic glucosides release HCN which inhibits cellular respiration. **Therapy:** Amygdalin
Banisteriopsis caapi (= *B. inebrians*), *B. muricata* and related species; Malpighiaceae	rain-forests of S America	bark and wood	β-carboline alkaloids; *N,N*-DMT (*B. muricata*) and saponins	II NT, MA	β-carboline alkaloids are strong MAO inhibitors and also stimulate serotonin receptors; induce sensory colourful illusions, hallucinations similar to LSD; mydriasis, salivation, vomiting, sweating, nausea, tremor, intestinal pains. **Therapy:** Indole alkaloids and *Banisteriopsis*
Baptisia australis (wild indigo), *B. leucantha*, *B. alba*, *B. tinctoria* and related species; Fabaceae	N America; orn. plant	all parts, especially fruits and seeds	cytisine, *N*-methylcytisine, and other quinolizidine alkaloids	II NT, MA, GI	Low doses induce vomiting and disturbance of GI tract; high doses cause burning sensation in mouth, salivation, perspiration, mydriasis, nausea and paralysis; also hallucinations and delirium. **Therapy:** Cytisine
Bassia scoparia (fireweed); Amaranthaceae (Chenopodiaceae)	America	all parts	triterpene saponins, oxalates, nitrates, alkaloids	II CT, IN; NT, AP	Triterpene saponins with cytotoxic properties; cause liver and kidney damage in livestock. **Therapy:** Saponins
Begonia gracilis and related species; Begoniaceae	S Africa; orn. plant	tubers	Ca²⁺-oxalate, cucurbitacins	II CT, IN, GI	Mucosal irritant; induces GI tract and kidney disturbance; vomiting, diarrhoea with blood. **Therapy:** Oxalic acid and cucurbitacins
Beloperone guttata, (= *Justicia brandegeeana*) Acanthaceae	Mexico, naturalised in Florida; orn. plant	all parts	Ca²⁺-oxalate, cucurbitacins	II CT, IN, GI	Mucosal irritant; induces GI tract and kidney disturbance; vomiting, bloody diarrhoea. **Therapy:** Oxalic acid
Berberis vulgaris (barberry) and related species; Berberidaceae	Europe, Asia, America; common orn. plant	all parts, especially bark	berberine, magnoflorine and other isoquinoline alkaloids	II CT, NT, GI, MP	Cytotoxic properties; nausea, vomiting, diarrhoea, kidney inflammation. High doses cause primary respiratory arrest and haemorrhagic nephritis. **Therapy:** Isoquinoline alkaloids and *Berberis*
Bersama abyssinica; Melianthaceae	E Africa	leaves, roots	unknown toxin	Ib CT, NT, AP	Livestock poisoning, fatal for cows; root extracts used for suicide in Zambia. **Therapy:** Cardiac glycosides
Berula erecta (water parsnips); Apiaceae	Europe, Africa, N America	aerial parts	possibly polyacetylenes or furanocoumarins	Ib CT, NT, AP	Strong livestock poison (kills cattle within hours). **Therapy:** Polyacetylenes

Scientific name (common name); Family	Occurrence	Hazardous plant parts	Active principle	Toxicity Class	Symptoms & Therapy
Blighia sapida (akee); Sapindaceae	W Africa; cultivated in W Indies and Asia	fruit walls, seeds with arils, unripe fruits	hypoglycin A (a NPAA)	Ib CT, GI	Hypoglycin A inhibits fatty acid metabolism and gluconeogenesis; especially poisonous to children; vomiting ("Jamaican vomiting disease"), hypoglycaemia, liver failure. **Therapy:** Non-protein amino acids
Boophone disticha; (= *Boophane disticha*) (bushman poison bulb); Amaryllidaceae	S Africa	bulbs	buphanidrine, lycorine, and other isoquinoline alkaloids	Ib NT, MA	Powerful analgesic, hallucinogen and neurotoxin; dizziness, restlessness, impaired vision, visual hallucinations, and finally coma and death; arrow poison of African bushmen. **Therapy:** Isoquinoline alkaloids and *Boophone*
Borago officinalis (borage); Boraginaceae	S Europe; cultivated herb	all parts; especially flowers	lycopsamine, intermedine, amabiline and related PAs	II CT, NT, MU	PAs have mutagenic and carcinogenic properties; substantial ingestion can cause hepatotoxic and neurotoxic disorders. **Therapy:** Senecionine
Boswellia sacra (frankincense tree) and related species; Burseraceae	Somalia, S Arabia	resin	essential oil with pinene, limonene, camphene, verbenone; boswellic acid	III MA, MP	Smoke has euphoric and stimulant effects; used traditionally as olibanum; anti-inflammatory activity. **Therapy:** Monoterpenes
Bowiea volubilis (climbing potato), *B. gariepensis*; Hyacinthaceae	S & E Africa	all parts, especially bulbs	scillaren-type bufadienolides: bovoside A, C, bowienine (alkaloid)	Ia CT, HP, GI, AP	Human and animal poisoning; symptoms of cardiac glycoside intoxication; similar to k-strophantin; nausea, vomiting, diarrhoea, salivation, spasmic pains, coronary arrest. **Therapy:** Cardiac glycosides
Brassica nigra (black mustard) and related species; Brassicaceae	Eurasia; widely cultivated	all parts, especially seeds	sinigrin and other glucosinolates; produces isothiocyanates upon hydrolysis	II-III CT, GI, IN	Skin irritation; pulmonary emphysema, respiratory distress, stomach pain; nausea, vomiting, diarrhoea; high doses: paralysis of CNS, cardiac and respiratory arrest; livestock poisoning. **Therapy:** Mustard oils
Brugmansia suaveolens (angel's trumpet), *B. arborea*, *B. aurea*, *B.* x *candida* (= *Methysticodendron amesianum*), *B. sanguinea*, *B.* x *insignis*, *B. versicolor* and related species; Solanaceae	C & S America; widly cultivated as orn. plant	all parts	scopolamine, hyoscyamine, tigloidine, 3-tigloyloxytropane 6β-ol and other tropane alkaloids (up to 0.4%)	Ia NT, MA, MP	mAChR antagonist, parasympatholytic properties; strongly hallucinogenic in various forms; intensive dreams; high doses: clonic spasms; strong heartbeat; tachycardia (>160 beats); mydriasis; inhibition of salivation; respiratory arrest; coma, death. **Therapy:** Hyoscyamine

Scientific name (common name); Family	Occurrence	Hazardous plant parts	Active principle	Toxicity Class	Symptoms & Therapy
Brunfelsia pauciflora; *B. uniflora, B. chiricaspi, B. grandiflora and related species;* Solanaceae	S & C America; widely cultivated orn. plant	all parts, especially roots	brunfelsamidine, manacine, manaceine (alkaloids), scopoletin	II CT, MA, MP	Psychoactive properties; used by S American Indians; abortifacient, fever medicine, ingestion induces salivation, weakness, paralysis of face, swollen tongue, distorted vision, trembling, nausea. **Therapy:** Pyrrolidine alkaloids
Bryonia dioica (bryony), *B. alba* and related species; Cucurbitaceae	Europe	all parts, especially fruits (red or black berries), roots	bryonin, bryonidin, bryonicin, and other cucurbitacins	Ib CT, IN, GI, MP	Skin irritant; nausea, vomiting, diarrhoea with blood, strong colic and spasms, kidney inflammation, tachycardia, respiratory arrest. **Therapy:** Cucurbitacins and *Bryonia*
Burttia prunoides; Connaraceae	NE Africa, Madagascar	roots, seeds	unknown substances, may be methionine sulfoxime	Ib CT, NT, AP	Used to kill rodents and dogs in Madagascar; ingestion causes convulsions, manic behaviour and coma. **Therapy:** Non-protein amino acids
Buxus sempervirens (box tree) and related species; Buxaceae	Europe; orn. tree	aerial parts	buxine, cyclobuxine and several related alkaloids	Ib CT, GI, NT, AP, IN	Contact dermatitis; initially exciting, later paralysing and hypotensive; nausea, vomiting, dizziness, diarrhoea, spasms, death by respiratory arrest; livestock poisoning. **Therapy:** *Buxus*
Caesalpinia gilliesii (bird-of-paradise), *C. pulcherrima, C. mexicana, C. vesicaria, C. bonduc;* Fabaceae	C & S America; orn. plant	mainly seeds	tannins, NPAAs	II–III CT, GI	Nausea, vomiting, abdominal pains, diarrhoea; some species are used as fish poison. **Therapy:** Flavonoids
Caladium bicolor and hybrids; Araceae	tropical America, Asia; orn. plant	all parts	Ca^{2+}-oxalate raphides	II, CT, IN, GI	Mucosal irritant; GI tract inflammation; severe swelling of throat and mouth, blister formation. **Therapy:** Oxalic acid
Calea ternifolia (= *C. zacatechichi*); Asteraceae	Mexico	aerial parts	germacranolides, caleicin, calein, and other sesquiterpene lactones	II MA	Dream and sleep inducing symptoms are apparent, also in animal experiments. **Therapy:** *Calea*

350

Scientific name (common name); Family	Occurrence	Hazardous plant parts	Active principle	Toxicity Class	Symptoms & Therapy
Calla palustris (water arum) and related species; Araceae	Europe, N America	all parts, especially leaves, berries, roots	pungent principle; needle-like Ca^{2+}-oxalate crystals; toxic peptides	II CT, IN, GI	Irritation, swelling and blistering in throat and mouth; burning sensation; disturbance of GI tract; in severe cases: cardiac attack and CNS paralysis. **Therapy:** Oxalic acid
Callilepis laureola (ox-eye daisy); Asteraceae	S Africa	roots	atractyloside (diterpene)	Ib CT, NT	Symptoms similar to strychnine; severe vomiting, abdominal pain, convulsions, and rapid progression into coma; death from acute liver and kidney damage. **Therapy:** *Callilepis*
Calophyllum inophyllum (mastwood); Clusiaceae	SE Asia; orn. plant in tropical America	seed kernels	inophyllum A-E, calophyllolide, calophynic acid (phenolics)	III GI	Nausea, vomiting, abdominal pains, diarrhoea; sap can cause eye inflammation. **Therapy:** Flavonoids
Calotropis procera (giant milkweed), *C. gigantea* (crown flower); Apocynaceae (Asclepiadoideae)	Africa, Asia, West Indies; orn. plant in tropics	latex, all parts	calotropin and other cardiac glycosides, triterpenes, Ca^{2+}-oxalate	Ib CT, IN, HP	Inflammation of skin and mucous membranes; typical cardiac glycoside poisoning, fatal in minute amounts, used as arrow poison. **Therapy:** Cardiac glycosides and oxalic acid
Caltha palustris (marsh marigold), *C. leptosepala*; Ranunculaceae	Europe, Asia, N America	all parts	ranunculin; produces protoanemonin upon hydrolysis; magnoflorine, triterpene saponins	II CT, IN, GI	GI tract and kidney disturbance; mutagenic properties; strong irritant of skin and mucous membranes. **Therapy:** Protoanemonin
Calycanthus floridus (allspice), *C. floridus*; Calycanthaceae	N America; orn. tree	all parts, especially seeds	calycanthine and other quinoline alkaloids	Ib NT	Alkaloids show strychnine-like properties; strong convulsions, muscular hyperactivity, heart trouble; poisoning of cattle and sheep has been recorded. **Therapy:** Indole alkaloids and *Strychnos*
Camellia sinensis (tea plant); Theaceae	S Asia; widely cultivated	leaves	purine alkaloids (up to 5%) (caffeine, theobromine, theophylline)	III MA, MP	Central stimulant; high doses are intoxicant with tinnitus, headache, dizziness, increased heartbeat, insomnia, unrest, nausea, diuresis. **Therapy:** Caffeine

Scientific name (common name); Family	Occurrence	Hazardous plant parts	Active principle	Toxicity Class	Symptoms & Therapy
Camptotheca acuminata (camptotheca, happy tree); Cornaceae	China; cultivated in India, Japan, USA	seeds, aerial parts	camptothecin and related indole alkaloids	Ib CT, GI, MP	Camptothecin inhibits DNA topoisomerase I; causes tiredness, asthenia, neutropenia (reduction of white blood cells), hair loss, somatitis, nausea, vomiting, diarrhoea and haemorrhagic cystitis. **Therapy:** *Camptotheca*
Canavalia ensiformis (jackbean); Fabaceae	Africa	seeds	concanavalin A (lectin)	II CT	Lethal for cattle in large quantities. **Therapy:** Abrin
Cannabis sativa (hemp) (= *C. indica, C. ruderalis*); Cannabaceae	Iran, India; widely cultivated	resin of female plant	tetrahydrocannabinol (THC) and other cannabinoids	III MA, MP	THC is responsible for euphoric and hallucinogenic activity; aphrodisiac, heart and CNS dysfunction at high doses. **Therapy:** *Cannabis*
Capparis fascicularis, C. spinosa, C. tomentosa; Capparidaceae	Africa, Europe	all parts, fruits	glucosinolates and isothiocyanates	III CT, GI	Substantial ingestion can cause GI tract disturbance; possibly fatal to livestock and humans. **Therapy:** Mustard oils
Capsicum annuum (chilli pepper), *C. frutescens* (tabasco pepper), *C. chinense*; Solanaceae	S & C America; widely cultivated	fruits, especially seeds	capsaicin and related capsaicinoids	III IN, MP	High doses cause burning sensation in mouth and throat; hyperthermia, chronic gastritis, kidney and liver disturbances, release of substance P from nerve cells; capsaicin inactivates sensory nerve cells. **Therapy:** *Capsicum*
Caragana arborescens; Fabaceae	Siberia; widely cultivated	all parts, seeds	seeds contain 2 lectins; 6% *L*-canavanine (a NPAA)	II–III CT, GI	Substantial ingestion can cause GI tract disturbance; animal poisoning has been reported. **Therapy:** Non-protein amino acids
Carissa macrocarpa (= *C. grandiflora*) (Natal plum), *C. bispinosa*; Apocynaceae	S Africa; orn. plant	aerial parts	cardiac glycosides, indole alkaloids	Ib–II CT, HP, GI	Substantial ingestion can cause GI tract, circulatory and cardial disturbance. **Therapy:** Cardiac glycoside
Carnegiea gigantea (giant cactus); Cactaceae	C America, Arizona	aerial parts, fruits	carnegine, gigantine, salsolidine and other phenylethylamines	II–III MA	Fruits have been used to make wine; alkaloids are mimics of dopamine and noradrenaline and could therefore be psychoactive and stimulant. **Therapy:** Bufotenine
Caryota mitis (fishtail palm), *C. urens*; Arecaceae	SE Asia; orn. plant in tropics	sap	Ca^{2+}-oxalate raphides; toxic peptide	II CT, IN	Sap is a strong irritant of skin and mucosal tissues; inflammation with oedema and blisters. **Therapy:** Oxalic acid

Scientific name (common name); Family	Occurrence	Hazardous plant parts	Active principle	Toxicity Class	Symptoms & Therapy
Cassia senna, *C. angustifolia*, *C. fistulosa* (golden shower), *C. occidentalis* (coffee senna); Fabaceae	Africa, Asia, America; sub-tropical orn. plant	aerial parts, leaves, fruits	sennoside A, B, chrysarobin, emodin, and other anthraquinones	III IN, GI, MP	Laxative; high doses cause intestinal pains, nausea, vomiting; emodin can cause urine discoloration (red or violet in alkaline urine). *C. occidentalis* can cause livestock poisoning. **Therapy:** Anthraquinones
Catalpa bignonioides, *C. ovata*; Bignoniaceae	N America; orn. tree	fruits	catalpin and other iridoid glycosides	II CT, NT, GI	Substantial ingestion induces disturbance of GI tract and CNS. **Therapy:** Terpenes and phenylpropanoids
Catha edulis (khat); Celastraceae	S Africa to Ethiopia, Arabian peninsula	leaves	cathinone, cathine (up to 0.9%); tannins	II–III MA	Central stimulant; widely used in Arabia and E Africa; causes excitations; reduces hunger feeling; causes disturbances of digestion, constipation and circulatory disorders. **Therapy:** Ephedrine and *Catha*
Catharanthus roseus (Madagascar periwinkle); Apocynaceae	Madagascar; orn. plant in tropics and sub-tropics	leaves	ajmalicine, catharanthine, vindoline and various others monoterpene indole alkaloids; vinblastine, vincristine	Ib–II CT, NT, MP	Plant causes severe disturbance of GI tract; dimeric indole alkaloids inhibit cell division by inhibiting microtubule formation, causes neurological disorders and affects all tissues with rapidly dividing cells; livestock poisoning. **Therapy:** Indole alkaloids and *Catharanthus*
Caulophyllum thalictroides (blue cohosh), and related species; Berberidaceae	N America	roots, seeds	N-methylcytisine, baptifoline and other quinolizidine alkaloids; saponins	Ib–II CT, NT, GI, MA	Low doses induce vomiting, diarrhoea and gastroenteritis; high doses cause burning sensation in mouth, induce salivation, perspiration, mydriasis, uterus contracting, nausea and paralysis; also hallucinations and delirium; respiratory arrest. **Therapy:** Cytisine
Cayaponia racemosa, *C. glandulosa*, *C. ophthalmica* and related species; Cucurbitaceae	tropical America	roots	cucurbitacins	Ib CT, GI	Strong purgative, nausea, vomiting, abdominal pains, diarrhoea. **Therapy:** Cucurbitacins
Celastrus scandens (bittersweet), *C. orbiculatus*; Celastraceae	N America, Asia	fruits	unknown GI irritant	III GI	Nausea, vomiting, abdominal pains, diarrhoea. **Therapy:** Saponins
Centaurium beyrichii (century), *C. calcosum*, *C. erythraea* and related species; Gentianaceae	N America, Europe	aerial parts	secoiridoids	II CT, AP	Secoiridoids with very bitter taste; livestock poisoning (sheep). **Therapy:** Terpenes and phenylpropanoids with aldehyde groups

Scientific name (common name); Family	Occurrence	Hazardous plant parts	Active principle	Toxicity Class	Symptoms & Therapy
Centella asiatica (Indian pennywort); Apiaceae	pan-tropical	aerial parts	hydrocotyline (an alkaloid); triterpenoid saponins	II CT, GI, MP	Irritation of skin and mucous membranes; substantial ingestion can cause GI tract disturbance. **Therapy:** Saponins.
Centrosema plumieri and related species; Fabaceae	C & S America	bark, seeds	canavanine (NPAA); isoflavones	II–III	Cuna Indians use the bark as fish poison.
Ceratophylla testiculata (bur buttercup); Ranunculaceae	N America	aerial part	ranunculin, protoanemonin	II CT, IN, GI, AP, MU	Protoanemonin is a strong irritant of skin and mucous membranes causing oedema, blistering and ulcerations; important animal poison. **Therapy:** Protoanemonin
Cerbera odollam, *C. venenifera*; Apocynaceae	India; SE Asia; Mada-gascar	seeds	tanghin and other cardiac glycosides	Ia CT, HP, IN	Often used for suicide and murder; typical cardiac glycoside intoxication. Can cause skin inflammation and blindness. **Therapy:** Cardiac glycosides and *Cerbera*
Cercocarpus montanus (mountain mahogany); Rosaceae	N America	aerial parts	cyanogenic glucosides	II CT, GI, AP	Symptoms of HCN poisoning; livestock poisoning has been reported. **Therapy:** Amygdalin
Cestrum parqui (green cestrum), *C. nocturnum* (night-blooming jessamine), *C. diurnum* (day-blooming jessamine), *C. aurantiacum*, *C. laevigatum* and related species; Solanaceae	S & C America; intro-duced in Africa, Asia and Europe	leaves, fruits	carboxyparquin, parquin (diterpenoids); solasonine, (alkaloids); trigogenin, yuccagenin and other steroid saponins; tropane alkaloids (hyoscyamine) and nicotine (*C. nocturnum*)	Ib–II CT, MA, AP	Slightly euphoric and hallucinogenic; high doses: fever, gastroenteritis. Cattle poisoning: salivation, watery eyes, colic, arched back, weakness, staggering gait and abdominal pain; vitamin D intoxication (*C. diurnum*). **Therapy:** *Cestrum*, hyoscyamine and pyrrolidine alkaloids
Chaenomeles japonica (flowering quince); Rosaceae	Asia; orn. plant	seeds	prunasin; 300 ppm HCN	II CT, GI	Ingestion seldom hazardous; only high doses could cause symptoms of HCN poisoning. **Therapy:** Amygdalin
Chaerophyllum temulentum; Apiaceae	Europe	leaves, fruits	alkaloids, polacetylenes	II CT, IN, GI	Substantial ingestion can cause GI tract and CNS disturbance. **Therapy:** Polyacetylenes
Chamaecyparis obtusa, *C. lawsoniana* and related species; Cupressaceae	Asia; N America; orn. plant	all parts	sabinene, thujone and other monoterpenes	Ib CT, GI, NT	Substantial ingestion can cause GI tract and CNS disturbance; strong cytotoxic effects, abortifacient. **Therapy:** Monoterpenes

Scientific name (common name); Family	Occurrence	Hazardous plant parts	Active principle	Toxicity Class	Symptoms & Therapy
Chelidonium majus (celandine); Papaveraceae	Europe, Asia, N America	all parts; reddish latex	chelidonine, chelerythrine, sanguinarine, berberine and other isoquinoline alkaloids	II CT, NT, GI, MP	Burning sensation in mouth and throat; nausea, vomiting, bloody diarrhoea; centrally sedative, spasmolytic and narcotic; low pulse; hypotension; cardiac arrest; chelidonine inhibits mitosis and is therefore cytotoxic. **Therapy:** Isoquinoline alkaloids and *Chelidonium*
Chenopodium ambrosioides var. *anthelminthicum* (American wormseed), *C. vulvaria* and other species; Amaranthaceae (Chenopodiaceae)	C & S America	aerial parts, seeds	ascaridol and other monoterpenes	Ib CT, NT, GI, MP	Ascaridol is the main toxin; high doses cause disturbances of CNS, unconsciousness, paralysis, hypotension, cerebral bleeding; inflammation of GI tract with spasms; abortifacient. **Therapy:** *Chenopodium ambrosioides*
Chenopodium quinoa (quinoa); Amaranthaceae (Chenopodiaceae)	S America; cultivated crop plant	seeds	triterpene saponins	II–III CT, GI	Triterpene saponins with cytotoxic properties; seed coats are very bitter because of triterpenes, which have to be removed before eating. **Therapy:** Saponins
Chimonanthus praecox (wintersweet); Calycanthaceae	China; common orn. plant	leaves	calycanthine, chimonanthine,	II CT, NT, GI, AP	Alkaloids show strychnine-like properties with strong convulsions, disturbance of GI tract, heart trouble; poisoning of cattle and sheep has been recorded. **Therapy:** Indole alkaloids and *Strychnos*
Chondrodendron tomentosum; Menispermaceae	S America	aerial parts	curarine, tubocurarine	Ia NT, MA, MP	Alkaloids inhibit nAChR; act as strong muscle relaxant, hypotensive, death from cardiac arrest. **Therapy:** Isoquinoline alkaloids and *Chondrodendron*
Chrysanthemum cinerariifolium (pyrethrum); Asteraceae	Balkans; widely cultivated	aerial parts, flowers	pyrethrin I and II, cinerin I, II, jasmolin I, II	II NT, MA, AP	Natural pyrethrins are used as insecticides; synthetic pyrethroids were derived from them; at high doses: headache, tinnitus, stomach pains, nausea, slow heart-beat, dyspnoea; synthetic derivates are more toxic. **Therapy:** Monoterpenes
Chrysanthemum vulgare (tansy) (= *Tanacetum vulgare*); Asteraceae	Europe, N Africa	aerial parts, especially flowers	thujone and other monoterpenes	II CT, NT, MA, MP	Thujone is a neurotoxin; at high doses the essential oil causes strong spasms, vomiting, gastroenteritis, convulsions, arrhythmia, mydriasis, uterine bleeding, miscarriage (abortifacient), kidney and liver disturbance, death through cardiac and respiratory arrest. **Therapy:** Monoterpenes and *Chrysanthemum*

355

Scientific name (common name); Family	Occurrence	Hazardous plant parts	Active principle	Toxicity Class	Symptoms & Therapy
Cicuta virosa (water hemlock), *C. bulbifera*, *C. douglasii*, *C. maculata*; Apiaceae	Europe, N Africa, N America	aerial parts, roots	cicutoxin, cicutol (polyacetylenes)	Ib CT, NT, GI	Burning sensation in mouth and throat, nausea, vomiting; stomach pains, large pupils, headache, tremors, epileptiform convulsions, delirium, death by respiratory arrest. **Therapy:** Polyacetylenes
Cinchona pubescens (quinine tree, Peruvian bark tree) and related species; Rubiaceae	S America	bark	quinine, quinidine and other quinoline alkaloids	II CT, NT, MP	Quinine is used to treat malaria, quinidine as an anti-arrhythmic drug; quinine impairs visual capacity; can cause fever, icterus and haematuria; vomiting, diarrhoea, dyspnoea, internal bleeding; respiratory arrest. **Therapy:** Quinoline alkaloids and *Cinchona*
Cinnamomum camphora (camphor tree); Lauraceae	China, SE Asia; widely cultivated	leaves	essential oil with camphor, safrole, terpineol, eugenol, cineole, pinene	II–III CT, NT, MP	High doses: excitation, intoxication similar to alcohol; dizziness, nausea, vomiting, inflammation of stomach mucosa; anxiety, inhibition of neuronal activity; nephritis. **Therapy:** Monoterpenes
Cissampelos perreira, *C. mucronata*, *C. capensis*; Menispermaceae	Africa	rhizomes, aerial parts	proaporphine alkaloids, such as glaziovine, insularine (a bisbenzyl-tetrahydroisoquinoline alkaloid)	Ib NT, GI	Modulators of cholenergic receptors; muscle relaxant properties; used as arrow poisons. **Therapy:** Isoquinoline alkaloids
Cissus quadrangularis; Vitaceae	SE Asia	aerial parts	Ca^{2+}-oxalate	II CT, IN	Irritation of mucous membranes and skin, burning sensation. **Therapy:** Oxalic acid
Citrullus colocynthis (bitter apple); Cucurbitaceae	W Africa, introduced in Mediter-ranean, Arabia, India	fruits, aerial parts	several cucurbitacins	Ib CT, GI, IN	Cytotoxic; strong purgative with profound irritation of GI tract and kidneys; strong abdominal pains; internal bleeding. **Therapy:** Cucurbitacins and *Citrullus*
Citrus aurantium (bitter orange) and related species; Rutaceae	Asia, Europe; widely cultivated	bitter peel	essential oil, limonene	II CT, NT, MP	Essential oil can cause violent colic and convulsions, large doses have led to death in children. **Therapy:** Monoterpenes

Scientific name (common name); Family	Occurrence	Hazardous plant parts	Active principle	Toxicity Class	Symptoms & Therapy
Clausena anisata; Rutaceae	Africa, Asia	aerial parts	essential oil with estragol	III CT, IN	Estragol has a terminal methylene group, which can bind to proteins; causes photodermatosis. **Therapy:** Myristicin
Claviceps purpurea (ergot), *C. paspali* and related fungi; Clavicipitaceae	Europe, Asia; grows on several grasses	fungal fruiting bodies (sclerotia)	several ergot alkaloids	Ib NT, MA, MP	Ergot alkaloid affects dopamine, noradrenaline and serotonin receptors, therefore inducing psychoactive, psychedelic and hallucinogenic effects; painful muscle contractions, epileptic convulsions; vasoconstriction can cause gangrene. **Therapy:** Ergot alkaloids and *Claviceps*
Cleistanthus collinus; Euphorbiaceae	India, Africa	all parts; leaves	heart poison, cleisthanin A,B (lignan glycoside)	Ia CT, HP, GI	Symptoms of cardiac glycoside poisoning; extracts have been used for poisoning of humans and animals. **Therapy:** Cardiac glycosides
Clematis recta, C. alpina, C. vitalba, C. microphylla, C. paniculata, C. virginiana and related species; Ranunculaceae	Europe; Asia, N America; orn. plant	all parts	ranunculin is converted enzymatically to protoanemonin (2 % DW)	II CT, IN, GI, AP, MU	Irritation of skin and mucous membranes with blistering and ulcerations, nausea, diarrhoea, intestinal bleeding, kidney disturbance; mutagenic. **Therapy:** Protoanemonin
Cleome gynandra (wild spider flower) and related species; Brassicaceae (Capparidaceae)	Asia; orn. plant	aerial parts	cyanogenic glucosides	Ib–II CT	HCN liberated from cyanogens can be poisonous (typical HCN intoxication): spasms, paralysis, respiratory arrest. **Therapy:** Amygdalin
Clibadium surinamense; Asteraceae	Guyana	leaves, seeds	terpenoids	II NT, AP	Narcotic, fish poison. **Therapy:** Sesquiterpenes
Clitocybe dealbata, C. cerussata, C. rivulosa, C. truncicola; Tricholomataceae	Europe, America	fruiting body	muscarine	II NT, MA	Muscarine is an agonist at mAChR; induces sweating, tears, salivation, bradycardia, hypotension; CNS dysfunction. **Therapy:** *Amanita muscaria*
Clivia miniata (bush lily, orange lily), *C. nobilis* and related species; Amaryllidaceae	S Africa; common orn. plant	all parts, especially rhizomes, leaves	lycorine, cliviamine, clivonine and cliviamartine, hippeastrine and other alkaloids	Ib–II NT, GI, MP	Salivation, vomiting, diarrhoea, leading to paralysis and collapse. **Therapy:** Lycorine
Clusia rosea (balsam apple); Clusiaceae	West Indies, Florida, Hawaii	viscous sap, fruits	GI irritant SM	II–III GI	Strong disturbance of GI tract with purging. **Therapy:** Anthraquinones

Scientific name (common name); Family	Occurrence	Hazardous plant parts	Active principle	Toxicity Class	Symptoms & Therapy
Codiaeum variegatum (ornamental croton); Euphorbiaceae	India, Malaysia; common orn. plant	all parts, sap, seeds	phorbol esters	Ib CT, IN, GI	Phorbol esters are strong irritants (see phorbol esters); the plant has caused gastrointestinal disturbance and contact dermatitis. **Therapy:** Phorbol esters
Coffea arabica (coffee tree), C. *canephora*, C. *liberica* and related species; Rubiaceae	Ethiopia; widely cultivated in tropics	leaves, especially seeds	caffeine (up to 1.5 %), chlorogenic acid (up to 7.5 %), trigonelline	III MA, GI, MP	Caffeine inhibits adenosine receptor; central stimulant; increases awareness and prevents sleepiness; high doses are intoxicating with tinnitus, headache, dizziness, increased heartbeat, insomnia, unrest, tachycardia. **Therapy:** Caffeine and *Coffea arabica*
Cola acuminata (cola tree), C. *nitida* and related species; Malvaceae (Sterculiaceae)	W Africa; widely cultivated	seeds	caffeine, theobromine (purine alkaloids) (up to 3.5 %)	III MA, GI, MP	See *Coffea arabica*
Colchicum autumnale (autumn crocus), C. *speciosum* and related species; Colchicaceae	Europe, Asia; orn. plant	all parts, especially seeds and bulbs	colchicine and related alkaloids	Ia CT, NT, MP, GI	Colchicine is a spindle poison; nausea, dizziness, burning of throat and stomach; purging, stomach pains, spasms, internal bleeding; strong diuresis; cardiovascular collapse, respiratory arrest. **Therapy:** Colchicine
Colocasia esculenta (= C. *antiquorum*) (taro), C. *gigantea*; Araceae	SE Asia, C America; widely cultivated	leaves, tuber	Ca^{2+}-oxalate raphides	II CT, IN, GI	Strong skin irritation; swollen and painful mucous membranes, burning sensation, disturbance of GI tract. **Therapy:** Oxalic acid
Colutea arborescens (bladder senna); Fabaceae	Mediterranean; orn. plant	leaves, seeds, inflated pods	in seeds: 5 % *L*-canavanine (NPAA) and lectins	II CT, GI	Canavanine is an antimetabolite; substantial ingestion causes diarrhoea and vomiting. **Therapy:** Non-protein amino acids
Conium maculatum (poison hemlock); Apiaceae	Europe, Asia, Africa, naturalised N America	all parts, seeds	coniine, conhydrine and other piperidine alkaloids	Ia NT, MP, MA	Strong neurotoxin, mental confusion, burning sensation in mouth and throat; vomiting, paralysis of muscles, low temperature, loss of sensation, convulsions, death through respiratory arrest. **Therapy:** *Conium*
Conocybe cyanopus, C. *kuehneriana*, C. *siligineoides*, C. *smithii*; Bolbitiaceae	Europe, Africa	fruiting body	psilocybin, baeocystine	II NT, MA	Alkaloids are serotonin receptor agonists; therefore psychoactive, psychedelic and hallucinogenic. **Therapy:** Bufotenine

358

Scientific name (common name); Family	Occurrence	Hazardous plant parts	Active principle	Toxicity Class	Symptoms & Therapy
Consolida regalis (larkspur), *C. ajacis, C. orientalis* and related species; Ranunculaceae	Europe, W Asia; orn. plants	all parts, especially seeds	delcosine, lycoctonine and other terpene alkaloids	Ia CT, NT, MA, AP	Alkaloids affect sodium channels and neuroreceptors; with sedating and dream-inducing properties; substantial ingestion causes nausea, gastroenteritis; arrhythmia, excitation, spasms, respiratory arrest. **Therapy:** Aconitine and *Delphinium*
Convallaria majalis (lily-of-the-valley); Ruscaceae (Convallariaceae)	Europe, Asia; naturalised N America	all parts, flowers, fruits	convallatoxin and other cardenolides; saponins	Ib CT, HP, GI, MP	Cardiac glycoside intoxication; nausea, gastrointestinal disturbance, diarrhoea, dizziness, hypertension, arrhythmia, coma, cardiac arrest. **Therapy:** Cardiac glycosides
Convolvulus pseudocantabrica (morning glory), *C. tricolor* and related species; Convolvulaceae	Europe; orn. plant	seeds	convolvamine (a tropane alkaloid); possibly ergot alkaloids	II–III MA	Tropane and ergot alkaloids have psychotropic properties that could explain a potential hypnotic and hallucinogenic effect reported from these plants. **Therapy:** Hyoscyamine and ergot alkaloids
Copelandia bispora, C. anomalus and related species; Bolbitiaceae	Europe, Africa	fruiting body	psilocybin	II NT, MA	Alkaloids are serotonin receptor agonists; therefore psychoactive, psychedelic and hallucinogenic. **Therapy:** Bufotenine
Coprinus atramentarius and related fungi; Agaricaceae	N America	fruiting body	coprine	Ib–II CT, NT, GI	Symptoms after 5–30 min; as if alcohol had been taken within 24 h. Symptoms include: hot and sweaty face, flushing of head, rapid and difficult breathing, tachycardia, violent headache, nausea, vomiting. **Therapy:** Non-protein amino acids
Corchorus olitorius; Malvaceae (Tiliacae)	tropical Africa	seeds	cardenolides (corchorin, helveticoside, evonoside)	Ia CT, HP, GI	Has been used as arrow poison. **Therapy:** Cardiac glycosides
Coriaria myrtifolia (myrtle-leaved sumac), *C. sarmentosa, C. thymifolia;* Coriariaceae	S Europe; N Africa; New Zealand; orn. plant America	shoots, roots, berries	coriamyrtin, coriamin, coriatin, tutin, hyaenanchin (sesquiterpene lactone); tannins	II NT, MA	GABA antagonist similar to picrotoxin; low doses increase blood pressure and respiration; high doses lead to spasms and seizures, salivation, vomiting, mydriasis and death; livestock poisoning. **Therapy:** Sesquiterpenes
Cornus sanguinea (dogwood) and related species; Cornaceae	Europe; orn. plant	fruits, leaves	aucubin (iridoid glucoside); tannins	III CT, GI	Ingestions of fruits seldom hazardous; may cause gastroenteritis when ingested in large quantities. **Therapy:** Terpenes and phenylpropanoids with aldehyde groups

359

Scientific name (common name); Family	Occurrence	Hazardous plant parts	Active principle	Toxicity Class	Symptoms & Therapy
Coronilla varia (= *Securigera varia*) (crown-vetch); Fabaceae	Europe	all parts, seeds	glycosides with cardiac activities (hyrcanoside, deglucohyrcanoside), glucose esters with nitropropionic acid	Ib CT, HP, GI	Glycosides appear to have similar activities as cardiac glycosides and cause similar symptoms; nitro compounds inhibit enzymes of citric acid cycle. Substantial ingestion causes vomiting, diarrhoea and abdominal pains. **Therapy:** Cardiac glycosides and non-protein amino acids
Corydalis cava, *C. aurea*, *C. caseana*, *C. flavula* and related species; Papaveraceae	Europe, N America; orn. plant	all parts, especially tubers (6% alkaloids)	bulbocapnine and other isoquinoline, protoberberine and aporphine alkaloids	II NT, MP, AP	The alkaloids interact with several neurotransmitter receptors; inhibition of active muscle activity; hypnotic, GI tract disturbance; livestock poisoning (sheep, cattle). **Therapy:** Isoquinoline alkaloids
Corynanthe pachyceras, *C. mayumbensis*; Rubiaceae	Africa	bark	rich in monoterpene indole alkaloids, corynanthine, yohimbine	Ib NT, MA	Extracts have been used as arrow poison; alkaloids interact with adrenergic neuroreceptors, inducing analgesic and aphrodisiac effects. Yohimbine is a central stimulant and can enhance general anxiety. Side effects include over-stimulation, irritability and anxiety attacks. **Therapy:** Indole alkaloids and *Pausinystalia johimbe*
Corynocarpus laevigatus (karaka nut); Corynocarpaceae	New Zealand; naturalised N America	seeds	karakin is cleaved to β-nitropropionic acid (NPAA)	II CT, NT AP	Nitropropionic acid inhibits succinate dehydrogenase, a central enzyme of the Krebs cycle. In animals: convulsions and cardiovascular collapse. **Therapy:** Non-protein amino acids
Coryphantha compacta and related species; Cactaceae	Mexico	aerial parts	hordenine, methyltyramine, tyramine and other β-phenylethylamines	II–III MA	Has been used by shamans as powerful psychoactive drug; the amines present can modulate dopamine and adrenergic receptors. **Therapy:** Bufotenine
Cotoneaster acuminatus, *C. horizontalis*, *C. simonsii*, *C. microphyllus*, *C. salicifolius*, *C. x watereri* and related species; Rosaceae	Europe; widely grown orn. plant	all parts, fruits	prunasin, amygdalin (cyanogenic glucosides)	II–III CT, GI	Cyanogenic glucosides release HCN, which is a strong inhibitor of cellular respiration; ingestion of red berries is often subject for consultations but is rarely hazardous; mostly GI disturbance. **Therapy:** Amagydalin
Cotyledon orbiculata (pig's ears); Crassulaceae	S Africa; orn. plant	all parts, especially leaves	orbicusides A, B and C and tyledoside C; bufadienolides	Ib CT, HP, GI, AP	Bufadienolides inhibit Na⁺, K⁺-ATPase and are therefore general heart and CNS poisons; causes GI tract disturbance, convulsions, and respiratory paralysis and death; mainly cause of animal poisoning. **Therapy:** Cardiac glycosides and *Cotyledon*

Scientific name (common name); Family	Occurrence	Hazardous plant parts	Active principle	Toxicity Class	Symptoms & Therapy
Courbonia glauca; Brassicaceae (Capparidaceae)	E Africa	all parts	glucosinolates (isothiocyanates)	Ib–II CT, NT	Mustard oils have pronounced cytotoxic activities; human fatalities have been reported. **Therapy:** Mustard oils
Crescentia cujete (calabash tree); Bignoniaceae	tropical America	leaves	cyanogenic glucosides	Ib CT, GI, AP	Cyanogenic glucosides release HCN, which is a strong inhibitor of cellular respiration; mostly GI disturbance; toxic to birds and mammals. **Therapy:** Amygdalin
Crinum bulbispermum (river lily); *C. longiflorum, C. amabile; C. asiaticum*; Amaryllidaceae	Africa, Asia, America; orn. plant	bulbs	rinine, powelline, lycorine, narcissine, crinamine or bulbispermine	Ib CT, IN, AP	Hypotensive, respiratory depressant; nausea, vomiting, dizziness, diarrhoea, kidney trouble and respiratory arrest. **Therapy:** *Crinum*
Crocus sativus (saffron); Iridaceae	Balkans, Asia	stigmas	picrocrocin is converted to safranal (the active component)	II NT, MA, MP	Similar to opium: psychedelic, narcotic, aphrodisiac; substantial ingestion causes excitation, headache, dizziness, vomiting, tachycardia, central paralysis, even death. **Therapy:** *Crocus sativus* and terpenes and phenylpropanoids
Crotalaria burkeana, *C. spartioides* (rattle box), *C. dura, C. globifera, C. incana, C. retusa, C. sagittalis, C. spectabilis, C. verrucosa, C. juncea, C. lanceolata* and related species; Fabaceae	Africa, America, Australia	all parts, especially seeds	pyrrolizidine alkaloids such as senecionine, monocrotaline, retrorsine and many others	II CT, NT, MU, AP	Mainly livestock poisoning; PAs are mutagenic and carcinogenic; neurotoxic, cause liver (veno-occlusive disease) and lung damage. **Therapy:** Senecionine
Croton tiglium (purging croton); Euphorbiaceae	India, SE Asia	seeds	TPA, phorbol esters; crotonide (purine alkaloid); a toxic lectin, crotin	Ia CT, IN, GI	Strong purgative; burning sensation in mouth; strong skin irritant; vomiting, unrest; strong diarrhoea; kidney trouble, dizziness, co-carcinogenic properties **Therapy:** Phorbol esters and *Croton*
Cryptostegia grandiflora (India rubber vine); *C. madagascariensis*; Apocynaceae (Asclepiadoideae)	India, West Indies, tropics	all parts	cardiac glycosides	Ib CT, HP	Symptoms of cardiac glycoside poisoning; human fatalities; livestock losses. **Therapy:** Cardiac glycosides
Cucumis africanus (wild cucumber), *C. myriocarpus*; Cucurbitaceae	S Africa	all parts, fruits	cucurbitacin B	Ib CT, GI, AP, MP	Cucurbitacins are very bitter and cytotoxic compounds, strong purgative, mainly animal poisoning. **Therapy:** Cucurbitacins

Scientific name (common name); Family	Occur-rence	Hazardous plant parts	Active principle	Toxicity Class	Symptoms & Therapy
Cucurbita lagenaria, see *Lagenaria siceraria*					
Cycas revoluta (sago cycas), *C. circinalis* and other cycads; Cycadaceae	S & E Asia	seeds, leaves, stems	cycasin, macrozamin (alkaloids); beta methylamino alanine (BMAA)	II CT, NT, GI, MU	BMAA is a neurotoxic NPAA (NMDA receptor agonist); cycasin is a methylating agent leading to mutations; ingestion causes internal bleeding; diarrhoea, nausea, vomiting, liver intoxication and neurodegenerative disorders (similar to amyotrophic lateral sclerosis); livestock poisoning. **Therapy:** *Encephalartos*
Cyclamen persicum (cyclamen), *C. purpurascens* and related species; Myrsinaceae (Primulaceae)	W Asia, Europe; orn. plants	all parts, especially bulbs	cyclamin and other triterpene saponins	II CT, GI	Substantial ingestion causes GI tract disturbance with nausea, vomiting, abdominal pains and spasms, sweating, circulatory disturbance, respiratory arrest; used to kill fish. **Therapy:** Saponins
Cydonia oblonga (quince); Rosaceae	Europe, Asia; worldwide cultivation	seeds	prunasin (cyanogenic glucoside); 500 ppm HCN	Ib–II CT, GI, AP	Cyanogenic glucosides release HCN, which is a strong inhibitor of cellular respiration; mostly GI disturbance. **Therapy:** Amygdalin
Cynanchum africanum, C. obtusifolium and *C. capense*; Apocynaceae (Asclepiadoideae)	S Africa	aerial parts	cynafoside B and F (pregnane glycosides)	Ib CT, NT, GI, AP	Symptoms of saponin intoxication; mostly animal poisoning: staggering gait and tremors, followed by spasms, convulsions and death. **Therapy:** Saponins
Cynodon dactylon (Bermuda grass); Poaceae	N America	aerial parts	cyanogenic glucosides	Ib–II CT, GI, AP	Cyanogenic glucosides release HCN, which is a strong inhibitor of cellular respiration; mostly GI disturbance; livestock poisoning. **Therapy:** Amygdalin
Cynoglossum officinale (hound's tongue) and related species; Boraginaceae	Europe, Asia; orn. plant	all parts, flowers	heliosupine, cynoglossine and other PAs	II CT, NT, MU, GI, AP	PAs are hepatotoxic, mutagenic and carcinogenic; substantial ingestion inhibits neuronal activities; paralytic; livestock poisoning. **Therapy:** Senecionine

362

Scientific name (common name); Family	Occurrence	Hazardous plant parts	Active principle	Toxicity Class	Symptoms & Therapy
Cytisus canariensis (= *Genista canariensis*) (Spanish broom); and related species with cytisine as main alkaloid; Fabaceae	Canary Islands; widely cultivated orn. plant	aerial parts, seeds	cytisine and other quinolizidine alkaloids	Ib NT, GI, MA	Cytisine is a strong nAChR agonist similar to nicotine; CNS active; psychedelic, euphoric, stimulant; high doses are very toxic (see *Laburnum*). **Therapy:** Cytisine
Cytisus scoparius (broom) (= *Sarothamnus scoparius*) and related species with sparteine/lupanine as main alkaloids; Fabaceae	Europe, Asia, common weed and orn. plant	all parts	sparteine and other quinolizidine alkaloids	II NT, GI, MP	Sparteine is an agonist at mAChR; substantial ingestion causes diuresis, uterus contractions, vomiting, diarrhoea, abdominal pain, tachycardia, cardiac irregularity; toxicity similar to nicotine. **Therapy:** Cytisine and *Cytisus*
Dalbergia nitidula, *D. stuhlmannii*; Fabaceae	E Africa	roots	dalbergin and other benzopyranoids; quinones	II–III CT, GI, NT	Appears to have caused human fatalities; can also cause contact dermatitis. **Therapy:** Hydroquinones
Daphne mezereum (mezereon), *D. cneorum*, *D. laureola*, *D. striata* and related species; Thymelaeaceae	Europe, Asia; in N America naturalised orn. plant	all parts, especially red berries	mezerein (phorbol ester), daphnin (coumarin glycoside)	Ia CT, IN, GI	Skin irritant; burning of throat and stomach, nausea, vomiting, gastroenteritis, internal bleeding, spasms, paralysis, kidney disturbance, bradycardia, coma, circulatory arrest. **Therapy:** Phorbol esters
Datisca glomerata (durang roots), *D. cannabina*; Datiscaceae	N America and Asia	all parts	cucurbitacins	Ib CT, GI, AP	Cucurbitacins are very bitter and cytotoxic compounds, strong purgative; mainly animal poisoning. **Therapy:** Cucurbitacins
Datura stramonium (thorn apple), *D. discolor*; *D. innoxia*, *D. metel*, *D. ferox*, *D. quercifolia*, *D. ceratocaula*, *D. wrightii* and related species; Solanaceae	Americas; worldwide as orn. plants and weeds	all parts, especially seeds and roots	hyoscyamine, scopolamine, atropine	Ia NT, MA, MP	Tropane alkaloids inhibit mAChR causing strong hallucinations; widely used as hallucinogen and aphrodisiac (see *D. stramonium*). **Therapy:** Hyoscyamine
Delphinium elatum (larkspur), *D. staphisagria*, *D. virescens*, *D. grandiflorum*, *D. ajacis* and related species; Ranunculaceae	Europe; N America, orn. plants	all parts, especially seeds	delphinine, nudicauline, staphisine, ajacine and other terpenoid alkaloids	Ia CT, NT, GI	Delphinine resembles aconitine in toxicity; inhibition of neuronal transmission (Na^+ channel opener); skin irritation, nausea, disturbance of GI tract and kidneys; dyspnoea; death from cardiac arrest; livestock poison. **Therapy:** Aconitine

363

Scientific name (common name); Family	Occurrence	Hazardous plant parts	Active principle	Toxicity Class	Symptoms & Therapy
Dendrocalamus asper (bamboo) and related species; Poaceae	Asia	young shoots	taxiphyllin and other cyanogenic glucosides	II CT, NT, GI	When shoots are crushed they release HCN which inhibits cellular respiration in mitochondria. Symptoms of HCN poisoning, HCN must be removed before consumption. **Therapy:** Amygdalin and *Dendrocalamus*
Derris elliptica (tuba root) and related species; Fabaceae	tropical Asia	roots	rotenone, ellipton, malacol and other isoflavones	Ib CT, NT, GI, AP	Blocks cellular respiration; GI tract disturbance; neurotoxin, inhibition of CNS and spinal cord: anaesthetic on tongue and throat; muscular arrest; used as insecticide and fish poison. **Therapy:** *Derris elliptica*
Desfontainia spinosa; Desfontainiaceae (Loganiaceae)	S & C. America	leaves, fruits	unknown	III MA	Has been used in S America as hallucinogen. **Therapy:** Bufotenine
Desmodium lasiocarpum; Fabaceae	America	seeds	β–carboline alkaloids, *N,N*-DMT	III NT, MA	Active agents modulate serotonin receptors and MAO; hallucinogenic properties, has been smoked as tobacco substitute. **Therapy:** Bufotenine
Dianella revoluta; D. nigra, D. caerulea; Hemerocallidaceae/Phormiaceae	Australia; orn. plant	aerial parts	stypandrol (a neurotoxic naphthoquinone)	II CT, NT, AP	Neurotoxic to nervus opticus and retina causing blindness in animals. **Therapy:** Quinones
Dicentra spectabilis (bleeding hearts); *D. cucullaria, D. canadensis, D. formosa* and related species; Papaveraceae	E Asia, N America; common orn. plant	all parts, especially roots	protopine, sanguinarine, chelerythrine and other isoquinoline, apomorphine and protoberberine alkaloids	II CT, IN, GI, NT, AP	Central sedative, spasmolytic and narcotic; dizziness; GI tract and kidney disturbance, arrhythmia, livestock poisoning. **Therapy:** Isoquinoline alkaloids
Dichapetalum cymosum (poison leaf); Dichapetalaceae	S Africa	aerial parts	fluoroacetic acid	Ia CT, GI, AP	Fluoroacetic acid is converted in the body to highly toxic fluorocitrate that inhibits the activity of aconitase enzymes in the TCA cycle and thus cellular respiration; mainly animal poisoning. **Therapy:** *Dichapetalum* and oxalic acid
Dictamnus albus (dittani); Rutaceae	Europe, Asia; orn. plant	aerial parts	dictamnine and other furanoquinoline alkaloids; monoterpenes	II CT, MU, NT, IN, GI	Furanocoumarins and alkaloids can intercalate or alkylate DNA; therefore mutagenic; plant sap is a strong skin irritant; when exposed to sunlight it can cause photodermatitis and contact dermatitis. **Therapy:** Coumarins and *Dictamnus*

Scientific name (common name); Family	Occurrence	Hazardous plant parts	Active principle	Toxicity Class	Symptoms & Therapy
Dieffenbachia seguine (= *D. maculata*) (dumb cane), *D. picta*; Araceae	tropical America; worldwide orn. plant	all parts	Ca^{2+}-oxalate raphides; cyanogenic glucosides, saponins, toxic peptides	Ib CT, IN, GI, NT	Raphides cause painful irritation by penetrating the mucous membranes of the mouth, throat or eyes; causes swelling of the mouth, tongue and face; GI tract and CNS disturbance. **Therapy:** Oxalic acid and *Dieffenbachia*
Digitalis purpurea (foxglove), *D. grandiflora, D. lanata, D. lutea* and related species; Plantaginaceae (Scrophulariaceae)	Europe; sometimes as orn. in N America and Africa	all parts	several cardenolides (purpurea glycoside, lanatoside, digitoxin, digoxin)	Ia CT, HP, GI, MP	Typical cardiac glycoside intoxication; vomiting, diarrhoea, gastroenteritis, severe headache, irregular heartbeat and pulse, convulsions, CNS disturbance, cardiac arrest, sudden death. **Therapy:** Cardiac glycosides
Dimorphandra gardneriana, D. mollis; Fabaceae	Brazil	fruits	unknown toxin; flavonoids	III AP	Insecticidal. **Therapy:** Flavonoids
Dimorphotheca cuneata (Karoo bietou), *D. nudicaulis, D. zeyheri* and *D. spectabilis;* Asteraceae	S Africa	aerial parts	linamarine and other cyanogenic glucosides	II CT, NT, GI, AP	Cyanogenic glucosides release HCN upon enzymatic hydrolysis; HCN blocks cellular respiration; mainly animal poisoning. **Therapy:** Amygdalin
Dioon edule; Cycadaceae	America	aerial parts	azoxyglucosides, such as cycasin	II CT, NT, GI, MU	Cycasin is a methylating agent leading to mutations; ingestion causes internal bleeding; diarrhoea, nausea, vomiting, liver intoxication and possibly neurodegenerative disorders; livestock poisoning. **Therapy:** *Encephalartos*
Dioscorea dregeana (wild yam), *D. bulbifera;* Dioscoreaceae	S Africa	aerial parts	dioscorine (piperidine alkaloid); steroidal saponins (glycosides of diosgenin)	Ib CT, NT, GI, MP	Used to treat hysteria, convulsions and epilepsy and to pacify psychotic patients; dioscorine is highly toxic to humans, causes liver damage and disturbance of GI tract. **Therapy:** Pyrrolidine alkaloids and *Dioscorea*
Diospyros mollis (ebony tree); Ebenaceae	Asia	aerial parts, fruits	diospyrol	II CT, IN	Inflammation of skin and mucous membranes. **Therapy:** *Rhus*
Dipladenia spp. (= *Mandevilla*); Apocynaceae	tropical America; orn. plant	all parts	melonine B and other indole alkaloids	II NT, GI	Substantial ingestion causes disturbance of GI tract and CNS. **Therapy:** Indole alkaloids

Scientific name (common name); Family	Occurrence	Hazardous plant parts	Active principle	Toxicity Class	Symptoms & Therapy
Diplopterys cabrerana (= *Banisteriopsis rusbyana*); Malpighiaceae	tropical S America	leaves, bark	*N,N*-DMT, *N*-methyltryptamine, 5-MeO DMT, bufotenine and β-carboline alkaloids	II NT, MA	Has been used as a hallucinogen, sometimes together with *Banisteriopsis caapi*. **Therapy:** Bufotenine and *Banisteriopsis*
Dirca palustris (leather wood); Thymelaeaceae	N America	all parts, bark	probably phorbol esters	II IN, GI	Inflammation of skin and mucous membranes; nauseating taste. **Therapy:** Phorbol esters
Dracunculus vulgaris; Araceae	Europe; orn. plant	aerial parts, fruits	aroin; Ca^{2+}-oxalates	Ib CT, IN, GI, NT	Mucosal irritant, disturbance of GI tract and CNS. **Therapy:** Oxalic acid and *Arum*
Drimia robusta, D. altissima; Hyacinthaceae	S Africa	bulbs	12β-hydroxyscillirosidin and other bufadienolides	Ib CT, HP, GI, MP	Inhibition of Na^+, K^+-ATPase causing heart glycoside poisoning; nausea, vomiting, diarrhoea, colic, haematuria, death by cardiac arrest; also skin irritant with blister formation; human and animal poisoning. **Therapy:** Cardiac glycosides
Dryopteris filix-mas (male fern); Dryopteridaceae	Europe, N Asia, N America	all parts, especially roots	albaspidin (a dimeric butanone phloroglucide)	II CT, GI, NT, AP	Nausea, vomiting, abdominal pain, diarrhoea, spasms, distorted vision, respiratory arrest. Even when the drug was taken against internal worms, human poisoning was not uncommon; today a known poison of sheep and cattle. **Therapy:** Sesquiterpenes
Duboisia myoporoides (corkwood duboisia), *D. bopwoodii* and related species; Solanaceae	Australia	leaves	scopolamine, hyoscyamine, tigloidine and other tropane alkaloids; nicotine and related pyridine alkaloids	Ib NT, MA, GI, MP	Has been used as hallucinogen by Australian aborigines; in combination with tobacco and opium; high doses: neurotoxic (see *Datura*); dizziness, tremor, excitation, mydriasis, nausea, vomiting, palpitations, even death. **Therapy:** Hyoscyamine and *Duboisia*
Duranta erecta (golden eye drop), *D. repens* (golden dew drop); Verbenaceae	America, SE Asia; orn. plant	leaves, fruits	durantoside I, II and III (iridoid glycosides); saponins; alkaloids (narcotine)	II CT; NT	The berries were said to have caused sleepiness, high temperature; dilated pupils, fast pulse, swelling of the lips and eyelids and convulsions. **Therapy:** Saponins
Ecballium elaterium (squirting cucumber); Cucurbitaceae	Mediterranean, Asia; orn. plant	fruits	cucurbitacin E, I	Ib CT, IN, GI, NT	Cucurbitacins are highly cytotoxic; strong purgative causing salivation, vomiting, inflammation of GI tract; abdominal pains, headache, tachycardia; internal bleeding; >0.6 ml juice can be lethal. **Therapy:** Cucurbitacins

Scientific name (common name); Family	Occurrence	Hazardous plant parts	Active principle	Toxicity Class	Symptoms & Therapy
Echium vulgare (viper's bugloss, blueweed; *E. plantagineum* (Patterson's curse) and related species; Boraginaceae	Europe, W Asia; worldwide as weeds	all parts, especially flowers	heliosupine and other PAs	II CT, NT, MU, AP	PAs are mutagenic and carcinogenic; hepatotoxic (veno-occlusive disease); intoxication of sheep has been recorded in Australia. **Therapy**: Senecionine
Encephalartos longifolius and related species; Zamiaceae	Africa	aerial parts	cycasin, macrozamin	II CT, NT, GI, MU	Cycasin is a methylating compound and therefore mutagenic and cytotoxic; liver poison. **Therapy**: *Encephalartos*
Enterolobium cyclocarpum; Fabaceae	Asia	seeds	albizziine (a NPAA); triterpenes	III GI, CT	Fish and cattle poisoning. **Therapy**: Non-protein amino acids
Ephedra sinica (ephedra), *E. distachya, E. gerardiana, E. americana, E. andina* and related species; Ephedraceae	Asia, Europe, America	all parts	l-ephedrine, d-pseudoephedrine, l,d-norephedrine, d-norpseudoephedrine (0.5–3.3 %)	II MA, MP, GI	Sympathomimetic, amphetamine-like activities (stimulant, euphoric); helps against fatigue and hunger; high doses induce heavy perspiration, activate breathing and muscle activity; insomnia; mydriasis, constipation; hypertension, arrhythmia, death. **Therapy**: Ephedrine
Epipremnum aureum (= *Pothos aureus, Raphidophora aurea, Scindapsus aureus*) (golden pothos); Araceae	tropics	all parts	calcium oxalate raphides; toxic peptides	Ib–II CT, IN, GI	Strong, painful irritation of skin and mucous membranes; GI tract disturbances. **Therapy**: Oxalic acid
Equisetum palustre (marsh horsetail), *E. fluviatile, E. hyemale, E. ramosissimum*; Equisetaceae	Europe, Asia, N America	aerial parts	palustrine, palustridine and other alkaloids (up to 0.3%); thiaminase (an enzyme), saponins	II CT, NT, GI, AP	Mainly known as animal poison for horses and cattle; animals show irritation, muscle tremor, staggering, falling down; death after complete exhaustion. **Therapy**: *Equisetum*
Eranthis hyemalis (winter aconite) and related species; Ranunculaceae	S Europe; orn. plant	all parts, especially tubers	eranthin A, B and other cardiac glycosides of the bufadienolide type	Ib CT, NT, GI,	Substantial ingestion causes symptoms of cardiac glycoside poisoning: nausea, vomiting, diarrhoea, colic, bradycardia, disturbed vision; dyspnoea; cardiac arrest. **Therapy**: Cardiac glycosides
Eriobotrya japonica (loquat); Rosaceae	Asia; widely cultivated	seeds	amygdalin (cyanogenic glucoside)	Ib CT, NT	Amygdalin releases HCN which inhibits cellular respiration; symptoms of HCN poisoning. **Therapy**: Amygdalin
Ervatamia coronaria (= *Tabernaemontana divaricarpa*) (crepe jasmin); Apocynaceae	India	aerial parts	indole alkaloids	Ib CT, NT, GI	Substantial ingestion causes disturbance of GI tract and CNS; can be fatal. **Therapy**: Indole alkaloids

367

Scientific name (common name); Family	Occurrence	Hazardous plant parts	Active principle	Toxicity Class	Symptoms & Therapy
Erysimum cheiri (= *Cheiranthus cheiri*) (wallflower); Brassicaceae	Mediterranean; common orn. plant	all parts, especially seeds	cheirotoxin, cheiroside A and other cardenolides; glucosinolates (glucocheirolin)	II CT, GI, HP	Substantial ingestion causes symptoms of cardiac glycoside intoxication with pronounced GI tract disturbance. **Therapy:** Cardiac glycosides
Erysimum crepidifolium, *E. diffusum*; Brassicaceae	C & S Europe	all parts, especially seeds	erysimoside, helveticoside and other cardiac glycosides (cardenolide type); up to 3.5% in seeds	Ib CT, NT, GI, AP	Substantial ingestion causes symptoms of cardiac glycoside poisoning with disturbance of GI tract and CNS; trembling, unrest, spasms, dyspnoea, cardiac arrest; animal poisoning, especially of geese. **Therapy:** Cardiac glycosides
Erythrina caffra (coral tree), *E. crista-galli; E. corallodendron, E. flabelliformis, E. berbacea, E. lysistemon, E. americana, E. berteroana* and several other species; Fabaceae	America, Africa; cultivated as orn. plant	all parts, especially seeds and bark	erythraline and erysotrine and several other *Erythrina* alkaloids	Ib NT, MA, MP	Alkaloids with neuromuscular blocking activities similar to curare but with oral toxicity; strongly hypertensive and muscle arresting; first symptoms are excitement, inconsistent talking, and tumbling as when drunk; hyperthermia, red skin and deep sleep that leads to coma and death; *Erythrina* has been used as hallucinogen in C America. **Therapy:** *Erythrina*
Erythronium dens-canis (dog's-tooth violet), *E. americanum* and related species; Liliaceae	N America; common orn. plant	all parts, especially bulbs	tulipalin, tuliposide A	III CT, IN, GI, AP	Gastrointestinal disturbance; serious skin allergen; animal poisoning. **Therapy:** Protoanemonin
Erythrophleum suaveolens (ordeal tree), *E. africanum, E. lasianthum, E. teysmannii, E. succirubrum*; Fabaceae	Africa; Asia	bark, aerial parts	cassaine, erythrophleine, and other diterpenoid alkaloids	Ia HP, GI, MP	Vomiting, diarrhoea, hypertension, positive inotropic, heart effects; death through respiratory and cardiac arrest. Also used for arrow poisons and ordeals. **Therapy:** Cardiac glycosides
Erythroxylum coca (coca plant), *E. novogranatense* and related species; Erythroxylaceae	S America	leaves	cocaine, cuskohygrine, truxilline and other tropane alkaloids	Ib NT, MA, MP	Strong central stimulant of sympathetic nervous system; aphrodisiac; Na^+ channel blocker and therefore local analgesic; reduction of salivation, hunger and thirst; tachycardia and enhanced sweating; mydriasis. **Therapy:** *Erythroxylum* and hyoscyamine
Eschscholzia californica (California poppy); Papaveraceae	N America; common orn. plant	all parts	protopine, cryptopine, sanguinarine, magnoflorine and other isoquinoline alkaloids; roots: up to 2.7%; leaves: up to 0.3%	II NT, MA, GI, MP	Mildly psychoactive; sedative; narcotic, spasmolytic and pain killing; uterus contracting; substantial ingestion causes burning sensation in mouth and throat; nausea, vomiting, bloody diarrhoea. **Therapy:** Isoquinoline alkaloids and *Chelidonium*

Scientific name (common name); Family	Occurrence	Hazardous plant parts	Active principle	Toxicity Class	Symptoms & Therapy
Eucalyptus globulus (blue gum), *E. cladocalyx* and related species; Myrtaceae	Australia; widely cultivated	aerial parts	essential oil with several monoterpenes, such as 1,8 cineol; *E. cladocalyx*: prunasin (a cyanogenic glucoside)	II–III CT, IN, GI, MP, AP	High doses cause CNS and GI disturbance (see monoterpenes); HCN poisoning through prunasin, especially in animals (sheep). **Therapy:** Monoterpenes and amygdalin
Euonymus europaeus (spindle tree), *E. americanus*, *E. occidentalis*, *E. atropurpureus*, *E. japonicus*, *E. alatus*, *E. verrucosus* and related species; Celastraceae	Europe, Asia, N America; orn. plant	all parts, especially fruits	*E. europaeus*: evobioside, evomonoside and other cardenolides; evonine and several alkaloids	Ib CT, GI, MP, HP	Irritation of GI tract; hallucinations, nausea, extensive vomiting, shock, hyperthermia, bloody diarrhoea, liver and kidney disturbance; arrhythmia, strong spasms, coma after 12 h, cardiac arrest. **Therapy:** Cardiac glycosides
Eupatorium cannabinum (hemp-agrimony), *E. perfoliatum*, *E. rugosum*; Asteraceae	Europe; N America; sometimes orn. plant	all parts	eupatoriopicrin, euparin (sesquiterpene lactones); several pyrrolizidine alkaloids	II CT, NT, MU	PAs have mutagenic and carcinogenic properties; they are transferred from milk to humans; tremor, delirium, death. **Therapy:** Senecionine
Euphorbia cyparissias (cypress spurge), *E. corollata*, *E. lactea*, *E. cotinifolia*, *E. myrsinitis*, *E. belioscopa*, *E. latbyris*, **E. tirucalli**, *E. ingens*, *E. pulcherrima* (Christmas flower), *E. milii* (crown of thorns), *E. marginata* (snow-on-the-mountain), *E. antiquorum* (Malayan spurge tree); Euphorbiaceae	Europe, Asia, Africa, America	all parts, especially latex and seeds	phorbol esters in latex; triterpenoids; *E. pulcherrima* without phorbol esters	Ib CT, IN, GI	Strong skin irritant (blister formation); burning irritation in mouth and throat; vomiting, stomach pain, purgative; bloody diarrhoea; arrhythmia, tinnitus, liver and kidney disturbances, coma; co-carcinogen. **Therapy:** Phorbol esters
Excoecaria velenifera, *E. agallocha* (blind-your-eyes), *E. oppositifolia*; Euphorbiaceae	E Africa, SE Asia	all parts, latex	phorbol esters; ellagic acid, tannins	Ib CT, IN, GI, AP	Burning irritation in mouth and throat; vomiting, stomach pain, purgative; bloody diarrhoea; arrhythmia, tinnitus, strong skin and eye irritant (blister formation; can cause blindness); camel poison. **Therapy:** Phorbol esters
Exogonium purga see Ipomoea purga					

Scientific name (common name); Family	Occurrence	Hazardous plant parts	Active principle	Toxicity Class	Symptoms & Therapy
Fabiana imbricata (false heath) and related species; Solanaceae	Chile; orn. plant	all parts	fabianine (a quinoline alkaloid); scopoletin and other coumarins; anthraquinones	II GI, MA, MP	Strong diuretic, antiseptic, general tonic; GI tract disturbance with diarrhoea; smoke with stimulant and euphoric activities. **Therapy:** *Fabiana*
Fagopyrum esculentum (buckwheat); Polygonaceae	Europe; widely cultivated	leaves	fagopyrin, a naphthodianthrone; rutin (flavonoid)	III IN, GI, AP	Naphthodianthrones can cause photodermatitis; substantial ingestion leads to GI tract disturbance. **Therapy:** Anthraquinones and *Hypericum*
Fagus sylvatica (beech); Fagaceae	Eurasia; widely cultivated	fruits	seeds contain up to 3 % oxalates and willardiine, a NPAA; saponins	III CT, GI	Substantial ingestion can cause GI tract and CNS disturbance; in earlier times, when large amounts of beech seeds were eaten, cases of poisoning had occurred. **Therapy:** Oxalic acid and non-protein amino acids
Fatsia japonica (Japanese fatsia); Araliaceae	Japan; common orn. plant	all parts	several triterpene saponins	II CT, NT, GI	Substantial ingestion causes disturbances of GI tract and kidneys. **Therapy:** Saponins
Ferula communis (giant fennel), *F. assa-foetida*; Apiaceae	Mediterranean	all parts	ferulenol in *F. communis*	Ib CT, GI, AP	Ferulenol is highly cytotoxic and stabilises microtubules, similar to taxol; ingestion causes disturbance of GI tract; known to cause animal poisoning ("ferulosis"). **Therapy:** *Taxus*
Ficaria verna (figwort) (= *Ranunculus ficaria*); Ranunculaceae	Europe, Asia	all parts, especially rhizomes	ranunculin is hydrolysed to bioactive protoanemonin	II–III CT, NT, GI	Substantial ingestion causes disturbances of GI tract and kidneys with nausea, diarrhoea, bloody vomiting; intestinal bleeding. **Therapy:** Protoanemonin
Ficus benjamina (weeping fig), *F. elastica*, *F. carica*, *F. pumila*; Moraceae	SE Asia; widely grown as orn. plant	latex	furanocoumarins, flavonoids, triterpenes, sesquiterpene glycosides, proteins	III IN, GI	Latex can cause photodermatosis; ingestion results in disturbance of GI tract; rarely hazardous. **Therapy:** Coumarins
Florestina tripteris; Asteraceae	C America	leaves	cyanogenic glucoside	Ib CT, NT, AP	HCN released from cyanogenic glucoside; symptoms of HCN intoxication; mainly livestock poisoning; fatal to sheep. **Therapy:** Amygdalin

Scientific name (common name); Family	Occurrence	Hazardous plant parts	Active principle	Toxicity Class	Symptoms & Therapy
Fritillaria imperialis (crown imperial), *F. meleagris*, *F. barellinii* and related species; Liliaceae	Asia; widely cultivated, orn. plant	especially bulbs	imperialine and other steroid alkaloids; tuliposide A; tulipalin A	II CT, NT, IN, GI	Steroidal alkaloids resemble aconitine in bioactivity; disturbances of GI tract and kidneys, vomiting, spasms, hypotension, cardiac arrest; contact allergen. **Therapy:** Aconitine
Fumaria officinalis (fumitory) and related species; Papaveraceae	Europe, Asia, N Africa	all parts	protopine, scoulerine, cryptopine, stylopine, and other isoquinoline alkaloids	II NT, MA, GI, MP	Mildly psychoactive; sedative; narcotic; substantial ingestion causes burning sensation in mouth and throat; nausea, vomiting, diarrhoea and hypotension. **Therapy:** Isoquinoline alkaloids and *Chelidonium*
Gagea lutea and related species; Liliaceae	Europe; orn. plant	all parts, especially bulbs	tulipalin, tuliposide A	III CT, IN, GI, AP	Substantial ingestion can lead to gastrointestinal disturbance; serious skin allergen; animal poisoning. **Therapy:** Protoanemonin
Galanthus nivalis (snowdrop) and related species; Amaryllidaceae	Europe, Asia; common orn. plant	mainly bulbs	galanthamine, lycorine and other alkaloids	II NT, GI, MP, AP	Galanthamine inhibits acetylcholine esterase; parasympathomimetic; causes nausea, vomiting, diarrhoea, small pupils; livestock poisoning. **Therapy:** Lycorine
Galerina steglichii; Cortinariaceae	Europe	fruiting body	psilocybin, psilocin, baeocystine	Ib NT, MA	Alkaloids are serotonin receptor agonists; therefore ingestion causes psychoactive, psychedelic and hallucinogenic effects. **Therapy:** Bufotenine
Galphimia glauca, (= *Thryallis*); Malpighiaceae	S & C America; orn. plant	leaves, branches	galphimine B (a triterpenoid)	III CT, GI	Ingestion of large amounts can cause GI tract disturbance; also, insecticidal. **Therapy:** Steroids and triterpenes
Gaultheria procumbens (wintergreen); Ericaceae	N America; orn. plant	aerial parts	gaultherin, a glycoside of methylsalicylate; methylsalicylate	II CT, GI, MP	High doses (30 g oil) lead to strong stomach and kidney disturbances; internal bleeding; death. **Therapy:** Monoterpenes and *Gaultheria*
Geigeria ornativa, G. aspera, G. filifolia; Asteraceae	S Africa	aerial parts	vermeeric acid is converted to vermeerin, the active component (sesquiterpene lactones)	II CT, GI, AP	Mostly animal poisoning; cough, diarrhoea or sometimes bloating, followed by stiffness and paralysis. Death may be caused by choking, respiratory paralysis, exhaustion from vomiting and purgation, heart failure or very often pneumonia. **Therapy:** Sesquiterpenes

Scientific name (common name); Family	Occurrence	Hazardous plant parts	Active principle	Toxicity Class	Symptoms & Therapy
Gelsemium sempervirens (yellow jessamine), *G. rigens*; Gelsemiaceae	N America; used as orn. plant	all parts, roots, nectar	gelsemine, sempervirine, and other indole alkaloids	Ia CT, NT, MA	Profuse sweating, paralysis of CNS; weakness, convulsions, reduced blood pressure and heart activity; mydriasis and visual disturbances, strychnine-like convulsions, death from respiratory arrest; *G. rigens* used for suicide by Asian women. **Therapy:** *Gelsemium*
Genista tinctoria (dyer's greenweed), *G. germanica* and related species; Fabaceae	Europe, Asia; sometimes orn. plant	all parts	cytisine, anagyrine and other quinolizidine alkaloids (0.3 %)	Ib NT, MA, GI, MP	Slightly psychoactive and hallucinogenic; cytisine is a nAChR agonist; diuretic, uterus contracting, causing abdominal pain, tachycardia, cardiac irregularity, vomiting, diarrhoea, dizziness; toxicity similar to nicotine. **Therapy:** Cytisine
Ginkgo biloba (maidenhair tree); Ginkgoaceae	China; widely cultivated	fruits	3-alkylbrenzcatechins (ginkgolic acid, cardanols); similar to urushiol (see *Anacardium*); 4-O-methylpyridoxine	III CT, NT, IN, GI, MP	The alkyl phenols are skin irritant contact allergens; 4-O-methylpyridoxine inhibits glutamate decarboxylase and thus the formation of GABA. Vomiting, diarrhoea and seizures are among the symptoms. **Therapy:** Phenolics
Girardinia heterophylla; Urticaceae	SE Asia	aerial parts	stinging hairs with histamine, acetylcholine, serotonin, formic acid, acetic acid	III IN	Painful itching with welt formation occurs at sites that were pricked by stinging hairs. **Therapy:** Oxalic acid
Glaucium flavum (yellow horned poppy) and related species; Papaveraceae	S Europe; orn. plant	all parts	glaucine, magnoflorine, protopine, sanguinarine and other isoquinoline alkaloids	II NT, MA, GI, MP	Psychoactive, narcotic, spasmolytic and pain killing, similar to opium; uterus contracting. Substantial ingestion causes GI tract and CNS disturbance. **Therapy:** Isoquinoline alkaloids and *Chelidonium*
Gliricidia sepium and related species; Fabaceae	America; introduced in tropical countries	roots, leaves, seeds, bark	canavanine (NPAA), benzopyranoids, flavonoids	III CT, AP	Bark and seeds used to kill rats and mice. **Therapy:** Non-protein amino acids
Gloriosa superba (= *G. rothschildiana*, *G. simplex*); Colchicaceae (Liliaceae)	Asia, Africa; used as orn. plant	all parts, especially bulb	colchicine, gloriosine, superbine	Ia CT, NT, GI	Colchicine is a spindle poison; nausea, dizziness, oropharyngeal pains, spasms, arrhythmia, internal bleeding; peripheral neuropathy; tachycardia, respiratory arrest. **Therapy:** Colchicine

Scientific name (common name); Family	Occurrence	Hazardous plant parts	Active principle	Toxicity Class	Symptoms & Therapy
Gluta usitata (= *Melanorrhoea usitata*) (theetsee, varnish tree); Anacardiaceae	SE Asia	aerial parts	alkyl phenols, ursuric acid	II CT, IN	Irritant on skin and mucosal tissues; causes severe inflammation with blister formation. **Therapy:** *Rhus*
Glyceria striata; Poaceae	N America	all parts	cyanogenic glucosides which release HCN	Ib–II CT, NT, GI, AP	Symptoms of HCN intoxication; livestock poisoning. **Therapy:** Amygdalin
Gnidia kraussiana (yellow heads), *G. burchellii* and related species; Thymelaeaceae	S Africa; a few orn. species	all parts, fruits	phorbol esters of daphnane type (kraussianin, gnidilatin, gnidilatidin), gnidicin, gnididin and gniditrin	Ia CT, IN, NT, GI, AP	Irritation of the nose and throat, coughing and sneezing, followed by headache and nausea; livestock poisoning: severe diarrhoea, weakness, fever, rapid, weak pulse, death. **Therapy:** Phorbol esters and *Gnidia*
Gomphocarpus fruticosus (milkweed, swanbush), (= *Asclepias*), *G. physocarpus* and related species; Apocynaceae (Asclepiadoideae)	S Africa; grown as orn. plant	all parts	cardiac glycosides	Ib CT, NT, GI, HP, AP	Typical cardiac glycoside intoxication; diuretic, sedative, irritation of GI tract with pronounced diarrhoea; cardiac arrest; livestock poisoning. **Therapy:** Cardiac glycosides
Gossypium herbaceum (cotton) and related species; Malvaceae	Asia; widely cultivated	especially seeds	gossypol, *p*-hemigossypolon and other quinones	II CT, NT, GI, AP	High doses lead to dyspnoea, disturbance of GI tract, internal bleeding, nephritis, central paralysis, spasms, cyanosis, even death; livestock poisoning. **Therapy:** Flavonoids and hydroquinones
Gratiola officinalis (hedge hyssop); Plantaginaceae (Scrophulariaceae)	Europe, Asia, N America	all parts	gratiogenin, gratioside and other tetracyclic triterpenes; cucurbitacin A	Ib CT, NT, GI, MP	Substantial ingestion can cause GI tract disturbance with nausea, vomiting, salivation, bloody diarrhoea, furthermore nephritis, respiratory and cardiac disturbances; death from respiratory arrest; abortifacient; toxicity similar to cardiac glycosides, to which there are structural similarities. **Therapy:** *Gratiola* and cardiac glycosides
Grevillea robusta; Proteaceae	Australia; garden tree	wood, leaves	resorcinol derivatives, grevillol (an alkylphenol)	III CT, IN, GI	Alkylphenols induce contact dermatitis (see hydroquinone); substantial ingestion causes GI tract disturbance. **Therapy:** Phenolics
Guaiacum officinale (lignum vitae); Zygophyllaceae	S & C America	wood, fruits	guajacol, guajaculen, saponins	II–III CT, GI	High doses lead to gastroenteritis with nausea and diarrhoea, palpitations, dizziness. **Therapy:** Phenolics
Gutierrezia sarothrae (= *Xanthocephalum*) (broom snakeweed); Asteraceae	S America	leaves	saponins	II–III CT, AP	Severely toxic for livestock; abortifacient. **Therapy:** Saponins

Scientific name (common name); Family	Occurrence	Hazardous plant parts	Active principle	Toxicity Class	Symptoms & Therapy
Gymnema sylvestre; Apocynaceae (Asclepiadoideae)	Africa	leaves	gymnemic acid and related glycosides	II CT, GI	Weakness, fever, diarrhoea, diuresis, hypoglycaemia; death by respiratory arrest. **Therapy:** Saponins
Gymnocladus dioicus (Kentucky coffee tree); Fabaceae	N America; widely cultivated	seeds, pulp	cytisine and other quinolizidine alkaloids	Ib–II NT, MA, GI	GI disorder, cytisine is nAChR agonist; irregular pulse, coma, fatal for livestock. **Therapy:** Cytisine
Gymnopilus aeruginosus, *G. validipes*, *G. braendlei*, *G. purpuratus* and related fungi; Cortinariaceae	world-wide	fruit body	psilocybin, psilocin, baeocystine	II NT, MA	Alkaloids are serotonin receptor agonists; therefore psychoactive, psychedelic, visionary and hallucinogenic. **Therapy:** Bufotenine
Gynura scandens; Asteraceae	E Africa	all parts	gynuramine and other pyrrolizidine alkaloids	II CT, NT, MU, GI	PAs have mutagenic and carcinogenic properties; hepatotoxicity, GI tract and CNS disturbance. **Therapy:** Senecionine
Gyromitra esculenta, *G. brunnea*, *G. californica*, *G. ambigua*, *G. fastigata*, *G. gigas*, *G. infula* and related fungi; Discinaceae	Europe	fruiting body	monomethyl hydrazine	Ia CT, NT, GI	Latent period 6–8 h; causes bloating, nausea, vomiting, bloody diarrhoea, muscle cramps, in severe cases: liver failure, high fever, coma, death after 2–4 days. **Therapy:** *Amanita phalloides*
Haemanthus coccineus (blood lily) and related species; Amaryllidaceae	S Africa; orn. plant	all parts	coccinine and an isomer, montanine (isoquinoline alkaloids)	II NT, MP	Weak hypotensive and convulsive agent. **Therapy:** Isoquinoline alkaloids
Hagenia abyssinica; Rosaceae	E Africa	flowers, fruits	kosotoxin, protokosin and other phloroglucinols	II CT, IN, NT, GI	Has been used to treat intestinal worms; paralysis of skeletal and cardiac muscles; disturbance of GI tract with nausea and vomiting; death by respiratory arrest. **Therapy:** Flavonoids
Halogeton glomeratus; Amaranthaceae (Chenopodiaceae)	N America	all parts	oxalates	II CT, GI	Substantial ingestion can cause GI tract and kidney disturbances. **Therapy:** Oxalic acid
Haplopappus heterophyllus (rayless goldenrod); Asteraceae	N America	all parts	tremetol	Ib–II CT, NT, GI	Tremetol is transferred into milk by animals that feed on the plant; human poisoning; fatal livestock poisoning. **Therapy:** Sesquiterpenes

Scientific name (common name); Family	Occurrence	Hazardous plant parts	Active principle	Toxicity Class	Symptoms & Therapy
Hedeoma pulegioides (false pennyroyal); Lamiaceae	N America	aerial parts	essential oil with pulegone and other monoterpenes	II CT, NT, GI	High doses have narcotic effects, cause irritation of GI tract and of internal organs; internal bleeding and even death. **Therapy:** Monoterpenes
Hedera helix (ivy), *H. canariensis* and related species; Araliaceae	Eurasia, N America; widely cultivated orn. plant	all parts, especially leaves, fruits	α-hederin and other triterpene saponins, sesquiterpenes; falcarinol (a polyacetylene)	II IN, GI, MP	Irritation of GI tract; nausea, vomiting, palpitations, eczema, dizziness, nervous depression, hyperthermia, death by respiratory arrest; mydriasis. Skin reactions include rashes, red, swollen skin, blisters, oedema and pain. **Therapy:** Saponins and polyacetylenes
Heimia salicifolia (sinicuiche); Lythraceae	Mexico	aerial parts	lythrine, cryogenine, heimine and other atypical quinolizidine alkaloids	II NT, MA	Psychoactive; stimulant, euphoric, auditory and visual hallucinogenic effects. **Therapy:** Cytisine
Helenium amarum and related species; Asteraceae	N America orn. plant	aerial parts	sesquiterpene lactones	II CT, NT, GI, AP	Substantial ingestion can cause disturbance of GI tract and CNS; livestock poisoning. **Therapy:** Sesquiterpenes
Heliotropium arborescens (garden heliotrope), *H. europaeum, H. angiospermum, H. indicum, H. curassavicum* and related species; Boraginaceae	S America, Europe, Asia; orn. plant	all parts	heliotrine and other pyrrolizidine alkaloids (0.9%)	II CT, NT, MU	PAs have mutagenic and carcinogenic properties; hepatotoxicity (veno-occlusive disease) and pulmonary hypertension, CNS disturbance; fatal livestock poisoning. **Therapy:** Senecionine
Helleborus viridis, H. niger (hellebore), *H. foetidus* and related species; Ranunculaceae	Europe, naturalised N America; common orn. plant	aerial parts	cardiac glycosides (bufadienolides): hellebrin; steroid saponins (helleborin), ranunculoside; alkaloids (celliamine, sprintillamine)	Ia CP, HP, GI, MP	Cardiac glycoside intoxication; vomiting, gastroenteritis with violent diarrhoea, delirium, convulsions, arrhythmia, death by respiratory arrest; alkaloids have similar properties as veratrine and aconitine; livestock poisoning. **Therapy:** Cardiac glycosides and aconitine
Hemerocallis spp. (daylilies); Hemerocallidaceae	Asia; orn. plant	aerial parts	stypandrol (=hemerocallin)	II CT, NT	Neurotoxic to nervus opticus and retina causing blindness in animals; animal poisoning. **Therapy:** Sesquiterpenes

Scientific name (common name); Family	Occurrence	Hazardous plant parts	Active principle	Toxicity Class	Symptoms & Therapy
Hepatica nobilis (liver leaf); Ranunculaceae	Europe; orn. plant	all parts	ranunculin is converted enzymatically to protoanemonin; magnoflorine (an aporphine alkaloid)	II CT, NT, IN, MP, MU	Substantial ingestion can cause GI tract and kidney disturbance; blistering and inflammation of skin. **Therapy:** Protoanemonin
Heracleum mantegazzianum (giant hogweed), *H. sphondylium*; Apiaceae	Caucasian mountains; introduced in many parts of the world	all parts, sap	8-methoxypsoralen and other furanocoumarins	II CT, IN, MU, GI	Furanocoumarins intercalate DNA and form crosslinks upon exposure to sunlight; strongly cytotoxic; leads to serious blister formation of the skin: FC are used in phytotherapy for psoriasis treatment. **Therapy:** Coumarins
Hesperis matronalis (dame's violet); Brassicaceae	Europe; used as orn. plant	all parts	glucosinolates (isothiocyanates) and cardenolides	II–III CT, NT, GI, HP	Substantial ingestion can cause skin irritation (mustard oils) and cardiac glycoside intoxication. **Therapy:** Mustard oils
Hieracium pilosella and related species; Asteraceae	Europe	all parts	coumarins, various flavonoids	III NT, MA	Shows psychoactive effects if large amounts are smoked. **Therapy:** Coumarins
Hippeastrum vittatum (Barbados lily), *H.* x *bortorum* hybrids; *H. equestre*; Amaryllidaceae	Peru; widely cultivated orn. plant	especially bulbs	lycorine, tazettine, haemanthamine, hippeastrine, galanthamine and other alkaloids	Ib CT, NT, GI	Ingestion can cause nausea, vomiting, diarrhoea; heavy sweating, kidney trouble, death in 2–3 h; has been used as arrow poison. **Therapy:** Lycorine
Hippobroma (= *Isotoma, Laurentia*) *longiflora* (star-of-Bethlehem); Campanulaceae	West Indies,. Hawaii, Guam	all parts	diphenyl lobelidol (alkaloid)	Ib NT, GI, AP	Disturbance of GI tract; symptoms of nicotine poisoning: hypertension, dilated pupils, sweating, seizures, coma, paralysis and death from respiratory failure; livestock poisoning. **Therapy:** Pyrrolidine alkaloids and *Lobelia*
Hippomane mancinella (manchineel tree); Euphorbiaceae	West Indies; Florida, Brazil	all parts, especially latex	physostigmine, brevifoline, hippomanine A, B; (=huratoxin)	Ib NT, IN, GI	Inhibits acetylcholine esterase, very aggressive on mucosal tissue (blister formation) or in eye (blindness); cystitis, vomiting, severe gastroenteritis; paralysis, death; phorbol esters are co-carcinogens; used as arrow poison. **Therapy:** Phorbol esters and indole alkaloids

Scientific name (common name); Family	Occurrence	Hazardous plant parts	Active principle	Toxicity Class	Symptoms & Therapy
Homalomena rubescens; Araceae	SE Asia	aerial parts	Ca²⁺-oxalate raphides	II CT, IN	Causes irritation by penetrating the mucous membranes of the mouth, throat or eyes; causes swelling of the mouth, tongue and face. **Therapy:** Oxalic acid
Homeria pallida (yellow tulp), *H. glauca, H. miniata* and related species; Iridaceae	S Africa; orn. plant	bulbs, aerial parts	1α-, 2α-epoxyscillirosidin and other bufadienolides	Ib CT, NT, GI, HP, AP	Ingestion can cause symptoms of cardiac glycoside poisoning with pronounced GI tract and cardiac disturbance; mainly animal poisoning. **Therapy:** Cardiac glycosides
Hoslundia opposita; Lamiaceae	trop. Africa	bark	essential oil with sesquiterpenes	II–III CT, NT	Ingestion of large amounts can be fatal for humans and sheep. **Therapy:** Sesquiterpenes
Humulus lupulus (hop plant); Cannabaceae	Europe; widely cultivated	flowers	humulone, lupulone, lupulin, essential oil	III MA, MP	Psychoactive: sedative effects; helps in case of nervousness; used for bittering beer. **Therapy:** Monoterpenes
Huperzia selago (fir club moss); Lycopodiaceae	Europe, Asia, America	all parts	lycopodine, arifoline, pseudoselagine and other alkaloids	II NT, MA, GI	Alkaloids affect cholinergic receptors; dizziness, vomiting; psychoactive; modulate sensations. **Therapy:** Cytisine
Hura crepitans (sand box tree); Euphorbiaceae	tropical America, Asia	all parts, especially latex and seeds	huratoxin and other phorbol esters; hurin and other lectins	Ib CT, IN, GI	Strong skin irritant (blister formation and contact dermatitis); can cause blindness; strong vomiting, severe gastroenteritis; phorbol esters are co-carcinogens; potent fish poison. **Therapy:** *Hura crepitans* and phorbol esters
Hyacinthoides non-scripta (bluebell); Hyacinthaceae	Europe; orn. plant	bulbs	oxalates, flavonoids	II GI,	Substantial ingestion can cause abdominal pain, diarrhoea, weak, slow pulse. **Therapy:** Oxalic acid
Hyaenanche globosa (= *Toxicodendron globosum*) (hyena poison); Euphorbiaceae	S Africa	fruits, seeds	sesquiterpene lactones such as tutin, mellitoxin, urushiol III and isodihydro-hyaenanchine	Ib CT, NT, GI	Ingestion can cause convulsions, delirium and coma, even death; animal and human poisoning. **Therapy:** *Hyaenanche*
Hydnocarpus kurzii (chaulmoogra tree); Achariaceae (Flacourtiaceae)	India, S Asia	oil	chaulmoogra acid, hydnocarpus acid, cyanogenic glucosides	II CT, NT, GI	In high doses: inflammation; ulcers, nausea, vomiting; spasms, hyperthermia, dyspnoea, cough, visual distortion, hypertensive. **Therapy:** Amygdalin

Scientific name (common name); Family	Occurrence	Hazardous plant parts	Active principle	Toxicity Class	Symptoms & Therapy
Hydrangea arborescens, *H. chinensis*, *H. macrophylla*; Hydrangeaceae	N America; Asia; common garden plant	rhizomes, leaves, flower buds	hydrangin (cyanogenic glucoside), saponins, quinazoline alkaloids (febrifugine)	II IN, GI, MA	High doses cause vertigo, gastroenteritis, dyspnoea, cerebral disturbances; common contact allergen; supposed to have mind-altering activities (active compound not known). **Therapy:** Amygdalin.
Hydrastis canadensis (goldenseal); Ranunculaceae	N America; orn. plant	all parts, especially rhizome	hydrastine, berberine and other isoquinoline alkaloids	Ib–II GI, NT, MA, MP	Toxic in large doses; GI tract disturbances; uterus contracting; vasoconstrictant; sympatholytic, CNS depressant, muscular arrest; hallucinations, delirium; cyanosis, vomiting; **Therapy:** Isoquinoline alkaloids
Hymenocallis declinata (spider lily), *H. caribaea*; Amaryllidaceae	N & tropical America; common orn. plant	bulbs	lycorine and related alkaloids	Ib–II CT, NT	Nausea, vomiting, abdominal spasms, diarrhoea, electrolyte imbalances. **Therapy:** Lycorine
Hymenoxys odorata, *H. richardsonii* (butterweed); Asteraceae	N America	all parts	sesquiterpene lactones	II CT, NT, GI, AP	Sesquiterpenes are cytotoxic; they can reduce glutathion levels in liver; mainly livestock poisoning. **Therapy:** Sesquiterpenes
Hyoscyamus niger (henbane), *H. muticus*, *H. albus*, *H. aurea*, *H. reticulatus* and related species; Solanaceae	Europe, N Africa, Asia, introduced into N America, Australia	all parts, roots, seeds	hyoscyamine, atropine, scopolamine and other tropane alkaloids	Ia NT, MA, MP	Tropane alkaloids block mAChR and are parasympatholytic; strong hallucinogens and aphrodisiac; high doses: mydriatic, cardiac stimulation, coma, death by respiratory arrest. **Therapy:** Hyoscyamine
Hypericum perforatum (St John's wort), *H. calycinum*, *H. aethiopicum* and related species; Hypericaceae / Clusiaceae	Europe, Asia, Africa, introduced to America, Australia; orn. plant	all parts	hypericin and other dianthrones, hyperforin, hyperoside	III IN, MA, MP	Skin irritant (light sensitisation); inhibit re-uptake of various neurotransmitters (used as antidepressant); can cause serotonin syndrome when used in conjunction with other serotonin-enhancing agents. **Therapy:** *Hypericum* and anthraquinones

Scientific name (common name); Family	Occurrence	Hazardous plant parts	Active principle	Toxicity Class	Symptoms & Therapy
Iberis amara (bitter candytuft); Brassicaceae	Europe; orn. plant	all parts; especially seeds	cucurbitacins; glucoiberin, glucocheirolin and other glucosinolates	II–III CT, GI, MP	In high doses: vomiting, gastroenteritis, bloody diarrhoea, abdominal pains, respiratory depressant, arrhythmia. **Therapy:** Mustard oils
Ilex aquifolium (holly), *I. vomitoria* (yaupon), *I. opaca* (American holly) and related species; Aquifoliaceae	Europe, N America, Japan; used as orn. plant	leaves and red berries	rutin, ursolic acid, amyrin, uvaol, other saponins; some theobromine	Ib–II CT, GI	Nausea, vomiting, diarrhoea, gastroenteritis; abdominal spasms; arrhythmia, paralysis, kidney trouble. *I. vomitoria* was used by Indians as an intoxicant. Cause of allergic reactions. **Therapy:** Saponins
Ilex cassine (black drink plant); Aquifoliaceae	N America	leaves	theobromine (0.3 %)	III MA	Central stimulant (see *Theobroma cacao*); has been used by American Indians. **Therapy:** Caffeine
Ilex paraguariensis (maté), *I. guayusa*; Aquifoliaceae	S America	leaves	caffeine (up to 1.6 %), theobromine (up to 0.45 %)	III MA, MP	Central stimulant; see *Coffea* and *Theobroma*. **Therapy:** Caffeine
Illicium anisatum (Japanese anise); Illiciaceae	Japan, Korea	fruits, seeds	anisatin and other sesquiterpene lactones; monoterpenes,	Ib CT, NT, GI	Substantial ingestion causes vomiting, diarrhoea, salivation; spasms, respiratory arrest, even death. **Therapy:** Sesquiterpenes
Indigofera endecaphylla (creeping indigo) and related species; Fabaceae	tropics	all parts	NPAAs	II CT, NT, AP	Mainly livestock poisoning. **Therapy:** Non-protein amino acids
Inocybe fastigiata, (= *I. rimosa*) *I. aeruginascens, I. coelestium, I. corydalina, I. erubescens, I. haematacta*; Cortinariaceae	Europe	fruiting body	psilocybin, baeocystine; muscarine (*I. erubescens*)	Ib NT, MA	Alkaloids are serotonin receptor agonists; therefore psychoactive, psychedelic and hallucinogenic; enhance perspiration, asthmatic breathing, rarely fatal. **Therapy:** Bufotenine
Iochroma fuchsioides (hummingbird flower); Solanaceae	Andean region	leaves, flowers	withanolide	II MA, GI, MP	Substantial ingestion may cause narcotic effects, vomiting, diarrhoea; possibly psychoactive. **Therapy:** *Withania*
Ipomoea purga (jalap bindweed); Convolvulaceae	C & S America	tubers	glycoretins, convolvulin, jalapin	Ib IN, GI	Formerly used as laxative; causing stomach pain, vomiting, diarrhoea, skin irritant. **Therapy:** Cucurbitacins

Scientific name (common name); Family	Occurrence	Hazardous plant parts	Active principle	Toxicity Class	Symptoms & Therapy
Ipomoea tricolor (morning glory) (= *I. violacea*), *I. carnea*, *I. crassicaulis*, *I. hederacea*, *I. muricata*, *I. nil*, *I. purpurea*; Convolvulaceae	Mexico; cultivated as orn. plant	seeds	ergometrine; lysergic acid amide and other derivatives of lysergic acid	Ib NT, MA	Hallucinogenic, similar to LSD but weaker; ingestion induces nausea, vomiting, weakness; high doses cause death; Mexican Indians used ololuqui (origin unresolved) for ritual and oracle purposes. **Therapy:** Ergot alkaloids and *Ipomoea tricolor*
Jatropha curcas (physic nut), *J. gossypifolia, J. integerrima, J. multifida, J. podagrica, J. cathartica* and related species; Euphorbiaceae	C America; cultivated in Africa, Asia; orn. plant	all parts, especially latex and seeds	curcin (a lectin); phorbol esters, curcanoleic acid	Ib CT, IN, GI	Strong skin irritant; strong and violent laxative; vomiting, delirium, dizziness, visual distortion, coma, death. **Therapy:** Phorbol esters
Juniperus communis (juniper); Cupressaceae	Europe; orn. plant	all parts, fruits	essential oil with pinene, terpineol	II–III CT, IN, GI, MP	Diuretic, mucosa and skin irritant; overdose can cause kidney and liver damage. **Therapy:** Monoterpenes
Juniperus sabina (savin), *J. virginiana, J. recurva*; Cupressaceae	Europe, Asia, N America; orn. plant	all parts; especially young twigs	3–5 % essential oil with sabinene and sabinylacetate, thujone, other monoterpenes	Ia CT, IN, NT, MA, AP	Strong skin irritant; ingestion causes nausea, excitation, arrhythmia, spasms, respiratory arrest; nephritis, bloody urine, liver intoxication; uterus contraction (therefore abortifacient); central paralysis; *J. recurva* possibly psychoactive; livestock poisoning. **Therapy:** Monoterpenes and *Juniperus*
Justicia pectoralis; Acanthaceae	C America	leaves	coumarins	III CT, MA	Leaves have been used as narcotic drug; sedative and hypnotic properties have been reported. **Therapy:** Coumarins
Kalanchoe lanceolata, *K. rotundifolia, K. blossfeldiana*; Crassulaceae	Africa; widely cultivated as orn. plant	aerial parts	lanceotoxin A, lanceotoxin B and hellebrigenin and other bufadienolides	II CT, GI, HP	Substantial ingestion causes GI tract disturbance and cardiac glycoside intoxication; mainly animal poisoning. **Therapy:** Cardiac glycosides
Kalmia latifolia (mountain laurel), *K. angustifolia, K. polifolia, K. microphylla* and related species; Ericaceae	N America; common orn. plant	leaves, nectar	andromedotoxin (=grayanotoxin) (diterpene); arbutin (quinone glucoside)	Ib NT, GI	Diterpenes activate Na^+ channels; they cause burning in mouth and throat, salivation, nausea, vomiting, diarrhoea, dizziness, slow pulse, hypotension, progressive paralysis, death; also honey derived from laurel can be toxic; sheep show severe poisoning when they feed on mountain laurel. **Therapy:** Andromedotoxin

Scientific name (common name); Family	Occurrence	Hazardous plant parts	Active principle	Toxicity Class	Symptoms & Therapy
Karwinskia bumboldtiana (buckthorn), *K. mollis*, *K. parviflora* and related species; Rhamnaceae	C&N America	fruits	dimeric anthracene derivatives (buckthorn toxins)	Ib–II NT, AP	Neurotoxins that cause demyelination of nerve cells; after latent period of several weeks: paralysis in humans, nausea, weakness, liver, lung and kidney damage; death in children and livestock. **Therapy:** Anthraquinone
Knowltonia bracteata; Ranunculaceae	Africa	aerial parts	ranunculin, protoanemonin	II CT, NT, IN, AP, MU	Severe skin and mucosal irritation with blisters and ulceration; internally: diarrhoea, abdominal pains, dizziness, seizures, tachycardia, nephritis, even death from respiratory and cardiac arrest. **Therapy:** Protoanemonin
Laburnum anagyroides (golden chain), *L. alpinum*, *L. x watereri* (a hybrid between former species); Fabaceae	Europe; widely cultivated orn. tree	all parts, especially seeds	cytisine and other quinolizidine alkaloids; 3.5 % in seeds	Ib NT, MA, GI	Slightly psychoactive and hallucinogenic; nAChR agonist; diuretic, uterus contracting, abdominal pain, tachycardia, hypotension; heart irregularity, vomiting, diarrhoea, dizziness, headache; delirium; toxicity similar to nicotine. **Therapy:** Cytisine
Lactuca virosa (prickly lettuce); Asteraceae	Eurasia, N America	all parts, especially latex	lactucin, lactucopicrin (sesquiterpene lactones)	Ib CT, NT, GI, MA, MP	Profound irritation of skin und mucous membranes; sedative properties (similarities with opium); pupil dilatation; dizziness, headache, visual distortions, sleepiness; heavy sweating; enhanced breathing and cardiac activity. **Therapy:** *Lactuca*
Lagenaria siceraria (bottle gourd, calabash); Cucurbitaceae	tropical Africa; widely cultivated	fruits, seeds	cucurbitacins	Ib CT, IN, GI	Skin irritant; nausea, vomiting, diarrhoea with blood, strong colic and spasms, kidney inflammation, tachycardia, respiratory arrest; ingestion has caused death in children. **Therapy:** Cucurbitacins
Lagochilus inebriens; Lamiaceae	Asia	aerial parts	lagochilin, flavonoids	III MA, MP	Has been used to treat nervousness, insomnia and stress; sedating properties. **Therapy:** Flavonoids
Lantana camara (lantana); Verbenaceae	tropical America; world-wide orn. plant	all parts, especially fruits	lantadene A, B, other triterpenes and sesquiterpenes	II CT, GI, AP	Mydriatic; dry mouth, vomiting, diarrhoea, cyanosis, dyspnoea, liver toxin; icterus, increased light sensitivity, flushed skin, respiratory failure; fatal to children; mainly animal poisoning. **Therapy:** Saponins

381

Scientific name (common name); Family	Occurrence	Hazardous plant parts	Active principle	Toxicity Class	Symptoms & Therapy
Laportea interrupta, L. bulbifera; Urticaceae	SE Asia	aerial parts	stinging hairs with histamine, acetylcholine, serotonin, formic acid, acetic acid	III IN	Painful itching with welt formation occurs at sites that were pricked by stinging hairs. **Therapy:** *Urtica dioica*
Larrea tridentata (creosote bush, chaparral leaf); Zygophyllaceae	N America	aerial parts	lignans, such as nordihydroguajaretic acid	Ib–II CT, IN, GI	Ingestion causes nausea, vomiting, diarrhoea, abdominal spasms; ingredients are cytotoxic and can induce renal cystitis and liver intoxication; contact allergen. **Therapy:** Phenolics
Lasiospermum bipinnatum, L. radiatum; Asteraceae	S Africa	aerial parts	dehydromyodesmone and dehydrongaione (furanosesquiterpenes)	II CT, NT, IN, AP	Mainly animal poisoning; photosensitivity, jaundice (icterus), colic, constipation and high temperature, lesions of the liver and lungs. **Therapy:** Sesquiterpenes
Lathyrus odoratus (sweet pea), *L. sativus, L. hirsutus, L. pusillus, L. sylvestris* and related species; Fabaceae	Europe; widely grown as orn. plant, America	seeds	β-aminopropionitrile, α-amino-β-oxalyl aminopropionic acid, homoarginine and related NPAAs	II CT, NT, AP	Cause of neurolathyrism; neurotoxic properties with muscle weakness, paralysis and psychic disturbances; more common in horses than in humans. **Therapy:** Non-protein amino acids
Latua pubiflora (witches tree); Solanaceae	Chile	all parts	hyoscyamine, atropine, scopolamine and other tropane alkaloids	Ia NT, MA,	Tropane alkaloids are mAChR antagonists and thus strongly parasympatholytic (see *Atropa, Datura*); highly hallucinogenic; the drug has been used by Chilean Indians. **Therapy:** Hyoscyamine
Lavandula angustifolia and related species (lavender); Lamiaceae	Europe; widely cultivated	aerial parts	essential oil with several monoterpenes; coumarins,	III CT, MA, MP	Essential oil is used to treat nervousness and insomnia. High doses can cause disturbance of liver and kidneys. **Therapy:** Monoterpenes
Ledum palustre see Rhododendron tomentosum					

Scientific name (common name); Family	Occurrence	Hazardous plant parts	Active principle	Toxicity Class	Symptoms & Therapy
Leontice leontopetalum; Berberidaceae	S Europe	roots, seeds	*N*-methylcytisine, baptifoline and other quinolizidine alkaloids; saponins	Ib–II CT, NT, GI, MA	Low doses induce vomiting, diarrhoea and gastroenteritis; high doses cause burning sensation in mouth; induce salivation; perspiration, mydriasis, uterus contracting; nausea and paralysis; also hallucinations and delirium; respiratory arrest. **Therapy:** Cytisine
Leonurus cardiaca (motherwort), *L. sibiricus*; Lamiaceae	Europe, Asia; orn.plant	aerial parts	alkaloids: leonuridine, leonurine, stachydrin; diterpenes: leosiberin, leosibericin	III MA, MP	Has been used as mild narcotic and sedative. **Therapy:** Terpenoids
Lepiota helveola and related fungi; Agaricaceae	N America	fruiting body	amanitins, phalloidins	Ia CT, NT, GI	Deadly poison; 6–24 h latent period before first symptoms occur. Diarrhoea, severe abdominal cramps, nausea, vomiting, liver and kidney failure; even death. **Therapy:** *Amanita phalloides*
Letharia vulpina (wolf lichen); Parmeliaceae	Europe	all parts of the lichen	vulpic acid	Ib NT, AP	Vulpic acid targets the CNS and leads to respiratory and circulatory arrest; has been used to kill wolves and foxes, hence the name. **Therapy:** *Letharia vulpina*
Leucaena leucocephala (= *L. glauca*); Fabaceae	tropical America, introduced in S Africa, Australia, Asia	aerial parts, young leaves, seeds	mimosine (NPAA)	II CT, AP	Mainly animal poisoning; delayed growth, oesophageal lesions, hair loss, goitrogenic effects. **Therapy:** Non-protein amino acids
Leucojum vernum (spring snowflake), *L. aestivum*; Amaryllidaceae	C Europe; common orn. plant	bulbs and aerial parts	lycorine, galanthamine, tazettine and other alkaloids	II CT, NT, GI	Galanthamine inhibits acetylcholine esterase, thus parasympathomimetic; ingestion causes nausea, vomiting, diarrhoea, small pupils. **Therapy:** Lycorine
Leucothoe davisiae (sweet bells); Ericaceae	N America	leaves, nectar (honey)	andromedotoxin (=grayanotoxin) (diterpene)	Ib NT, GI	Diterpenes activate Na^+ channels; burning in mouth and throat; salivation; nausea, vomiting, diarrhoea, dizziness; slow pulse, hypotension, weakness, progressive paralysis, convulsions, death. **Therapy:** Andromedotoxin
Ligustrum vulgare (privet), *L. lucidum, L. japonicum* and related species; Oleaceae	Europe, Asia, America; orn. plant	aerial parts, fruits	ligustrin (= syringin), oleuropein and other secoiridoid glycosides	II CT, GI	Nausea, vomiting, dizziness, headache, diarrhoea, gastroenteritis, convulsions, circulatory arrest; fatal livestock poisoning (with neurotoxicity). **Therapy:** Terpenes and phenylpropanoids

Scientific name (common name); Family	Occurrence	Hazardous plant parts	Active principle	Toxicity Class	Symptoms & Therapy
Lilium tigrinum and other lilies; Liliaceae	orn. plant	all parts, especially bulbs	tulipalin, tuliposide A	III CT, IN, GI, AP	Substantial ingestion can cause gastrointestinal disturbance; serious skin allergen; mainly animal poisoning. **Therapy:** Protoanemonin
Linum usitatissimum (flax), *L. catharticum* (purging flax), *L. lewisii*, *L. neomexicanum*, *L. rigidum* and related species; Linaceae	Europe, America; widely cultivated	seeds, seedlings, roots	linamarin (a cyanogenic glucoside) in seeds and seedlings; podophyllotoxin in roots; saponins	II CT, NT, GI, MP, AP	HCN liberated from linamarin can be poisonous (typical HCN intoxication): spasms, paralysis, respiratory arrest. Podophyllotoxin is cytotoxic and purgative; livestock poisoning. **Therapy:** Amygdalin and flavonoids
Lippia rehmannii, *L. javanica* and *L. scaberrima*; Verbenaceae	S Africa; cultivated	leaves	triterpenoids: rehmannic acid, better known as lantadene A; icterogenic acid	II CT, NT, GI, AP	Mainly animal poisoning; liver damage as in *Lantana camara*. **Therapy:** Saponins
Liriodendron tulipifera (tulip tree, yellow poplar); Magnoliaceae	N America; widely cultivated as orn. tree	all parts	glaucine, liriodenine and other alkaloids; cyanogenic glucosides	II CT, NT, GI	Substantial ingestion causes disturbance of GI tract, CNS and of circulation. **Therapy:** Isoquinoline alkaloids
Lithospermum officinale and related species; Boraginaceae	Europe, Asia; introduced in N America	all parts, seeds	pyrrolizidine alkaloids; lithospermic acid and other phenolics	II CT, GI, MU, NT	PAs have mutagenic and carcinogenic properties; hepatotoxic and neurotoxic upon substantial ingestion. **Therapy:** Senecionine
Lobelia inflata (Indian tobacco) *L. tupa*, *L. dortmanna*, *L. cardinalis*, *L. siphilitica* and related species; Campanulaceae	America; common orn. plants	aerial parts	lobeline, lelobine, lobinine and other piperidine alkaloids	Ib NT, MA, MP	Lobeline is an agonist at nAChR; psychoactive, enhances respiration at low doses; leads to respiratory arrest at high doses; emetic and GI tract disturbance, followed by symptoms of nicotine poisoning: bradycardia, hypertension, arrhythmia, convulsions, anxiety, death through respiratory arrest. **Therapy:** Pyrrolidine alkaloids and *Lobelia*

Scientific name (common name); Family	Occurrence	Hazardous plant parts	Active principle	Toxicity Class	Symptoms & Therapy
Lolium temulentum (rye grass, darnel); Poaceae	Europe, Asia, introduced into most parts of the world	all parts	temuline, loline and other pyrrolizidine alkaloids (fungal products; composition depends on the infecting fungus)	Ib–II NT, MA, AP, GI	Alkaloids are psychoactive; ingestion causes symptoms of CNS disturbance with drunkenness and dizziness, hence the name, staggering, tremor, headache, visual distortion, sleepiness, death via respiratory arrest; poisoning better known from horses, pigs and poultry. **Therapy:** Senecionine and pyrrolidine alkaloids
Lonchocarpus violaceus (violet lancepod), *L. floribundus*, *L. rariflorus*, *L. sericeus*; Fabaceae	C America; cultivated	bark, seeds	rotenone (isoflavone); longistyline A, B, C, D which resemble kavalactones in structure (stilbenes)	Ib CT, NT, MA	Neurotoxin: anaesthetic for tongue and throat; induces vomiting, salivation, sweating, dizziness, trembling, muscular arrest; inhibition of CNS and spinal cord; abortifacient; used as fish poison. **Therapy:** *Derris elliptica*
Lonicera xylosteum (honey suckle), *L. caprifolium*, *L. nigra*, *L. periclymenum*, *L. tartarica* and related species; Caprifoliaceae	Europe; naturalised in N America; common as orn. plant	berries	xylostein, cyanogenic glucosides, saponins	III CT, GI, NT	Vomiting and GI tract disturbance; older reports describe further symptoms: psychoactivity, hyperthermia, dilated pupils, tachycardia, pollakisuria, nephritis, respiratory arrest. **Therapy:** Terpenes and phenylpropanoids
Lophopetalum toxicum; Celastraceae	Philippines	bark	cardenolides	Ib CT, NT, GI, HP	Symptoms of cardiac glycoside intoxication; extracts were used as arrow poison. **Therapy:** Cardiac glycosides
Lophophora williamsii (peyote cactus); Cactaceae	Mexico; orn. plant	aerial parts	mescaline and other phenylethylamines (4.5–7 %)	Ib–II NT, MA, MP	Ingestion causes psychedelic hallucinations (similar to LSD) with visions, creative dreams and loss of reality; aphrodisiac; high doses lead to GI disorder, hypotension, bradycardia, respiratory depression, vasodilatation, paralysis. **Therapy:** *Lophophora*
Lotononis laxa, *L. carnosa*, *L. involucrata*, *L. fruticoides* and related species; Fabaceae	S Africa	aerial parts, seeds	prunasin (cyanogenic glucoside), senecionine, integerrimine (PAs), lupanine and other QA	II CT, NT, GI, AP, MU	Cyanogenic glucoside releases HCN upon enzymatic hydrolysis, an inhibitor of cellular respiration; PAs are hepatotoxic, DNA alkylating and therefore mutagenic and carcinogenic. **Therapy:** Amygdalin, senecionine and cytisine

Scientific name (common name); Family	Occurrence	Hazardous plant parts	Active principle	Toxicity Class	Symptoms & Therapy
Lupinus albicaulis, L. densiflorus, L. sericeus, L. caudatus; Fabaceae	N America	all parts, seeds	cytisine, anagyrine and other quinolizidine alkaloids	Ia–Ib NT, MA, MU, AP	Anagyrine causes malformations in livestock (crooked calf disease); cytisine is a strong nAChR agonist and therefore neurotoxic similar to nicotine. **Therapy:** Cytisine
Lupinus polyphyllus (lupin), *L. nootkatensis, L. albus, L. angustifolius, L. luteus, L. perennis, L. hirsutissimus* and related species; Fabaceae	Europe, America; widely cultivated as orn. and crop plants	all parts, especially seeds	lupanine, sparteine and related quinolizidine alkaloids	Ib–II NT, MP, AP	Lupanine and other QAs are modulators of nAChR and mAChR and also inhibit Na^+ channels; cause salivation, vomiting, problems to swallow, hyperthermia, arrhythmia, mydriasis, excitement and delirium, paralysis, death through respiratory arrest. **Therapy:** Cytisine
Lycium barbarum (box thorn), *L. chinense, L. europaeum, L. pallidum, L. ruthenicum, L. turcomanicum, L. carolinianum, L. halmifolium* and related species; Solanaceae	Europe, Asia; other species America; orn. plant	all parts, red berries	hyoscyamine and other alkaloids are probably not present (wrong information in literature); steroidal saponins, withanolides	III CT, IN, AP	Thorns may produce an irritant dermatitis; animal poisoning has been observed with some species. **Therapy:** Saponins
Lycopersicon esculentum (tomato) (= *Solanum lycopersicum*) and related species; Solanaceae	S America; cultivated worldwide	aerial parts, except red fruits	steroidal glycoalkaloids such as tomatine in leaves and green fruits	II CT, NT, GI, AP	Steroidal alkaloids are cytotoxic and neurotoxic when ingested in high amounts; cause disturbance of GI tract; also livestock poisoning. **Therapy:** *Solanum*
Lycopodium clavatum (clubmoss) *L. annotinum* and related species; Lycopodiaceae	cosmopolitan	all parts	lycopodine, clavatine and related quinolizidine alkaloids	II NT, MA, MP	Neurotoxin similar to curare; induces vomiting, dizziness, unconsciousness; it is also psychoactive and modulates sensations. **Therapy:** Cytisine and indole alkaloids
Lycoris africana (= *L. aurea*) (golden spider lily), *L. radiata, L. squamigera;* Amaryllidaceae	E Asia; common orn. plant	bulbs	lycorine and related alkaloids	Ib CT, NT, GI	Lycorine inhibits protein biosynthesis; high doses are neurotoxic and lead to strong disturbances of GI tract (vomiting, diarrhoea, abdominal spasms). **Therapy:** Lycorine

Scientific name (common name); Family	Occurrence	Hazardous plant parts	Active principle	Toxicity Class	Symptoms & Therapy
Macleaya cordata (plume poppy); Papaveraceae	E Asia; common orn. plant	all parts	protopine, berberine, sanguinarine and other isoquinoline alkaloids	II NT, MA, GI	Burning sensation in mouth and throat; alkaloids affect several neuroreceptors; substantial ingestion causes disturbance of CNS and GI tract; bloody diarrhoea, arrhythmia, low pulse; hypotension; cardiac arrest. **Therapy:** Isoquinoline alkaloids and *Sanguinaria*
Maclura pomifera (Osage orange) and related species; Moraceae	pan-tropical	latex	toxicarioside; cardenolides	Ia CT, NT, HP, GI	Cardiac glycosides inhibit Na^+, K^+-ATPase and are therefore strong neuronal and muscular blockers; used as arrow poison. **Therapy:** Cardiac glycosides
Macrozamia miquelii and related cycads; Zamiaceae	America; orn. plant	all parts	azoxyglycosides, such as cycasin	II CT, NT, GI, MU	Cycasin is a methylating compound and therefore mutagenic and cytotoxic; liver poison; livestock poisoning. **Therapy:** *Encephalartos*
Mahonia aquifolium (Oregon grape) and related species; Berberidaceae	N America; widely grown as orn. plant	all parts, especially bark	berberine, oxyacanthine, berbamine and other isoquinoline alkaloids	II CT, NT, GI, MP	Cytotoxic properties (used to treat psoriasis); substantial ingestion can cause nausea, vomiting, diarrhoea, tachycardia; kidney inflammation; mature fruits are rarely toxic. **Therapy:** *Chelidonium* and isoquinoline alkaloids
Maianthemum bifolium (May lily); Ruscaceae (Convallariaceae)	Europe, N Asia	all parts, especially fruits	probably cardiac glycosides	II CT, GI, HP	At high doses: cardiac glucoside intoxication; nausea, gastrointestinal disturbance, diarrhoea, arrhythmia, dizziness, hypertension, coma. **Therapy:** Cardiac glycosides
Malus domestica (apple) and related species; Rosaceae	Europe; cultivated worldwide	seeds	prunasin (cyanogenic glucoside)	Ib–II CT	When seeds are crushed they release HCN, which is a strong poison; inhibits cellular respiration; high doses: vomiting, flushed face, increased respiration; headache, respiratory and cardiac arrest. **Therapy:** Amygdalin
Mammea americana (mammee apple); Clusiaceae	West Indies	seeds	coumarins	II CT, NT, AP	Highly toxic to insects and vertebrates. **Therapy:** Coumarins
Mandragora officinarum (mandrake), *M. autumnalis*, *M. caulescens* (= *Anisodus bumilis*), *M. turcomanica* and related species; Solanaceae	Mediterranean, Asia	mainly roots	hyoscyamine, atropine, scopolamine and other tropane alkaloids	Ia NT, MA, GI, MP	Tropane alkaloids are mAChR antagonists with parasympatholytic properties; psychedelic; hallucination in various forms; used as hallucinogen since antiquity. High doses: clonic spasms; strong heart beat; tachycardia (>160 beats); inhibition of salivation; respiratory arrest; coma. **Therapy:** Hyoscyamine

387

Scientific name (common name); Family	Occur-rence	Hazardous plant parts	Active principle	Toxicity Class	Symptoms & Therapy
Manihot esculenta (tapioca, cassava); Euphorbiaceae	C America; cultivated as food plant in tropics	tubers, leaves	linamarin, lotaustralin; cyanogenic glucosides	Ib–II CT	When roots are crushed they release HCN which inhibits cellular respiration in mitochondria. Symptoms of HCN poisoning. May be responsible for nutritional neuropathies in Africa. **Therapy:** Amygdalin
Marsdenia condurango; Apocynaceae (Asclepiadoideae)	Africa	aerial parts, bark	condurangin, steroid glucosides	II–III CT, GI	The drug has been used in phytomedicine to stimulate the secretion of saliva and gastric juices; substantial ingestion can cause disturbance of GI tract and CNS. **Therapy:** Saponins
Melaleuca alternifolia (tea tree) and related species; Myrtaceae	Australia; widely cultivated	aerial parts	essential oil with several monoterpenes, such as terpinene-4-ol;	III CT, NT, IN, MP	Has been used to treat infections; high doses can cause CNS and GI disturbance (see thujone). **Therapy:** Monoterpenes
Melia azedarach (Persian lilac, chinaberry); Meliaceae	Asia; widely cultivated as orn. tree	all parts; especially fruits and bark	several triterpenes (melinoon and melianol), also kulinone, kulacton, meliantriol, meliatoxins A_1, A_2, B_1 and B_2; alkaloids (azaridine)	Ib CT, NT, GI, AP	In high doses fruits can cause nausea, vomiting, diarrhoea, thirst, cold perspiration, spasms, even death; also used as natural insecticide with pronounced livestock toxicity. **Therapy:** Saponins
Melianthus comosus (honeyflower), *M. major* and related species; Melianthaceae	S Africa; orn. plants	aerial parts	melianthugenin, melianthusigenin and hellebrigenin 3-acetate and other bufadienolides	Ib CT, NT, GI, HP	Ingestion causes symptoms of cardiac glycoside poisoning; severe human and livestock intoxication; even honey from the plant and parasitising mistletoes become poisonous. **Therapy:** Cardiac glycosides
Melilotus albus (sweet clover), *M. officinalis, M. indica*; Fabaceae	Europe; America, now almost cosmopolitan	aerial parts	coumarins; from dicoumarols upon drying and silage formation	III CT, AP, MP	Mainly animal poisoning; dicoumarol is an anticoagulant leading to internal bleeding (haemorrhages and haematomas); sudden death in cattle. The drug is used in phytomedicine to treat the symptoms of venous and lymphatic insufficiency. **Therapy:** Coumarins
Melochia pyramidata; Malvaceae (Sterculiaceae)	Nicaragua	aerial parts	meloquinine (a pyridone alkaloid)	II NT, AP	Animal poisoning with paralysis, bradycardia, hypotension. **Therapy:** Pyrrolidine alkaloids
Melothria scabra; Cucurbitaceae	America	fruits, roots	cucurbitacins	Ib CT, GI, AP	Cucurbitacins are strongly cytotoxic; substantial ingestion causes severe disturbance of GI tract; mainly livestock poisoning.

Scientific name (common name); Family	Occurrence	Hazardous plant parts	Active principle	Toxicity Class	Symptoms & Therapy
Menispermum canadense (moonseed) and related species; Menispermaceae	N America	aerial parts, especially fruits	dauricine and related isoquinoline alkaloids	Ib NT, HP	Dauricine inhibits cardiac K⁺ channels and thus causes arrhythmia; excitation of CNS leading to seizures and neuromuscular arrest; reported to be fatal for children. **Therapy:** Isoquinoline alkaloids
Mentha pulegium (pennyroyal); Lamiaceae	Europe, Asia; cultivated	all parts	pulegone, piperitone and other monoterpenes	II CT, GI, NT	Slightly narcotic, intoxicating; high doses: vomiting, diarrhoea, hypertension, central paralysis, abortifacient; death through respiratory arrest. **Therapy:** Monoterpenes
Menziesia ferruginea; Ericaceae	America	all parts	andromedotoxin (=grayanotoxin) (diterpene)	Ib NT, GI	Diterpenes activate Na⁺ channels; burning in mouth and throat; salivation; nausea, vomiting, diarrhoea, dizziness, slow pulse, hypotension, progressive paralysis, death. **Therapy:** Andromedotoxin
Mercurialis annua (annual mercury), *M. perennis*; Euphorbiaceae	Europe, Asia	all parts	saponins, aliphatic amines (methylamine, trimethylamine)	III CT, AP	Saponins are cytotoxic; in animals: gastroenteritis with kidney and liver damage. **Therapy:** Saponins
Mesembryanthemum crystallinum (ice plant) and related plants; Mesembryanthemaceae (Aizoaceae)	S Africa; common orn. plant in tropics and subtropics	all parts	mesembrine (alkaloids), oxalic acid; tannins	III CT, NT, MA	Mesembrine has similar but weaker properties than cocaine; psychoactive; sedative; substantial ingestion causes disturbance of GI tract and kidneys. **Therapy:** *Mesembryanthemum tortuosum*
Mesembryanthemum tortuosum (=*Sceletium tortuosum*) (sceletium); Aizoaceae (Mesembryanthemaceae)	S Africa	aerial parts	mesembrine, mesembrinine and other alkaloids	III CT, NT, MA	Mesembrine has similar but weaker properties than cocaine; psychoactive; sedative; substantial ingestion causes disturbance of GI tract. **Therapy:** *Mesembryanthemum tortuosum*
Microcycas calocoma; Zamiaceae	Cuba	all parts	azoxyglycosides, such as cycasin	II CT, NT, GI, MU	Cycasin is a methylating compound and therefore mutagenic and cytotoxic; liver poison; livestock poisoning. **Therapy:** *Encephalartos*
Mimosa hostilis, (= *M. tenuiflora*), *M. scabrella*; Fabaceae	C & S America	bark	*N,N*-DMT, 5-hydroxytyramine, β-carboline alkaloids, triterpene saponins, Ca²⁺-oxalates	II NT, MA	Alkaloids stimulate serotonin receptors causing brief but strong hallucinogenic effects. **Therapy:** Bufotenine

Scientific name (common name); Family	Occurrence	Hazardous plant parts	Active principle	Toxicity Class	Symptoms & Therapy
Mitragyna speciosa; Rubiaceae	SE Asia,	leaves	mitragynine, ajmalicine, mitraversine, corynanthedine and other monoterpene indole alkaloids	II MA, MP	Stimulating like cocaine, sedating like opium; increases excitability of sympatic autonomous nerve system, excitation of motoric CNS centres. Locally used to treat pains. **Therapy:** Indole alkaloids
Momordica charantia (balsam pear), *M. balsamica*; Cucurbitaceae	Asia, C America, Gulf Coast states; orn. plant	seeds, outer rind of ripe fruits	cucurbitacins, toxic lectins (momordin)	Ib CT, GI, MP	Disagreeable bitter taste; severe vomiting and diarrhoea, momordin inhibits ribosomal protein biosynthesis in the same way as abrine (see abrine). **Therapy:** Cucurbitacins
Monstera deliciosa (ceriman, delicious monster); Araceae	Mexico; world-wide as orn. plant	roots, aerial parts	Ca^{2+}-oxalate raphides; toxic peptides	II CT, IN, GI	High doses: skin irritant; blister formation; severe swelling of throat and mouth, burning sensation in mouth; disturbance of GI tract and internal bleeding; possibly cardiac arrhythmia; CNS disturbance; spasms; low body temperature. **Therapy:** Oxalic acid
Moraea polystachya (blue tulp), *M. buttonii* (formerly *M. rivularis*), *M. spatbulata* (= *M. spathacea*), *M. umguiculata* (= *M. tenuis*), *M. stricta* (= *M. trita*), *M. graminicola* and *M. carsonii*; Iridaceae	S Africa	bulbs, aerial parts	16β-formyloxybovo-genin A and other bufadienolides	Ib CT, NT, HP, GI	Ingestion causes cardiac glycoside poisoning with pronounced disturbance of GI tract, CNS and heart; mainly animal intoxications. **Therapy:** Cardiac glycosides
Mucuna pruriens (= *Dolichos pruriens*) (cow itch); Fabaceae	cultivated in tropics of Asia, Africa, America	seeds	*N,N*-DMT, 5-MeO-DMT, bufotenine, serotonin; histamine; nicotine (alkaloids); irritating compounds mucunine, prurierine	II CT, IN, MA	The amines are agonist of 5HT-R causing psychedelic, hallucinogenic, aphrodisiac effects; the colourful seeds are sometimes used in necklaces. **Therapy:** Bufotenine
Mundulea sericea (cork bush); Fabaceae	Africa, Asia	bark, leaves, roots	deguelin and tephrosin, possibly rotenone and other rotenoids (isoflavonoids)	Ib CT, NT, GI, AP	Rotenone blocks cellular respiration and is thus a strong toxin; fish and insect poison; human and animal poisoning reported. **Therapy:** Downy slvosia

Scientific name (common name); Family	Occurrence	Hazardous plant parts	Active principle	Toxicity Class	Symptoms & Therapy
Myoporum insulare; M. laetum (Ngaio tree); Myoporaceae	S Pacific, Australia, New Zealand; introduced S California	leaves, fruits	dehydrongaione, ngaione; (furano-sesquiterpenoids)	II, CT, NT, GI, AP	Ingestion causes disturbance of GI tract and CNS; liver damage and photodermatosis; deadly livestock poisoning reported. **Therapy:** Sesquiterpenes
Myristica fragrans (nutmeg tree); Myristiceae	SE Asia; cultivated in tropics	seeds	essential oil: myristicin (4%), eugenol, safrole, sabinene and other phenylpropanoids and terpenoids	II CT, NT, MA, MP	Myristicin is psychoactive similar to amphetamines (MDMA), aphrodisiac; high doses induce gastroenteritis, heart palpitations, unconsciousness, hyperhidrosis, haematuria, even death. **Therapy:** Myristicin
Myroxylon balsamum (tolu balsam tree); Fabaceae	C & S America	all parts; tolu balsam	benzoic acid, cinnamic acid and their esters	II CT, IN, GI, MP	Tolu balsam is used mainly in cough and throat syrups; substantial ingestion causes disturbance of GI tract and CNS, allergenic. **Therapy:** Myristicin
Narcissus pseudonarcissus (daffodil), *N. cyclaminens, N. jonquilla, N. poeticus, N. tarzetta* and related species; Amaryllidaceae	Europe, Asia; widely cultivated orn. plant	all parts, especially bulbs	lycorine, haemanthamine, narciclasine, tazettine and other isoquinoline alkaloids; Ca²⁺-oxalate crystals	Ib–II CT, NT, IN, GI	Ca^{2+}-oxalate and alkaloids cause skin irritation and inflammation (contact dermatitis); alkaloids cause nausea, vomiting, diarrhoea, abdominal spasms, heavy perspiration, and even death. **Therapy:** Lycorine and oxalic acid
Narthecium ossifragum (bog asphodel); Nartheciaceae	N Europe	all parts	steroidal saponins, nartheside A, B; tannins (10%)	II CT, IN, GI, AP	Nartheside A and B are converted into the reactive narthogenins, which have similar properties as protoanemonin and tulipalin; saponin with haemolytic and cytotoxic properties; substantial ingestion causes hepatitis, oedema; severe poisoning in sheep; toxins induce secondary photosensitisation. **Therapy:** Protoanemonin and saponins
Naucleopsis spp.; Moraceae	S America	latex	toxicariosides; cardenolides	Ia CT, NT, HP, GI	Cardiac glycosides inhibit Na⁺, K⁺-ATPase causing symptoms of cardenolide intoxication; has been used as arrow poison. **Therapy:** Cardiac glycosides
Nepeta cataria (catnip); Lamiaceae	Europe; orn. plant	aerial parts	essential oil with nepetalactone	III CT, NT, MA	Nepetalactone has reactive aldehyde groups which can bind proteins; psychoactive for cats and mice; it has been speculated that it induces mind-altering properties when smoked or ingested. **Therapy:** Terpenes and phenylpropanoids with aldehyde groups

Scientific name (common name); Family	Occurrence	Hazardous plant parts	Active principle	Toxicity Class	Symptoms & Therapy
Nerium oleander (oleander), (= N. indicum); Apocynaceae	Mediterranean; widely cultivated as orn. plant	all parts; nectar, even honey	oleandrine and several other cardenolides	Ia CT, HP, GI, MP	Typical symptoms of cardiac glycoside poisoning; tongue and throat become numb, nausea, vomiting, bloody diarrhoea, spasms, arrhythmia, bradycardia, dilated pupils, dyspnoea, blue lips and hands; respiratory arrest, death can occur after 2–3 h. **Therapy:** Cardiac glycosides
Neurolaena lobata; Asteraceae	West Indies	leaves, stems	sesquiterpene lactones	II CT, NT	Mainly used as fish poison and insecticide. **Therapy:** Sesquiterpenes
Nicandra physaloides and related species; Solanaceae	S America; widely cultivated	all parts; roots	withanolides (such as nicandrenone), roots with pyrrolidine alkaloids	III NT	Slightly intoxicating; ingestion of fruits rarely hazardous. **Therapy:** Withania
Nicotiana glauca (tree tobacco); Solanaceae	S America; introduced into N America, S Europe, W Asia, S Africa	all parts	anabasine, nornicotine and other piperidine alkaloids	Ib NT, MA	Anabasine is an agonist at AChR; salivation, perspiration, spasms of eyelids and lips, dyspnoea, cyanosis; mydriasis, bleeding in most internal organs, respiratory arrest. **Therapy:** Pyrrolidine alkaloids and N. glauca
Nicotiana tabacum (tobacco), N. rustica, N. sylvestris, N. attenuata, N. longiflora and related species; Solanaceae	C America; widely used and cultivated; orn. plants	all parts	nicotine, nornicotine and other pyridine alkaloids: 0.04 to 4 % in N.tabacum; up to 7.5 % in N.rustica	Ib NT, MA, MP	Nicotine is an agonist at nAChR; causes psychedelic feelings and excitation; burning in mouth and throat; nausea, vomiting, disturbance of GI tract, vasoconstriction, hypertension, mydriasis, arrhythmia, seizures, collapse, respiratory arrest. **Therapy:** Pyrrolidine alkaloids and N. tabacum
Nierembergia hippomanica; Solanaceae	S America; orn. plant	all parts	brunfelsamidine	II NT, MA, GI	Ingestion can induce salivation, nausea, weakness, paralysis of face, swollen tongue, distorted vision, trembling, arrhythmia; human and animal intoxication. **Therapy:** Brunfelsia
Nolina texana; Agavaceae	N America	flowers	saponins	II CT, GI	Serious livestock poisoning. **Therapy:** Saponins
Notholaena sinuata (jimmy fern); Pteridaceae	America	all parts	triterpenes	II CT, NT, AP	Mainly animal poisoning; high mortality in sheep. **Therapy:** Pteridium

Scientific name (common name); Family	Occurrence	Hazardous plant parts	Active principle	Toxicity Class	Symptoms & Therapy
Notholirion spp.; Liliaceae	Asia; orn. plant	all parts, especially bulbs	tulipalin, tuliposide A	III CT, IN, GI, AP	Ingestion can cause gastrointestinal disturbance; serious skin allergen; animal poisoning. **Therapy:** Protoanemonin
Nuphar lutea (yellow waterlily); Nymphaeaceae	Europe, Asia; orn. plant	all parts	nupharine, desoxynupharidine and other simple quinolizidine alkaloids	II NT, MA	Alkaloids probably affect acetylcholine receptors, therefore psychoactive; similar to atropine and papaverine; spasmolytic, hypotensive. **Therapy:** Cytisine
Nymphaea alba (white waterlily), *N. caerulea, N. ampla*; Nymphaeaceae	Europe, Asia, America, Africa	all parts	nupharine, nymphaline, aporphine alkaloids	II NT, MA	Alkaloids probably affect acetylcholine receptors, therefore psychoactive: intoxicant; slightly hallucinogenic; similar to atropine and papaverine; spasmolytic, hypotensive. **Therapy:** Cytisine
Oenanthe aquatica (water dropwort), *O. crocata, O. sarmentosa*; Apiaceae	Europe, Asia; introduced or native to N America	all parts	oenanthotoxin (polyacetylene); essential oils	Ib CT, NT, GI, IN	Polyacetylenes bind to proteins and inhibit them; inflammation and blister formation in mouth; inflammation of GI tract; vertigo, coma, seizures with bloody foam, mydriasis, bradycardia, loss of short-term memory; poisonous to humans, cattle, horses and pigs. **Therapy:** Polyacetylenes
Ormosia coccinea and related species; Fabaceae	America	all parts, seeds	ormosanine, sparteine and related quinolizidine alkaloids	Ib–II NT, MP, AP	Lupanine and other QAs are modulators of nAChR and mAChR and also inhibit Na+ channels; salivation, vomiting, problems to swallow, hyperthermia, arrhythmia, mydriasis, excitement and delirium, paralysis, death through respiratory arrest. **Therapy:** Cytisine
Ornithogalum thyrsoides (star-of-Bethlehem), *O. umbellatum, O. conicum, O. ornithogaloides, O. prasinum, O. tenellum, O. longibracteatum* and related species; Hyacinthaceae	Europe, Africa; introduced to N America; orn. plant	all parts, bulbs, flowers	convallatoxin and other cardenolides in bulbs; prasinoside G and other steroid glycosides	Ib–II CT, NT, HP	Symptoms of cardiac glycoside poisoning; nausea, gastrointestinal disorders; abdominal pain and convulsions, heart failure and even death; severe diarrhoea that may last up to three weeks; human and animal poisoning. **Therapy:** Cardiac glycosides
Oxalis acetosella (wood sorrel), *O. pes-caprae, O. corniculata* and related species; Oxalidaceae	Europe, Africa, Asia, N America	all parts	Ca^{2+}-oxalates (up to 1.3 %)	III CT, IN, AP	Substantial ingestion can cause irritation of mucosa and GI tract; bradycardia, hypotension, spasms, central paralysis, circulatory arrest, even death; cause of human and animal poisoning. **Therapy:** Oxalic acid

Scientific name (common name); Family	Occur-rence	Hazardous plant parts	Active principle	Toxicity Class	Symptoms & Therapy
Pachycereus pecten-aboriginum; Cactaceae	Mexico, N Ame-rica	aerial parts	carnegine, alsolidine, heliamine and other β-phenylethylamines	III MA	Psychoactive; similar to mescaline-containing cacti. **Therapy:** Bufotenine
Pachypodium succulentum, *P. lamerei* and related species; Apocynaceae	Madagas-car; S Africa; common orn. plant	all parts	probably glycosides with digitalis properties	Ib–II CT, GI, HP	Probably symptoms of cardiac glycoside poisoning. **Therapy:** Cardiac glycosides
Pachyrrhizus erosus (yam bean); Fabaceae	SE Asia; natura-lised C & N America	mature fruit pods, seeds	rotenone, pachyrrhizin, erosenin; saponins (pachysaponin A, B)	Ib–II CT, GI	Rotenone is an inhibitor of the mitochondrial respiratory chain; disturbance of GI tract; symptoms of rotenone poisoning with multiple organ failure; fatal for humans, fish and insects. **Therapy:** *Derris elliptica*
Pachysandra terminalis (Japanese spurge); Buxaceae	Japan; widely cultivated	all parts	cyclobuxine and other steroid alkaloids	Ib CT, NT, GI, IN	Initially exciting; later paralysing and hypotensive; induces nausea, vomiting, dizziness, diarrhoea, spasms, respiratory arrest; contact dermatitis. **Therapy:** Solanine
Pachystigma pygmaeum, *P. thamnus, P. schumanniana;* Rubiaceae	S Africa	aerial parts	a polyamine	II NT, AP	Mainly animal poisoning; heart failure in livestock. **Therapy:** Cardiac glycosides
Palicourea marcgravii; Rubiaceae	Brazil	leaves	monofluoro acetic acid; fluoro fatty acids, N-methyltyramine	Ib CT, AP	Monofluoro acetic acid inhibits TCA cycle and thus energy metabolism; several cases of livestock poisoning known. **Therapy:** Oxalic acid
Panaeolus cyanescens, *P. africanus, P. cambodginiensis,* *P. castaneifolius, P. fimicola,* *P. papilonaceus, P. sphintrinus,* *P. subbalteatus, P. tropicales* and related species; Bolbitiaceae	Asia, Europe, Africa	fruiting body	psilocin, psilocybin, baeocystine	II NT, MA	Alkaloids are serotonin receptor agonists; therefore psychoactive, psychedelic and hallucinogenic. **Therapy:** Bufotenine and *Psilocybe*
Pancratium trianthum, *P. maritimum;* Amaryllidaceae	Europe; orn. plant	bulbs	trispheridine, tacettine, lycorine	Ib–II CT, NT, MA, GI	Alkaloids cause nausea, vomiting, diarrhoea, abdominal spasms, heavy perspiration, and even death; possibly slightly hallucinogenic. **Therapy:** Lycorine
Papaver bracteatum, P. orientale and related species; Papaveraceae	Asia; orn. plant	aerial parts, latex	thebaine, oripavine and other isoquinoline alkaloids	Ib–II CT, NT, GI	Thebaine causes spasms similar to strychnine; substantial ingestion causes disturbance of GI tract and CNS. **Therapy:** Morphine and *Datura stramonium*

Scientific name (common name); Family	Occur-rence	Hazardous plant parts	Active principle	Toxicity Class	Symptoms & Therapy
Papaver rhoeas and other poppy species; Papaveraceae	Europe, N Africa, Asia; weed	all parts, especially latex	rhoeadine, protopine, berberine, and other isoquinoline alkaloids	II CT, NT, GI, AP	Substantial ingestion causes vomiting, spasms, abdominal pains and CNS disturbance; in animals: central excitation, gastroenteritis, unrest, epileptiform spasms, unconsciousness. **Therapy:** Morphine and isoquinoline alkaloids
Papaver somniferum (opium poppy); Papaveraceae	Asia; widely cultivated	latex (opium)	papaverine, morphine, codeine, thebaine	Ib NT, GI, MA, MP	Opium: sedating; induces happiness and euphoria; inhibits CNS and respiratory centre; nausea, vomiting, hypothermia, respiratory arrest; morphine is a powerful painkiller. **Therapy:** Morphine and *Papaver*
Paris quadrifolia (herb Paris); Melanthiaceae (Trilliaceae)	Europe, W Asia	all parts, especially fruits	paridin, aristyphnin, and other steroidal saponins	II–III CT, NT, GI	Saponins are haemolytic and cytotoxic when absorbed; sensory irritation; nausea, small pupils, nephritis, CNS disturbance; respiratory arrest. **Therapy:** Saponins
Parthenocissus quinquefolium (Virginia creeper); Vitaceae	N America; orn. plant	fruits	oxalates	II–III CT	Plant might have caused poisoning and death of children. **Therapy:** Oxalic acid
Passiflora incarnata (passion flower), *P. caerulea*, *P. quadrangularis* and related species; Passifloraceae	S America; widely cultivated	all parts (except edible seed arils)	leaves contain cyanogenic glucosides; flavonoids; harman alkaloids	III MA, MP	Mild sedative; if harman alkaloids are present, then hallucinogenic effects likely; cyanogenic glucosides release HCN which block respiratory chain in mitochondria. **Therapy:** Indole alkaloids
Paullinia cupana (guarana) and related species; Sapindaceae	S America; cultivated in Brazil	seeds, aerial parts	seeds with caffeine (3–8%); saponins	III MA, MP	Central stimulant; high doses cause intoxication with headache, dizziness, palpitations, insomnia, unrest, vomiting, diarrhoea, diuresis. **Therapy:** Caffeine
Pausinystalia yohimbe (yohimbe); Rubiaceae	W & C Africa	bark	yohimbine, ajmalicine, corynanthine, and other monoterpene indole alkaloids	Ib NT, MA, GI, MP	Yohimbine is a sympatholytic inducing intoxication with hallucinations; used as aphrodisiac (similar to Viagra); high doses: salivation, increased respiration, diarrhoea, hypotension, cardiac arrest; death. **Therapy:** Indole alkaloids and *Pausinystalia*
Peddiea africana (poison olive), *P. volkensii*; Thymelaeaceae	S Africa	roots, aerial parts	*Peddiea* factor A$_1$ and other diterpenes (phorbolesters of daphnane-type)	Ia CT, NT, IN, GI	Ingestion of phorbol esters induces strong irritation of mucous membranes, causing severe disturbance of GI tract and CNS; skin irritant; has been used for murder. **Therapy:** Phorbol esters

Scientific name (common name); Family	Occurrence	Hazardous plant parts	Active principle	Toxicity Class	Symptoms & Therapy
Pedilanthus tithymaloides (slipper flower); Euphorbiaceae	Mexico; common orn. plant in warm climates	all parts, latex	phorbol esters	Ib CT, IN, GI	Strong skin and mucosal irritant; severe diarrhoea and GI tract disturbance. **Therapy:** Phorbol esters
Peganum harmala (African rue); Zygophyllaceae	Asia, N Africa	seeds, aerial parts, roots	harman alkaloids, in seeds and roots; quinoline alkaloids (peganine, vasicine) in aerial parts	Ib NT, MA, MP	β-carboline alkaloids are agonist at serotonin receptors and inhibit MAO; intoxicant with hallucinogenic, antidepressive, aphrodisiac effects; induces happiness. **Therapy:** *Peganum*
Pelecyphora aselliformis (hatchet cactus); Cactaceae	Mexico	aerial parts	hordenine, anhalidine, traces of mescaline	III MA	Psychoactive, similar to but weaker than *Lophophora williamsii*. **Therapy:** *Lophophora*
Pentalinon (= *Urechites*) *luteum* (yellow night shade); Apocynaceae	America	leaves, seed pods	urechitoxin and other cardiac glycosides	Ib CT, NP, HP	Symptoms of cardiac glycoside poisoning; overdose causes heart failure. **Therapy:** Cardiac glycosides
Pernettya mucronata, P. furens; Ericaceae	Chile, New Zealand; orn. plant	aerial parts, berries, nectar	andromedotoxin; sesquiterpenes	Ib CT, NT, MA	Psychoactive, dream inducing; visions; burning sensation in mouth; salivation, dizziness, vomiting, diarrhoea, spasms, hypotonia, bradycardia, paralysis, respiratory arrest. **Therapy:** Andromedotoxin
Petasites hybridus (butterbur) and related species; Asteraceae	Europe, Asia, introduced in N America	all parts	petasin and other sesquiterpenes; senkirkine and other pyrrolizidine alkaloids	II–III CT, NT, MU, MP	PAs have mutagenic and carcinogenic properties (especially in liver); sesquiterpenes are sedating and used for treating migraine and allergic rhinitis. **Therapy:** Senecionine and sesquiterpenes
Petroselinum crispum (parsley); Apiaceae	Europe & W Asia; widely cultivated	aerial parts, roots	apiol, myristin and other phenylpropanoids; apiin, bergapten, psoralen and other furanocoumarins	III CT, IN	Apiol increases contraction of smooth muscles of gut, bladder and uterus; has been used traditionally as abortifacient; high doses cause liver cirrhosis; bloody diarrhoea, bleeding of GI tract, nephritis; furanocoumarins are skin irritants and cause photodermatitis; myristicin has psychoactive properties (see *Myristica fragrans*). **Therapy:** Myristicin
Peucedanum galbanum (= *Noto bubon galbanum*) (blister bush); Apiaceae	S Africa	all parts	psoralen, xanthotoxin, bergapten and other furanocoumarins	II CT, MU, IN	Furanocoumarins intercalate DNA and form crosslinks upon exposure to sunlight; strongly cytotoxic; leads to serious blister formation of the skin when plant sap comes into contact with skin and sunlight. **Therapy:** Coumarins and *Heracleum*

Scientific name (common name); Family	Occurrence	Hazardous plant parts	Active principle	Toxicity Class	Symptoms & Therapy
Peumus boldus (boldo); Monimiaceae	Chile	all parts	boldine, isocorydine, reticuline and other isoquinoline alkaloids	II NT, MP	High doses cause paralysis, hallucinations and CNS disturbance. **Therapy:** Isoquinoline alkaloids
Phacelia crenulata, P. minor, P. tanacetifolia; Hydrophyllaceae / Boraginaceae	N America; introduced into Europe	all parts	prenylated hydroquinones (geranylbenzoquinone, geranylhydroquinone)	III CT, IN	Geranylbenzoquinone can alkylate proteins and make them allergenic; known contact allergen in America. **Therapy:** Hydroquinone
Phalaris arundinacea (reed grass), *P. aquatica, P. canariensis, P. minor, P. tuberosa*; Poaceae	Europe, Asia, America	aerial parts	*N,N*-DMT, 5-MeO-DMT, gramine	III NT, MA	Tryptamine derivates are agonists at serotonin receptors; therefore psychoactive; psychedelic, hallucinogenic. **Therapy:** Bufotenine
Phaseolus lunatus (lima bean); Fabaceae	America; widely cultivated	seeds	phaseolunatin (cyanogenic glucoside)	II CT, NT, GI	Cyanogenic glucosides release HCN; substantial ingestion causes HCN poisoning; small tropical varieties may be fatal even when cooked. **Therapy:** Amygdalin
Phaseolus vulgaris (French bean), *P. coccineus* (scarlet runner bean); Fabaceae	tropical America; widely cultivated	uncooked fruits	phasin (a lectin), (is destroyed by heating)	II–III CT, IN, GI	Substantial ingestion of raw fruits cause vomiting, gastroenteritis, diarrhoea, haemorrhagic abdominal pain and spasms. **Therapy:** Abrin and *Phaseolus vulgaris*
Philodendron selloum (= *P. bipinnatifidum*), *P. scandens* and related species; Araceae	tropics; common orn. plant	aerial parts	Ca²⁺-oxalate crystals	II–III CT, IN, GI	Substantial ingestion can cause disturbance of GI tract; inflammation of mouth and throat. **Therapy:** Oxalic acid
Phoradendron serotinum (American mistletoe), *P. quadrangulare, P. tomentosum, P. villosum*; Santalaceae	N & C America	all parts	phoratoxin, a toxic lectin	Ib–II CT, GI	Substantial ingestion can cause severe gastroenteritis and spasms and even multiple organ failure. **Therapy:** Abrin
Phragmites australis (reed) (= *P. communis*) (reed); Poaceae	cosmopolitan	roots	*N,N*-DMT, 5-MeO-DMT, gramine	III NT, MA	Tryptamine derivates are agonists at serotonin receptors; therefore psychoactive; psychedelic, hallucinogenic. **Therapy:** Bufotenine
Physalis alkekengi (Japanese lantern plant); *P. peruviana, P. crassifolia*; Solanaceae	America; common orn. plant	all parts; unripe berries (red berries are edible)	steroidal glycoalkaloids; physalin A–V	III CT, GI, NT, MA	Green parts are rich in alkaloids that cause GI tract disturbance; nausea, cold perspiration and even delirium and hallucinations; cause of many consultations in poison centres; however, ingestion of red berries is rarely hazardous. **Therapy:** Solanine

Scientific name (common name); Family	Occurrence	Hazardous plant parts	Active principle	Toxicity Class	Symptoms & Therapy
Physostigma venenosum (calabar bean), *P. mesoponticum*; Fabaceae	W Africa, E Africa	seeds	physostigmine, eseramine, physovenine and other indole alkaloids	Ia NT, MA, MP	Physostigmine inhibits acetylcholine esterase causing sweating, trembling, tachycardia, inflammation of mucosal tissues, respiratory and cardiac arrest. **Therapy:** *Physostigma*
Phytolacca americana (= *P. decandra*) (pokeweed), *P. acinosa* (= *P. esculenta*), *P. dioica, P. dodecandra, P. octandra*; Phytolaccaceae	N America, Asia; introduced into Europe and Africa	roots, leaves	lectins; phytolaccatoxin (triterpene saponins)	II CT, GI	Vomiting, diarrhoea, stomach cramps, weakened pulse, in severe cases: breathing difficulty, convulsions; death; used as molluscicide. **Therapy:** Saponins
Pieris japonica (lily-of-the-valley-bush), *P. floribunda*; Ericaceae	E Asia; N America; common orn. plant	leaves, nectar (honey)	phlorizin and derivates; andromedotoxin; saponins	II CT, NT, GI	Burning in mouth, salivation, vomiting, spasms, diarrhoea; also CNS effects: headache, weakness, blurred vision, convulsions, death by respiratory arrest. **Therapy:** Andromedotoxin
Pilocarpus jaborandi (jaborandi), *P. alavardoi* and related species; Rutaceae	Brazil; S & C America	leaves	pilocarpine, pilosine and related imidazol alkaloids	Ib NT, MP, AP	Pilocarpine is an agonist at mAChR; substantial activation of glandular secretions; diarrhoea, uterus contractions (therefore abortifacient); disturbance of heart and circulation; animal poisoning. **Therapy:** *Pilocarpus*
Pinus ponderosa (yellow pine); Pinaceae	N America	needles	essential oil; isocupressic acid and its esters	III CT, AP	If pregnant cows feed on *P. ponderosa* needles, they often abort ("pine needle abortion"). **Therapy:** Monoterpenes
Piper auritum; Piperaceae	C & S America	leaves	safrole and several other phenylpropanoids and monoterpenes	III NT, MA, MU	Stimulant; safrole is mutagenic. **Therapy:** Myristicin
Piper betle (betel pepper); Piperaceae	SE Asia	leaves	eugenol, chavicol and monoterpenes	III CT, NT, MP	Stimulates senses, carminative, antibiotic; used together with *Areca catechu*. **Therapy:** Myristicin
Piper cubeba (cubebs); Piperaceae	Sunda islands, E Asia;	fruits	cubebin, cubebenic acid	II CT, NT, IN	High doses cause inflammation of urinary tract, bladder and kidney pains, albinuria, tachycardia, nausea, vomiting, diarrhoea; disturbance of CNS, aphrodisiac. **Therapy:** Monoterpenes

Scientific name (common name); Family	Occurrence	Hazardous plant parts	Active principle	Toxicity Class	Symptoms & Therapy
Piper methysticum (kava kava); Piperaceae	Poly-nesia	rhizomes	kavain, methysticin, yangonin, and other kawa lactones	III NT, MA, MP	Psychoactive; relieves from anxiety and sorrow; anticonvulsive, muscle relaxant, narcotic, high doses cause nausea, vomiting, diarrhoea. **Therapy:** Piper
Piptadenia peregrina, see Anadenanthera; Fabaceae					
Piptoporus betulinus and related fungi; Fomitopsidaceae	Eurasia	fruiting body	several indole alkaloids	II–III NT, MA	Mushroom with weak hallucinogenic and psychedelic properties. **Therapy:** Bufotenine
Piscidia piscipula (Jamaica dogwood), P. guaricensis; Fabaceae	West Indies, Asia, C America	bark, leaves, roots	jamaicine, ichthynone, piscerythrone, rotenone and other isoflavones	Ib CT, NT, GI, AP	Rotenone blocks mitochondrial respiration; high doses cause vomiting, salivation, perspiration, dizziness, trembling and cardiac arrest; used as fish and arrow poison. **Therapy:** Derris
Plumeria alba (frangipani), P. rubra, P. obtusa; Apocynaceae	tropical America; orn. plant	latex	indole alkaloids	II–III IN, NT	Skin irritant; toxic when ingested in large quantities. **Therapy:** Indole alkaloids
Pluteus cyanopus, P. nigriviridis and related species; Pluteaceae	Europe	fruiting body	psilocybin, baeocystine	II NT, MA	Alkaloids are serotonin receptor agonists; therefore psychoactive, psychedelic and hallucinogenic. **Therapy:** Bufotenine and Psilocybe
Podophyllum peltatum (may apple), P. hexandrum, P. pleianthum and related species; Berberidaceae	N America, Asia; orn. plant	all parts (except fruits)	podophyllotoxin, peltatin, and other lignans	Ib CT, GI	Inhibits cell division (spindle poison); initial oropharyngeal pains; strong purgative; vomiting, haemorrhagic gastroenteritis; respiratory stimulation, peripheral neuropathy, coma or death by respiratory arrest. **Therapy:** Flavonoids and P. peltatum
Polygonatum multiflorum (Solomon's seal), P. odoratum, P. verticillatum and related species; Ruscaceae (Convallariaceae)	Europe, Asia; N America orn. plant	all parts, especially fruits	acetidine 2 carboxylic acid (NPAA); steroidal saponins (diosgenin as aglycone)	II–III CT, NT, GI	Only high doses toxic: causing nausea, vomiting, diarrhoea, cytotoxicity, heart and CNS disturbance. **Therapy:** Saponins and NPAAs
Polygonum punctatum (water smartweed) and related species; Polygonaceae	N America	aerial parts	Ca^{2+}-oxalate	II–III CT, IN, AP	Typical symptoms of calcium oxalate intoxication; poisonous to humans, fatal to livestock. **Therapy:** Oxalic acid

Scientific name (common name); Family	Occur-rence	Hazardous plant parts	Active principle	Toxicity Class	Symptoms & Therapy
Primula obconica (top primrose), *P. vera, P. elatior, P. denticulata* and related species; Primulaceae	Asia, Europe, America; common orn. plant	all parts	primin (a prenylated benzoquinone) in trichomes of stems and flowers of *P. obconica* (see hydroquinone); triterpene saponins in rhizomes of other species	III CT, IN, MP	*P. obconica:* skin irritant; contact dermatitis; quinones can bind to proteins and alter their activity. Primroses with saponins can cause disturbance of GI tract when consumed in larger quantities. **Therapy:** Hydroquinone and saponins
Prunus laurocerasus (laurel cherry), *P. dulcis* (bitter almond), *P. armeniaca, P. avium, P. domestica, P. padus, P. persica, P. serotina; P. spinosa, P. virginiana* and related species; Rosaceae	Europe, Asia, N America; widely cultivated	all parts, especially seeds	amygdalin, prunasin (cyanogenic glucosides); especially high concentration in seeds (5–8 %)	Ib–II CT, NT, GI	When seeds are crushed they release HCN which is a strong respiratory poison; high doses with HCN poisoning symptoms: burning sensation in throat, sweating, abdominal pains, vomiting, red face, salivation, convulsions, respiratory and cardiac arrest. **Therapy:** Amygdalin
Psilocybe mexicana, P. azurescens, P. baeocystis, P. bohemica, P. caerulescens, P. caerulipes, P. coprophila, P. cubensis, P. cyanescens, P. hoogshagenii, P. natalensis, P. pelliculosa, P. semilanceata, P. strictipes, P. stuntzii and many related species; Strophariaceae	Europe, America, Asia	fruiting body	psilocybin, psilocin, baeocystine	II NT, MA	Alkaloids are serotonin receptor agonists; therefore psychoactive, psychedelic and hallucinogenic. **Therapy:** Bufotenine and *Psilocybe*
Psychotria (= *Cephaelis*) *ipecacuanba* (ipecac) and related species; Rubiaceae	S & C America	root	emetine and cephaeline (alkaloids)	Ib CT, GI, MP	Emetine is a strong protein biosynthesis inhibitor and DNA intercalator; ingestion causes nausea, vomiting; gastroenteritis (intestinal mucosa start bleeding), diarrhoea, dyspnoea, muscle arrest; death by cardiac arrest. **Therapy:** *Psychotria*
Psychotria viridis, P. carthaginensis, P. colorata, P. poeppigiana, P. psychotriaefolia; Rubiaceae	S America; cultivated in California, Hawaii	leaves	N,N-DMT, MMT, 2-methyltetrahydro-β-carboline	III NT, MA	Tryptamine derivatives are serotonin receptor agonists; hallucinogenic, have been used by Amazonian Indians as intoxicant, together with ayahuasca. **Therapy:** Bufotenine

Scientific name (common name); Family	Occurrence	Hazardous plant parts	Active principle	Toxicity Class	Symptoms & Therapy
Ptelea trifoliata (hop tree); Rutaceae	N America	all parts	quinoline alkaloids, furanocoumarins	III CT, NT, IN, MU	Compounds can intercalate DNA; skin irritant; substantial ingestion can cause disturbance of GI tract and CNS. **Therapy:** Quinoline alkaloids and coumarins
Pteridium aquilinum (bracken fern); Dennstaedtiaceae	cosmopolitan	all parts, young shoots	cyanogenic glucosides; ptaquiloside (sesquiterpene); thiaminase	Ib CT, NT, GI, MU, AP	Ptaquiloside is a strong mutagen and causes stomach and bladder cancer (mostly cattle, also humans); thiaminase destroys vitamin B_1 leading to CNS disturbances in animals. **Therapy:** *Pteridium* and sesquiterpenes
Pteronia pallens; Asteraceae	S Africa	aerial parts	unknown	II CT, AP	Mainly animal poisoning; chronic or acute liver damage, heavy breathing, fast and weak pulse, general weakness, apathy and jaundice. **Therapy:** Sesquiterpenes
Pulsatilla vulgaris (pasque flower), *P. pratensis* and related species; Ranunculaceae	Europe, N Asia; orn. plant	all parts	ranunculin is converted enzymatically to protoanemonin, saponins	Ib–II IN, CT, AP, MP, MU	GI tract and kidney disturbance; mutagenic; blistering and inflammation of skin; livestock poisoning. **Therapy:** *Pulsatilla* and protoanemonin
Punica granatum (pomegranate); Lythraceae (Punicaceae)	Asia, introduced into Mediterranean; common orn. plant	bark, leaves, young fruits	pseudopelletierine, pelletierine and other piperidine alkaloids (up to 0.7%); tannins, oxalate	II CT, NT, GI	Alkaloids affect acetylcholine receptors; high doses cause nausea, vomiting, disturbance of GI tract, internal bleeding, CNS disturbance, bradycardia, convulsions, death by respiratory arrest. **Therapy:** *Punica* and pyrrolidine alkaloids
Pyracantha coccineus (firethorn); Rosaceae	Europe; orn. plant	seeds	amygdalin, 120 ppm HCN/FW	III CT, GI	Rarely hazardous, although ingestion of fruits a frequent cause of consultations in poison centres. Only ingestion of large quantities of seeds cause GI tract disturbance. **Therapy:** Amygdalin
Pyrus communis (pear); Rosaceae	Europe; cultivated worldwide	seeds	prunasin (cyanogenic glucoside)	III CT, GI	When seeds are crushed they release HCN which is a strong poison; inhibits cytochrome oxidase in mitochondria; ingestion of large amounts of seeds: vomiting, flushed face, increased sweating; respiratory and cardiac arrest. **Therapy:** Amygdalin

Scientific name (common name); Family	Occurrence	Hazardous plant parts	Active principle	Toxicity Class	Symptoms & Therapy
Quassia amara (quassia) and related species; Simaroubaceae	S America	wood	bitter triterpenes; quassin and other quassinoids, also β-carboline alkaloids	II–III CT	Quassinoids and canthinone alkaloids show antibacterial, antifungal, antiviral, antiparasitic (amoebae, trypanosomes, worms, insects), antitumour and positive inotropic activities. β-Carboline and canthinone alkaloids inhibit cAMP-phosphodiesterase. Substantial ingestion can cause GI tract and CNS disturbance. **Therapy:** Saponins
Quercus robur (oak), *Q. petraea, Q. glauca, Q. douglasii* and related species; Fagaceae	Europe, N America; widely cultivated	bark, fruits	high amounts of tannins and other polyphenols; digallic acid, ellagitannins such as castalagin	III CT, GI, AP	Mainly animal poisoning; feeding on leaves and seeds causes gastroenteritis and severe kidney damage. **Therapy:** Flavonoids
Ranunculus acris (buttercup), *R. bulbosus, R. sceleratus, R. multifidus* and related species; Ranunculaceae	Europe, Asia, N America; a worldwide weed	all parts	ranunculin is enzymatically converted to the active protoanemonin	Ib–II CT, IN, MU, GI	Severe skin and mucosal irritation with blisters and ulceration; internally: diarrhoea, abdominal pains, tinnitus, headache, dizziness, seizures, tachycardia, nephritis, even death from respiratory and cardiac arrest. **Therapy:** Protoanemonin
Rauvolfia serpentina (Indian snakeroot), *R. caffra, R. rosea* and related species; Apocynaceae	India, SE Asia; Africa	all parts, especially roots	reserpine, yohimbine, rescinnamine, serpentine, ajmaline, tetraphylline and other monoterpene indole alkaloids	Ib NT, MA, MP	Reserpine blocks the uptake of neurotransmitters into synaptic vesicles; sedating CNS and smooth muscles; hypotension, bradycardia, cardiac arrest; used as arrow and fowl poison. **Therapy:** Indole alkaloids
Rhamnus catbartica (buckthorn), *R. frangula, R. purshiana;* Rhamnaceae	Europe, N Africa N America	bark, berries	frangulin, barbaloin, and related anthraquinones; saponins	II CT, GI, MP	Powerful laxative: nausea, vomiting, bloody diarrhoea, abdominal pains, even nephritis and collapse; probably mutagenic. **Therapy:** Anthraquinones
Rhazya stricta; Apocynaceae	Asia; orn. plant	leaves	indole alkaloids	II CT, NT, GI	Mainly animal poisoning; substantial ingestion causes GI tract and CNS disturbance. **Therapy:** Indole alkaloids
Rheum rhabarbarum (= R x hybridum) (rhubarb), *Rheum rhaponticum, R. palmatum;* Polygonaceae	Europe, Asia; widely grown food plant	all parts, roots	high oxalate levels, rhein, aloe emodin, chrysophanol (anthraquinone glycosides)	III CT, GI, MP	Laxative; roots are edible; leaves cause vomiting, stomach pain, internal bleeding, nephritis, breathing difficulty; can be fatal; livestock poisoning. **Therapy:** Oxalic acid

Scientific name (common name); Family	Occurrence	Hazardous plant parts	Active principle	Toxicity Class	Symptoms & Therapy
Rhododendron ponticum, R. chrysanthum, R. simsii, R. indicum, R. ferrugineum and related species; Ericaceae	Asia, Europe, N America; common orn. plant	all parts, nectar (honey)	rhododendrin, andromedotoxin (=grayanotoxin I); triterpenes, arbutin	Ib–II CT, NT, GI	Burning in mouth, salivation, vomiting, spasms, diarrhoea, also CNS effects: headache, weakness, blurred vision, convulsions, death by respiratory arrest. **Therapy:** Andromedotoxin
Rhododendron tomentosum (= Ledum palustre) (wild rosemary); Ericaceae	Europe, Asia, N America	aerial parts	arbutin, ledol and other sesquiterpenes	II CT, NT, MA, GI	Hypnotic and sedating (similar to beer); high doses can cause gastroenteritis, nephritis, spasms and progressive paralysis; hypotensive, abortifacient. **Therapy:** Sesquiterpenes and Rhododendron tomentosum
Rhodomyrtus macrocarpa (finger cherry); Myrtaceae	SE Asia, Australia	aerial parts	saponins	II CT, GI	In case of contact with eyes, temporary or permanent blindness can result. **Therapy:** Phorbol esters
Rhodotypos scandens (= R. tetrapetala, R. kerrioides) (jetbead); Rosaceae	N America; orn. plant	fruits	probably amygdalin	II CT, GI	Probably mild HCN poisoning due to cyanogenic glucosides. **Therapy:** Amygdalin
Rhus toxicodendron (= Rhus radicans, Toxicodendron radicans) (poison ivy), R. diversiloba (= T. diversilobum), R. toxicarium (= T. pubescens), R. vernix and related species; Anacardiaceae	N America; orn. plant	all parts	urushiol (=toxicodendrin); (3.3 % in leaves, 1.6 % in twigs, 3.6 % in unripe fruits)	Ia CT, IN, GI	Strongly irritant; blisters and ulceration of skin and mucosal tissues; nausea, vomiting, strong colic, bloody diarrhoea, dizziness, excitation, nephritis, haematuria. **Therapy:** Rhus
Rhus typhina (staghorn sumach); Anacardiaceae	N America; introduced Europe; orn. plant	all parts, especially latex	ellagic acid	III CT, GI	Ingestion of substantial amounts leads to disturbance of GI tract; skin irritant. **Therapy:** Flavonoids
Rhynchosia pyramidalis; Fabaceae	Mexico, Caribbean; orn. plant	seeds	unknown alkaloids	II NT, MA	Probably psychoactive; used in Mexico as narcotic, poison and intoxicant. **Therapy:** Pyrrolidine alkaloids

Scientific name (common name); Family	Occurrence	Hazardous plant parts	Active principle	Toxicity Class	Symptoms & Therapy
Ricinus communis (castor oil plant); Euphorbiaceae	originally India & Africa; widely cultivated as orn. plant	seeds	ricin (a lectin); ricinine (pyridine alkaloid); ricinolic acid (fatty acid)	Ia CT, IN, GI, MP	Oil has been used as a laxative; ricin is very toxic and inhibits protein biosynthesis; parenteral application can cause life-threatening multisystem organ failure; ingestion causes nausea, bloody vomiting, bloody diarrhoea, nephritis, liver damage, convulsions, tachycardia, circulatory arrest; also poisonous for animals (press cake). **Therapy:** Abrin and *ricinus*
Robinia pseudoacacia (black locust) and related species; Fabaceae	N America; introduced to Europe, Asia	all parts, especially roots, bark, fruits	robin (a lectin); tannins	Ib CT, GI	The lectin has agglutinating properties and is cytotoxic, causes nausea, vomiting, diarrhoea; sleepiness, mydriasis, seizures, abdominal pains; parenteral application can cause life-threatening multisystem organ failure; toxic for cattle and horses. **Therapy:** Abrin
Rourea glabra, R. volubilis; Connaraceae	tropics	seeds	rourinoside and rouremin	Ib NT	Very poisonous to carnivores. **Therapy:** Phenolics
Rubia tinctorum (common madder); Rubiaceae	Mediterranean; cultivated	roots	anthraquinones such as lucidin, rubiadin	II–III CT, GI, MP	Substantial ingestion causes GI tract disturbance: nausea, vomiting, bloody diarrhoea, abdominal pains; nephritis, probably mutagenic. **Therapy:** Anthraquinones
Rudbeckia hirta (black-eyed susan), *R. laciniata*; Asteraceae	America; orn. plant	all parts	polyacetylenes and thiophenes	II CT, IN, GI	Mainly livestock poisoning. **Therapy:** Sesquiterpenes
Rumex crispus (curled dock), *R. acetosa* and related species; Polygonaceae	Europe, Asia, N America	all parts	tannins; physcion, aloe-emodin (anthraquinones); high oxalate levels	II–III CT, IN, GI, AP	Mainly GI tract disturbance with nausea, vomiting, bloody diarrhoea, abdominal pains; nephritis, mostly animal poisoning; probably mutagenic. **Therapy:** Oxalic acid
Ruscus aculeatus (butcher's broom); Ruscaceae (Asparagaceae)	Europe	all parts, fruits	steroidal saponins	III CT, GI	Cytotoxic, gastrointestinal disturbance; no serious intoxication in humans. **Therapy:** Saponins
Ruta graveolens (garden rue) and related species; Rutaceae	S Europe; orn. plant	all parts	bergapten, psoralen (furanocoumarins); kokusagenine, skimmianine, rutamine, dictamnine (quinoline alkaloids)	Ib–II CT, NT, GI, MU, MP	Furanocoumarins and alkaloids intercalate and alkylate DNA; strong skin and mucosal irritant, blister formation, itching; internally: visual distortions, salivation, gastroenteritis, abortifacient, haematuria. **Therapy:** *Ruta* and coumarins

Scientific name (common name); Family	Occurrence	Hazardous plant parts	Active principle	Toxicity Class	Symptoms & Therapy
Ryania speciosa; Salicaceae (Flacourtiaceae)	America	aerial parts	ryanodine and other indole alkaloids	Ib NT, GI, AP	Ryanidine inhibits an IP$_3$ receptor at the sarcoplasmatic reticulum of heart cells and blocks calcium efflux which is needed to activate muscle contraction; violent GI tract poison, very toxic to vertebrates and insects. **Therapy:** Aconitine
Saccharomyces cerevisiae (yeast); Saccharomycetaceae	worldwide		used to make beer or wine from sugar containing plants (e.g. *Vitis vinifera*)	III NT, MA	Alcohol modulates GABA receptors; explaining the psychoactive effects; perhaps the metabolically formed acetaldehyde can react with biogenic amines to form psychoactive alkaloids. **Therapy:** Ethanol
Salvia divinorum (magic mint); Lamiaceae	Mexico	leaves	salvinorin A & B	II NT, MA	Sap of the plant is ingested or dried leaves are smoked; unpleasant psychoactivity; psychedelic, hallucinogenic. **Therapy:** Bufotenine or ergot alkaloids
Samanea saman see Albizia saman					
Sambucus ebulus (dwarf elder), *S. racemosa, S. australis, S. canadensis, S. pubens*; Adoxaceae (Caprifoliaceae)	Europe, Asia, N America	all parts, especially fruits	ebuloside and other iridoid glucosides; lectins;	III CT, GI, AP	Burning of throat, nausea, vomiting, bloody diarrhoea, dizziness, headache, visual problems, cardiac disturbance, even death; animal poisoning has been observed. **Therapy:** Terpenes and phenylpropanoids
Sambucus nigra (elderberry); Adoxaceae (Caprifoliaceae)	Europe, Asia, N America	all parts, roots, unripe fruits (harmless if ripe and cooked)	sambunigrin (cyanogenic glucoside); lectins	III CT, GI, MP	HCN is released upon hydrolysis of sambunigrin; high doses lead to vomiting and diarrhoea. **Therapy:** Amygdalin
Sanguinaria canadensis (bloodroot); Papaveraceae	N America; orn. plant	leaves, roots	sanguinarine, berberine, protopine and other isoquinoline alkaloids	Ib–II CT, NT, MP	Inhibits acetylcholine esterase and thus enhances the activity of acetylcholine; activates smooth muscles; vomiting, diarrhoea, respiratory repressing, shock, coma. **Therapy:** Isoquinoline alkaloids and *Sanguinaria*
Sansevieria trifasciata (mother-in-law's tongue) and related species; Agavaceae	Africa; orn. plant	all parts	saponins	II–III CT, GI	Saponins have cytolytic and cytotoxic properties; ingestion can cause vomiting, nausea and diarrhoea. **Therapy:** Saponins

Scientific name (common name); Family	Occurrence	Hazardous plant parts	Active principle	Toxicity Class	Symptoms & Therapy
Santalum album (white sandalwood); Santalaceae	Asia	wood, essential oil	α and β-santalol	II CT, GI, IN	Substantial ingestion can cause nephritis, disturbances of GI tract; burning sensation in stomach, dyspepsia, haematuria; skin irritant. **Therapy:** Oxalic acid and terpenoids
Sapindus saponaria, S. drummondii, S. rarak (soap nut tree); Sapindaceae	S & N America; SE Asia	fruits	triterpene saponins	II CT, GI, IN	Saponins have haemolytic and cytotoxic properties; pronounced GI disturbances: nausea, vomiting, diarrhoea, abdominal spasms; dermatitis; fish poison in Venezuela. **Therapy:** Saponins
Saponaria officinalis (soapwort); *S. vaccaria*; Caryophyllaceae	Europe; orn. plant	all parts	triterpene saponins	III CT, GI, AP	Saponins are cytotoxic and disturb membrane fluidity; high doses cause nephritis, disturbances of GI tract; livestock poisoning. **Therapy:** Saponins
Sarcococca saligna; Buxaceae	Asia; orn. plant	aerial parts	steroidal alkaloids	Ib CT, NT, GI	Ingestion can cause disturbance of GI tract and CNS. **Therapy:** *Buxus* and solanine
Sarcostemma viminale; Apocynaceae (Asclepiadoideae)	Africa to Asia	aerial parts	sarcovimicide A and other steroid glycosides	II CT, NT, GI, AP	Mainly animal poisoning: hypersensitivity, seizures and paralysis, fish poison. **Therapy:** Saponins
Sartwellia laveriia; Asteraceae	N America	all parts	essential oil with sesquiterpenes	II–III CT, GI, AP	Serious liver damage in livestock. **Therapy:** Sesquiterpenes
Sassafras albidum (= *S. officinale*) (sassafras); Lauraceae	N America	wood, essential oil	safrole, pinene, eugenol, myristicin	II CT, MA, MU, MP	High doses cause nephritis, unconsciousness, weakness, liver disturbance; CNS stimulant (similar to the amphetamine MDMA); aphrodisiac; safrole becomes active in the liver and can alkylate DNA; mutagenic and carcinogenic; skin irritant. **Therapy:** Myristicin
Sauromatum venosum (= *S. guttatum*) (voodoo lily); Araceae	Africa	all parts, fruits	aroin, Ca^{2+}-oxalates	Ib–II CT, GI	Ingestion can cause disturbance of GI tract and CNS. **Therapy:** Oxalic acid
Scadoxus puniceus (red paint brush), *S. multifloris*; Amaryllidaceae	S Africa; cultivated as orn. plant	all parts	haemanthamine, haemanthidine and other isoquinoline alkaloids	Ib–II CT, NT, GI, MP, AP	Substantial ingestion causes disturbance of GI tract and CNS; hypotensive, convulsive; animal poisoning. **Therapy:** *Scadoxus*

Scientific name (common name); Family	Occurrence	Hazardous plant parts	Active principle	Toxicity Class	Symptoms & Therapy
Schefflera actinophylla (= *Brassaia actinophylla*) (umbrella tree); Araliaceae	Australia, Java, New Guinea; orn. plant in subtropics	leaves, sap	falcarinol (polyacetylene); Ca^{2+}-oxalate raphides	II CT, IN, GI	Strong irritant of skin and mucous membranes causing oedema and blistering; may produce contact dermatitis; vomiting, disturbance of GI tract. **Therapy:** Oxalic acid
Schinus terebinthifolius (Brazilian pepper tree), *S. molle* (pepper tree) and related species; Anacardiaceae	C & S America; cultivated in Africa, Europe, Asia	fruits, essential oil	(15:1)-cardanol; α- & β-phellandrene, limonene, myrcene, α-pinene and other terpenoids	II CT, NT, GI	(15:1)-cardanol is a strong mucosal and skin irritant, headache, swollen lids; GI tract disturbances; should not be used as a spice in larger quantities. **Therapy:** Saponins
Schizanthus pinnatus (poor man's orchid) and related species; Solanaceae	S America; orn. plant	aerial parts	tropane alkaloids: esters of 6-hydroxytropine, tiglic, angelic and senecioic acid	Ib–II NT, MA	Tropane alkaloids inhibit mAChR, therefore parasympatholythic; high doses cause hallucinations and CNS disturbance (see *Atropa, Datura*). **Therapy:** Hyoscyamine
Schizozygia caffaeoides; Apocynaceae	E Africa	aerial parts	schizozygine and other indole alkaloids	II NT, GI	Substantial ingestion causes disturbance of GI tract and CNS. **Therapy:** Indole alkaloids
Schoenocaulon drummondii (green lily), *S. texanum*, *S. officinale* (sabadilla); Melanthiaceae	America	all parts, especially seeds	cevadine, veratridine, sabadine and other steroid alkaloids	Ia CT, NT, MA, GI	The alkaloids activate Na$^+$ channels and are thus nerve and muscle poisons; furthermore burning in mouth, nausea, diarrhoea, excitation of CNS, tantrum, hypothermia, bradycardia, dyspnoea, convulsions, even death; strong skin irritant. **Therapy:** Aconitine
Scilla bifolia (squill), *S. rigidifolia, S. natalensis, S. siberica, S. peruviana, S. hispanica* and related species; Hyacinthaceae	Europe, Africa, Asia, America; orn. plant	all parts, especially bulbs and seeds	saponins, proscillaridin A and other bufadienolides	Ib–II CT, NT, HP, GI	Saponins are haemolytic and cytotoxic; substantial cardiac glycoside toxicity (if cardenolides are present); substantial GI tract disturbance. **Therapy:** Cardiac glycosides
Scopolia carniolica (scopolia) and related species; Solanaceae	E Europe	all parts	hyoscyamine, scopolamine and other tropane alkaloids (up to 0.8 %)	Ib NT, MA, GI, MP	mAChR antagonist with parasympatholytic properties; hallucinations in various forms; clonic spasms; strong heart beat; tachycardia (>160 beats); mydriasis; inhibition of salivation; death from respiratory arrest. **Therapy:** Hyoscyamine

Scientific name (common name); Family	Occurrence	Hazardous plant parts	Active principle	Toxicity Class	Symptoms & Therapy
Scrophularia aquatica (water figwort); Scrophulariaceae	Europe	all parts	unknown; cardiac or iridoid glycosides (?)	II AP, GI	Only animal poisoning has been reported; causes violent diarrhoea in livestock. **Therapy:** Anthraquinones
Securidaca longipedunculata (violet tree); Polygalaceae	tropical Africa	aerial parts, essential oil	saponins; securinine (indole alkaloid); methylsalicylate	II CT, GI, MP	Substantial ingestion can cause disturbance of GI tract and CNS; fatal if overdosed; widely used in African medicine; the bark has been used as an ingredient of arrow poison and as ordeal poison. **Therapy:** Saponins
Sedum acre (wall pepper); Crassulaceae	Europe, N Africa; orn. plant	all parts	sedamine, sedamine, sedinine and other piperidine alkaloids	II CT, IN, NT, GI	mAChR agonist with strong pungent taste, hence the name; severe irritation of mouth and throat; induces salivation, vomiting, intestinal spasms, paralysis and even respiratory arrest. **Therapy:** Pyrrolidine and piperidine alkaloids
Selenicereus (= *Cereus*) *grandiflorus* (night-blooming cactus); Cactaceae	Mexico; widely cultivated	all parts	betacyanins, flavonoids, steroidal glycosides	III IN, GI	High doses cause arrhythmia, burning sensation in mouth, nausea, vomiting, diarrhoea. **Therapy:** Saponins
Semecarpus anacardium (= *Anacardium orientale*) (kidney bean); Anacardiaceae	India; cultivated	fruits	cardol, catechols, anarcardic acid, anacardol	II CT, IN, GI	Skin and mucosal irritant, contact dermatitis, blister formation; gastroenteritis and paralysis of jaws, used as fish poison. **Therapy:** *Rhus*
Senecio jacobaea (tansy ragwort), *S. latifolius* (including *S. sceleratus*), *S. douglasii, S. vulgaris, S. vernalis, S. retrorsus, S. isatideus, S. burchellii* and other ragwort species; Asteraceae	Eurasia, America, Africa; common weed	all parts, especially flowers	senecionine and other pyrrolizidine alkaloids	II CT, MU, NT, MP	PAs are hepatotoxic (veno-occlusive disease), alkylate DNA and are therefore mutagenic and carcinogenic; inhibits peripheral nerves; important animal poison. **Therapy:** Senecionine and *Senecio*
Sesbania punicea, *S. grandiflora* (rattlebox), *S. drummondii, S. vesicaria*; Fabaceae	America, introduced into Africa; orn. plant	aerial parts, seeds	sesbanimide A and other piperidine alkaloids, saponins	II CT, NT, GI, AP	Mainly animal poisoning; nausea, vomiting, diarrhoea, rapid pulse, respiratory failure. **Therapy:** Pyrrolidine alkaloids
Sida acuta; Malvaceae	pan-tropical	aerial parts	ephedrine, cryptolenine	II–III NT, MA	Has been smoked as marijuana substitute; CNS stimulant. **Therapy:** *Cannabis sativa*

Scientific name (common name); Family	Occurrence	Hazardous plant parts	Active principle	Toxicity Class	Symptoms & Therapy
Simmondsia chinensis (jojoba); Simmondsiaceae	N America; widely cultivated	seeds	simmondsine, nitrile glycosides	II CT, NT, AP	Seeds are used for oil production; press cake causes animal poisoning; 750 mg/kg are lethal for rats. **Therapy:** Amygdalin
Sium latifolium (greater water parsnip), *S. suave* and related species; Apiaceae	Europe, N America	all parts, fruits	essential oil (7%) in fruits with limonene, pinene; polyacetylenes	II–III CT, IN, NT, GI	Polyacetylenes attack proteins; cause gastroenteritis with nausea, vomiting, diarrhoea, weakness, bradycardia and muscle paralysis. **Therapy:** Polyacetylenes
Skimmia japonica; Rutaceae	Asia; orn. plant	aerial parts	5-methoxypsoralen; dictamine, skimmianine (quinoline alkaloids)	II CT, MU, IN, GI	Skimmianine and other quinoline alkaloids are DNA intercalating compounds which exhibit mutagenic and cytotoxic properties; substantial ingestion causes inflammation of GI tract and CNS disturbance. **Therapy:** *Dictamnus*
Smodingium argutum (African poison ivy); Anacardiaceae	S Africa	aerial parts	3-(8,11 heptadecadienyl) catechol and other alkylphenols	Ib–II CT, IN, GI	Alkylphenols cause dermatitis through the ease with which they penetrate the upper skin layers; they are oxidised to highly reactive quinines which react with proteins in the skin and mucous membranes; cause severe allergic reactions, including rashes and blistering. **Therapy:** Hydroquinones and *Rhus*
Solandra maxima, *S. grandiflora* (chalice vine), *S. brevicalyx*, *S. guerrerensis*, *S. guttata* and related species; Solanaceae	tropical America and Mexico; orn. plant	aerial parts	hyoscyamine, atropine and other tropane alkaloids	Ib NT, MA, MP	mAChR antagonist; parasympatholytic properties; cause hallucinatious in various forms; tonic-clonic spasms; strong heartbeat; tachycardia (>160 beats); mydriasis; inhibition of salivation; respiratory arrest; coma. **Therapy:** Hyoscyamine
Solanum glaucophyllum, *S. malacoxylon*; Solanaceae	S America	aerial parts	1,25 dihydroxy-cholecalciferol	III CT, AP	Vitamin D₃ agonists; causes calcification of heart, liver and lung; animal poisoning in Argentina. **Therapy:** Solanine
Solanum tuberosum (potato), *S. dulcamara, S. carolinense, S. pyracanthum, S. rostratum, S. pseudocapsicum, S. incanum, S. sodomeum, S. nigrum*; Solanaceae	Americas, Africa, Europe; widely cultivated	green parts (red berries are usually not hazardous)	green fruits and leaves contain steroidal glyco-alkaloids, such as solanine, soladulcidine, solanine, solasodine etc.; saponins	Ib–II CT, NT, GI, AP, MP	Disturbance of GI tract, vomiting, spasms, internal bleeding, salivation, trembling, restlessness, headache, delirium, fever and coma. In severe cases death may occur through respiratory arrest. Livestock poisoning. **Therapy:** Solanine and *Solanum*

409

Scientific name (common name); Family	Occurrence	Hazardous plant parts	Active principle	Toxicity Class	Symptoms & Therapy
Sophora secundiflora (mescal bean), *S. tomentosa* and related species; Fabaceae	America; orn. plant	all parts, especially seeds	cytisine and other quinolizidine alkaloids; isoflavones	Ib NT, MA, GI	Alkaloids activate acetylcholine receptors; psychoactive, used by Indians in Mexico as hallucinogen; vomiting, diarrhoea, abdominal pain, diuretic, uterus contracting, hypotension; heart irregularity, toxicity similar to nicotine, death through respiratory failure. **Therapy:** Cytisine and *Sophora*
Sorbus aucuparia (mountain ash) and related species; Rosaceae	Europe; orn. tree	fruits	parasorbic acid (reactive lactone); formed from the glycoside parasorboside	III CT, IN	Although a frequent cause for consultations in poison centres, actual intoxication is quite rare; substantial ingestion of raw fruits can cause disturbance of GI tract. **Therapy:** Protoanemonin
Sorghum bicolor (grain sorghum, millet) and related species; Poaceae	Africa; widely cultivated	aerial parts	dhurrin (cyanogenic glucoside), nitrate	II CT, NT, GI; AP	Cyanogenic glucosides release HCN upon enzymatic hydrolysis, which is a strong inhibitor of cellular respiration; HCN poisoning; mainly animal poisoning. **Therapy:** Amygdalin
Spartium junceum (Spanish broom); Fabaceae	Europe; widely cultivated as orn. plant	all parts, especially flowers and seeds	N-methylcytisine, cytisine and other quinolizidine alkaloids	Ib NT, MA, GI	Affect nicotinic acetylcholine receptors (agonists); mildly psychoactive and hallucinogenic; toxicity similar to nicotine (see *Laburnum*). **Therapy:** Cytisine
Spathiphyllum floribundum (peace lily); Araceae	S America; orn. plant	all parts, fruits	toxic peptides; cyanogenic glucosides; saponins, Ca^{2+}-oxalate	Ib CT, IN, GI	Skin irritant; blister formation; burning sensation in mouth; disturbance of GI tract with internal bleeding; cardiac arrhythmia; CNS disturbance. **Therapy:** Oxalic acid
Spigelia anthelmia (= *Anthelmia quadriphylla*) (West Indian pinkroot), *S. marilandica*; Loganiaceae (Strychnaceae)	tropical America	all parts	spigeline, actinidine (alkaloids); benzoyl-choline, 3,3-dimethyl-acryloylcholine	Ib NT, GI	High doses cause nausea, vomiting, mydriasis, dyspnoea, followed by strychnine-like convulsions; anthelmintic; has been used for murder. **Therapy:** Indole alkaloids
Spirostachys africana (tamboti); Euphorbiaceae	Africa	all parts, latex	diterpenes, such as stachenone, stachenol	Ia CT, IN, GI	Phorbol esters activate protein kinase C and are co-carcinogens; strong skin and mucous membrane irritant; induce blisters on skin and mucosal tissues. **Therapy:** Phorbol esters

Scientific name (common name); Family	Occurrence	Hazardous plant parts	Active principle	Toxicity Class	Symptoms & Therapy
Sprekelia formosissima (= *Amaryllis formosissima*) (Aztec lily); Amaryllidaceae	C America; orn. plant	bulbs	haemanthamine and other Amaryllidaceae alkaloids	II CT, NT, GI	Ingestion causes GI tract and CNS disturbance; strong emetic. **Therapy:** Lycorine
Stephania japonica (= *S. bernandiifolia*); Menispermaceae	SE Asia	tubers	picrotoxin (sesquiterpene)	Ib NT, MA, MP	Stimulates motoric elements of CNS, occurrence of clonic and tonic spasms, headache; vomiting; sleepiness; coma. **Therapy:** Sesquiterpenes
Sternbergia lutea (fall daffodil); Amaryllidaceae	S Europe, Asia; orn. plant	bulbs	lycorine, haemanthamine, tazettine and other isoquinoline alkaloids	II CT, NT, GI	Alkaloids cause diarrhoea, vomiting, heavy perspiration, and even death. **Therapy:** Lycorine
Stictocardia tiliafolia; Convolvulaceae	Panama	seeds	ergot alkaloids (0.3 %) such as agroclavine, ergine, chanoclavine I; lysergic acid, ergometrine and others	Ib–II NT, MA, GI	Strongly psychedelic (4–8 seeds), similar to LSD (see *Ipomoea*). **Therapy:** Ergot alkaloids
Stillingia treculeana; Euphorbiaceae	tropical America	aerial parts	cyanogenic glucosides	Ib CT, NT, AP	Substantial ingestion causes HCN poisoning, sheep toxin. **Therapy:** Amygdalin
Strongylodon macrobotrys (jade vine); Fabaceae	Philippines	aerial parts	cytisine, hydroxynorcytisine	II NT, MA	Cytisine is a nAChR agonist; the plant is said to cause CNS disturbance with excitation, disorientation and delirium. **Therapy:** Cytisine
Strophanthus gratus, *S. kombe, S. hispidus, S. caudatus, S. sarmentosus, S. speciosus;* Apocynaceae	Africa, Asia; orn. plant	all parts, especially seeds	k-strophantoside, k-strophantin, cymarin, strophantidol, periplocymarin and other cardenolides	Ia CT, NT, GI, HP, MP	Inhibition of Na^+, K^+-ATPase; blocking neuronal and cellular transport activities; nausea, vomiting, bloody diarrhoea, convulsions, myosis, coma, cardiac arrest; death; used for murder and as arrow poison. **Therapy:** Cardiac glycosides and *Strophanthus*
Strychnos nux-vomica (nux vomica), *S. ignatii, S. spinosa, S. madagascariensis, S. benningsii, S. castelanei, S. crevauxii;* Loganiaceae (Strychnaceae)	Asia; cultivated in Africa and on Hawaii	aerial parts, especially fruits and seeds	strychnine, brucine, colubrine and other monoterpene indole alkaloids	Ia CT, NT, MA, MP	Strychnine inhibits glycine receptor (Cl–channel), an inhibitory neurotransmitter, CNS active; causes spasms and convulsions, salivation, death from respiratory arrest. **Therapy:** *Strychnos*

411

Scientific name (common name); Family	Occurrence	Hazardous plant parts	Active principle	Toxicity Class	Symptoms & Therapy
Strychnos toxifera (curare); Loganiaceae (Strychnaceae)	S America	all parts	toxiferine	Ia NT, MP	Inhibits nAChR; used in surgery to paralyse muscles; death from cardiac arrest; famous arrow poison. **Therapy:** *Chondrodendron*
Stypandra glauca (= *S. imbricata*) (blind grass); Hemerocallidaceae	Australia	aerial parts	stypandrol	II CT, NT, AP	Neurotoxic to nervus opticus and retina causing blindness in animals. **Therapy:** Sesquiterpenes
Styphnolobium japonicum (pagoda tree); Fabaceae	Japan; cultivated in Asia, Africa, Europe	aerial parts, fruits, seeds	sophoricoside and other isoflavonoids, rutin; lectins (in seeds), stizolamine (a pyrazine alkaloid; no QA)	III CT, AP	Mainly animal poisoning; phytoestrogenic effects; lectin intoxication. **Therapy:** Flavonoids
Suckleya suckleyana; Amaranthaceae (Chenopodiaceae)	N America	all parts	cyanogenic glucosides	Ib–II CT, NT, AP	Only animal intoxication; substantial ingestion causes HCN poisoning of livestock. **Therapy:** Amygdalin
Symphoricarpos albus (snowberry), *S. occidentalis*, *S. orbiculatus*; Caprifoliaceae	N America; common orn. plant	berries	saponins, traces of chelidonine	III GI	Nausea, vomiting, diarrhoea, abdominal pains, sleepiness; skin irritant. **Therapy:** Saponins
Symphytum officinale (comfrey), *S. x uplandicum*, *S. tuberosum* and related species; Boraginaceae	Europe, Asia; orn. plant	all parts, roots	symphytine, echimidine and other pyrrolizidine alkaloids; allantoin	III CT, NT, MU, AP, MP	PAs are hepatotoxic, mutagenic and carcinogenic; substantial ingestion inhibits peripheral nerves and causes GI tract disturbance; important animal poison. **Therapy:** Senecionine and *Symphytum*
Symplocarpus foetidus (skunk cabbage); Araceae	N America	leaves, rhizome	Ca^{2+}-oxalate raphides, toxic peptides	Ib CT, IN, GI	Painful and burning irritation and inflammation of mucous membranes; GI tract disturbance. **Therapy:** Oxalic acid
Synadenium cupulare (dead-man's tree), *S. grantii*; Euphorbiaceae	Africa	all parts, latex	12-O-tigloyl-4-deoxyphorbol-13-isobutyrate and several other tigliane-type diterpene esters of the 4-deoxyphorbol type	Ia CT, IN, GI, AP	Phorbol esters activate protein kinase C and are co-carcinogens; strong skin irritant; induce blisters and ulceration on skin and mucosal tissues; mainly animal poisoning. **Therapy:** Phorbol esters and *Synadenium*
Tabebuia impetiginosa (pau d'arco); Bignoniaceae	tropical America; orn. plant	bark	lapachol and other naphthoquinones	III CT, IN, GI	Naphthoquinones have irritating properties on mucosa and skin; substantial ingestion causes cytotoxic effects and GI tract disturbance. **Therapy:**

412

Scientific name (common name); Family	Occurrence	Hazardous plant parts	Active principle	Toxicity Class	Symptoms & Therapy
Tabernaemontana divaricata (crepe jasmine) and related species; Apocynaceae	Asia	bark of roots and stem	indole alkaloids, vobasine, ibogamine	Ib NT, MA, MP	Narcotic, psychedelic and hallucinogenic properties. **Therapy:** *Tabernanthe iboga*
Tabernanthe iboga (iboga), *T. dichotoma, T. sananbo*; Apocynaceae	tropical Africa, Asia, S America	roots, seeds	roots: ibogaine, voacangin, tabernanthine; seeds: catharanthine, voaphylline	Ib NT, MA, MP	Narcotic, psychedelic and hallucinogenic properties. **Therapy:** *Tabernanthe iboga*
Tagetes lucida, T. erecta (marigold) and related species; Asteraceae	America; common orn. plant	all parts, roots	thiophenes; essential oil with terpenoids (including substances similar to salvinorin A)	II–III CT, IN, MA	Thiophenes are reactive polyacetylenes and can cause allergic reactions; slightly psychoactive stimulant; plant has been used by Aztecs as fumigant or added to balche, a holy drink. **Therapy:** Polyacetylenes
Tamus communis (black bryony, murrain berries); Dioscoreaceae	Europe, W Asia, N Africa	all parts, especially roots, berries	Ca²⁺-oxalate raphides; steroid saponins	III CT, GI	Burning in mouth and throat, vomiting, diarrhoea, gastroenteritis, skin irritant; ingestion of high amounts can be fatal. **Therapy:** Oxalic acid
Tanacetum vulgare, see *Chrysanthemum vulgare*					
Tanaecium nocturnum; Bignoniaceae	tropical America	aerial parts, root bark	cyanogenic glucosides	II CT, MA, AP	Cyanogenic glucosides release HCN upon enzymatic hydrolysis; HCN is a strong inhibitor of cellular respiration; has been used as a snuff for ritual purposes and as hallucinogen by S American Indians; cattle poisoning. **Therapy:** Amygdalin
Taxus baccata (English yew), *T. brevifolia* (Californian yew), *T. canadensis, T. cuspidata, T. floridana*; Taxaceae	Europe, Asia, N America	all parts (except red aril of fruits)	taxin A, B, C; taxicin I, II (*T. baccata*), taxol (*T. brevifolia*); cyanogenic glucosides	Ia CT, NT, MP	Taxol stabilises microtubules and is used in cancer therapy; taxins inhibit K⁺ and Ca²⁺ channels. Symptoms include salivation, vomiting, painful diarrhoea, disturbance of circulation and heart activity; death through respiratory and circulatory arrest; has been used as arrow poison. **Therapy:** *Taxus*

Scientific name (common name); Family	Occurrence	Hazardous plant parts	Active principle	Toxicity Class	Symptoms & Therapy
Tephrosia vogelii, T. cinerea; Fabaceae	Africa, America	leaves	rotenone and other rotenoids (isoflavonoids)	Ib CT, NT, GI, AP	Blocks cellular respiration; GI tract disturbance; neurotoxin, inhibition of CNS and spinal cord: anaesthetic on tongue and throat; muscular arrest; used as insecticide and fish poison. **Therapy:** *Derris elliptica*
Terminalia sericea, T. superba, T. oblongata; Combretaceae	E Africa, Australia	roots, leaves, fruits	high levels of tannins (16 %); punicalagin	II–III CT, GI, AP	Substantial ingestion causes GI tract and kidney disturbances; mainly animal poisoning. **Therapy:** Flavonoids
Tetrapterys multiglandulosa; Malpighiaceae	Brazil	aerial parts	procyanidins	II–III CT, GI, AP	Substantial ingestion causes GI tract and kidney disturbances; mainly animal poisoning. **Therapy:** Flavonoids
Thalictrum minus (meadow-rue) and related species; Ranunculaceae	Europe; orn. plant	all parts	isoquinoline and protoberberine alkaloids, such as berberine; furthermore cyanogenic glucosides	II NT, GI, MP	Substantial ingestion causes disturbance of GI tract and CNS; hypotensive, cytotoxic. **Therapy:** *Chelidonium* and amygdalin
Thamnosma texana; Rutaceae	C America	aerial aprts	furoquinoline alkaloids	II CT, MU, NT, IN, GI	Furoquinoline alkaloids can intercalate or alkylate DNA; therefore mutagenic; plant sap is a strong skin irritant; when exposed to sunlight it can cause photodermatitis and contact dermatitis; animal poisoning. **Therapy:** Quinoline alkaloids and *Dictamnus*
Thapsia garganica (Spanish turpeth root); Apiaceae	N Africa, W Mediterranean	aerial parts	sesquiterpene lactones, thapsigargin	II CT, NT, GI	Ingestion can cause disturbance of GI tract and CNS; strong human and animal poison. **Therapy:** Sesquiterpenes
Theobroma cacao (cacao); Malvaceae (Sterculiaceae)	tropical America; cultivated in Africa, Asia	seeds	theobromine (1.45 %), caffeine (0.05 %), tannins	III MA, GI, MP	Purine alkaloids inhibit adenosine receptors; central stimulant; increases awareness and prevents sleepiness; high doses are intoxicating with tinnitus, headache, dizziness, increased heartbeat, insomnia, unrest, tachycardia; has been used as aphrodisiac. **Therapy:** Caffeine and *Coffea arabica*
Thesium lineatum (poison bush), *T. namaquense;* Santalaceae	S Africa	aerial parts	thesiuside, a cardiac glycoside	Ib CT, NT, HP, GI, AP	Ingestion causes cardiac glycoside intoxication; mainly animal poisoning. **Therapy:** Cardiac glycosides

Scientific name (common name); Family	Occurrence	Hazardous plant parts	Active principle	Toxicity Class	Symptoms & Therapy
Thevetia peruviana (yellow oleander); Apocynaceae	C & S America; common orn. plant	all parts, seeds	thevetin, neriifolin and other cardenolides	Ib CT, HP, GI, MP	Typical symptoms of cardiac glycoside poisoning; tongue and throat become numb, nausea, vomiting, spasms, pronounced arrhythmia, respiratory arrest, bradycardia, dilated pupils, dyspnoea, even death. **Therapy:** Cardiac glycosides
Thiloa glaucocarpa; Combretaceae	Brazil	leaves	high tannin content; ellagitannin, castalagin, stachyurin; saponins	II CT, GI, AP	Saponins can enhance tannin toxicity synergistically; mostly animal poisoning. **Therapy:** Saponins and flavonoids
Thuja occidentalis (arbor-vitae), *T. orientalis*, *T. koraiensis*, *T. standishii*; Cupressaceae	N America; China; common orn. plant	all parts	essential oil with thujone	Ib CT, NT, MA, MP, AP	Ingestion of plant material or its essential oil cause cytotoxic effects (abortifacient) and induce severe disturbance of GI tract, CNS and internal organs. **Therapy:** Monoterpenes and *Thuja*
Tribulus terrestris (devil's thorn), *T. zeyheri*, *T. cistoides*; Zygophyllaceae	Africa	all parts	diosgenin and other triterpenoid saponins; harman and norharman (β-carboline alkaloids)	II CT, NT, IN, AP	Saponins are cytotoxic; harman alkaloids can intercalate DNA and are especially active upon illumination; mainly animal poisoning; causing photosensitivity, with distinctive swelling of the head, lips, cheeks and ears. **Therapy:** Saponins and indole alkaloids
Trichocereus pachanoi (San Pedro cactus), *T. peruvianus*, *T. bridgesii*, *T. macrogonus*, *T. terscheckii*, *T. werdermannianus*; Cactaceae	S America	aerial parts	mescaline (up to 2%); hordenine, tyramine and other phenylethylamines	Ib NT, MA, MP	Ingestion causes psychedelic hallucinations (similar to LSD) with visions, creative dreams and loss of reality; aphrodisiac; high doses lead to GI disorder, hypotension, bradycardia, respiratory depression, vasodilatation, paralysis. **Therapy:** *Lophophora*
Tricholoma muscarium and related fungi; Tricholomataceae	Japan	fruiting body	ibotenic acid, muscimol, muscarine	II NT, MA	Muscimol is a strong parasympatholytic, psychoactive and hallucinogenic agent, causing inebriation, manic behaviour, delirium, deep sleep. **Therapy:** *Amanita muscaria*
Trichosanthes integrifolia; Cucurbitaceae	SE Asia	seeds	cucurbitacins	Ib CT, IN	Skin irritant; nausea, vomiting, excitation, dizziness, diarrhoea with blood, strong colic, spasms, kidney inflammation, tachycardia, respiratory arrest. **Therapy:** Cucurbitacin

Scientific name (common name); Family	Occurrence	Hazardous plant parts	Active principle	Toxicity Class	Symptoms & Therapy
Triglochin maritima (sea arrowgrass), *T. palustris*; Juncaginaceae	Europe, Asia, N America	all parts	taxiphyllin, triglochinin and other cyanogenic glucosides	II CT, GI, AP	HCN is released upon hydrolysis of cyanogenic glucosides; high doses lead to vomiting and diarrhoea; animal poisoning. **Therapy:** Amygdalin
Trillium erectum (birthroot); Melanthiaceae (Trilliaceae)	N America, Asia; orn. plant	all parts, especially roots	steroidal saponins (trillin, trillarin, diosgenin; aglycones chlorogenin, nologenin)	II CT, GI	Saponins are haemolytic and cytotoxic (abortifacient); substantial ingestion causes vomiting, diarrhoea and CNS disturbance. **Therapy:** Saponins
Trollius europaeus (globeflower) and related species; Ranunculaceae	Europe, Asia; orn. plant	all parts	ranunculin is enzymatically converted to the active protoanemonin	Ib–II CT, IN, MU, GI	Severe skin and mucosal irritation with blisters and ulceration; internally: diarrhoea, abdominal pains, tinnitus, headache, dizziness, seizures, tachycardia, nephritis, even death from respiratory and cardiac arrest. **Therapy:** Protoanemonin
Tulipa gesneriana (tulip) and hybrids; Liliaceae	S Europe, Asia; common orn. plant	all parts, especially bulbs	tulipalin, tuliposide A	III CT, IN, AP	Substantial ingestion causes gastrointestinal disturbance; plant sap induces serious skin allergies; animal poisoning. **Therapy:** Protoanemonin
Turbina corymbosa (Ololiuqui); Convolvulaceae	Mexico, C & S America	seeds, leaves, roots	ergot alkaloids: ergine, erginine, chanoclavine	Ib NT, MA, GI	Ergot alkaloids are hallucinogenic, similar to LSD but weaker; ingestion induces nausea, vomiting, weakness; high doses cause death. Mexican Indians used ololiuqui (origin unresolved) for ritual and oracle purposes. **Therapy:** Ergot alkaloids and *Turbina*
Turnera diffusa (damiana); Passifloraceae/Turneraceae	C & S America	aerial parts	essential oil with 1,8-cineole, pinene, sesquiterpenes, arbutin, caffeine	III CT, NT, MA	Supposed to be a stimulant similar to *Cannabis* with aphrodisiac properties. **Therapy:** Monoterpenes
Tylecodon wallichii, *T. grandiflorus*; Crassulaceae	S Africa	aerial parts	cotyledoside and other bufadienolides	Ib CT, GI, HP, AP	Ingestion causes typical cardiac glycoside poisoning; human and animal intoxication. **Therapy:** Cardiac glycosides
Typhonium trilobatum; Araceae	SE Asia	aerial parts	Ca^{2+}-oxalate raphides	II CT, IN	Severe swelling of throat and mouth, possible asphyxiation. **Therapy:** Oxalic acid
Ulex europaeus (gorse) and related species; Fabaceae	Europe; introduced worldwide	all parts, seeds	cytisine and other quinolizidine alkaloids	II NT, MA	Affects nicotinic acetylcholine receptors (agonists); inducing uterus contractions, vomiting, diarrhoea, abdominal pain, tachycardia, cardiac arrhythmia. **Therapy:** Cytisine

Scientific name (common name); Family	Occurrence	Hazardous plant parts	Active principle	Toxicity Class	Symptoms & Therapy
Urginea maritima (= *Drimia maritima*) (red squill), *U. sanguinea*, *U. altissima*, *U. physodes*; Hyacinthaceae	S Europe, S Africa	all parts, bulbs	scillaren A and other bufadienolides; urginin (*U. altissima*) and physodine A (*U. physodes*)	Ib CT, HP, GI, MP	Inhibition of Na+, K+-ATPase causing heart glycoside poisoning; nausea, vomiting, diarrhoea, colic, haematuria, death by cardiac arrest; also skin irritant with blister formation; human and animal poisoning. **Therapy:** Cardiac glycosides
Urtica dioica (nettle) and related species; Urticaceae	Europe; world-wide introduced weed	aerial parts	stinging hairs with acetylcholine, histamine, serotonin	III CT, IN, NT	Painful itching with welt formation occurs at sites that were pricked by stinging hairs. **Therapy:** *Urtica*
Vaccinium uliginosum (northern bilberry); Ericaceae	circumpolar Europe, Asia, N America	fruits	tannins, flavonoids, arbutin (without psychoactivity)	III GI, MA	Said to have hallucinogenic properties; might be infested by fungi that produce hallucinogens; has been used to make wine; also an additive to other drinks (including *Amanita muscaria*). **Therapy:** Ethanol
Valeriana officinalis (valerian) and related species; Valerianaceae	Europe, Asia, America	roots	didrovaltrate and other valepotriates (iridoids)	III CT, MU, MA, MP	Known medicinal plant with sedating properties; high doses cause headache, unrest, insomnia, arrhythmia, even central paralysis, cardiac arrest; valepotriates are reactive substances that can alkylate DNA (therefore mutagenic). **Therapy:** Monoterpenes
Veratrum album (white hellebore); *V. viride*, *V. nigrum*, *V. californicum*, *V. parvifolium*, *V. tenuipetalum* and related species; Melanthiaceae	Europe, Asia, N America	all parts	protoveratrine A & B, germerine, cyclopamine	Ia NT, IN, AP, MP, MA	Alkaloids activate Na+-channels; cyclopamine causes malformation (cyclopian eye); hallucinogenic, heart and neurotoxin; death by respiratory and cardiac arrest, skin irritant; animal poisoning. **Therapy:** Aconitine
Viburnum opulus (guelder rose), *V. lantana*; Adoxaceae (Caprifoliaceae)	Europe, Asia; common orn. plant	bark, fruits	viburnin, amyrin, oxalates	III CT, GI	Substantial ingestion of plant material can cause disturbance of GI tract and CNS; with vomiting, diarrhoea, nausea, excitation, convulsions, dyspnoea. **Therapy:** Saponins and oxalic acid
Vicia cracca (tufted vetch), *V. ervilia*, *V. villosa* and related species; Fabaceae	Europe, Asia, Africa, America	leaves, seeds	glycosides of Pyrimidines; NPAAs	II–III CT, IN, GI, AP	Mostly animal poisoning; the glycosides appear to cause light sensitivity disease "hairy vetch poisoning" with hair loss, itching, conjunctivitis, diarrhoea, nephritis. **Therapy:** Non-protein amino acids

Scientific name (common name); Family	Occurrence	Hazardous plant parts	Active principle	Toxicity Class	Symptoms & Therapy
Vicia faba (broad bean); Fabaceae	cultivated worldwide	seeds	vicioside, convicin (glycosides of pyrimidines), lectins	II–III CT, GI, AP	In persons with GPDG deficiency, symptoms of favism can be caused, even death. More common in S Europe: symptoms include headache, dizziness, vomiting, fever, anaemia, favism; potential livestock poison. **Therapy:** Non-protein amino acids
Vinca minor (periwinkle), *V. major*; Apocynaceae	Europe; orn. plant	all parts	vincamine, eburnamenine and other monoterpene indole alkaloids	II CT, NT, GI, MP	Substantial ingestion can causes GI tract, heart and CNS disturbances; has been used as medicinal plant to enhance glucose supply to CNS. **Therapy:** Indole alkaloids
Vincetoxicum hirundinaria and related species; Apocynaceae (Asclepiadoideae)	Europe, Asia,	all parts, especially rhizome	vincetoxin (a mixture of steroid glycosides), tylophorine	II CT, NT, GI	Substantial ingestion cause GI tract and CNS disturbance; salivation, vomiting, diarrhoea, convulsions, paralysis of muscles, respiratory arrest. **Therapy:** Saponins
Viola odorata (violet) and related species; Violaceae	Europe; common orn. plant	rhizomes, seeds	flavonoids, triterpene saponins	II–III CT, GI, MP	Substantial ingestion causes GI tract and CNS disturbance ; severe gastroenteritis, restlessness, respiratory and circulatory depression. **Therapy:** Saponins
Virola calophylla, V. elongata (= *V. tbeiodora*), *V. calophylloidea, V. cuspidata* and related species; Myristicaceae	S & C. America	resin, inner bark	*N,N*-DMT, 5-MeO-DMT and other trypta-mines; 6-methoxy-harmaline and other β-carboline alkaloids	III NT, MA	Tryptamines are 5-HT-R agonists; β-carboline alkaloids are MAO inhibitors; therefore psychoactive; powder has been used by Indians as a hallucinogenic snuff. **Therapy:** Bufotenine
Viscum album (mistletoe) and related species; Santalaceae (Viscaceae)	Europe, Asia, intro-duced in N Ame-rica	all parts	viscotoxins, lectins, SM depend on host plant	II CT, GI, MP	Viscotoxins and lectins are responsible for cytotoxic and hypotensive effects; substantial GI tract disturbance: nausea, vomiting, diarrhoea, abdominal pains, convulsions. **Therapy:** Abrin and *Viscum*
Vitis vinifera (grapevine); Vitaceae	Mediter-ranean; cultivated world-wide	fruits	mature grapes have more than 20% sugar, which is fermented to ethanol	III MA	Ethanol is an agonist at GABA-receptors; induces the well-known symptoms of alcohol intoxication. **Therapy:** Ethanol
Wisteria sinensis (Chinese wisteria), *W. floribunda* (Japanese wisteria); Fabaceae	Asia, America; orn. plant	all parts, especially bark, fruits and seeds	wistarine; haemagglutinating lectins	II CT, GI	Nausea, vomiting, gastroenteritis, diarrhoea, abdominal pains, mydriasis, circulatory disturbance; dangerous for children when more than 2–4 seeds are ingested. **Therapy:** Non-protein amino acids

Scientific name (common name); Family	Occurrence	Hazardous plant parts	Active principle	Toxicity Class	Symptoms & Therapy
Withania somnifera (winter cherry, ashwagandha); Solanaceae	Africa, Asia; cultivated as orn. plant	roots, aerial parts	withaferin A, withanolides (steroid lactones)	III NT, MA, MP	Plants have been used medicinally as sedative, sleep inducing; antistress tonic. **Therapy:** Saponins
Xanthium strumarium (cocklebur), X. spinosum, X. sibiricum; Asteraceae	S America; introduced to Asia, Europe, Africa	seeds, seedlings, leaves	xanthinine, xanthinine; carboxyatractyloside (sesquiterpene lactones)	II CT, IN, AP	Mainly animal poison (anorexia, haemorrhagic, dizziness, weakness, dyspnoea; cardiac arrest); also human fatalities. **Therapy:** Xanthium
Xanthosoma violacea (malanga), X. sagittifolium; Araceae	tropical America; cultivated in America	leaves	Ca²⁺-oxalates raphides	Ib–II CT, IN, GI	Painful burning and swelling of throat and mouth; GI tract disturbance. **Therapy:** Oxalic acid
Xysmalobium undulatum (uzara); Apocynaceae (Asclepiadoideae)	S Africa	roots	uzarin, asclepioside, and other cardenolides	Ib CT, NT, HP, MP	Inhibits gut motility; high doses inhibit smooth muscles generally; i.v. application results in typical cardiac glycoside intoxication, even death. **Therapy:** Cardiac glycosides
Zamia integrifolia (coontie), Z. furfuracea, Z. pumila; Zamiaceae	C America	all parts	cycasin (azoglycoside), releases methylazoxymethanol	II CT, MU, GI	Nausea, vomiting, abdominal spasms, visual distortions, lethargy and coma; mutagen and carcinogen; cattle poisoning. **Therapy:** Encephalartos
Zantedeschia aethiopica (calla lily); Araceae	S Africa; common orn. plant	all parts	probably Ca²⁺-oxalates	III CT, IN, GI	Burning sensation in mouth and throat; nausea, vomiting. **Therapy:** Oxalic acid
Zephyranthes atamasco (zephyr lily, atamasco), Z. carinata (rain lily); Amaryllidaceae	N America	bulbs	lycorine and related alkaloids	Ib–II CT, NT, AP	Substantial ingestion causes nausea, vomiting, abdominal spasms, diarrhoea. Poisonous for cattle and horses. **Therapy:** Lycorine
Zigadenus brevibracteatus (death camas), Z. fremontii, Z. muscaetoxicum, Z. venenosus, Z. paniculatus, Z. nuttallii and other species; Melanthiaceae	N America	bulbs	zygadenine, zygacine; protoveratrine and related steroidal alkaloids	Ia CT, NT, MA	The alkaloids activate Na⁺ channels and thus block neuronal activity; neurological and cardiac symptoms predominate besides GI tract disturbance; salivation, extreme thirst, strong vomiting, diarrhoea, headache, dizziness, muscular weakness, cardiac irregularities, convulsions, coma, death. **Therapy:** Aconitine

419

Glossary of chemical, medical, pharmacological and toxicological terms

ABC transporter – an important ATP-driven transporter at biomembranes (such as p-glycoprotein), which pumps out lipophilic xenobiotics

Abiotic – not associated with living organisms

Abortifacient – a substance that causes abortion

Absolute lethal concentration (LC$_{100}$) – lowest concentration of a substance, which kills 100% of test organisms or species under defined conditions. This value is dependent on the number of organisms used in its assessment

Absorbed dose – amount of a substance absorbed into an organism or into organs and tissues of interest

Absorption – active or passive uptake of a substance through the skin or a mucous membrane; substances enter the bloodstream and are transported to other organs. Chemicals can also be absorbed by breathing or swallowing

Accepted risk – probability of suffering disease or injury which is accepted by an individual

Accidental exposure – unintended contact with a substance resulting from an accident

Accumulation – successive additions of a substance to a target organism, or organ, which leads to an increasing amount of the chemical in the organism or organ

Acetylcholine – a neurotransmitter that binds to nicotinic (nACh-R) or muscarinic (mACh-R) receptors

Acetylcholine esterase inhibitor – a substance that inhibits cholinesterase (AChE) and thus the breakdown of acetylcholine to acetate and choline

Acidosis – pathological condition in which the pH of body fluids is below normal and therefore the pH of blood falls below the reference value of 7.4

Acute – a symptom or condition that appears suddenly (and lasts for a short period; minutes or hours)

Acute toxicity test – experimental animal study to determine which adverse effects occur in a short time (usually up to 14 d) after a single dose of a substance or after multiple doses given in up to 24 h

Adaptogen – a substance with a non-specific action that causes improved resistance to physical and mental stress

Addiction – physical or psychological dependence on a substance for the sake of relief, comfort, stimulation, or exhilaration, which it affords; often with craving when the drug is absent

Additive – a substance that is added to a mixture (typically for taste, colour or texture)

Additivity – the effect of a combination of two or more individual substances is equivalent to the sum of the expected individual responses

Adduct – new chemical species AB, formed by direct combination of two separate molecular entities A and B

Administered dose – a defined quantity given orally or parenterally

Adenoma – an abnormal benign growth of glandular epithelial tissue

Adenocarcinoma – malignant tumour originating in glandular epithelium or forming recognisable glandular structures

Adenylylcyclase – enzyme of signal transduction; catalyses the formation of the second messenger cAMP from ATP

Administration – application of a known amount of a chemical to an organism in a reproducible manner and by a defined route

Adrenaline – the hormone that binds to adrenergic receptors; causes the "fight or flight" response

Adrenergic (sympathomimetic) – a substance that binds to adrenergic neuroreceptors and produces an effect similar to the normal impulses (caused by adrenaline, noradrenaline) of the sympathetic nervous system; antagonists are sympatholytics

Adsorption – enrichment of substances on surfaces

Adulterant – an undesirable ingredient found in a commercial product

Adulteration – non-allowed substitution of a substance or materials in a drug or food by another, usually being inactive or toxic

Aetiology – science dealing with the cause or origin of disease

Aglycone – the non-sugar part of a glycoside (after removal of the sugar part)

Agonist – substance that binds and activates a cellular receptor

AIDS – acquired immunodeficiency syndrome, a condition (weakened immune system) caused by HIV (a retrovirus)

Albuminuria – presence of albumin from blood plasma in the urine

Alcoholic extract – soluble fraction of plant material obtained after extraction with ethanol

Alkaloid – a chemical substance containing nitrogen as part of a heterocyclic ring structure; often highly toxic or mind-altering

Alkylating agent – reactive secondary metabolite, which introduces an alkyl substituent into DNA, proteins or other molecules

Allergen – an antigenic substance that triggers an allergic reaction (hypersensitivity)

Allergy – a hypersensitivity to allergens (often pollen or reactive secondary metabolite) that causes rhinitis, urticaria, asthma and contact dermatitis

Alopecia – loss of hair

Alzheimer's disease – see dementia

Ames test – in vitro test for mutagenicity using mutant strains of the bacterium *Salmonella typhimurium*. The test can be carried out in the presence of a given microsomal fraction (S-9) from rat liver to allow metabolic transformation of mutagen precursors to active derivatives

Amino acid – chemical substances that form the building blocks of proteins ("proteinogenic amino acids")

Amoebiasis – a (sub-)tropical protozoan infection with *Entamoeba histolytica*

Amoebicidal – a substance that kills amoebae

Anaemia – reduced number of red blood cells in the blood, often causing pallor and fatigue

Anaesthetic – a substance that causes localised or general loss of feeling or sensation; general anaesthetics produce loss of consciousness; local anaesthetics render a specific area insensible to pain

Analeptic – a substance that stimulates the central nervous system

Analgesic – a substance that relieves pain without loss of conscience

Anaphrodisiac – a substance that reduces sexual desire

Anaphylactic shock – a severe, life-threatening form of a general allergic reaction to an antigen or hapten to which a person has previously been sensitised

Anemia (see anaemia)

Aneuploid – missing or extra chromosomes or parts of chromosomes

Angina pectoris – severe pain in the chest

Anoxia – total absence of O_2; refers sometimes to a decreased oxygen supply in tissues

Antagonist – inhibitor at cellular receptors; blocks the activity of an endogenous ligand; reverses or reduces the effect modulated by an agonist

Antibody – specific protein produced by the immune system (an immunoglobulin molecule), which can bind specifically to an antigen or hapten which induced its synthesis

Anthelmintic – a substance that kills or expels intestinal worms

Anthraquinones – secondary metabolites with an anthracene skeleton; anthrones show strong laxative effects

Anti-arrhythmic – a substance that counteracts irregular heartbeat

Anti-asthmatic – a substance that alleviates the spasms of asthma

Antibacterial – a substance that kills or inhibits the growth of bacteria

Antibiotic – a substance that kills or inhibits the growth of microorganisms

Anticholinergic – a substance that blocks the parasympathetic nerve impulse

Anticoagulant – a substance that prevents blood from clotting

Anticonvulsant – a substance that prevents or relieves convulsions

Antidepressant – a substance that alleviates depression

Antidiabetic – a substance that prevents or alleviates diabetes

Antidiuretic – a substance that prevents or slows urine formation

Antidote – a substance that counteracts the effect of a potentially toxic substance

Anti-emetic – a substance that prevents vomiting

Antifungal – a substance that kills or inhibits the growth of fungi

Antigen – substance which induces the immune system to produce a specific antibody or specific cells

Antihistamine – a substance that improves allergic symptoms by blocking the action of histamine

Antihydrotic – a substance that reduces perspiration

Antihypertensive – a substance that reduces high blood pressure

Anti-inflammatory – a substance that causes

symptomatic relief of inflammation

Antimetabolite – a substance structurally similar to a metabolite, which competes with it or replaces it, and thus prevents or reduces its normal function

Antimicrobial – a substance that kills or inhibits the growth of microorganisms

Antimitotic – a substance that prevents or inhibits cell division (mitosis)

Antimycotic – see antifungal

Anti-oedemic – a substance that prevents swelling

Antioxidant – a substance that is able to protect cells or counteract the damage caused by oxidation and free oxygen radicals (reactive oxygen species, ROS)

Antiparasitic – a substance that kills parasites

Antiphlogistic – a substance that prevents inflammation

Antipruritic – a substance that alleviates or prevents itching

Antipyretic – a substance that alleviates fever

Antirheumatic – a substance that relieves the symptoms of rheumatism

Antiseptic – a substance that stops or inhibits infection

Antispasmodic – a substance that reduces muscular spasms and tension

Antitumour – a substance that counteracts tumour formation or tumour growth

Antitussive – a substance that reduces the urge to cough

Anuria – the inability to urinate

Anxiety – symptoms of fear not caused by any danger or threat

Aperitif – a drink that stimulates the appetite

Aphasia – loss or impairment of the power of speech or writing, or of the ability to understand written or spoken language or signs, due to a brain injury, disease or drugs

Aphrodisiac – a substance that increases sexual desire

Apnoea – cessation of breathing

Apoptosis – programmed cell-death leading to a progressive fragmentation of DNA and disintegration of cells without causing inflammation

Aqueous extract – soluble fraction of plant material obtained after extraction with water

Aromatherapy – the medicinal use of aroma substances by inhalation, bath, massage, etc.

Arrhythmia – any deviation from the normal rhythm of the heartbeat

Arteriosclerosis – accumulation of fatty deposits in the blood vessels causing them to narrow and harden, resulting in heart disease or stroke

Arthritis – inflammation of joints

Asphyxiant – substance that blocks the transport or use of oxygen by living organisms

Asthenia – diminishing strength and energy, weakness

Asthma – chronic respiratory disease characterised by bronchoconstriction, excessive mucus secretion and oedema of the pulmonary alveoli, resulting in difficulty in breathing out, wheezing, and cough

Astringent – a substance (often tannins) that reacts with proteins in wounds, on the surface of cells or membranes, resulting in a protective layer and causing contraction

Ataxia – loss of muscle coordination leading to unsteady or irregular walking or movement

Atherosclerosis – pathological condition; changes of arterial walls that lead to arteriosclerosis

Atrophy – wasting away of the body or of an organ or tissue

Autonomic nervous system – that part of the nervous system that regulates the heart muscle, smooth muscles and glands; it comprises the sympathetic nervous system and the parasympathetic nervous system

Autopsy – post-mortem examination of the organs and body tissue to determine cause of death

Ayurvedic medicine – traditional Indian medicine

Bacteriostatic – a substance that prevents the multiplication of bacteria

Bactericide – substance that kills bacteria

Bacterium – a microorganism consisting of a single cell surrounded by a cell wall; DNA is circular; bacteria do not have internal membrane systems or a nucleus

Base pairing – complexation of the complementary pair of polynucleotide chains of nucleic acids by means of hydrogen bonds between complementary purine and pyrimidine bases, adenine (A) with thymine (T) or uracil, cytosine (C) with guanine (G)

Benign – not cancerous, not malignant since tumour does not form metastases and has still positional control. Or disease without persisting harmful effects

Benzodiazepine receptor – binding site for benzodiazepines at the GABA receptor; target for several sedatives and tranquillisers

Bidesmosidic saponins – saponins with 2 sugar chains

Bile – a bitter fluid excreted by the liver via

...he gall bladder that helps to digest fats

Biliary dyskinesia – inability to secrete bile

Bilirubin – orange-yellow pigment ($C_{33}H_{36}O_6N_4$), a breakdown product of haem-containing proteins (haemoglobin, myoglobin, cytochromes), it is excreted in the bile by the liver

Bioaccumulation – when harmful substances enter the ecosystem, some of them move through the food chain, by one organism eating another. In the end, substances accumulate, sometimes in high concentration in the top consumer or predator (often humans)

Bioactivation – any metabolic conversion of a xenobiotic to a more toxic derivative (e.g. pyrrolizidine alkaloids)

Bioassay – an experiment in which test organisms are exposed to varying concentrations of a substance. The response of the test organism is determined as a function of experimental conditions

Bioavailability – amount of a substance that is available for pharmacological or toxicological response after absorption

Biomembrane – permeation barrier around every cell or cellular compartments consisting of phospholipids, cholesterol and membrane proteins. The integrity of its biomembrane is a requisite for any cell

Biopsy – excision of a small piece of living tissue for microscopic or biochemical examination and diagnosis

Bitter tonic – a substance that promotes appetite and digestion by stimulating the secretion of digestive juices

Blood-brain barrier – blood vessels of the brain are covered with especially tight endothelial tissues, so that only few substances can enter the brain

Bradycardia – pulse under 60 beats per minute

Bronchitis – inflammation of the mucous membranes of the bronchial tubes

Bronchodilatory – a substance that expands air passage through the bronchi and reduces bronchial spasm

Bruise – a non-bleeding injury to the skin

Cachexia – weight loss due to chronic illness or prolonged emotional stress

Cancer – various types of malignant cells that multiply out of control

Carcinogen – a substance that may cause cancer

Carcinogenicity – a complex multistage process leading to abnormal cell growth and cell differentiation. During initiation cells undergo mutations, during promotion mutated cells are stimulated (e.g. by co-carcinogens) to progress to cancer

Carcinoma – malignant growth of epithelial cells

Cardiac glycoside – a steroidal glycoside that increases the strength or rhythm of the heartbeat

Cardiotonic – a substance that has a strengthening or regulating effect on the heart

Cardiotoxic – harmful to heart cells

Carminative – a substance that reduces flatulence

Catabolism – process of breakdown of complex molecules into simpler ones, often providing biologically available energy in form of ATP

Catalyst – a compound that speeds up the rate of a reaction

Catarrh – inflammation of mucous membranes

Catechol-*O*-methyltransferase (COMT) – enzyme which inactivates neurotransmitters with a phenolic OH group (dopamine, noradrenaline, serotonin) through methylation

Cathartic – laxative, purgative

Chemotherapy – treatment of cancer with chemical substances

Chiral – if the mirror image of a substance is not superimposable, it is called chiral

Cholagogue – a substance that stimulates the flow of bile from the gall bladder; distinction between cholekinetics and choleretics

Cholekinetic – a substance that stimulates the release of bile by contraction of the gall bladder and bile ducts

Choleretic – a substance that stimulates the liver to produce bile

Cholesterol – the most common steroid (fat-like material) found in the human body; important for membrane fluidity and as a precursor for steroid hormones; high cholesterol levels are associated with an increased risk of coronary diseases

Cholinesterase inhibitor – a substance that inhibits the action of cholinesterase (AChE); AChE catalyses the hydrolysis of choline esters: a cholinesterase inhibitor causes hyperactivity in parasympathetic nerves

Chromosomal aberration – abnormal chromosome number or structure

Chronic – occurring over a long period of time (>1 year)

Chronic exposure – continued exposure over an extended period of time

Chronic ailment – a condition that extends over a long period

Chronotoxicology – science of the influence of biological rhythms on the toxicity of substances

Cirrhosis – liver disease defined by increased fibrous tissue, with loss of functional liver cells, and increased resistance to blood flow through the liver portal

CNS – central nervous system

Co-carcinogen – a substance that amplifies the effect of a carcinogen

Colic – abdominal pains, caused by muscle contraction of an abdominal organ, accompanied by nausea, vomiting and perspiration

Concentration – the amount of a given substance in a given volume of air or liquid

Concentration–response curve – graph of the relation between exposure concentration and the magnitude of the resultant biological change

Conjugate – derivative of a substance formed by its combination with chemicals such as acetic acid, glucuronic acid, glutathione, glycine, sulphuric acid, etc.

Conjunctiva – the mucous membranes of the eyes and eyelids

Conjunctivitis – inflammation of conjunctiva

Contact dermatitis – inflammatory condition of the skin resulting from dermal exposure to an allergen or an irritating chemical

Constipation – lack of bowel movement leading to prolonged passage times of faeces

Contaminant – any kind of adverse substance that contaminates water, air or food

Contraindication – condition that makes some particular treatment improper or undesirable

Cortex – dried bark

Covalent bond – a bond created between 2 atoms when they share electrons

Crohn's disease – chronic inflammation of the intestinal tract

Cumulative effect – mutually enhancing effects of repeated doses of a harmful substance

Cutaneous – relating to the skin

Cyanogenic glucoside (CG) – secondary metabolite that is activated upon wounding, releasing the toxin HCN

Cyanosis – bluish coloration, especially of the skin and mucous membranes and fingernail beds; occurs when oxygenation is deficient and reduced haemoglobin is abundant in the blood vessels

Cytochrome P-450 – important haemoprotein which has the task to hydroxylate many endogenous and exogenous substrates (which are later conjugated and excreted). The term includes a large number of isoenzymes which are coded for by a superfamily of genes

Cyclooxygenase – key enzyme of prostaglandin biosynthesis

Cystitis – inflammation of the bladder

Cytoplasm – basic compartment of the cell (surrounded by the plasma membrane) in which nucleus, endoplasmic reticulum, mitochondria and other organelles are imbedded

Cytotoxic – a substance that is toxic to cells i.e. damages cell structure or function

Decoction – watery extract obtained by boiling

Decongestant – a substance that removes mucus from the respiratory system and opens the air passages so that breathing becomes easier

Dementia – loss of individually acquired mental skills; Alzheimer's disease is a severe form of dementia

Dependence – psychic craving for a drug or other substance which may or may not be accompanied by a physical dependency

Depression – psychic disturbance, often associated with low concentrations of dopamine and noradrenaline

Dermal – referring to the skin

Dermal absorption – absorption through the skin

Dermatitis – inflammation of skin (e.g. by contact dermatitis)

Detergent – a substance capable of dissolving lipids

Detoxification – biochemical modification which make a toxic molecule less toxic

Developmental toxicity – adverse effects on the embryo or growing foetus

Diabetes mellitus – abnormally high blood sugar levels caused by lack of insulin

Diaphoretic – a substance that increases sweating (profuse perspiration)

Diarrhoea – abnormally frequent discharge of watery stool (more than 3 times per day)

Diffusion – the process by which molecules migrate through a medium and spread out evenly

Disulphide bridge – a bond between two SH-groups, e.g. in a protein

Diuresis – discharge of urine

Diuretic – a substance that increases the volume of urine

DNA – desoxyribonucleic acid, the biomolecule in cells that stores the genetic information; composed of 2 complementary nucleic acid strands bonded by G-C and A-T pairs

Dose – the amount of a substance to which a person or test organism is exposed. The effective dose depends on body weight

Dosage – dose expressed as a function of the organism being dosed and time, for example mg/kg body weight/day

Dose–response curve – graph of the relationship between dose and the degree of changes produced

Drug – term for a therapeutic agent, but also commonly employed for abused substances

Drug-resistance – having a (often acquired) resistance against a drug, by developing modified targets, increasing the degradation of an active compound or by exporting it out of a cell

Dysentery – inflammation of the colon; often caused by bacteria (shigellosis) or viruses, accompanied by pain and severe diarrhoea

Dysmenorrhoea – abnormal or painful menstruation

Dyspepsia – indigestion

Dysplasia – abnormal development of an organ or tissue

Dyspnoea – difficult breathing

Dysuria – painful urination

Eczema – acute or chronic inflammation of the skin with redness, itching, papules, vesicles, pustules, scales, crusts or scabs

Effective concentration (EC) – EC_{50} is the median concentration that causes 50% of maximal response

Effective dose (ED) – ED_{50} is the median dose that causes 50% of maximal response

Electronegative – atoms that draw electrons of a bond toward itself (e.g. oxygen)

Embryotoxicity – any toxic effect on the conceptus as a result of prenatal exposure during the embryonic stages of development: including malformations, malfunctions, altered growth, prenatal death and altered postnatal function

Emesis – vomiting

Emetic – a substance causing vomiting

Emollient – a substance that soothes and softens the skin

Endocrine – pertaining to hormones or to the glands that secrete hormones directly into the bloodstream

Endoplasmic reticulum – Endomembrane system in which proteins are modified post-translationally

Endorphins – peptides made by the body with similar actions as morphine

Endothelia – layer of cells lining the inner surface of blood and lymphatic vessels

Enteritis – inflammation of the intestines

Enzyme – protein that catalyses a chemical reaction, e.g. the hydrolysis of acetylcholine

Epidemiology – science that studies the occurrence and causes of health conditions in human populations; scientists try to find out whether a factor (e.g., nutrient, contaminant) is associated with a given health effect

Epigastric – referring to the upper-middle region of the abdomen

Epilepsy – chronic brain condition characterised by seizures and loss of consciousness

Epileptiform – occurring in severe or sudden spasms, as in convulsion or epilepsy

Epithelia – cell layer covering the internal and external surfaces of the body

Epitope – any part of a molecule that carries an antigenic determinant

Ergot – a fungus (*Claviceps purpurea*) that infects grasses (especially rye) and produces pharmacologically active alkaloids

Ergotism – poisoning by eating ergot-infected grain

Erythema – redness of the skin produced by congestion of the capillaries

Essential oil (= volatile oil) – mixture of volatile terpenoids and phenylpropanoids responsible for the taste and smell of many plants, especially spices

Estrogen (oestrogen) – a female sex hormone

Ethnobotany – study how different human cultures use plants for medicinal purposes

Excretion – elimination of chemicals or drugs from the body, mainly through the kidney and the gut. Volatile compounds may be eliminated by exhalation. In the GI tract elimination may take place via the bile, the shedding of intestinal cells and transport through the intestinal mucosa

Expectorant – a substance that increases mucous secretion or its expulsion from the lungs; distinction between secretolytics and secretomotorics

Exposure – contact with a substance by swallowing, breathing, or directly through skin or eye. We distinguish between short-term and long-term exposure.

Extract – a concentrated preparation (semi-liquid, solid or dry powder) of the soluble fraction of plant material

Familial Mediterranean fever – a condition with recurrent attacks of fever and pain

Febrifuge – a substance that reduces fever

Febrile – relating to fever

First-pass effect – biotransformation of a chemical in the liver (after absorption from the intestine and before it reaches the systemic circulation)

Flatulence – accumulation of excessive gas

in the intestines

Fluid extract – an alcohol-water extract concentrated to the point where, e.g. 1 ml equals 1 g of the original herb

Folium – dried leaves

Food allergy – hypersensitivity reaction to chemicals in the diet to which a person has previously been exposed and sensitised

Forced diuresis – clinical method of stimulating diuresis, with the aim of achieving increased clearance of a toxic substance in urine

Frame-shift mutation – point mutation deleting or inserting of one or two nucleotides in a gene: this shifts the normal reading frame and causes the formation of functionless proteins

Free radical – an unstable form of oxygen molecule that can damage cells

Gallstone – a solid or semi-solid body in the gall bladder or bile duct

Gargle – a fluid used as throat wash

Gastritis – inflammation of the stomach

Gastroenteritis – inflammation of the gastrointestinal tract (stomach and intestine), associated with nausea, pain and vomiting

Genetic toxicity – damage to DNA by a mutagen, causing mutations and altered genetic expression (mutagenicity). Non-repaired mutations in somatic cells are inherited to daughter cells whereas mutations in germ cells can reach the next generation

Gingivitis – inflammation of the gums

Glaucoma – an eye disease characterised by increased intra-ocular pressure

GLC – high resolution gas-liquid chromatography (a technique used to analyse volatile chemical compounds and extracts)

Glucosinolate – secondary metabolite that becomes activated upon wounding of a plant releasing active isothiocyanate

Glycoside – a chemical substance that yields at least one simple sugar upon hydrolysis

Gout – increased uric acid level in blood and sporadic episodes of acute arthritis

GRAS – abbreviation for "generally regarded as safe", the status given to foods and herbal medicines by the American Food and Drug Administration (FDA)

Gravel – small concretions in the bladder or kidney

Haematuria – blood in the urine

Haematoma – local accumulation of clotted blood

Glycoprotein – protein that carries sugar groups

Haemodialysis – removal of toxins from the blood through dialysis, using an artificial kidney (allowing the diffusion of toxins from the blood)

Haemolysis – the disruption of red blood cells and release of haemoglobin in blood

Haemoperfusion – removal of toxins from the blood with the aid of a column of charcoal or adsorbent resin

Haemorrhage – profuse bleeding

Haemorrhagic nephritis – blood in the urine

Haemorrhoids (= piles) – painful and swollen anal veins

Haemosorption perfusion – passage of a patient's blood through a set of columns filled with a haemosorbent (activated charcoal, ion-exchange resin, etc.): the purpose of the operation is to remove a toxic substance from the organism, particularly in an emergency

Haemostyptic – a substance that reduces or stops bleeding

Hallucinogen – a substance that induces the perception of objects that are not actually present

Hazard – capability of an agent to cause adverse effects

Hepatitis – inflammation of the liver

Hepatotoxic – toxic to the liver

Herbalist – a person with experience in herbal medicine and / or herbal therapy

Herpes simplex – localised infection on the lips or genitalia caused by the herpes virus

HIV – human immunodeficiency virus that causes AIDS

Homeopathy – a medicine system using minute amounts of substances that cause in a healthy person the same effect (symptoms) than those caused by the condition under treatment

Hormone – a substance released into the bloodstream that affects organ systems elsewhere in the body

HPLC – high performance liquid chromatography (a technique used to analyse chemical compounds and extracts)

Hydrogen bond – an attraction between a hydrogen atom (H) on an electronegative atom (mostly N or O)

Hydrolysis – breaking down a molecule by addition of water; in cells hydrolases (glucosidases, lipases, DNAses) catalyse this reaction

Hydrophilic substance – a substance soluble in water but not in oil

Hydrophobic substance – a substance that is repelled by water, but soluble in lipids

Hyperaemia (hyperemia) – abnormal blood

...ccumulation is a localised part of the body

Hyperlipidemia – characterised by enhanced lipid levels in the blood; triglycerides >160 mg/100 ml) and cholesterol (>260 mg/ 00 ml)

Hyperplasia – abnormal growth of normal cells in a tissue or organ

Hypersensitivity – allergic reaction of a person to a chemicals to which they had been exposed previously

Hypertension – high blood pressure; >140/90 mm Hg)

Hypertonic solution – abnormally high salt levels having a higher osmotic pressure than blood or another body fluid

Hypertrophy – abnormal increase in size of an organ (cell numbers remain constant)

Hypnotic – a substance that induces sleep

Hypoglycaemic – abnormally low level of blood sugar

Hypothermia – low body temperature

Hypotonia/hypotension – low blood pressure; (< 105/60 mm Hg)

Hypoxia – abnormally low oxygen content or tension in the body

Icterus – jaundice; deposition and retention of bile pigment in the skin

Immune stimulant – a substance capable of improving the immune system

Immunosuppression – reduction of the immune response

Incidence rate – measure of the frequency of new events occurring in a population

Inflammation – localised swelling, redness and pain as a result of an infection or injury

Influenza (flu) – an acute and highly contagious disease caused by viruses that infect mucous membranes of the respiratory tract

Ingestion – swallowing (eating or drinking) of chemicals. After ingestion chemicals can be absorbed from the GI tract into the bloodstream and distributed throughout the body

Inhalation – exposure to a substance through breathing it; if taken up from the lungs, a substance can enter the bloodstream

Inhibitory concentration (IC) – IC_{50} is the median concentration that causes 50% inhibition

Inotropic – a substance that stimulates the contraction of muscles, e.g. of the heart

Inorganic agents – material that consists of elements or inorganic compounds

Insomnia – inability to sleep

Insulin – a hormone made in the pancreas that controls the level of glucose in the blood

Intercalation – planar and lipophilic compounds can intercalate between base stacks of DNA; this leads to frame shift mutations (resulting in inactive proteins)

Intoxication – term has two meanings: 1. Poisoning with clinical signs and symptoms and 2. Drunkenness from ethanol-containing beverages or other compounds affecting the CNS

In vitro – in the laboratory or test tube

In vivo – in a living animal or human

Irritant – a substance causing irritation of the skin, eyes or respiratory system

Ion channel – membrane protein that can form water-containing pores so that mineral ions can enter or leave cells

Ionic bond – a bond created when 2 atoms trade electrons and then attract each other due to their opposing charges (e.g. NH_3^+-groups form ionic bonds to COO^--groups)

Iridoids – a subgroup of monoterpenoids, with iridoid glucosides, secoiridoids and secologanin

Isothiocyanate – secondary metabolites released from glucosinolates upon hydrolysis; exhibits strong skin irritating properties

Itch (= pruritis) – skin irritation

Jaundice – yellow coloration of skin and mucosa; caused by abnormally high level of bile pigments in the blood

Lacrimator – substance which can irritate the eyes and causes tear formation

Lactation – production and secretion of milk by female mammary glands

Latency – time period from the first exposure to a substance until the appearance of biological effects

Lavage – irrigation or washing out of the stomach, intestine or the lungs

Laxative – substance that causes evacuation of the intestinal contents

Lethal – deadly; fatal; causing death

LD_{50} – lethal dose that kills 50% of the individuals in a test group

LD_{100} – lethal dose that kills all (100%) of the individuals in a test group

Leukaemia – malignant disease of the blood-forming organs

Leukopenia – low white blood cell count

Ligand – substance that binds to a receptor in a specific way like a key in a lock

Liniment – ointment for topical application

Lipid – a substance soluble in non-polar solvents; insoluble in water

Lipophilic (= hydrophobic) – a substance soluble in oil or a non-polar solvent

Lowest-observed-adverse-effect-level (LOAEL) – lowest concentration or amount of a substance which causes an adverse effect

Lowest-observed-effect-level (LOEL) – lowest concentration or amount of a substance which causes any biological effects

Lysosome – cytoplasmic organelle containing hydrolytic enzymes

Maceration – preparation made by soaking plant material

Malaise – slight feeling of bodily discomfort

Malaria – a parasitic disease caused by *Plasmodium* parasites; it is transmitted by mosquitoes

Malignant – a disease which gets progressively worse and result in death if not treated, or a cancer with uncontrolled growth and metastasis

Mania – emotional disturbance characterised by an expansive and elated state (euphoria), rapid speech, flight of ideas, decreased need for sleep, distractability, grandiosity, poor judgement and increased motor activity

MAO inhibitor – inhibitor of monoamine oxidase that degrades the neurotransmitters adrenaline, noradrenaline, dopamine and serotonin

Mastitis – inflammation of the breast

Materia medica – the various materials (from plants, animals or minerals) that are used in medicine (healing)

Maximum tolerable dose (MTD) – highest amount of a substance that does not kill test animals (LD_o)

MDR – multiple drug resistance; caused by overexpression of p-glycoprotein, an important ATP-driven transporter at biomembranes, which pumps out lipophilic xenobiotics

Melanoma – a tumour of skin and mucosa arising from the pigment producing cells

Menopause – permanent cessation of menstruation caused by decreased production of female sex hormones

Metastasis – movement of cancer cells from one part of the body to another, starting new tumours in other organs

Microtubules – linear tubular structures of higher cells, formed from tubulin dimers; essential for cell division and vesicular transport processes

Migraine – recurrent condition of severe pain in the head accompanied by other symptoms (nausea, visual disturbance)

Mineralocorticoid – the steroid of the adrenal cortex (aldosteron) that regulates salt metabolism

Minimum lethal dose (LD_{min}) – lowest amount of a substance that may cause death

Miosis – abnormal contraction of the pupil to less than 2 mm

Mitochondria – important compartment of eukaryotic cells; site of the Krebs cycle and respiratory chain (production of ATP); mitochondria have their own DNA, replication, transcription and ribosomes

Mitogen – chemical which induces mitosis and cell proliferation

Mitosis – cell division

Monoamine oxidase – the enzyme that catalyses the removal of amine groups (e.g. dopamine, noradrenaline)

Monodesmosidic saponins – saponins with one sugar chain

Morbidity – any form of "sickness", "illness" and "morbid condition"

Mucilage – solution of viscous (slimy) substances (usually polysaccharides) that form a protective layer over inflamed mucosal tissues

Mucosa – mucous tissue layer on the inside of the respiratory or gastrointestinal tract

Mucus – clear, viscose secretion formed by mucous membranes

Multiple sclerosis – disorder of the central nervous system caused by a destruction of the myelin around axons in the brain and spinal cord that lead to various neurological symptoms

Mutagenic – a substance that induces genetic mutations; resulting in alterations or loss of genes or chromosomes

Myalgia – non-localised muscle pain

Mycotoxin – toxin produced by a fungus

Mydriasis – dilation of the pupil of the eye

Na^+, K^+-ATPase – important ion pump of animal cells; pumps Na^+ out of the cell and K^+ into the cell; is inhibited by cardiac glycosides

Narcotic – a substance that produces insensibility or stupor, combined with a sense of well-being, or more specifically an opioid, any natural or synthetic drug that has morphine-like activity

Necrosis – death of cells or tissue, usually accompanied by an inflammation

Neoplasm – new and abnormal formation of a tumour by fast cell proliferation

Nephritis – kidney inflammation, accompanied by proteinuria, haematuria, oedema and hypertension

Nephrotoxic – a substance that is harmful to the kidney

Neuralgia – severe pain along nerve ends

Neuritis – inflammation of nerves

Neuron – nerve cell

Neuropathy – general term for diseases of the central or peripheral nervous system

Neurotoxin – substance with adverse effects on the central and peripheral nervous system; such as transient modulation of mood or performance of CNS

Neurotransmitter – signal compounds in synapses of neurons that help to convert an electric signal into a chemical response ; important neurotransmitters are acetylcholine, noradrenaline, adrenaline, dopamine, serotonin, histamine, glycine, GABA, glutamate, endorphins and several other peptides

Neurovesicle – small vesicles in the presynapse that are filled with neurotransmitters

No-observed-adverse-effect-level (NOAEL) – greatest concentration or amount of a substance, which causes no detectable adverse alteration

Non-protein amino acid (NPAA) – secondary metabolite that is an analogue of a proteinogenic amino acid; if incorporated into proteins, the latter are usually inactivated

Noxious substance – harmful substance

Nutritive – nourishing, nutritious

Nycturia – nightly urge to urinate

Nystagmus – involuntary, rapid, rhythmic movement of the eyeball

Oedema (edema) – swelling of tissue due to an accumulation of fluids, often caused by kidney or heart failure

Ointment – semisolid medicinal preparation that is used topically

Oleum – nonvolatile oil; fat

Oliguria – elimination of a small amount of urine in relation to fluid intake

Ophthalmic – relating to the eye

Opium – dried latex of *Papaver somniferum* with several alkaloids, especially morphine

Oral – by mouth (p.o.)

Organic agent – molecules made of carbon and hydrogen atoms, often containing oxygen and nitrogen in addition

Organelle – nanomachine or membrane embraced compartment within a cell that has a specialised function, e.g., ribosome, peroxisome, lysosome, Golgi apparatus, mitochondrion and nucleus

Osteoporosis – a reduction in bone mass, resulting in fractures

Otitis – inflammation of the ear

Oxidation – a reaction that adds oxygen atoms to a molecule or when electrons are lost to a molecule

Oxytocic – speeding up of parturition

Palpitation – noticible regular or irregular heartbeat

Paraesthesia – abnormal sensation, as burning or prickling

Paralysis – loss or impairment of muscle activity

Parasympathetic nervous system – that part of the nervous system that slows the heart rate, increases intestinal (smooth muscle) and gland activity, and relaxes sphincter muscles

Parasympatholytic – anticholinergic substance that induces effects resembling those caused by interruption of the parasympathetic nerve

Parasympathomimetic – cholinomimetic substance that induces effects resembling those caused by stimulation of the parasympathetic nervous system

Parenteral administration – administration of medicinal substances by injection (i.v. = intravenous; i.m. = intramuscular; s.c. = subcutaneous) or intravenous drip

Paresis – weak paralysis

Parkinsonism – one of several neurological disorders manifesting in unnaturally slow or rigid movements

Pathogen – a microorganism that may cause disease

Percutaneous – absorption through the skin

Periodontitis – inflammation of the area around a tooth

Peristalsis – waves of involuntary contraction in the digestive system

Peritoneal dialysis – clinical procedure of artificial detoxication in which a toxic substance from the body is absorbed into a liquid that has been pumped into the peritoneum

Pesticide – substance (such as fungicide, insecticide, herbicide) used to kill agricultural pests and pathogens

p-gp – p-glycoprotein; an important ATP-driven transporter at biomembranes, which pumps out lipophilic xenobiotics

Pharmaceuticals – drugs, medical products, medicines, or medicaments

Pharmacodynamics – the study of how active substances work in the body; e.g., whether they bind to a receptor

Pharmacogenetics – the study of the influence of hereditary genetic factors on the effects of drugs on individual organisms

Pharmacognosy – the study of herbal drugs, their identification, properties and uses

Pharmacokinetics – the study of how active substances are absorbed, moved, distributed, metabolised and excreted

Pharmacology – the study of the nature, properties and uses of drugs (see pharmacodynamics, pharmacokinetics); includes the study of endogenous active compounds

Pharmacopoeia (Pharmacopeia) – an official and authoritative book or publication listing all the various drugs that may be used

Phase 1 reaction – enzymic modification of a xenobiotic or drug by oxidation, reduction, hydrolysis, hydration, dehydrochlorination or other reactions

Phase 2 reaction – conjugation of a substance, or its metabolites from a phase 1 reaction, with endogenous hydrophilic molecules, making them more water-soluble that may be excreted in the urine or bile

Phenotype – the expressed structural and functional characteristics which depend on genotype and environmental conditions

Phorbolester – diterpene from Euphorbiaceae and Thymelaeaceae, resembling diacylglycerol in structure and therefore activates protein kinase C

Phosphodiesterase – enzyme of signal transduction; inactivates cAMP or cGMP

Phospholipase C – enzyme of signal transduction; splits inositol phosphates to IP3 and diacylglycerol (DAG)

Phospholipids – phosphorylated lipids that are building blocks of cell membranes

Phytomedicine – see phytotherapy

Photo-irritation – skin inflammation due to light exposure, caused by metabolites which were formed in the skin by photolysis

Photosensitisation – increasing sensitivity to sunlight

Phototoxicity – adverse effects of compounds activated by light exposure (e.g. furanocoumarins)

Phytotherapy – application of plant drugs or products derived from them to cure diseases or to relieve their symptoms

Phytotoxicity – adverse effects of compounds to plants

Piscicide – toxins used to kill fish

Placebo – drug preparation without active ingredients, which cannot be distinguished from the original drug; used in placebo controlled clinical trials

Plasma (blood plasma) – liquid part of blood which surrounds the blood cells and platelets

Pneumonitis – inflammation of the lung

Point mutation – exchange of a single base pair in DNA

Poison – toxicant that causes immediate death or illness even at very low doses

Potentiation – the response to a combination of active substances is greater than expected from the individual compounds (synergism)

Poultice – a semisolid mass of plant materials in oil or water applied to the skin

ppb – parts per billion, 1 µg in 1 kg

ppm – parts per million, 1 mg in 1 kg

Precursor – substance from which another molecule is formed

Prescription drug – a drug that requires prescription from a physician

Procarcinogen – compound which has to be metabolised before it can induce a tumour

Prodrug – a substance that is converted to its active form within the body

Prophylactic – a substance that prevents disease

Prostaglandins – a group of physiological active substances within tissues that cause stimulation of muscles and numerous other metabolic effects; important for inflammation processes

Prostatitis – bacterial infection of the prostate (also see benign prostate hyperplasia)

Protein kinases – enzymes that phosphorylate other proteins which become activated or inactivated by this modification; important are protein kinase A and protein kinase C

Proteinuria – excretion of excessive amounts of protein in the urine

Pruritis – itching

Psoriasis – inherited skin condition caused by an enhanced growth of dermal cells resulting in the production of dandruff

Psychosis – mental disorder characterised by personality changes and loss of contact with reality

Psychotropic – a substance that affects the mind or mood

Pulmonary – referring to the lungs

Purgative – see laxative

Pyretic (pyrogen) – a substance that induces fever

Rate – frequency of events during a specified time interval

Reactive oxygen species, ROS – see antioxidant

Receptor – protein (often a membrane protein) that has a binding site for another molecule (« ligand »); important for signal transduction in cells

Relaxant – a substance that reduces tension

Renal – referring to the kidneys

Repellent – compounds used to repel herbivores or predators

Replication – duplication of DNA prior to cell division

Reproductive toxicology – adverse effects of chemicals on the embryo, foetus, neonate and prepubertal mammal and the adult reproductive and neuro-endocrine systems

Resin – amorphous brittle substance resulting from a plant secretion

Re-uptake inhibitor – inhibitors of transporters for the neurotransmitters dopamine, noradrenaline and serotonin at presynaptic and vesicle membrane

Rheumatism – general term referring to painful joints

Rhinitis – inflammation of the mucosa of the nose

Rhizoma – dried rhizomes; underground stem

Risk – probability of manifestation of a hazard under specific conditions

Rodenticide – pesticide used to kill rodents

Rubefacient – a substance (counter-irritant) that causes reddening of skin

Saponins – glycosides of triterpenes and steroids; the aglycone is usually lipophilic, whereas the saponins are amphiphilic with detergent properties; distinguished are monodesmosidic saponins with 1 sugar chain and bidesmosidic saponins with 2 sugar chains

Saturated fats – fats with fatty acids without double bonds (animal fats, coconut oil, etc.)

Saturnism – intoxication caused by lead

Scar tissue – new cell growth following injury

Sclerosis – excessive hardening of an organ due to growth of fibrous tissue

Secondary metabolite – chemical substances of plants (usually of low molecular weight) with a high structural diversity that are used as defence or signal compounds by the plants producing them. Several secondary metabolites have a restricted occurrence in the plant kingdom. In contrast, primary metabolites are essential and present in all plants

Secretolytic – a substance that leads to a better solubilisation of mucus and favours its discharge

Sedative – a substance that calms down the nerves (tranquilliser)

Serum (blood) – water-soluble protein fraction of the blood that remains after clotting

Sesquiterpene lactone – terpene with 15 C atoms; its exocyclic methylene group can bind to SH-groups of proteins or glutathione

SH-group – functional group in proteins that can form disulphide bonds with other proteins

Side effect – activity of a drug other than that desired for beneficial pharmacological effect

Solvent abuse (solvent sniffing) – inhalation (or drinking) of volatile solvents, with the intent to become intoxicated (dangerous for liver and CNS)

Soporific – a substance that induces or promotes sleep

Spasm – involuntary contraction of muscles

Special extract – extract that is enriched in the active principle whereas unwanted compounds have been discarded

Steam distillation – a method of selectively extracting volatile compounds and oil from plant material by boiling or steaming in water, followed by condensation

Stomachic – a substance that promotes appetite and digestion

Styptic – a substance applied externally to stop bleeding by contracting the tissue and blood vessels

Subacute – form of repeated exposure or administration not long enough to be called "long-term" or "chronic"

Substrate – a molecule that binds to an enzyme and then undergoes a chemical reaction

Sympathetic nervous system – that part of the nervous system that accelerates the heart rate, constricts blood vessels and raises blood pressure

Sympatholytic – substance that blocks signal transmission from the adrenergic (sympathetic) postganglionic fibres to effector organs

Sympathomimetic – substance that induces effects resembling those of impulses transmitted by the postganglionic fibres of the sympathetic nervous system

Symptom – subjective perception of a disease or of effects induced by a toxin

Synapse – neurons are connected with other neurons or target tissues via synapses where the action potential is converted into a chemical signal (neurotransmitter)

Syndrome – summary of symptoms occurring together; it often characterises a particular disease-like state

Synergistic – the phenomenon that the combined effect of two or more substances is greater than the sum of their individual effects (see potentiation)

Syrup – a sugary solution intended for oral administration (such as cough syrups)

Tachycardia – pulse over 100 beats per minute

Tachypnoea – fast breathing

Tannins – secondary metabolites with several phenolic OH-groups, that can form hydrogen and ionic bonds with proteins, thereby altering their conformation; distinguished are gallotannins and catechol tannins, which derive from catechin or epicatechin

Target – any component of the human body that can be affected by a drug

TCM – traditional Chinese medicine; oldest therapeutic system of mankind that is still in use and esteemed

Tea – an infusion made by pouring boiling water over a measure quantity of dried plant material and leaving it for a while to steep

Teratogen – a substance given to pregnant women that causes abnormal growth and malformations in an embryo

Terpenoids – a very large group of secondary metabolites, including monoterpenes (with 10 carbons), sesquiterpenes (15 C), diterpenes (20 C), triterpenes (30 C), steroids (27 or less), tetraterpenes (40 C)

Testosterone – the male sex hormone

Tetanic – referring to tetanus, characterised by tonic muscle spasms

Therapeutic margin – difference between toxic and therapeutic doses

Tincture – an extract of medicinal plant material made with alcohol (ethanol) or a mixture of alcohol and water

Tinnitus – noise in the ear (ringing, buzzing, roaring or clicking)

Tonic – a substance that maintains or restores health and vigour (usually taken over a lengthy period); also term for tension, especially muscular tension

Topical application – external application (on the skin)

Toxicant – chemicals producing adverse biological effects of any nature

Toxicology – science that studies toxins and their effects in humans or animals

Toxification – metabolic activation of a potentially toxic substance

Toxin – a harmful biogenic substance or agent causing injury in living organisms as a result of physicochemical interactions

Transcription – process of copying the base sequence of a gene into mRNA

Translation – process of copying the base sequence of mRNA into the amino acid sequence of proteins in the ribosome

Transporter – a membrane protein that catalyses the transport of a molecule from one side of a biomembrane to the other side

Tuber – dried tubers

Tuberculosis – a bacterial disease caused by *Mycobacterium tuberculosis* that affects the lungs and other organs (often chronic and fatal i not treated with antibiotics)

Tumour (tumor) – an abnormal growth o tissue (benign or malignant)

Ulcer – an inflamed lesion on the skin or o a mucous membrane

Unsaturated fats – fats with fatty acids wit double bonds between the carbon atoms

Uptake – absorption of a chemical into th body, a cell, or into the body fluids by passag through membranes

Urinary calculi – concretions in the urethra

Urticaria – transient reaction of the ski characterised by smooth, slightly elevate patches that are redder or paler than the sur rounding skin; often with severe itching

Vasoconstrictor – a substance that causes narrowing of the blood vessels and reduce blood flow

Vasodilator – a substance that causes an in crease in the internal diameter of blood ves sels and enhanced blood flow

Venom – animal toxin usually employed fo self-defence or predation; mostly delivered b a bite or sting

Venous tone – the firmness of the walls o veins

Ventricular fibrillation – irregular heartbea characterised by uncoordinated contraction of the ventricle

Vermifuge – a substance that kills or expel intestinal worms

Vertigo – dizziness

Vesicant – compounds producing blisters o the skin

Vesicle – a subcellular container in the cel to store chemicals, e.g. neurotransmitters ar stored in neurovesicles

Virus – infectious complex of macromol ecules that contain their genetic informatio either as DNA or RNA; viruses need hos cells for replication and the formation of nev viral particles

Virustatic – a substance that inhibits th multiplication of viruses

Volatile oil – various terpenoids that evapo rate easily (they add taste and smell to man plants)

Vomit – the expulsion of matter from th stomach via the mouth

Withdrawal effect – negative feelings follow ing withdrawal of a drug on which a perso has become dependent

Xenobiotic – foreign compound absorbe by animals or humans

Further reading

Acamovic, T, Steward CS, Pennycott, TW (2004) Poisonous Plants and Related Toxins. CABI, Wallingford

Alberts A, Mullen P (2000) Psychoaktive Pflanzen. Pilze und Tiere. Kosmos Verlag, Stuttgart

Ayensu ES (1978) Medicinal Plants of West Africa. Reference Publications Inc., Algonac

Ayensu ES (1981) Medicinal Plants of the West Indies. Reference Publications Inc., Algonac

Balick MJ, Cox PA (1996) Drogen, Kräuter und Kulturen. Spektrum Akademischer Verlag, Heidelberg

Bellamy D, Pfister A (1992) World Medicine – Plants, Patients and People. Blackwell Publishers, Oxford

Bindon P (1996) Useful Bush Plants. Western Australian Museum, Perth

Blaschek W, Hänsel R, Keller K, Reichling J, Rimpler G, Schneider G (1998) Hagers Handbuch der Pharmazeutischen Praxis. 5th ed. Vol 2 und 3. Springer Verlag, Berlin

Boulos L (1983) Medicinal Plants of North Africa. Reference Publications Inc., Algonac

Bresinsky A, Besl H (1985) Giftpilze. Wissenschaftliche Verlagsgesellschaft, Stuttgart

Brown RG (2002) Dictionary of Medicinal Plants. Ivy Publishing House, Raleigh

Bruneton J (1999) Pharmacognosy, Phytochemistry, Medicinal Plants. 2nd ed., Intercept, Hampshire

Bruneton J (1999) Toxic Plants Dangerous to Humans and Animals. Intercept, Hamphire

Burger A, Wachter H (1998) Hunnius Pharmazeutisches Wörterbuch. 8th ed. Walter de Gruyter, Berlin

Burrows GM, Tyrl RJ (2006) Handbook of Toxic Plants of North America. Wiley-Blackwell, Oxford

Chevallier A (2001) Das große Lexikon der Heilpflanzen. Dorling Kindersley, London

Daunderer M (1995) Lexikon der Pflanzen- und Tiergifte. Ecomed, Hamburg

Dewick, PM (2001) Medicinal Natural Products. Wiley, Chichester

Duke JA (1992) Database of Phytochemical Constituents of GRAS Herbs and Other Economic Plants. CRC Press, Boca Raton

Duke JA (2002) Handbook of Medicinal Herbs. 2. Aufl. CRC Press, Boca Raton

Duke JA, Foster S (1998) A Field Guide to Medicinal Plants and Herbs of Eastern and Central North America. Houghton Mifflin, Boston

Erhardt W, Götz E, Bödeker N, Seybold S (Hrsg) (2002) Zander Handwörterbuch der Pflanzennamen. 17th ed., Ulmer Verlag, Stuttgart

Fattorusso E, Taglialatela-Scafati (2008) Modern Alkaloids. Structure, Isolation, Synthesis and Biology. Wiley-VCH, Weinheim

Frohne D, Pfänder HJ (2004) Giftpflanzen. 5th ed., Wissenschaftliche Verlagsgesellschaft, Stuttgart

Goodman LS, Gilman AG, Limbird LE, Hardman JG, Goodman Gilman A (2001) The Pharmacological Basis of Therapeutics. 10th ed., McGraw-Hill, London

Hänsel R, Keller K, Rimpler H, Schneider G (1992–1994) Hagers Handbuch der Pharmazeutischen Praxis. 5th ed., vol 4–6. Springer Verlag, Berlin

Hänsel R, Sticher O (2007) Pharmakognosie – Phytopharmazie. 8th ed., Springer Verlag, Heidelberg

Harborne JB, Baxter H (1993) Phytochemical Dictionary – A Handbook of Bioactive Compounds from Plants. Taylor & Francis, London

Harborne, JB, Baxter, H & Moss, GP (1997). Dictionary of Plant Toxins. John Wiley & Sons, Chichester.

Hardman JG, Limbird LE, Molinoff PB, Ruddon RW, Goodman Gilman A (1998) Pharmakologische Grundlagen der Arzneimitteltherapie. 9th ed., The McGraw-Hill Co., London

Hausen BM, Vieluf, IK (1997) Allergiepflanzen. Handbuch und Atlas. Nikol, Hamburg

Hegnauer R (1962–2001) Chemotaxonomie der Pflanzen. Vols 1–11. Birkhäuser Verlag, Basel

Heinrich M (2001) Ethnopharmazie und Ethnobotanik. Wissenschaftliche Verlagsgesellschaft, Stuttgart

Hocking GM (1997) A Dictionary of Natural Products. Plexus Publishing, Medford

Hüllmann H, Mohr K (1999) Pharmakologie und Toxikologie. 14th ed., Thieme, Stuttgart

Isaacs J (1987) Bush Food. Lansdowne Publishing, Sydney

Iwu MM (1993) Handbook of African Medicinal Plants. CRC Press, Boca Raton

Jastorf B, Störmann R, Wölcke U (2003) Struktur-Wirkungs-Denken in der Chemie. Universitätsverlag Aschenbeck & Isensee,

Bremen

Kapoor LD (2000) CRC Handbook of Ayurvedic Medicinal Plants. CRC Press, Boca Raton

Kayser O, Quax W (2007) Medical Plant Biotechnology. From basic research to industrial applications. Wiley-VCH, Weinheim

Kellerman TS, Coetzer JAW, Naudé TW (1988) Plant Poisonings and Mycotoxicoses of Livestock in Southern Africa. Oxford University Press, Cape Town

Kletter C, Kriechbaum M (2001) Tibetan Medicinal Plants. Medpharm, Stuttgart

Kokwaro JO (1976) Medicinal Plants of East Africa. East Africa Literature Bureau, Nairobi

Langenheim JH (2003) Plant Resins – Chemistry, Evolution, Ecology and Ethnobotany. Timber Press, Portland

Lassak EV, McCarthy T (2001) Australian Medicinal Plants. JB Books, Marleston

Leung AY, Foster S (1996) Encyclopedia of Common Natural Ingredients Used in Food, Drugs, and Cosmetics. 2nd ed. John Wiley & Sons, New York

Lewin L (1998) Phantastica: A Classic Survey on the Use and Abuse of Mind-Altering Plants. Park Street Press

Lewis WH, Elvin-Lewis MPF (1977) Medical Botany – Plants Affecting Man's Health. John Wiley & Sons, New York

Mabberley DJ (1997) The Plant Book – A Portable Dictionary of the Vascular Plants. 2nd ed., Cambridge University Press, Cambridge

Macias FA, Garcia Galindo JC, Molinillo JMG, Cutler HG (2004) Allelopathy – Chemistry and mode of action of allelochemicals. CRC Press, Boca Raton

Mann J (1992) Murder, Magic and Medicine. Oxford University Press, Oxford

Martindale W (1993) Extra Pharmacopoeia. 30th ed., The Pharmaceutical Press, London

Martinetz D, Lohs K, Janzen J (1989) Weihrauch und Myrrhe. Wissenschaftliche Verlagsgesellschaft, Stuttgart

Mebs D (1989) Gifte im Riff. Toxikologie und Biochemie eines Lebensraumes. Wissenschaftliche Verlagsgesellschaft, Stuttgart

Mebs D (1992) Gifttiere. Wiss. Verlagsgesellschaft, Stuttgart

Merck (2006). The Merck Index. 14th ed. Merck, Rahway.

Moerman DE (1998) Native American Ethnobotany. Timber Press, Portland

Moll E, Moll G (1989) Poisonous Plants. Struik, Cape Town

Mors W, Rizzini CT, Pereira NA, DeFilipps R (2000) Medicinal Plants of Brazil. Latino

Herbal Press, Spring Valley

Morton JF (1981) An Atlas of Medicinal Plants of Middle America. Charles C. Thomas, Springfield

Müller-Ebeling C, Rätsch C, Storl W-D (1999) Hexenmedizin. AT Verlag, Aarau

Munday J (1988) Poisonous Plants in South African Gardens and Parks. Delta Books, Johannesburg

Mutschler E, Geisslinger G, Kroemer HK, Schäfer-Korting M (2001) Arzneimittel wirkungen – Lehrbuch der Pharmakologie und Toxikologie. 8th ed., Wissenschaftlich Verlagsgesellschaft, Stuttgart

Nelson LS, Shih RD, Balick MJ (2007) Handbook of Poisonous and Injurious Plants. 2nd ed., Springer Verlag, New York

Neuwinger HD (1994) Afrikanische Arzneipflanzen und Jagdgifte. Wissenschaftliche Verlagsgesellschaft, Stuttgart

Neuwinger HD (1996) African Ethnobotany, Poisons and Drugs. Chapman & Hall, London

Neuwinger HD (2000) African Traditional Medicine – A Dictionary of Plant Use and Applications. Medpharm Scientific Publishers, Stuttgart

Newall CA, Anderson LA, Phillipson JD (1996) Herbal Medicine – A Guide for Health Care Professionals. The Pharmaceutical Press, London

Oliver-Bever B (Hrsg) (1986) Medicinal Plants of Tropical West Africa. Cambridge University Press, Cambridge

Panter K, Wierenga T, Pfister J (2007) Poisonous Plants: Global Research and Solutions. CABI, Wallingford

Perrine D (1996) The Chemistry of Mind-Altering Drugs: History, Pharmacology, and Cultural Context. American Chemical Society

Poyala G (2003) Biochemical Targets of Plant Bioactive Compounds. Taylor & Francis, London

Rai M, Carpinella MC (2006) Advances in Phytomedicine. Naturally Occurring Bioactive Compounds. Elsevier, Amsterdam

Rätsch C (1998) Enzyklopädie der psychoaktiven Pflanzen – Botanik, Ethnopharmakologie und Anwendung. Wissenschaftlich Verlagsgesellschaft, Stuttgart

Rätsch C (2005) The Encyclopedia of Psychoactive Plants: Ethnopharmacology and It's Applications. Park Street Press

Rimpler H (1999) Biogene Arzneistoffe. Deutscher Apotheker Verlag, Stuttgart

Roberts MF, Wink M (1998) Alkaloids. Biochemistry, Ecology and Medicinal Applications. Plenum Press, New York

Roth I, Lindorf H (2002) South American

Medicinal Plants. Springer, Berlin

Roth L, Daunderer M, Kormann K (1994) Giftpflanzen – Pflanzengifte. Ecomed Verlagsgesellschaft, Landsberg

Roth L, Frank H, Kormann K (1990) Giftpilze – Pilzgifte. Ecomed Verlagsgesellschaft, Landsberg

Russo E (2001) Handbook of Psychotropic Herbs: A Scientific Analysis of Herbal Remedies for Psychiatric Conditions. Haworth Press

Schultes RE, Hofmann A (1987) Pflanzen der Götter. 2nd ed., Hallwag Verlag, Bern

Seeger R, Neumann HG (2005) Giftlexikon. Deutscher Apotheker Verlag, Stuttgart

Seigler DS (1995) Plant Secondary Metabolism. Kluwer Academic Publishers, Boston

Siegel RK (2005) Intoxication: The Universal Drive for Mind-Altering Substances. Park Street Press

Small E, Catling PM (1999) Canadian Medicinal Crops. NRC Research Press, Ottawa

Spinella M (2001) The Psychopharmacology of Herbal Medicine: Plant Drugs That Alter Mind, Brain, and Behavior. MIT Press, Cambridge

Stuart D (2004) Dangerous garden – the quest for plants to change our lives. Harvard University Press, Cambridge

Swerdlow JL (2000) Nature's Medicine Plants that Heal. National Geographic, Washington

Tang W, Eisenbrand G (1992) Chinese Drugs of Plant Origin. Springer Verlag, Berlin

Teuscher E, Lindequist U (1994) Biogene Gifte – Biologie, Chemie, Pharmakologie. 2nd ed., Gustav Fischer Verlag, Stuttgart

Teuscher E, Melzig MF, Lindequist U (2004) Biogene Arzneimittel. 6th ed., Wissenschaftliche Verlagsgesellschaft, Stuttgart

Tyler VE (1993) The Honest Herbal. 3rd ed., Pharmaceutical Products Press, New York

Tyler VE (1994) Herb of Choice. The Therapeutic Use of Phytomedicinals. Pharmaceutical Products Press, New York

Vahrmeijer J (1981) Poisonous Plants of Southern Africa that Cause Stock Losses. Tafelberg Publishers, Cape Town

Van Wyk B-E, Gericke N (2000) People's Plants – A Guide to Useful Plants of Southern Africa. Briza Publications, Pretoria

Van Wyk B-E, Van Heerden F, Van Oudtshoorn B (2002) Poisonous Plants of South Africa. Briza Publications, Pretoria

Van Wyk B-E, Van Oudtshoorn B, Gericke N (1997) Medicinal Plants of South Africa. Briza Publications, Pretoria

Van Wyk B-E, Wink M. (2004) Medicinal Plants of the World. Briza Pretoria

Van Wyk, B-E (2005) Food Plants of the World. Briza Publications, Pretoria

Verdcourt B, Trump EC (1969) Common Poisonous Plants of East Africa. Collins, London

Von Koenen E (2001) Medicinal, Poisonous, and Edible Plants in Namibia. Klaus Hess Publishers, Windhoek & Göttingen

Wagner H, Bladt S (1995) Plant Drug Analysis – A Thin Layer Chromatography Atlas. Springer Verlag, Berlin

Wagner H, Vollmar A, Bechthold A (2007) Pharmazeutische Biologie 2. Biogene Arzneistoffe und Grundlagen von Gentechnik und Immunologie. 7th ed., Wissenschaftliche Verlagsgesellschaft, Stuttgart

Watt JM, Breyer-Brandwijk MG (1962) The Medicinal and Poisonous Plants of Southern and Eastern Africa. 2nd ed. Livingstone, London

Wiart C (2006) Ethnopharmacology of Medicinal Plants: Asia and the Pacific. Humana Press

Wichtl M (2002) Teedrogen und Phytopharmaka. 4th ed., Wissenschaftliche Verlagsgesellschaft, Stuttgart

Williamson EM (2002) Major Herbs of Ayurveda. Churchill Livingstone, London

Williamson EM, Okpako DT, Evans FJ (1996) Selection, Preparation and Pharmacological Evaluation of Plant Material. John Wiley & Sons, New York

Wink M (1999) Biochemistry of Plant Secondary Metabolism. Annual Plant Reviews, Vol 2. Sheffield Academic Press, Sheffield

Wink M (1999) Functions of Plant Secondary Metabolites and their Exploitation in Biotechnology. Annual Plant Reviews, Vol 3. Sheffield Academic Press, Sheffield

Wink M (2000) Interference of alkaloids with neuroreceptors and ion channels. In Bioactive natural products. Atta-Ur-Rahman ed.), pp. 3–129, Vol 11, Elsevier, Amsterdam

Wink M (2007) Molecular modes of action of cytotoxic alkaloids – From DNA intercalation, spindle poisoning, topoisomerase inhibition to apoptosis and multiple drug resistance. In – "The Alkaloids, (G. Cordell, ed.), Academic press, Vol. 64, 1–48, 2007 Elsevier, Amsterdam

Wolters B (1994) Drogen, Pfeilgift und Indianermedizin. Urs Freund Verlag, Greifenberg

Wolters B (1996) Agave bis Zaubernuß. Urs Freund Verlag, Greifenberg

435

Acknowledgements

The authors wish to thank the following institutions and persons:
Briza Publications and the production team, especially Eben van Wyk and Jacqueline Huisma·

Dr Coralie Wink for translating the book into German and her help in refining and editing t
English version.

Heidelberg University, Germany, and University of Johannesburg, South Africa, for providi
the authors with institutional support. Some of the plants in this book were photographed
the Botanical Gardens in Darmstadt, Düsseldorf, Frankfurt, Hamburg, Heidelberg, Ke
Kirstenbosch, Mainz, Marburg, Montpellier, Pretoria, Rostock, Stellenbosch, Strasbourg, Vien
and Worcester.

Persons and institutions who have contributed photographs (see below), or provided suppc
or encouragement: Siegmar Bauer (Medpharm Scientific Publishers, Stuttgart), Ted Cc
(Heidelberg), Luc Legal, Adrian and Leonie Wink (Dossenheim).

PHOTOGRAPHIC CONTRIBUTIONS

All photographs were taken by Prof. Dr Michael Wink (360) except those from other photographe
listed alphabetically below. The photographs are numbered according to page number and positic
(alphabetically from top to bottom and from left to right).

Prof. Dr W. Barthlott (University of Bonn, Germany): 148b,c; 137b; **José María de Jesús
Almonte** (HUMO Herbarium of the University of Morelos, Mexico) 76a,b; 231a; **Thomas
Brendler** (PlantaPhile, Berlin): 116a, 189a; **Dr. B. von Daake** (Botanical Garden Marburg,
Germany) 59a, 108a; **Wilmer A. Diaz** (Jardín Botanico del Orinoco; Venezuela) 50a,b; 66a;
Prof. Dr P. Endress (University of Zürich) 157, 222a; **Dr Robin Foster** (The Field Museum,
Chicago) 85a,b; **Fulvio Castillo Suarez** (Mexico) 23b, 194a; **Dr. G Gerlach** (Bot. Garden
Munich, Germany) 63a,b; **D. Greig** (Australia) 116b; **Dr Michael Hasler** (Frankfurt, German
87a , 121a, 166a; **Prof. Dr H. Hilger** (Free University of Berlin, Germany) 138a; **Bruce Holst**
(Marie Selby Botanical Gardens, Sarasota, FL, USA) 141b; **Niels Jacobsen,** (Wilderness, Sout
Africa) 16g; **Fredi Kasparek** (Herten, Germany) 149 a/b; **Thomas Lehr** (Hofheim, Germany
47b; **Dr Russell J. Molyneux** (ARS/USDA Albany, USA) 254a,b; **Rolando Pérez** (STRI,
Panamá) 141b; **Prof. Dr F. Oberwinkler** (University Tübingen, Germany) 194b; **Prof. Dr Jen
Rohwer** (University Hamburg, Germany) 66bc; 189b, 231b; **Dr Patricia M. Tilney** (University
of Johannesburg, South Africa): 226b; **Tracey Slotta** (USDA-NRCS PLANTS Database, USA
187a; **Prof. Dr Fanie R van Heerden** (University KwaZulu Natal, South Africa) 142a; **Frits
van Oudtshoorn** (Nylstroom, South Africa) 219a,b; **Prof. Dr B.-E. van Wyk** (University of
Johannesburg, South Africa) 11b,h; 21a,b,c,d; 23c, 29 d,e; 31c; 34, 39b; 40a; 41 a,b; 42b; 44b;
46a,b; 49; 53c; 54a,b; 56a,b; 57a; 58a; 60b; 64a; 67a,b; 68a,b; 70a; 71a; 75a,b; 77a,b; 79a,b; 80b;
81a; 82a, 83; 84a,b; 88a,b; 89; 91; 92a,b; 94a,b; 99a,b; 100b; 101a,b; 102a; 104b; 106a,c; 107a;
109; 110; 111a; 114a,b; 115a,b; 119a,b; 120a; 122b; 123a; 124a,b; 125a; 128a,b; 129; 131a,b; 132a;
134a,b; 135b; 142b; 143a,b;144a,b; 145c 146a,b; 148a,d; 151a,b; 153a,b; 158a; 164; 165b; 168a;
169b; 170a; 172; 173a,b; 175a; 176b; 178a; 181a,b; 182a,b; 183a; 184a,b; 185a; 190b; 193a,b; 195a,
199c; 202a; 206a; 211a,b; 214; 216a; 218a,b; 220; 221a,b; 222a; 223a,b; 226b; 227a,b; 228a; 232a
234b; 237c; 239a; 309a; 333c; **Eben van Wyk** (Sannieshof, South Africa) 168b; **Teodor van
Wyk** (Johannesburg, South Africa): 215b; **Dr Karan Vasisht** (Panjab University, Chandigarh,
India) 200a; **Prof. Y. Zu** (Harbin, China) 76b.

436

Index

The index refers to plant and substance names. Page numbers in **bold print** indicate main entries and/or illustrations.

445

458